Quantum Processes in Semiconductors

Fourth Edition

B. K. Ridley FRS
Professor of Physics
University of Essex

CLARENDON PRESS · OXFORD
1999

OXFORD
UNIVERSITY PRESS

Great Clarendon Street, Oxford OX2 6DP

Oxford University Press is a department of the University of Oxford.
It furthers the University's objective of excellence in research, scholarship,
and education by publishing worldwide in

Oxford New York
Athens Auckland Bangkok Bogotá Buenos Aires Calcutta
Cape Town Chennai Dar es Salaam Delhi Florence Hong Kong Istanbul
Karachi Kuala Lumpur Madrid Melbourne Mexico City Mumbai
Nairobi Paris São Paulo Singapore Taipei Tokyo Toronto Warsaw

with associated companies in Berlin Ibadan

Oxford is a registered trade mark of Oxford University Press
in the UK and in certain other countries

Published in the United States
by Oxford University Press Inc., New York

First published 1982
Second edition 1988
Third edition 1993
Fourth edition 1999

A catalogue record for this book is available from the British Library

Library of Congress Cataloging in Publication Data
(Data available)

ISBN 0 19 850580 9 (hbk)
ISBN 0 19 850579 5 (pbk)

Typeset by Newgen Imaging Systems (P) Ltd., Chennai, India

Printed in India by Thomson Press Ltd.

Preface to the Fourth Edition

This new edition contains three new chapters concerned with material that is meant to provide a deeper foundation for the quantum processes described previously, and to provide a statistical bridge to phenomena involving charge transport. The recent theoretical and experimental interest in fundamental quantum behaviour involving mixed and entangled states and the possible exploitation in quantum computation meant that some account of this should be included. A comprehensive treatment of this important topic involving many-particle theory would be beyond the scope of the book, and I have settled on an account that is based on the single-particle density matrix. A remarkably successful bridge between single-particle behaviour and the behaviour of populations is the Boltzmann equation, and the inclusion of an account of this and some of its solutions for hot-electrons was long overdue. If the Boltzmann equation embodied the important step from quantum statistical to semi-classical statistical behaviour, the drift-diffusion model completes the trend to fully phenomenological description of transport. Since many excellent texts already cover this area I have chosen to describe only some of the more exciting transport phenomena in semiconductor physics such as those involving a differential negative resistance, or involving acoustoelectric effects, or even both, and something of their history.

A new edition affords the opportunity to correct errors and omissions in the old. No longer being a very assiduous reader of my own writings, I rely on others, probably more than I should, to bring errors and omissions to my attention. I have been lucky, therefore, to work with someone as knowledgable as Dr N.A. Zakhleniuk who has suggested an update of the discussion of cascade capture, and has noted that the expressions for the screened Bloch–Gruneisen regime were for 2-D systems and not for bulk material. The update and corrections have been made, and I am very grateful for his comments.

My writing practically always takes place at home and it tends to involve a mild autism that is not altogether sociable, to say the least. Nevertheless, my wife has put up with this once again with remarkable good humour and I would like to express my appreciation for her support.

Thorpe-le-Soken, 1999 B.K.R.

Preface to the Third Edition

One of the topics conspicuously absent in the previous editions of this book was the scattering of phonons. In a large number of cases phonons can be regarded as an essentially passive gas firmly anchored to the lattice temperature, but in recent years the importance to transport of the role of out-of-equilibrium phonons, particularly optical phonons, has become appreciated, and a chapter on the principal quantum processes involved is now included. The only other change, apart from a few corrections to the original text (and I am very grateful to those readers who have taken the trouble to point out errors) is the inclusion of a brief subsection on exciton annihilation, which replaces the account of recombination involving an excitonic component. Once again, only processes taking place in bulk material are considered.

Thorpe-le-Soken B.K.R.
December 1992

Preface to the Second Edition

This second edition contains three new chapters—'Quantum processes in a magnetic field, Scattering in a degenerate gas, and Dynamic screening'—which I hope will enhance the usefulness of the book. Following the ethos of the first edition I have tried to make the rather heavy mathematical content of these new topics as straightforward and accessible as possible. I have also taken the opportunity to make some corrections and additions to the original material—a brief account of alloy scattering is now included—and I have completely rewritten the section on impact ionization. An appendix on the average separation of impurities has been added, and the term 'third-body exclusion' has become 'statistical screening', but otherwise the material in the first edition remains substantially unchanged.

Thorpe-le-Soken 1988 B.K.R.

Preface to the First Edition

It is a curious fact that in spite of, or perhaps because of, their overwhelming technological significance, semiconductors receive comparatively modest attention in books devoted to solid state physics. A student of semiconductor physics will find the background theory common to all solids well described, but somehow all the details, the applications, and the examples—just those minutiae which reveal so vividly the conceptual cast of mind which clarifies a problem—are all devoted to metals or insulators or, more recently, to amorphous or even liquid matter. Nor have texts devoted exclusively to semiconductors, excellent though they are, fully solved the student's problem, for they have either attempted global coverage of all aspects of semiconductor physics or concentrated on the description of the inhomogeneous semiconducting structures which are used in devices, and in both cases they have tended to confine their discussion of basic physical processes to bare essentials in order to accommodate breadth of coverage in the one and emphasis on application in the other. Of course, there are distinguished exceptions to these generalizations, texts which have specialized on topics within semiconductor physics, such as statistics and band structure to take two examples, but anyone who has attempted to teach the subject to postgraduates will, I believe, agree that something of a vacuum exists, and that filling it means resorting to research monographs and specialist review articles, many of which presuppose a certain familiarity with the field.

Another facet to this complex and fascinating structure of creating, assimilating, and transmitting knowledge is that theory, understandably enough, tends to be written by theoreticians. Such is today's specialization that the latter tend to become removed from direct involvement in the empirical basis of their subject to a degree that makes communication with the experimentalist fraught with mutual incomprehension. Sometimes the difficulty is founded on a simple confusion between the disparate aims of mathematics and physics—an axiomatic formulation of a theory may make good mathematical sense but poor physical sense—or it may be founded on a real subtlety of physical behaviour perceived by one and incomprehended by the other, or more usually it may be founded on ignorance of each other's techniques, of the detailed analytic and numerical approximations propping up a theory on the one hand, and of the detailed method and machinery propping up an experimental result on the other. Certainly, experimentalists cannot avoid being theoreticians from time to time, and they have to be aware of the basic theoretical structure of their subject. As students of physics operating in an area where physical intuition is more important than logical deduction they are not likely to appreciate a formalistic account of that basic structure even though it may possess elegance. Intuition functions on rough approximation rather than rigour, but too few accounts of theory take that as a guide.

This book, then, is written primarily for the postgraduate student and the experimentalist. It attempts to set out the theory of those basic quantum-mechanical processes in homogeneous semiconductors which are most relevant to applied semiconductor physics. Therefore the subject matter is concentrated almost exclusively on electronic processes. Thus no mention is made of phonon–phonon interactions, nor is the optical absorption by lattice modes discussed. Also, because I had mainstream semiconductors like silicon and gallium arsenide in mind, the emphasis is on crystalline materials in which the electrons and holes in the bands obey non-degenerate statistics, and little mention is made of amorphous and narrow-gap semiconductors. Only the basic quantum-mechanics is discussed; no attempt is made to follow detailed applications of the basic theory in fields such as hot electrons, negative-differential resistance, acousto-electric effects, etc. To do that would more than triple the size of the book. The theoretical level is at elementary first- and second-order perturbation theory, with not a Green's function in sight; this is inevitable, given that the author is an experimentalist with a taste for doing his own theoretical work. Nevertheless, those elementary conceptions which appear in the book are, I believe, the basic ones in the field which most of us employ in everyday discussions, and since there is no existing book to my knowledge which contains a description of all these basic processes I hope that this one will make a useful reference source for anyone engaged in semiconductor research and device development.

Finally, a word of caution for the reader. A number of treatments in the book are my own and are not line-by-line reproductions of standard theory. Principally, this came about because the latter did not exist in a form consistent with the approach of the book. One or two new expressions have emerged as a by-product, although most of the final results are the accepted ones. Where the treatment is mine, the text makes this explicit.

Colchester 1981 B.K.R.

Contents

1. Band structure of semiconductors

1.1. The crystal Hamiltonian

FOR an assembly of atoms the classical energy is the sum of the following:

(a) the kinetic energy of the nuclei;

(b) the potential energy of the nuclei in one another's electrostatic field;

(c) the kinetic energy of the electrons;

(d) the potential energy of the electrons in the field of the nuclei;

(e) the potential energy of the electrons in one another's field;

(f) the magnetic energy associated with the spin and the orbit.

Dividing the electrons into core and valence electrons and leaving out magnetic effects leads to the following expression for the crystal Hamiltonian:

$$H=\sum_{l}\frac{\mathbf{p}_l^2}{2M_l}+\sum_{l,m}U(\mathbf{R}_l-\mathbf{R}_m)+\sum_{i}\frac{\mathbf{p}_i^2}{2m}+\sum_{i,l}V(\mathbf{r}_i-\mathbf{R}_l)+\sum_{i,j}\frac{e^2/4\pi\epsilon_0}{|\mathbf{r}_i-\mathbf{r}_j|} \tag{1.1}$$

where l and m label the ions, i and j label the electrons, $\mathbf{p}$ is the momentum, M is the ionic mass, m is the mass of the electron, $U(\mathbf{R}_l-\mathbf{R}_m)$ is the interionic potential, and $V(\mathbf{r}_i-\mathbf{R}_l)$ is the valence–electron–ion potential.

The Schrödinger equation determines the time-independent energies of the system:

$$H\Xi=E\Xi \tag{1.2}$$

where H is now the Hamiltonian operator.

1.2. Adiabatic approximation

The mass of an ion is at least a factor of $1{\cdot}8\times10^3$ greater than that of an electron, and for most semiconductors the factor is well over 10^4. For comparable energies and perturbations ions therefore move some 10^2 times slower than do electrons, and the latter can be regarded as instantaneously adjusting their motion to that of the ions. Therefore the total wavefunction is approximately of the form

$$\Xi=\Psi(\mathbf{r},\mathbf{R})\Phi(\mathbf{R}) \tag{1.3}$$

where $\Phi(\mathbf{R})$ is the wavefunction for all the ions and $\Psi(\mathbf{r},\mathbf{R})$ is the wavefunction for all the electrons instantaneously dependent on the ionic position.

The Schrödinger equation can be written

$$\Psi(\mathbf{r},\mathbf{R})H_{\mathrm{L}}\Phi(\mathbf{R})+\Phi(\mathbf{R})H_{\mathrm{e}}\Psi(\mathbf{r},\mathbf{R})+H'\Psi(\mathbf{r},\mathbf{R})\Phi(\mathbf{R})=E\Psi(\mathbf{r},\mathbf{R})\Phi(\mathbf{R}) \quad (1.4)$$

where

$$H'\Psi(\mathbf{r},\mathbf{R})\Phi(\mathbf{R})=H_{\mathrm{L}}\Psi(\mathbf{r},\mathbf{R})\Phi(\mathbf{R})-\Psi(\mathbf{r},\mathbf{R})H_{\mathrm{L}}\Phi(\mathbf{R}) \quad (1.5)$$

$$H_{\mathrm{L}}=\sum_{l}\frac{\mathbf{p}_l^2}{2M_l}+\sum_{l,m}U(\mathbf{R}_l-\mathbf{R}_m) \quad (1.6)$$

$$H_{\mathrm{e}}=\sum_{i}\frac{\mathbf{p}_i^2}{2m}+\sum_{i,l}V(\mathbf{r}_i-\mathbf{R}_l)+\sum_{i,j}\frac{e^2/4\pi\epsilon_0}{|\mathbf{r}_i-\mathbf{r}_j|}. \quad (1.7)$$

The relative contribution of H' is of the order m/M_l. The adiabatic approximation consists of neglecting this term. In this case eqn (1.4) splits into a purely ionic equation

$$H_{\mathrm{L}}\Phi(\mathbf{R})=E_{\mathrm{L}}\Phi(\mathbf{R}) \quad (1.8)$$

and a purely electronic equation

$$H_{\mathrm{e}}\Psi(\mathbf{r},\mathbf{R})=E_{\mathrm{e}}\Psi(\mathbf{r},\mathbf{R}). \quad (1.9)$$

1.3. Phonons

Provided that the ions do not move far from their equilibrium positions in the solid their motion can be regarded as simple harmonic. If the equilibrium position of an ion is denoted by the vector $\mathbf{R}_{l0}$ and its displacement by $\mathbf{u}_l$, the Hamiltonian can be written

$$H_{\mathrm{L}}=\sum_{l}\frac{\mathbf{p}_l^2}{2M_l}+\sum_{l,m}D_{l,m}(\mathbf{R}_l-\mathbf{R}_m)\mathbf{u}_l\mathbf{u}_m+H_{\mathrm{L0}}(\mathbf{R}_{l0})+H'_{\mathrm{L}} \quad (1.10)$$

where $D_{lm}(\mathbf{R}_l-\mathbf{R}_m)$ is the restoring force per unit displacement, $H_{\mathrm{L0}}(\mathbf{R}_{l0})$ is an additive constant dependent only on the equilibrium separation of the ions, and H'_{L} represents the contribution of anharmonic forces. The displacements can be expanded in terms of the normal modes of vibration of the solid. The latter take the form of longitudinally polarized and transversely polarized acoustic waves plus, in the case of lattices with a basis, i.e. more than one atom per primitive unit cell, longitudinally and transversely polarized 'optical' modes. (See Section 3.9 for an account of the theory for long-wavelength acoustic modes.) Ionic motion therefore manifests itself in terms of travelling plane waves

$$\mathbf{u}(\omega,\mathbf{q})=\mathbf{u}_0\exp\{\mathrm{i}(\mathbf{q}.\mathbf{r}-\omega t)\} \quad (1.11)$$

which interact weakly with one another through the anharmonic term H'_{L}. Figure 1.1 shows the typical dispersion relation between ω and $\mathbf{q}$.

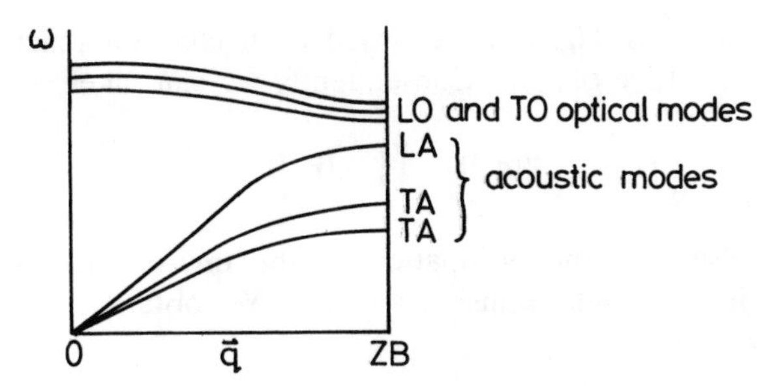

FIG. 1.1. Dispersion of lattice waves.

The energy in a mode is given by

$$E(\omega, \mathbf{q}) = \{n(\omega, \mathbf{q}) + \tfrac{1}{2}\}\hbar\omega \tag{1.12}$$

where $n(\omega, \mathbf{q})$ is the statistical average number of phonons, i.e. vibrational quanta, excited. At thermodynamic equilibrium $n(\omega, q) = n(\omega)$ is given by the Bose–Einstein function for a massless particle

$$n(\omega) = \frac{1}{\exp(\hbar\omega/k_B T) - 1}\,. \tag{1.13}$$

The following points should be noted.

(1) The limits of $\mathbf{q}$ according to periodic boundary conditions are $2\pi/Na$ and the Brillouin zone boundary, where N is the number of unit cells of length a along the cavity.

(2) The magnitude of a wavevector component is $2\pi l/Na$, where l is an integer. The curves in Fig. 1.1 are really closely spaced points.

(3) An impurity or other defect may introduce localized modes of vibration in its neighbourhood if its mass and binding energy are different enough from those of its host.

(4) For long-wavelength acoustic modes $\omega = v_s q$. For others it is often useful to approximate their dispersion by $\omega =$ *constant.*

1.4. The one-electron approximation

If the electron–electron interaction is averaged we can regard any deviation from this average as a small perturbation. Thus we replace the repulsion term as follows:

$$\sum_{i,j} \frac{(e^2/4\pi\epsilon_0)}{|\mathbf{r}_i - \mathbf{r}_j|} = H_{e0} + H_{ee} \tag{1.14}$$

where H_{e0} contributes a constant repulsive component to the electronic energy and H_{ee} is a fluctuating electron–electron interaction which can be regarded as small. If H_{ee} is disregarded each electron reacts independently with the lattice of ions. Consequently we can take

$$\Psi(\mathbf{r}, \mathbf{R}) = \prod_i \psi_i(\mathbf{r}_i, \mathbf{R}) \tag{1.15}$$

with the proviso that the occupation of the one-electron states is in accordance with the Pauli exclusion principle. We obtain the one-electron Schrödinger equation

$$H_{ei}\psi_i(\mathbf{r}_i\mathbf{R}) = E_{ei}\psi_i(\mathbf{r}_i, \mathbf{R}) \tag{1.16}$$

where

$$H_{ei} = \frac{\mathbf{p}_i^2}{2m} + \sum_l V(\mathbf{r}_i - \mathbf{R}_l). \tag{1.17}$$

This Hamiltonian still depends on the fluctuating position of ions, and it is useful to reduce the Hamiltonian into one that depends on the interaction with the ions in their equilibrium positions with the effect of ionic vibrations taken as a perturbation. Thus we take

$$H_{ei} = H_{e0i} + H_{ep} \tag{1.18}$$

$$H_{e0i} = \frac{\mathbf{p}_i^2}{2m} + \sum_l V(\mathbf{r}_i - \mathbf{R}_{l0}) \tag{1.19}$$

where the H_{ep} is the electron–phonon interaction. The electronic band structure is obtained from (dropping the subscripts i and e)

$$\left\{\frac{\mathbf{p}^2}{2m} + \sum_l V(\mathbf{r} - \mathbf{R}_{l0})\right\}\psi(\mathbf{r}) = E\psi(\mathbf{r}). \tag{1.20}$$

1.5. Bloch functions

In the case of a perfectly periodic potential the eigenfunction is a Bloch function:

$$\psi_{n\mathbf{k}}(\mathbf{r}) = u_{n\mathbf{k}}(\mathbf{r})\exp(i\mathbf{k}.\mathbf{r}) \tag{1.21}$$

$$u_{n\mathbf{k}}(\mathbf{r} + \mathbf{R}) = u_{n\mathbf{k}}(\mathbf{r}) \tag{1.22}$$

where $\mathbf{R}$ is a vector of the Bravais lattice, n labels the band and $\mathbf{k}$ is the wavevector of the electron in the first Brillouin zone (Fig. 1.2). From eqns (1.21) and (1.22) if follows that

$$\psi_{n\mathbf{k}}(\mathbf{r} + \mathbf{R}) = \psi_{n\mathbf{k}}(\mathbf{r})\exp(i\mathbf{k}.\mathbf{R}). \tag{1.23}$$

If a macroscopic volume V is chosen whose shape is a magnified version

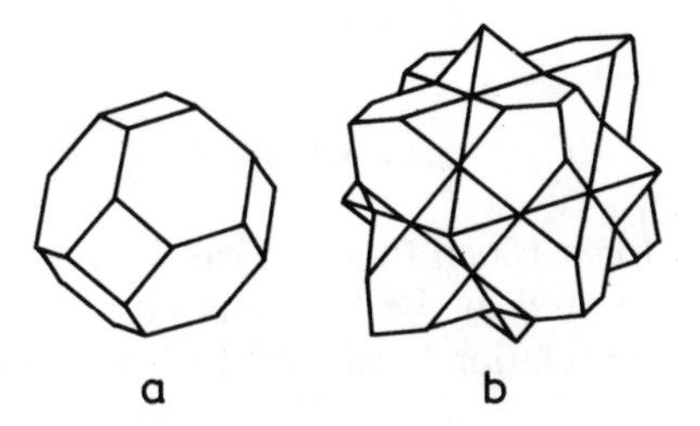

FIG. 1.2. The first and second zones for a face-centred cubic lattice. The first has half the volume of the cube that is determined by extending the six square faces. The second has the same volume as this cube.

of the primitive cell, then we can apply the periodic boundary condition

$$\psi_{n\mathbf{k}}(\mathbf{r}+N\mathbf{a})=\psi_{n\mathbf{k}}(\mathbf{r}) \tag{1.24}$$

where $\mathbf{a}$ is a vector of the unit cell and N is the number of unit cells along the side of V in the direction of $\mathbf{a}$. This decouples the properties of the wavefunction from the size of a crystal, provided the crystal is macroscopic. Equations (1.23) and (1.24) constrain $\mathbf{k}$ such that

$$\exp(\mathrm{i}\mathbf{k}.N\mathbf{a})=1.$$

Therefore

$$\mathbf{k}.N\mathbf{a}=2\pi n \tag{1.25}$$

where n is an integer. In terms of reciprocal lattice vectors $\mathbf{K}$, defined by

$$\mathbf{K}_i.\mathbf{a}_j=2\pi\,\delta_{ij} \tag{1.26}$$

the electronic wavevector assumes the values

$$\mathbf{k}=\frac{n_1}{N_1}\mathbf{K}_1+\frac{n_2}{N_2}\mathbf{K}_2+\frac{n_3}{N_3}\mathbf{K}_3. \tag{1.27}$$

Thus the volume of an electronic state in $\mathbf{k}$-space is given by

$$\Delta k_1\,\Delta k_2\,\Delta k_3=(2\pi)^3/V. \tag{1.28}$$

If $\mathbf{q}$ is any vector that satisfies the periodic boundary conditions then the wavefunction can be written generally as an expansion in plane waves:

$$\psi(\mathbf{r})=\sum_{\mathbf{q}} c_{\mathbf{q}}\exp(\mathrm{i}\mathbf{q}.\mathbf{r}). \tag{1.29}$$

This general expansion can be related to the Bloch form by putting $\mathbf{q}=\mathbf{k}-\mathbf{K}$ where $\mathbf{k}$ is not necessarily confined to the first Brillouin zone:

$$\psi_{\mathbf{k}}(\mathbf{r})=\exp(\mathrm{i}\mathbf{k}.\mathbf{r})\sum_{\mathbf{K}} c_{\mathbf{k}-\mathbf{K}}\exp(-\mathrm{i}\mathbf{K}.\mathbf{r}) \tag{1.30}$$

and thus

$$u_{\mathbf{k}}(\mathbf{r}) = \sum_{\mathbf{K}} c_{\mathbf{k}-\mathbf{K}} \exp(-\mathrm{i}\mathbf{K}.\mathbf{r}). \tag{1.31}$$

Yet another form for a Bloch function can be formed out of functions $\phi_n(\mathbf{r}-\mathbf{R})$ which are centred at the lattice points $\mathbf{R}$. These are known as Wannier functions. The relation between Bloch and Wannier functions is

$$\psi_{n\mathbf{k}}(\mathbf{r}) = \sum_{\mathbf{R}} \phi_n(\mathbf{r}-\mathbf{R}) \exp(\mathrm{i}\mathbf{k}.\mathbf{R}). \tag{1.32}$$

This is a useful formulation for describing narrow energy bands when the Wannier function can be approximated by atomic orbitals in the tight-binding approximation.

Since the Bloch functions are eigenfunctions of the one-electron Schrödinger equation they are orthogonal to one another, *viz.*

$$\int \psi^*_{n'\mathbf{k}'} \psi_{n\mathbf{k}} \, \mathrm{d}\mathbf{r} = \delta_{n'n} \, \delta_{\mathbf{k}'\mathbf{k}} \tag{1.33}$$

with

$$\psi_{n\mathbf{k}} = \frac{1}{V^{1/2}} u_{n\mathbf{k}}(\mathbf{r}) \exp(\mathrm{i}\mathbf{k}.\mathbf{r}). \tag{1.34}$$

1.6. Nearly-free-electron model

When the periodic potential is very weak the valence electron is almost free, and hence

$$E_{\mathbf{k}} \approx \hbar^2 k^2/2m. \tag{1.35}$$

In the cases of semiconductors with diamond and sphalerite structure there are two atoms in each primitive cell and eight valence electrons. Therefore there have to be four valence bands with two electrons of opposing spin in each state. By allowing $\mathbf{k}$ to extend beyond the first zone, we can work out the total width of the four valence bands by equating it with the Fermi energy E_{F} for a free-electron gas of the same density as the valence electrons. Observations of soft X-ray emission confirm that the width of the valence band in these semiconductors is indeed close to E_{F}. Thus it is reasonable to assume that the valence electrons are almost free, and eqn (1.35) is a good approximation to the energy provided we take into account the effect of the lattice.

Restricting $\mathbf{k}$ to the first Brillouin zone (Fig. 1.2) we obtain

$$E_{\mathbf{k}} \approx \hbar^2 q^2/2m \tag{1.36}$$

$$\mathbf{q} = \mathbf{k} + \mathbf{K}. \tag{1.37}$$

The first band is obtained for $\mathbf{K}=(0,0,0)2\pi/a$, and is obviously parabolic. At the zone boundary there is an energy gap in general. The second band is obtained from the smallest non-zero reciprocal lattice vectors, which are $\mathbf{K}=(1,1,1)2\pi/a$ and its cubic fellows (e.g. $(-1,1,1)2\pi/a$) and $\mathbf{K}_2=(2,0,0)2\pi/a$ and its cubic fellows (e.g. $(0,2,0)2\pi/a$)). At the zone boundary along the $\langle 100\rangle$ direction $q=K_2/2=2\pi/a$ and $k=-K_2/2=-2\pi/a$. As q increases k moves towards zero, reaching it when $q=K_2=4\pi/a$. At the zone boundary along the $\langle 111\rangle$ direction $q=K_1/2=\sqrt{3}\pi/a$ and $k=-K_1/2=-\sqrt{3}\pi/a$. As q increases, k moves to zero, reaching it when $q=K_1=2\sqrt{3}\pi/a$. The band continues to be parabolic in both directions, except close to the zone boundaries.

The first and second bands are parabolic directions because the appropriate reciprocal lattice vector simply subtracts from q. Bands 3 and 4 are not that simple because $\mathbf{K}_1$ and $\mathbf{K}_2$ are neither parallel nor anti-parallel in this case. The region in reciprocal lattice space which contains the first four Brillouin zones is the Jones zone (Fig. 1.3).

Bands 1 and 2 reach the surface of the Jones zone at the points (2, 0, 0) and (1, 1, 1). Bands 3 and 4 are associated with combinations of $\mathbf{k}$, $\mathbf{K}_1$, and $\mathbf{K}_2$ which keep $\mathbf{q}$ close to the zone boundary for all $\mathbf{k}$. The smallest $\mathbf{q}$ corresponds to the centre of a face $\mathbf{q}=\mathbf{K}_1-\mathbf{K}_2/2$ ($q=2\sqrt{2}\pi/a$). With $\mathbf{k}$ along the $\langle 100\rangle$ direction the band is described by $\mathbf{q}\doteq\mathbf{K}_1-\mathbf{k}$. When $\mathbf{k}=0$, $\mathbf{q}=\mathbf{K}_1$ ($q=2\sqrt{3}\pi/a$). Thus q changes by an amount $\sqrt{3}-\sqrt{2}$ in units of $2\pi/a$ as $\mathbf{k}$ sweeps through the zone in the $\langle 100\rangle$ direction, and hence the energy changes very little with $\mathbf{k}$. This band is far from being free-electron-like. The other band is also flat, for again q changes comparatively little with $\mathbf{k}$ because $\mathbf{k}$ is more or less perpendicular to the reciprocal vector.

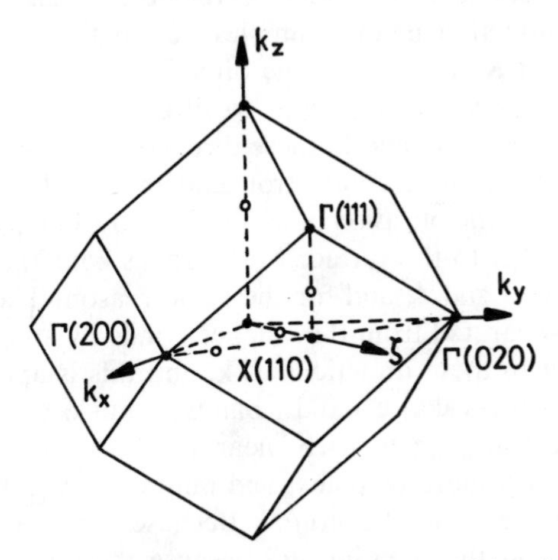

FIG. 1.3. The Jones zone for face-centred cubic crystals containing eight electrons per cell includes the first four Brillouin zones.

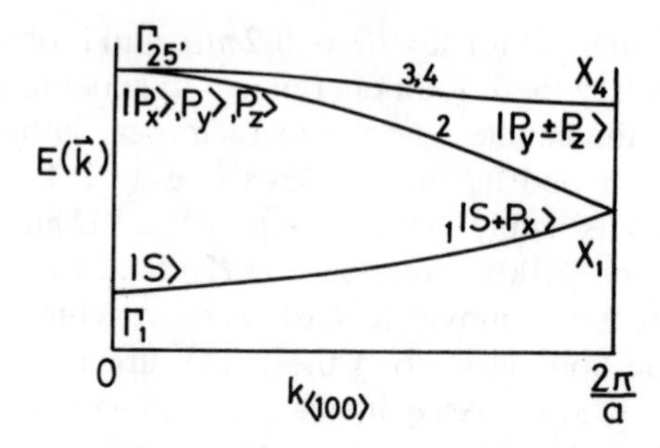

FIG. 1.4. The general form of the valence band. The symmetry symbols and orbitals are appropriate for diamond and silicon.

The free-electron model predicts different energies at the $\mathbf{q}$ corresponding to the corners of the Jones zone. We have already seen that $\mathbf{q} = \mathbf{K}_1$ at one. Along the cube-edge direction $\mathbf{q} = \mathbf{K}_2$ ($q = 4\pi/a$), which has a greater magnitude. Thus the energy of the latter is greater on the free-electron model. The periodic potential with its cubic symmetry (in the case under discussion) alters that picture to one in which these energies are identical. This common energy is the maximum attained at the surface of the Jones zone, which is a free-electron-like result since the corners are indeed furthest from the zone centre. The minimum energy for bands 3 and 4 appearing at $\mathbf{k} = \mathbf{K}_2/2$, corresponding to the middle of the face which is the nearest point of the surface to the zone centre, is equally free electron like. The valence band has the general form depicted in Fig. 1.4.

Pushing the free-electron model beyond the Jones zone in order to describe the conduction band is simplest, as in the case of the valence band, when $\mathbf{k}$ and $\mathbf{K}$ are in the same direction. In bands 3 and 4 and in bands 5 and 6 this will correspond to directions perpendicular to the surface of the Jones zone, and in these directions the mass of the electron will be near that for the free electron and the bands will be parabolic. However, in the principle directions ⟨100⟩ and ⟨111⟩, for which $\mathbf{k}$ lies more or less parallel to the surface of the Jones zone, bands 5 and 6 will be flat like bands 3 and 4, and for the same reason. They might also be expected to be separated from the valence bands by an energy gap which remains constant in these directions of $\mathbf{k}$, and this is approximately true. However, unlike the valence bands, bands 5 and 6 are not degenerate. Moreover, their form, particularly near the corners and edges of the Jones zone, is much more complex, and minima at $\mathbf{k} = 0(\Gamma)$, $\mathbf{k} = \mathbf{K}_2/2(\mathrm{X})$, and $\mathbf{k} = \mathbf{K}_1/2(\mathrm{L})$ are general features. Because the minimum energy of bands 3 and 4 is at the X point and because the energy gap is roughly constant it is expected that the lowest minimum of the conduction band also lies at X, but this is only roughly correct and holds only for semiconductors involving elements in the early rows of the periodic table.

1.6.1. Group theory notation

Many features of band structure depend upon symmetry, and in particular the symmetry of the cubic lattice. The special points in the Brillouin zone of a simple lattice are shown in Fig. 1.5 (see also Figs. 1.6 and 1.7). The symmetry types at various points as illustrated by simple basis functions are given in Table 1.1. The representations for the Γ group are given in Table 1.2.

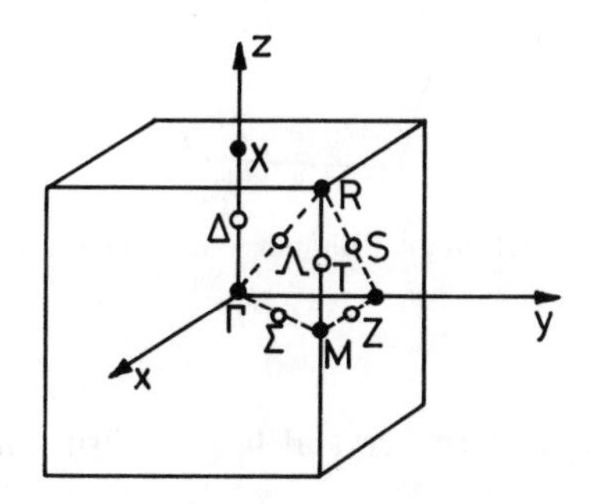

FIG. 1.5. Special points in the Brillouin zone of a simple cubic lattice.

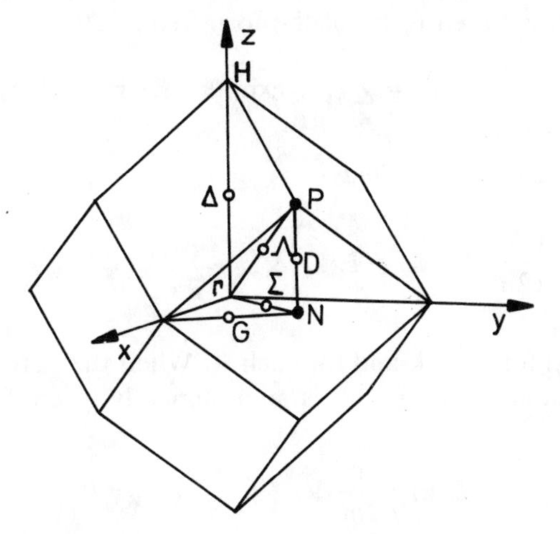

FIG. 1.6. Special points in the Brillouin zone of a body-centred cubic lattice.

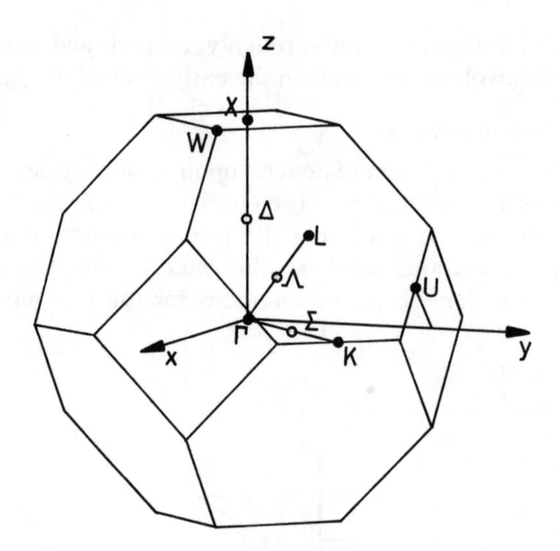

FIG. 1.7. Special points in the Brillouin zone of a face-centred cubic lattice.

1.7. Energy gaps

The periodic potential of the crystal (eqn (1.20)) can be expanded in a Fourier series:

$$\sum_{l} V(\mathbf{r}-\mathbf{R}_l) = \sum_{\mathbf{K}'} V_{\mathbf{K}'} \exp(\mathrm{i}\mathbf{K}'.\mathbf{r}) \tag{1.38}$$

and a constant potential can be added to make $V_{\mathbf{K}'} = 0$ when $\mathbf{K}' = 0$. If the Bloch function is taken to be of the form (eqn 1.30)

$$\psi_{\mathbf{k}} = \sum_{\mathbf{K}} c_{\mathbf{k}-\mathbf{K}} \exp\{\mathrm{i}(\mathbf{k}-\mathbf{K}).\mathbf{r}\} \tag{1.39}$$

the Schrödinger equation yields†

$$\left\{\frac{\hbar^2}{2m}(\mathbf{k}-\mathbf{K})^2 - E(\mathbf{k})\right\} c_{\mathbf{k}-\mathbf{K}} + \sum_{\mathbf{K}'} V_{\mathbf{K}'-\mathbf{K}} c_{\mathbf{k}-\mathbf{K}'} = 0 \tag{1.40}$$

which will hold for each **k** and for each **K**. When the periodic potential is weak the solution of eqn (1.40) is approximately given by

$$E(\mathbf{k}) \approx \frac{\hbar^2}{2m}(\mathbf{k}-\mathbf{K})^2 \qquad c_{\mathbf{k}-\mathbf{K}} \neq 0 \tag{1.41}$$

† $\sum_{\mathbf{K},\mathbf{K}'} V_{\mathbf{K}'} c_{\mathbf{k}-\mathbf{K}} \exp\{\mathrm{i}(\mathbf{k}-\mathbf{K}+\mathbf{K}').\mathbf{r}\} = \sum_{\mathbf{K},\mathbf{K}'} V_{\mathbf{K}'-\mathbf{K}} c_{\mathbf{k}-\mathbf{K}'} \exp\{\mathrm{i}(\mathbf{k}-\mathbf{K}).\mathbf{r}\}$

TABLE 1.1
Symmetry types and basis functions of a cubic lattice

Points	Notation		Basis functions
Γ, R	Γ_1 [a]	or A_{1g} [b]	1
	Γ_2	A_{2g}	$x^4(y^2-z^2)+y^4(z^2-x^2)+z^4(x^2-y^2)$
	Γ_{12}	E_g	$z^2-\frac{1}{2}(x^2+y^2)$, x^2-y^2
	$\Gamma_{15'}$	T_{1g}	$xy(x^2-y^2)$, $yz(y^2-z^2)$, $zx(z^2-x^2)$
	$\Gamma_{25'}$	T_{2g}	xy, yz, zx
	Γ'_1	A_{1u}	$xyz\{x^4(y^2-z^2)+y^4(z^2-x^2)+z^4(x^2-y^2)\}$
	Γ'_2	A_{20}	xyz
	$\Gamma_{12'}$	E_u	$xyz\{z^2-\frac{1}{2}(x^2+y^2)\}$, $xyz(x^2-y^2)$
	Γ_{15}	T_{1u}	x, y, z
	Γ_{25}	T_{2u}	$z(x^2-y^2)$, $x(y^2-z^2)$, $y(z^2-x^2)$
X, M	X_1		1
	X_2		x^2-y^2
	X_3		xy
	X_4		$xy(x^2-y^2)$
	X_5		yz, zx
	$X_{1'}$		$xyz(x^2-y^2)$
	$X_{2'}$		xyz
	$X_{3'}$		$z(x^2-y^2)$
	$X_{4'}$		z
	$X_{5'}$		x, y
Δ, T	Δ_1		1
	Δ_2		x^2-y^2
	Δ_3		xy
	Δ_4		$xy(x^2-y^2)$
	Δ_5		x, y
Σ, S	Σ_1		1
	Σ_2		$z(x-y)$
	Σ_3		z
	Σ_4		$x-y$
Z	z_1		1
	z_2		yz
	z_3		y
	z_4		z
Λ	Λ_1		1
	Λ_2		$xy(x-y)+yz(y-z)+zx(z-x)$
	Λ_3		$x-z$, $y-z$

From Kittel 1963.
[a] Notation of Bouckaert, Smoluchowski, and Wigner 1936.
[b] Chemical notation: g stands for even function, u for odd function, under inversion.

TABLE 1.2
Γ *Group representation with spin*

Without spin	$\Gamma_1\Gamma_2\Gamma_{12}$	$\Gamma_{15'}$	$\Gamma_{25'}$	$\Gamma'_1\Gamma'_2\Gamma_{12'}$	Γ_{15}	Γ_{25}
With spin	$\Gamma_6\Gamma_7\Gamma_8$	$\Gamma_6+\Gamma_8$	$\Gamma_7+\Gamma_8$	$\Gamma'_6\Gamma'_7\Gamma'_8$	$\Gamma'_6+\Gamma'_8$	$\Gamma'_7+\Gamma'_8$

Γ_6 and Γ_7 are doublets; Γ_8 is a quartet.

or

$$E(\mathbf{k}) \neq \frac{\hbar^2}{2m}(\mathbf{k}-\mathbf{K})^2 \qquad c_{\mathbf{k}-\mathbf{K}} \approx 0. \tag{1.42}$$

In the non-degenerate case, when eqn (1.41) is satisfied for only one $\mathbf{K}$, say $\mathbf{K}=\mathbf{K}_a$, a more accurate solution is obtained by noting that for $\mathbf{K}=\mathbf{K}_b$, where $\mathbf{K}_b \neq \mathbf{K}_a$,

$$c_{\mathbf{k}-\mathbf{K}_b} = \frac{\sum_{\mathbf{K}'} V_{\mathbf{K}'-\mathbf{K}_b} c_{\mathbf{k}-\mathbf{K}'}}{E_a - (\hbar^2/2m)(\mathbf{k}-\mathbf{K}_b)^2} \approx \frac{V_{\mathbf{K}_a-\mathbf{K}_b} c_{\mathbf{k}-\mathbf{K}_a}}{E_a - E_b} \tag{1.43}$$

where $E_a = \hbar(\mathbf{k}-\mathbf{K}_a)^2/2m$ and $E_b = \hbar^2(\mathbf{k}-\mathbf{K}_b)^2/2m$. Putting this back in eqn (1.40) gives

$$E(\mathbf{k}) = E_a + \sum_{\mathbf{K}'} \frac{V_{\mathbf{K}'-\mathbf{K}_a} V_{\mathbf{K}_a-\mathbf{K}'}}{E_a - E'} = E_a + \sum_{\mathbf{K}'} \frac{|V_{\mathbf{K}'-\mathbf{K}_a}|^2}{E_a - E'}. \tag{1.44}$$

The effect of energy states lying above E is to depress $E(\mathbf{k})$, and the effect of those lying below E_a is to raise $E(\mathbf{k})$, i.e. energy levels tend to repel one another.

In the degenerate case, when eqn (1.41) is satisfied for a given eigenvalue $E(\mathbf{k})$ by, say, $\mathbf{K}=\mathbf{K}_a=\mathbf{K}_b$, neglecting all coefficients except those for $\mathbf{K}_a$ and $\mathbf{K}_b$ gives

$$\begin{aligned} \{E_a - E(\mathbf{k})\} c_{\mathbf{k}-\mathbf{K}_a} + V_{\mathbf{K}_b-\mathbf{K}_a} c_{\mathbf{k}-\mathbf{K}_b} &= 0 \\ \{E_b - E(\mathbf{k})\} c_{\mathbf{k}-\mathbf{K}_b} + V_{\mathbf{K}_a-\mathbf{K}_b} c_{\mathbf{k}-\mathbf{K}_a} &= 0, \end{aligned} \tag{1.45}$$

whence, with $K = |\mathbf{K}_b - \mathbf{K}_a|$,

$$\{E_a - E(\mathbf{k})\}\{E_b - E(\mathbf{k})\} - |V_K|^2 = 0.$$

Therefore

$$E(\mathbf{k}) = \tfrac{1}{2}(E_a + E_b) \pm \tfrac{1}{2}\{(E_a+E_b)^2 - 4(E_a E_b - |V_K|^2)\}^{1/2} \tag{1.46}$$

which for $E_a = E_b = E_0$ becomes

$$E(\mathbf{k}) = E_0 \pm |V_K| \tag{1.47}$$

corresponding to an energy gap of magnitude $2|V_K|$.

When the unit cell consists of two atoms, which are not necessarily identical, the potential can be denoted

$$V(\mathbf{r}-\mathbf{R}_{l0}) = V_1(\mathbf{r}-\mathbf{R}_{l1}) + V_2(\mathbf{r}-\mathbf{R}_{l2}) \tag{1.48}$$

where $\mathbf{R}_{l0}$, $\mathbf{R}_{l1}$, and $\mathbf{R}_{l2}$ are the position vectors of the centre of the cell, atom 1, and atom 2 respectively. If $\mathbf{R}_{l1} = \mathbf{R}_{l0} + \mathbf{r}_c$ and $\mathbf{R}_{l2} = \mathbf{R}_{l0} - \mathbf{r}_c$, where $\mathbf{r}_c$ is the covalent bond length (half the interatomic distance), eqn (1.48) becomes

$$V(\mathbf{r}-\mathbf{R}_{l0}) = V_1(\mathbf{r}-\mathbf{r}_c-\mathbf{R}_{l0}) + V_2(\mathbf{r}+\mathbf{r}_c-\mathbf{R}_{l0}) \tag{1.49}$$

and so, instead of eqn (1.38), we can put

$$\sum_l V(\mathbf{r}-\mathbf{R}_{l0}) = \sum_{\mathbf{K}'} \{V_{1\mathbf{K}'} \exp(-i\mathbf{K}'.\mathbf{r}_c) + V_{2\mathbf{K}'} \exp(i\mathbf{K}'.\mathbf{r}_c)\}\exp(i\mathbf{K}'.\mathbf{r})$$

$$= \sum_{\mathbf{K}'} (V^S_{\mathbf{K}'} \cos \mathbf{K}'.\mathbf{r}_c + iV^A_{\mathbf{K}'} \sin \mathbf{K}'.\mathbf{r}_c)\exp(i\mathbf{K}'.\mathbf{r}) \tag{1.50}$$

where a division has been made into a symmetric part $V^S_{\mathbf{K}'} = (V_{1\mathbf{K}'} + V_{2\mathbf{K}'})/2$ and an antisymmetric part $V^A_{\mathbf{K}'} = (V_{2\mathbf{K}'} - V_{1\mathbf{K}'})/2$. The latter will be zero if the atoms are identical. The energy gap is therefore given by

$$E_g^2 = 4(|V^S_K \cos Kr_c|^2 + |V^A_K \sin Kr_c|^2). \tag{1.51}$$

Thus the energy gap can be regarded as having a symmetric or homopolar component and an antisymmetric or polar component. A division in this way is useful for understanding how the covalent and polar aspects of binding influence the electronic structure of semiconductors, as we shall see.

The eigenfunctions for the two bands (eqn 1.39), at the band-edges, are of the form

$$\psi_{a,b\mathbf{k}} = \{c_{a\mathbf{k}} \exp(-i\mathbf{K}_a.\mathbf{r}) \pm c_{b\mathbf{k}} \exp(-i\mathbf{K}_b.\mathbf{r})\}\exp(i\mathbf{k}.\mathbf{r}) \tag{1.52}$$

with $|c_{a\mathbf{k}}| = |c_{b\mathbf{k}}|$. In the case of the direct gap between valence and conduction bands we can take $\mathbf{K}_a = 0$, $\mathbf{k} \to \mathbf{k} + \mathbf{k}_F$, where $\mathbf{k}$ is now restricted to the first Brillouin zone, $\mathbf{k}_F$ is the Fermi-level vector for the free-electron valence band, and $\mathbf{K}_b \approx 2\mathbf{k}_F$, which in effect assumes the Jones zone to be a sphere of radius k_F. In this approximation (isotropic model)

$$\psi_{a,b\mathbf{k}} = \{c_{a\mathbf{k}} \exp(i\mathbf{k}_F.\mathbf{r}) \pm c_{b\mathbf{k}} \exp(-i\mathbf{k}_F.\mathbf{r})\}\exp(i\mathbf{k}.\mathbf{r}). \tag{1.53}$$

1.8. Spin–orbit coupling and orbital characteristics

In the crystal Hamiltonian of eqn (1.1) we left out magnetic energy. Even in non-magnetic semiconductors the contribution of magnetic energy, although small compared with purely electrostatic components, is not negligible in relation to the energy gaps. The source of this energy in non-magnetic materials is the interaction between spin and orbit, as described by the one-electron Hamiltonian

$$H_{so} = \frac{\hbar^2}{4m^2c^2}[\nabla V(\mathbf{r}-\mathbf{R}_{l0}) \times \mathbf{p}].\boldsymbol{\sigma} \tag{1.54}$$

where $\mathbf{p}$ is the momentum operator and $\boldsymbol{\sigma}$ is the spin operator. As in the case of a free atom the effect of spin–orbit coupling is to remove orbital degeneracies.

The degeneracies occurring in the band structure of semiconductors near the principle energy gap are those associated with bands 2, 3, and 4 at $\mathbf{k}=0$, and between bands 3 and 4 elsewhere in the zone. In order to discover how spin–orbit coupling splits these degeneracies we need to know the orbital characteristics associated with these bands. For diamond, sphalerite, and wurzite type crystals the constituent atoms in their free state possess valence electrons in the atomic $|s\rangle$ and $|p\rangle$ states, and we expect the valence and indeed the lower conduction bands to have $|s\rangle$- and $|p\rangle$-like orbital characteristics. (In the case of heavier atoms $|d\rangle$ states are also involved, but we ignore this complication for the present.) The chemical picture of bonding in these lattices involves the production of sp^3 hybridized bonds directed towards the corners of a regular tetrahedron. The two atoms in the unit cell contribute their $|s\rangle$ and $|p\rangle$ orbitals in bonding combinations to give the valence band in which the electron density is high between the atoms, or in anti-bonding combinations to give the conduction band in which the electron density tends to be high at, rather than between, the atoms. We can therefore associate bonding $|s\rangle$ orbital characteristics with the lowest valence band and bonding $|p\rangle$ orbital characteristics with bands 2, 3, and 4 at $\mathbf{k}=0$, and also anti-bonding $|s\rangle$ orbital characteristics with the conduction band, again at $\mathbf{k}=0$. Orbital characteristics can be assigned to high symmetry points in the zone on the basis of symmetry (Table 1.3). (Note that the Bloch function in eqn (1.53), which is that for the two-band isotropic model, evidently has a bonding $|s\rangle$ character or an anti-bonding $|s\rangle$ character.) At other points of the zone the orbitals become a mixture of $|s\rangle$ and $|p\rangle$ characteristics.

Returning to the question of spin–orbit splitting, we can write eqn (1.54) as follows for a spherically symmetric field:

$$H_{so}=\frac{\hbar^2}{4m^2c^2}\frac{1}{r}\frac{dV}{dr}(\mathbf{r}\times\mathbf{p}).\boldsymbol{\sigma}=\lambda\mathbf{L}.\mathbf{S} \tag{1.55}$$

where $\mathbf{L}$ is the orbital angular momentum operator, $\mathbf{S}$ is the spin angular momentum operator, and $\lambda=(dV/dr)/2m^2c^2r$. Since $\mathbf{L}$ and $\mathbf{S}$ combine vectorially to give a total angular momentum $\mathbf{J}$ we can write

$$\mathbf{J}^2=\mathbf{L}^2+\mathbf{S}^2+2\mathbf{L}.\mathbf{S} \tag{1.56}$$

whence

$$\begin{aligned}\langle\mathbf{L}.\mathbf{S}\rangle&=\tfrac{1}{2}\langle\mathbf{J}^2-\mathbf{L}^2-\mathbf{S}^2\rangle\\&=\frac{\hbar^2}{2}\{j(j+1)-l(l+1)-s(s+1)\}.\end{aligned} \tag{1.57}$$

TABLE 1.3
Orbital character of Bloch functions

Symmetry point	Symmetry symbol	Orbital character
Γ	Γ_1	$\|s\rangle$
($\mathbf{k}=0$)	Γ_{15}	$\|p_x\rangle, \|p_y\rangle, \|p_z\rangle$
L, Λ	L_1	$\|s\rangle$
($\mathbf{k}=\mathbf{k}_{\langle 111\rangle}$)	L_1	$\frac{1}{\sqrt{3}}\|p_x+p_y+p_z\rangle$
	L_3	$\left\{\begin{array}{l}\frac{1}{\sqrt{6}}\|p_x+p_y-2p_z\rangle \\ \frac{1}{\sqrt{2}}\|p_x-p_y\rangle\end{array}\right.$
X, Δ	X_1	$\frac{1}{\sqrt{2}}(\|s\rangle_A+\|p_x\rangle_B)$ $\begin{pmatrix}\text{Atom A}\\ \text{Atom B}\end{pmatrix}$
($\mathbf{k}=\mathbf{k}_{\langle 100\rangle}$)	X_3	$\frac{1}{\sqrt{2}}(\|s\rangle_B+\|p_x\rangle_A)$
	X_5	$\frac{1}{\sqrt{2}}\|p_y-p_z\rangle$
		$\frac{1}{\sqrt{2}}\|p_y+p_z\rangle$

From Bassani 1966.

At the top of the valence band are three degenerate $|p\rangle$-like bands. Thus $l=1$ and $j=\frac{3}{2}$ or $\frac{1}{2}$, since $j=l\pm s$, and $s=\frac{1}{2}$ for one electron. For $j=\frac{1}{2}$, $\langle\mathbf{L}.\mathbf{S}\rangle=-\hbar^2$, and for $j=\frac{3}{2}$, $\langle\mathbf{L}.\mathbf{S}\rangle=+\hbar^2/2$. Thus the states are split by an amount Δ_0 proportional to $\frac{3}{2}\hbar^2$, the double degenerate $j=\frac{3}{2}$ state moving up $\Delta_0/3$ and the single $j=\frac{1}{2}$ state moving down by $2\Delta_0/3$ (Fig. 1.8). The upper pair remain degenerate at $\mathbf{k}=0$ and each member retains an orbital characteristic which is $|p\rangle$ like with, for $\mathbf{k}\neq 0$, lobes directed

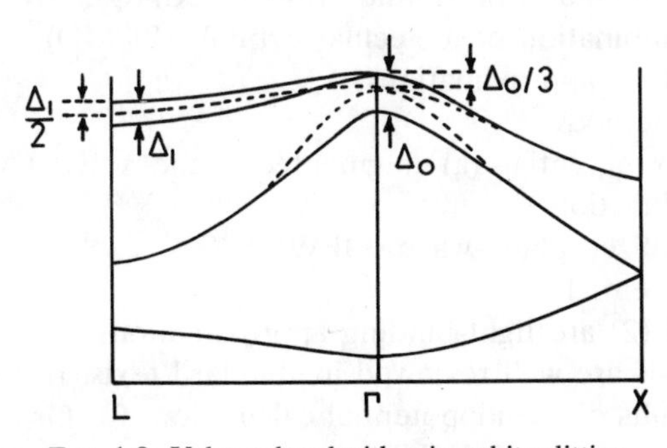

FIG. 1.8. Valence band with spin–orbit splitting.

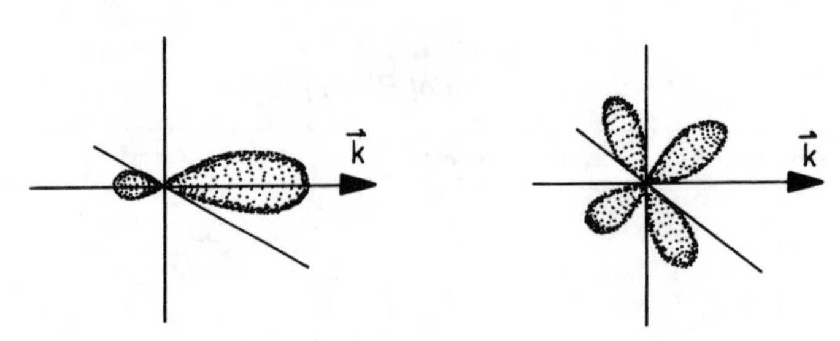

FIG. 1.9. Orbital characters of valence bands.

along mutually perpendicular directions which are themselves perpendicular to the **k** direction. However, the split-off band assumes an orbital which is $|p\rangle$ like with a lobe along the **k** direction when $\mathbf{k} \neq 0$ (Fig. 1.9). When $\mathbf{k} \neq 0$ the degeneracy of the upper bands is removed by spin–orbit splitting, except for **k** along the cube edge. The splitting is maximum along the $\langle 111 \rangle$ (Λ) directions and at L it is denoted by Δ_1. If we take $|p\rangle$ orbitals at the L point we obtain an identical splitting to that at Γ, but since we have only two $|p\rangle$ states and not three we must weight that splitting by a factor of $\frac{2}{3}$. Thus $\Delta_1 = 2\Delta_0/3$, and one band moves up $\Delta_1/2$ and the other down by $\Delta_1/2$ (Fig. 1.8). (In directions other than $\langle 100 \rangle$ and $\langle 111 \rangle$ the degeneracy is removed not only by spin–orbit splitting but by the much larger splitting of the cubic field.)

In silicon the lowest conduction band at Γ is triply degenerate like the top of the valence band and is split only weakly by the spin–orbit interaction. In other semiconductors the conduction band in the vicinity of the gap exhibits non-degenerate valleys at Γ, L, and Δ or X, all of which have an orbital character which is $|s\rangle$ like.

1.9. Band structures

Electronic band structures have been calculated using various models:

(1) linear combination of atomic orbitals (LCAO);
(2) linear combination of molecular orbitals (LCMO);
(3) free-electron approximations;
(4) cellular methods;
(5) muffin-tin potential (a) augmented plane wave (APW) and (b) Green's function;
(6) orthogonalized plane-wave (OPW);
(7) pseudopotential.

Models (1) and (2) are tight-binding approximations.

These methods are well reviewed in standard texts. In Figs. 1.10–1.12 we give the results of pseudopotential calculations for Group IV elemental, III–V compound, and II–VI compound semiconductors. Observed

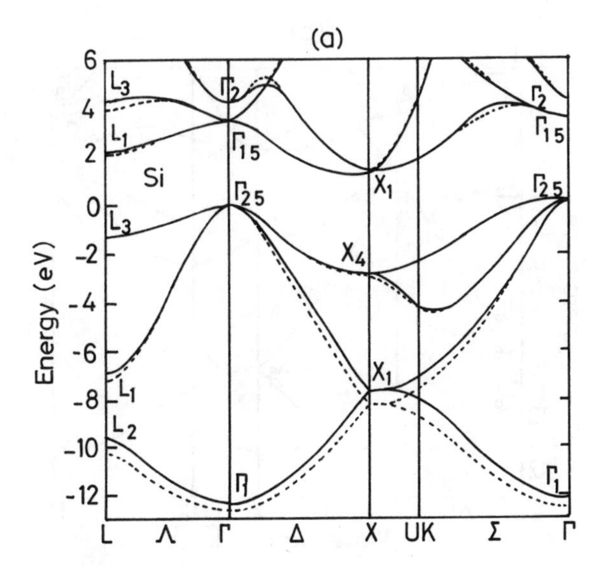

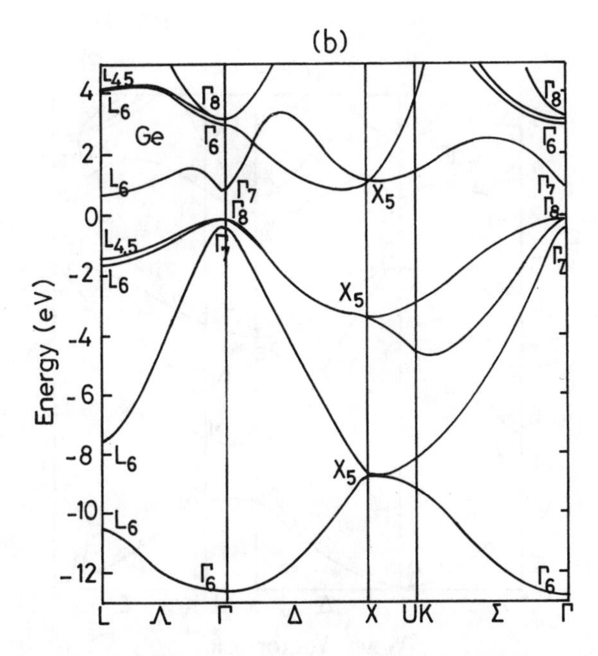

FIG. 1.10. Band structures for (a) silicon, (b) germanium, and (c) α-tin. In the case of silicon two results are presented: the non-local pseudopotential (solid curve) and the local pseudopotential (dotted curve). (From Chelikowsky and Cohen 1976.)

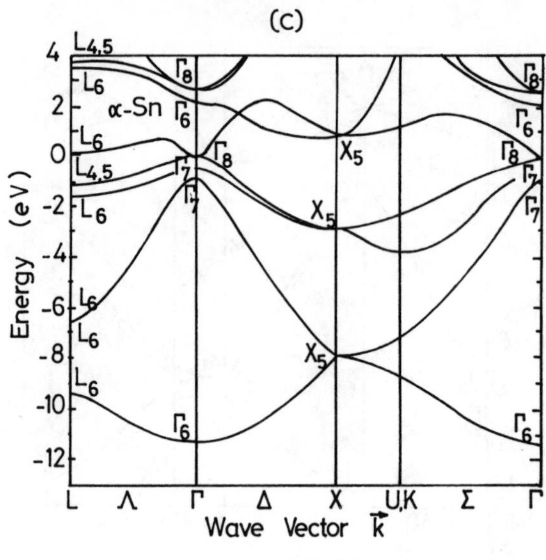

FIG. 1.10. (c).

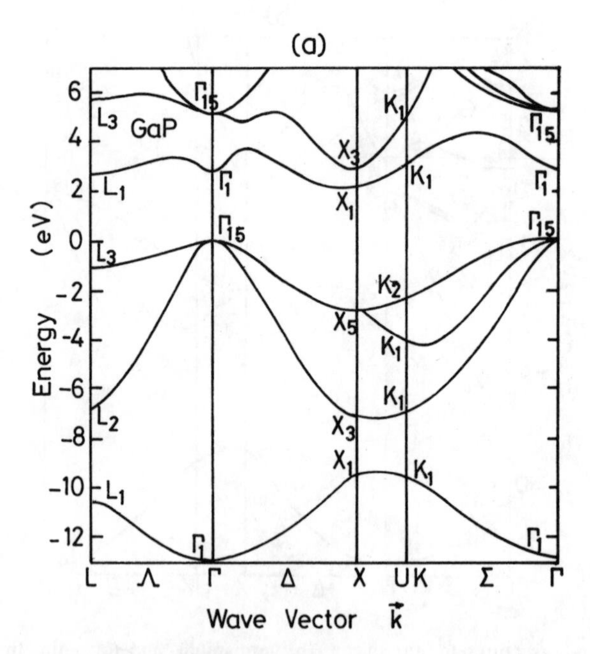

FIG. 1.11. Band structures of III–IV compounds: (a) GaP; (b) GaAs; (c) GaSb; (d) InP; (e) InAs; (f) InSb. (From Chelikowsky and Cohen 1976.)

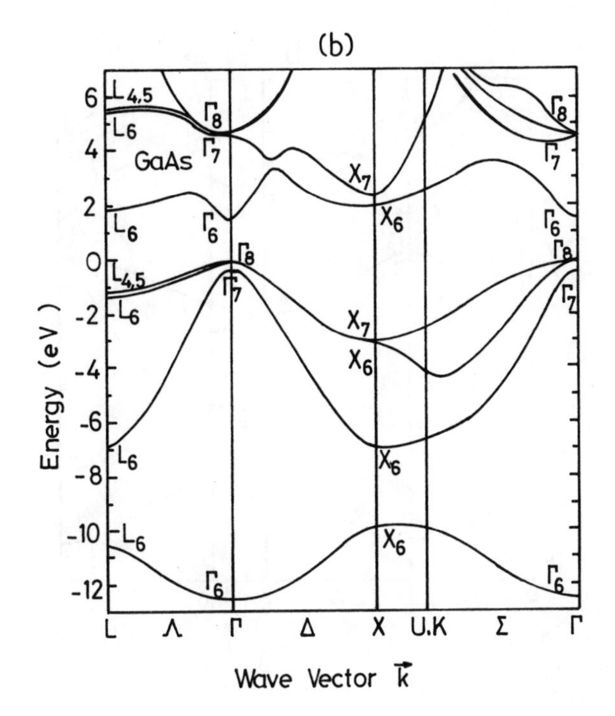

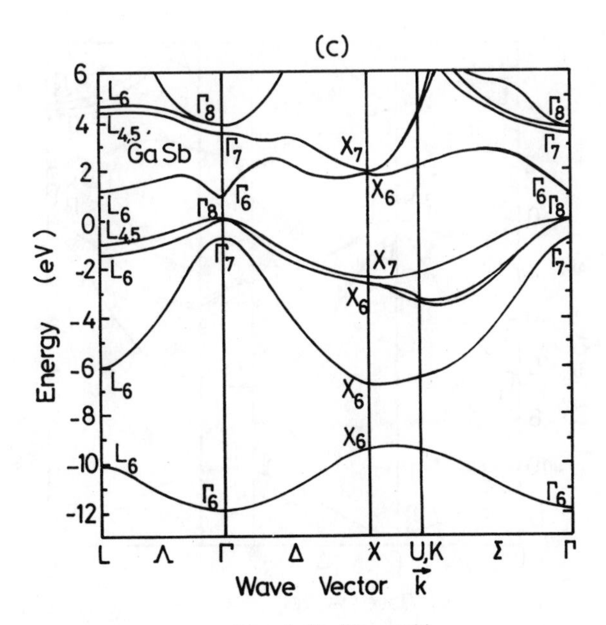

FIG. 1.11. (b) and (c).

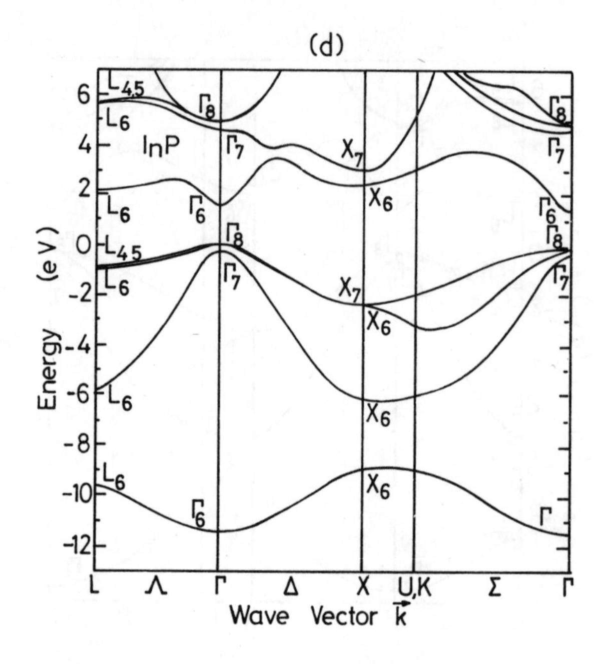

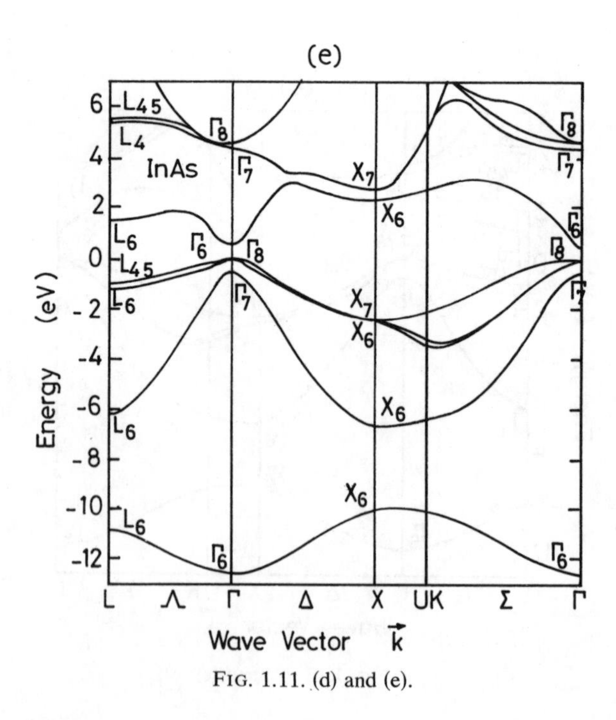

FIG. 1.11. (d) and (e).

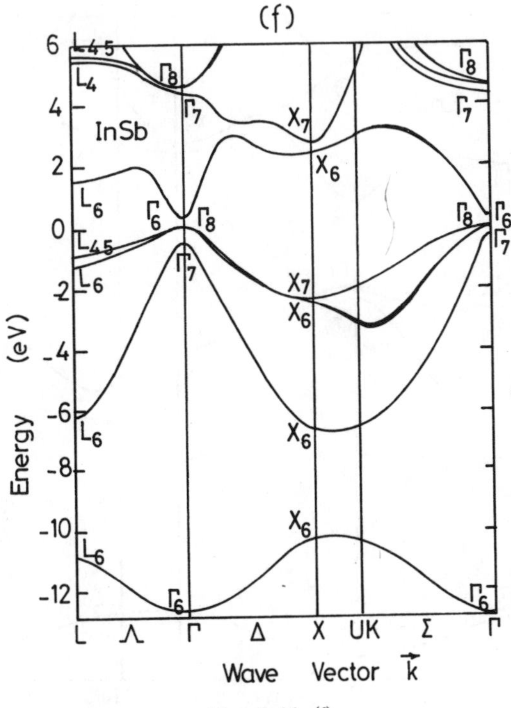

FIG. 1.11. (f).

energy gaps at 300 K between the top of the valence band and the Γ, L, and X valleys in the conduction band for several Group IV and III–V compound semiconductors are depicted in Fig. 1.13.

The major points to be noted are as follows:

(1) light atoms tend to have the X valley lowest in agreement with the nearly-free-electron model;
(2) heavy atoms tend to have small energy gaps;
(3) polar materials tend to have larger energy gaps than non-polar materials;
(4) the energy gaps of polar materials tend to be direct gaps, i.e. at the Γ point.
(5) germanium is peculiar in having the lowest conduction valley at L.

1.10. Chemical trends

A semi-quantitative account of chemical trends in band structure can be given on the basis of a model developed by Phillips (1968, 1973) and Van Vechten (1969) which is very much in the spirit of Pauling's discussion of the nature of the chemical bond (Pauling 1960).

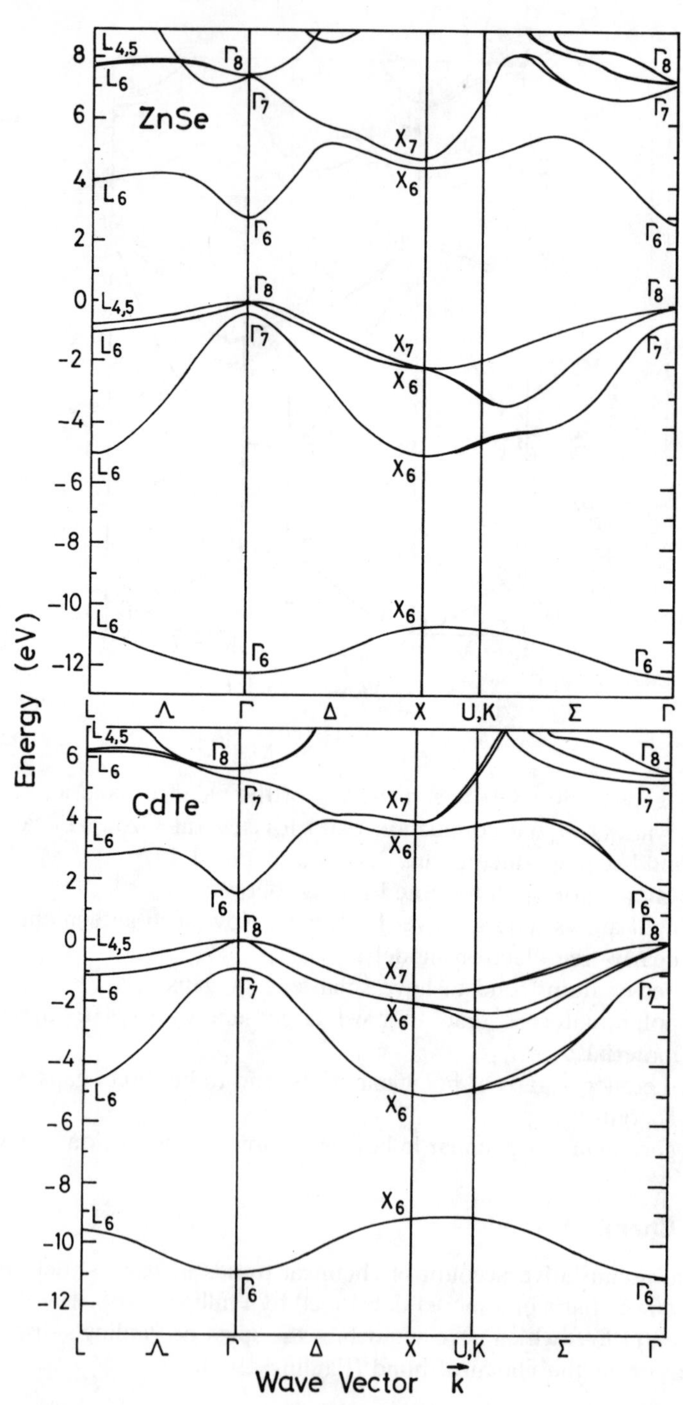

FIG. 1.12. Band structures for (a) ZnSe and (b) CdTe. (From Chelikowsky and Cohen 1976.)

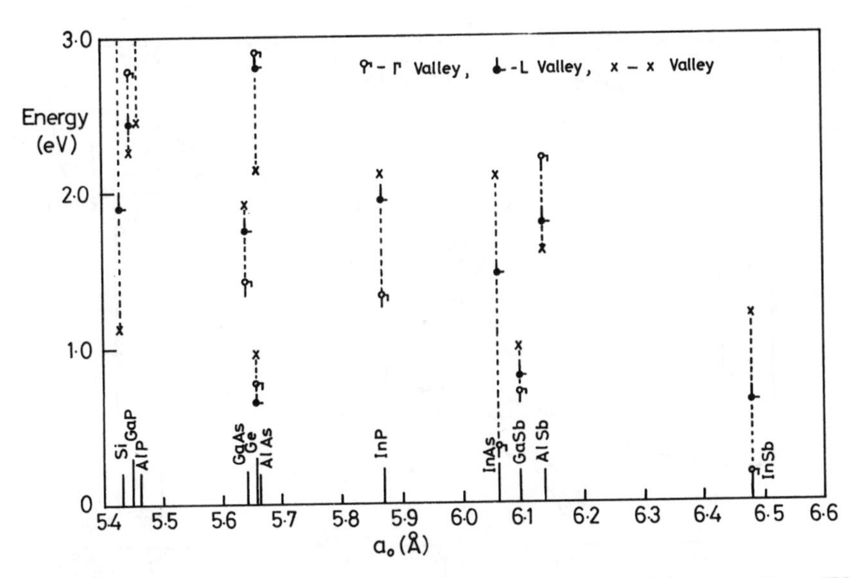

FIG. 1.13. Energies of Γ, L, and X valleys above the valence band for some Group IV elements and III–IV compounds at 300 K: ○, Γ valley; ●, L valley; ×, X valley.

The energy gaps which are of most importance in semiconductors are those between the valence band and the valleys in the conduction band at the centre of the zone (Γ) and at the zone edge in the ⟨111⟩ direction (L) and in the ⟨100⟩ direction (X). These gaps are labelled E_0, E_1, and E_2 respectively (Fig. 1.14).

The absolute levels below the vacuum level are fixed by the valence band energy at X, i.e. E_{X4}. This point is distinctive as being at the centre of a face in the Jones zone, and is an obvious candidate to be the central strut about which the band structure is supported.

Increasing polarity causes bands to become narrower, and in the limit to go to the free-atomic level. This level is therefore one which is independent of polarity and about which the valence band broadens. It turns out that E_{X4} is within 5 per cent of the ionization potential for the free atom in the cases of silicon and germanium, and so is a good choice to be the polarity-independent pivot for the band structure. It is therefore assumed that

$$E_{X4} = E_{X4}(\mathrm{Si})\left\{\frac{(d_A d_B)^{1/2}}{d_{\mathrm{Si}}}\right\}^{-s} \tag{1.58}$$

where $E_{X4}(\mathrm{Si})$ is the silicon gap and d_{Si} is its covalent interatomic distance (2·35 Å). Since E_{X4} is observed to be close to the free-atom ionization energy for silicon and germanium, it is taken to be a property of the row of the periodic table to which the constituent atoms belong. Thus d_A and

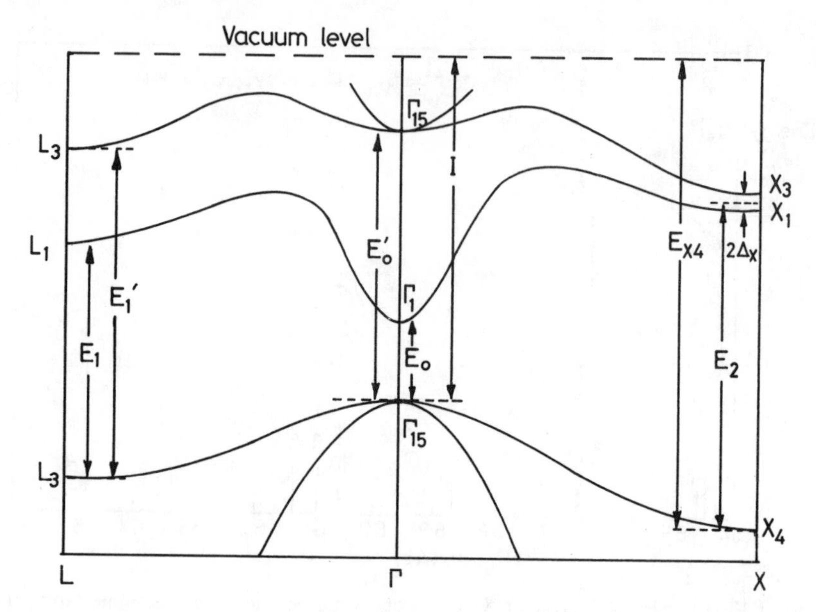

FIG. 1.14. Principal energy gaps (ignoring spin–orbit splitting).

d_B are the interatomic distances of the Group IV elements in rows A and B, to which the constituent atoms belong, and these distances are referred to as 'normal' covalent interatomic distances to distinguish them from the actual ones found in compounds AB.

The other basic strut is the ionization energy *I*. The structure of energy gaps implied by the nearly-free-electron picture (eqn (1.51)) suggests that an energy gap can be considered to be formed of two components, a symmetric or homopolar part and an antisymmetric or polar part. Thus

$$I^2 = I_h^2 + C^2 \tag{1.59}$$

where I_h is the homopolar ionization energy and C is a polar contribution which is dependent on the difference between the electron affinities of the two atoms in the unit cell. Values for C have been given by Van Vechten (1969). The purely covalent part is taken to depend upon the observed nearest neighbour d as follows:

$$I_h = I(\mathrm{Si})(d/d_{\mathrm{Si}})^{-s}. \tag{1.60}$$

The absolute level of the valence band at L is simply assumed to be the arithmetic mean of the energies at Γ and X, i.e.

$$E_{L1} = (I + E_{X4})/2 \tag{1.61}$$

All energy gaps are then taken to be of the form of eqn (1.59) provided that no d-band intrudes. A filled d-band affects E_0 and E_1 because the conduction band states with which these gaps are associated have $|s\rangle$ orbitals which penetrate the atomic core, while the $|p\rangle$ orbitals of the valence states do not, and thus the gaps are sensitive to the presence of a full d-shell. However, E_0' and E_1', which involve only $|p\rangle$ orbitals which do not penetrate the core, are not affected. Although the orbital character is $|s\rangle$-like for the X_1 and X_3 bands, it is only half as strongly so as Γ_1 or L_1, and the effect of the d-band mixing is deemed negligible for E_2. Thus for the unaffected gaps we have equations of the form

$$E_2^2 = E_{2h}^2 + C^2 \tag{1.62}$$

but for E_0 and E_1,

$$E_i^2 = (E_{ih}^2 + C^2)\{1-(D_{av}-1)(\Delta E_i/E_{ih})\}^2 \qquad i = 0, 1 \tag{1.63}$$

$$D_{av} = (v_A D_A + v_B D_B)/8 \tag{1.64}$$

where v_A and v_B are the valencies of the two atoms, $D_A = 1$ for rows up to and including silicon, $D_A = 1{\cdot}25$ for the row containing germanium, and $D_A = 1{\cdot}46$ for the row containing tin. Like the E_{ih}, the ΔE_i depend upon the nearest-neighbour distance in the usual way:

$$E_{ih} = E_{ih}(\mathrm{Si})\left(\frac{d}{d_{\mathrm{Si}}}\right)^{-s} \qquad \Delta E_i = \Delta E_i(\mathrm{Si})\left(\frac{d}{d_{\mathrm{Si}}}\right)^{-s}. \tag{1.65}$$

Finally, the splitting of the conduction band at X in polar compounds is taken to be simply related to the polar component of energy C as follows: (Fig. 1.14)

$$\Delta_X = 0{\cdot}071C\ \mathrm{eV}. \tag{1.66}$$

The lower minima X_1 correspond to Bloch functions with anti-nodes at the anion site. Since the anion is more electronegative (Table 1.4) it is more attractive for electrons. The higher minima X_3 correspond to Bloch functions with anti-nodes at the cation.

TABLE 1.4
Phillips' electronegativities

I	II	III	IV	V	VI	VII
Li 1·00	Be 1·50	B 2·00	C 2·50	N 3·00	O 3·50	F 4·00
Na 0·72	Mg 0·95	Al 1·18	Si 1·41	P 1·64	S 1·87	Cl 2·10
Cu 0·79	Zn 0·91	Ga 1·13	Ge 1·35	As 1·57	Se 1·79	Br 2·01
Ag 0·57	Cd 0·83	In 0·99	Sn 1·15	Sb 1·31	Te 1·47	I 1·63
Au 0·64	Hg 0·79	Tl 0·94	Pb 1·09	Bi 1·24		

From Phillips 1973.

TABLE 1.5
Basic energies for predicting chemical trends

Parameter	Value for Si (eV)	Exponent S
I_h	5·17	1·308
E_{X_4}	8·63	1·43
E_{0h}	4·10	2·75
E_{1h}	3·60	2·22
E_{2h}	4·50	2·382
E'_{0h}	3·40	1·92
E'_{1h}	5·90	1·67
ΔE_0	12·80	5·07
ΔE_1	4·98	4·97

From Van Vechten 1969.

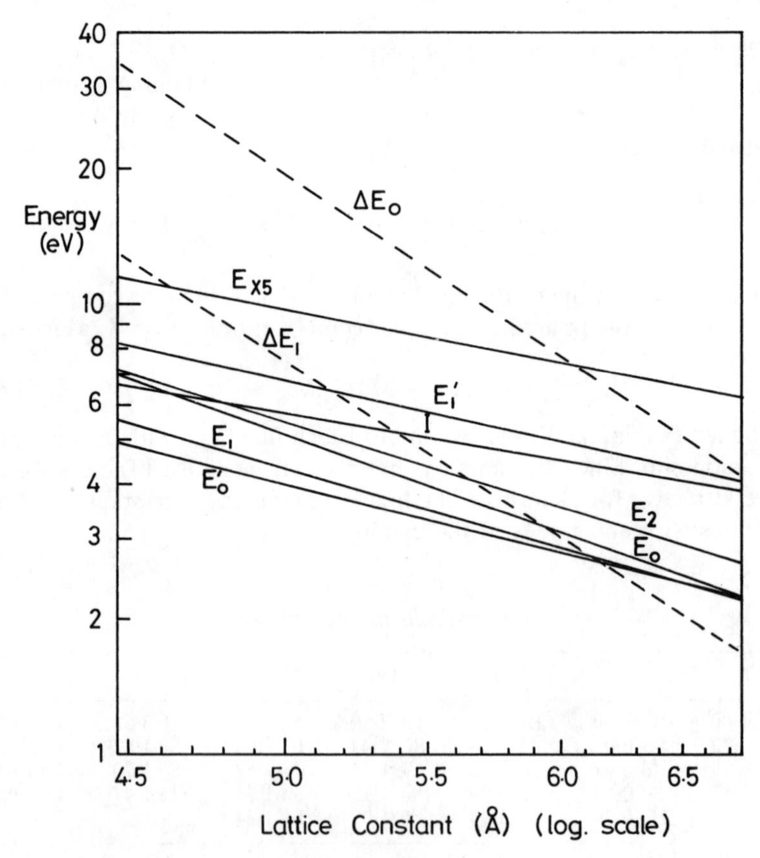

FIG. 1.15. Homopolar energies. (Data from Van Vechten 1969.)

The foregoing equations provide a reasonably successful basis for understanding the sources of variation of band structure and for predicting roughly the principal energies. Table 1.5 gives the various quantities which appear, and these are plotted in Fig. 1.15. Table 1.6 gives the observed and calculated parameters for semiconductors with diamond or sphalerite structure.

Figure 1.16 shows the homopolar energy gaps between the top of the valence band and each of the conduction-band valleys. These are distinguished notationally from the gaps described above by the subscript g (although $E_g = E_0$). These gaps apply directly only to silicon, germanium, and α-tin. For silicon $D = 1$ and the smallest gap is the indirect one to X. For germanium $D = 1{\cdot}25$ and d-shell mixing has reduced the gaps at Γ and L just enough to make the indirect gap to L the smallest. For α-tin $D = 1{\cdot}46$ and the direct gap undercuts the gap at L. Figure 1.17 shows a similar set of curves for a polar material with $C = 3$. The polar interaction enhances the gaps appreciably even in the case of strong d-shell mixing.

Spin–orbit splitting is readily incorporated. Following the discussion in Section 1.8 we merely subtract $\Delta_0/3$ from the 'unprimed' gaps (E_0, E_1, I) at Γ and L, and $2\Delta_0/3$ from the 'primed' gaps (E_0', E_1').

TABLE 1.6

Band-structure parameters of diamond and sphalerite structures

Semi-conductor	a_0 (Å)	C (eV)	D	Band gaps without spin–orbit splitting: $E_{g\Gamma}$ (eV) Calc.	$E_{g\Gamma}$ (eV) Obs.	E_{gL} (eV) Calc.	E_{gL} (eV) Obs.	E_{gX} (eV) Calc.	E_{gX} (eV) Obs.	Δ_0 (eV)
C	3·567	0	1·00	13·04		5·77		5·48	5·48	0·006
Si	5·431	0	1·00	4·10	4·08	1·87		1·04	1·13	0·044
Ge	5·657	0	1·25	0·96	0·89	0·61	0·76	0·84	0·96	0·29
Sn	6·489	0	1·46	0·13	0·1	0·18	0·3	0·35		0·48
BP	4·538	0·68	1·00	6·76		2·88		1·81	2·0	
BAs	4·777	0·38	1·11							
AlP	5·4625	3·14	1·00		4·6		3·7		2·4	0·06
AlAs	5·6611	2·67	1·11		3·05		2·9		2·26	0·29
AlSb	6·1355	3·10	1·19	2·67	2·5	2·39	2·0	2·15	1·87	0·75
GaP	5·4495	3·30	1·11	2·85	2·77	2·75	2·5	3·05	2·38	0·127
GaAs	5·6419	2·90	1·235	1·55	1·55	1·89	1·86	2·37	2·03	0·34
GaSb	6·094	2·10	1·325	1·00	0·99	1·17	1·07	1·36	1·30	0·80
InP	5·868	3·34	1·19	1·45	1·37	2·25	2·0	2·93	2·1	0·11
InAs	6·058	2·74	1·325	0·56	0·5	1·45	1·77	2·14	2·1	0·38
InSb	6·478	2·10	1·425	0·39	0·5	1·01	0·91	1·40	1·47	0·82
ZnS	5·4093	6·20	1·08	4·37	3·80	5·72		6·96		0·07
ZnSe	5·6676	5·60	1·175	3·37	2·9	4·26		5·82		0·43
ZnTe	6·101	4·48	1·235	2·72	2·56	3·64		4·26		0·93
CdTe	6·477	4·90	1·303	1·89	1·80	3·40		4·32		0·92

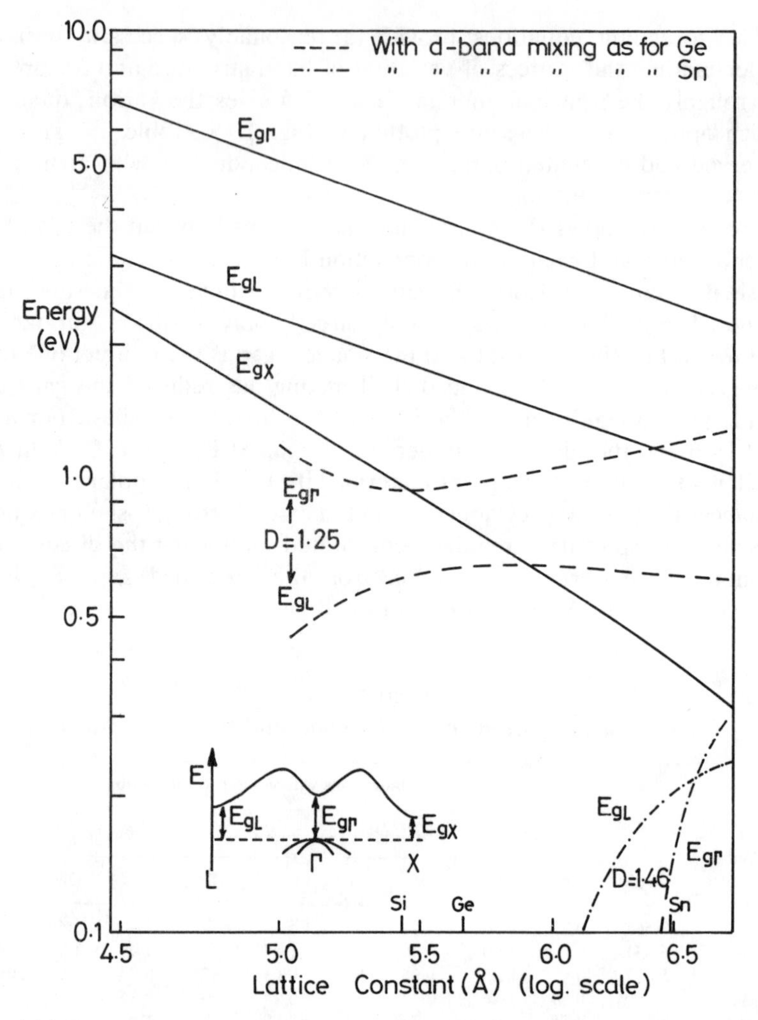

FIG. 1.16. Homopolar energy gaps: broken curves, with d-band mixing as for germanium; chain curves, with d-band mixing as for tin.

1.11. k . p perturbation and effective mass

In practice the important regions of the band structure are those which are most commonly populated by the mobile excitations of the crystal: electrons in the lowest conduction-band valley and holes at the top of the valence band. If we know the solution of the one-electron Schrödinger equation at these points in the Brillouin zone it is possible to obtain solutions in the immediate neighbourhood by regarding the scalar product

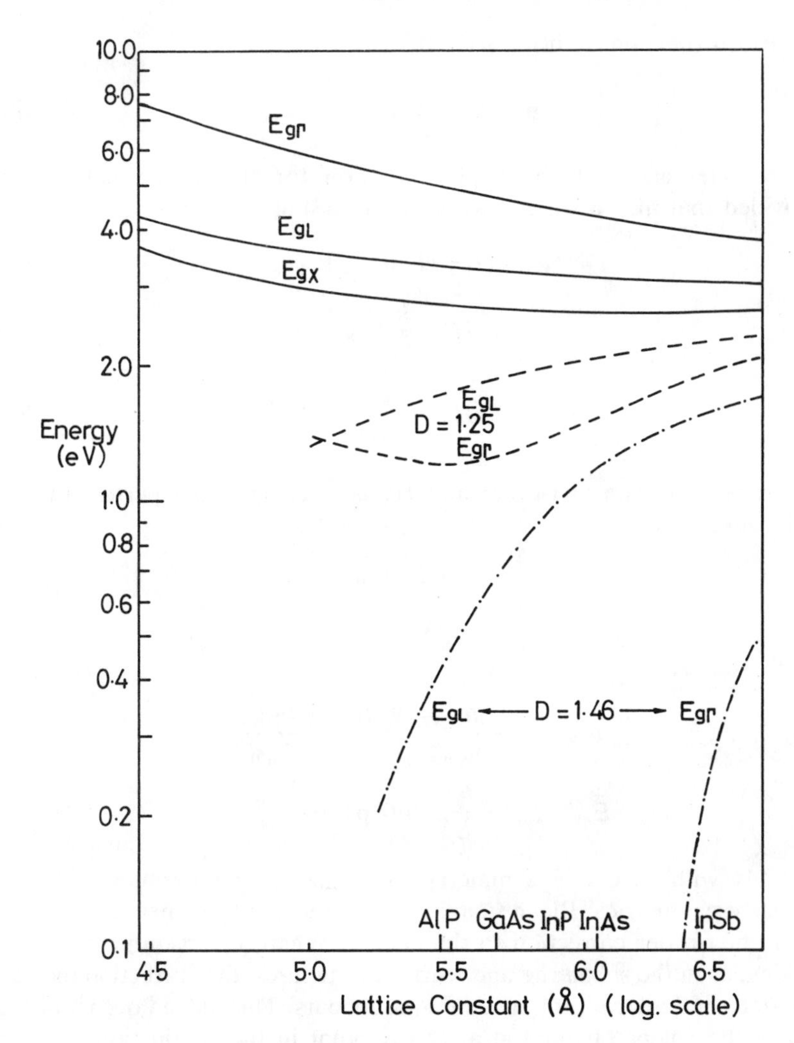

FIG. 1.17. Energy gaps in polar material ($C = 3$).

k.p, where **k** is the wavevector measured from the Brillouin zone point, as a perturbation.

Suppose that the eigenvalues and Bloch functions are known for all bands at the centre of the zone (Γ). The Schrödinger equation

$$\left\{\frac{\mathbf{p}^2}{2m}+\sum_l V(\mathbf{r}-\mathbf{R}_{l0})\right\}u_{n\mathbf{k}}(\mathbf{r})\exp(\mathrm{i}\mathbf{k}.\mathbf{r})=E_{n\mathbf{k}}u_{n\mathbf{k}}(\mathbf{r})\exp(\mathrm{i}\mathbf{k}.\mathbf{r}) \quad (1.67)$$

can be transformed to an equation containing only the periodic part of the Bloch function by using $\mathbf{p}=-i\hbar\nabla$:

$$\left\{\frac{1}{2m}(\mathbf{p}+\hbar\mathbf{k})^2+V(\mathbf{r})\right\}u_{n\mathbf{k}}=E_{n\mathbf{k}}u_{n\mathbf{k}} \tag{1.68}$$

where $V(\mathbf{r})$ is merely a simpler notation for the periodic potential. Provided that $\hbar\mathbf{k}\ll\mathbf{p}$ we can write eqn (1.68) as

$$(H_0+H_1+H_2)u_{n\mathbf{k}}=E_{n\mathbf{k}}u_{n\mathbf{k}} \tag{1.69}$$

$$H_1=\frac{\hbar}{m}\mathbf{k}.\mathbf{p} \tag{1.70}$$

$$H_2=\frac{\hbar^2\mathbf{k}^2}{2m} \tag{1.71}$$

and regard H_1 as a first-order and H_2 as a second-order perturbation.

To zero order

$$\begin{aligned}u_{n\mathbf{k}}&=u_{n0}\\ E_{n\mathbf{k}}&=E_{n0}.\end{aligned} \tag{1.72}$$

To first order

$$\begin{aligned}u_{n\mathbf{k}}&=u_{n0}+\frac{\hbar}{m}\sum_{m\neq n}\frac{\mathbf{k}.\langle m0|\,\mathbf{p}\,|n0\rangle u_{m0}}{E_{n0}-E_{m0}}\\ E_{n\mathbf{k}}&=E_{n0}+\frac{\hbar}{m}\mathbf{k}.\langle n0|\,\mathbf{p}\,|n0\rangle.\end{aligned} \tag{1.73}$$

Crystals with inversion symmetry, like silicon and germanium, have similarly symmetrical Bloch functions. Since **p** is antisymmetric there will be no first-order correction to the energy. Where inversion symmetry is lacking, as in the sphalerite and wurtzite structures, the correction may be non-zero and a term proportional to **k** appears. This indeed occurs at the top of the valence band and at the X point in the conduction band in these materials, and results in a shift of the extremum in **k** (Fig. 1.18).

However, there is always a first-order correction to the Bloch function. For example, the Bloch functions of the conduction band and valence band are respectively $|s\rangle$ like and $|p\rangle$ like and have a strong momentum matrix element connecting them. The second-order correction is not as urgent for the Bloch function as it is for the energy.

To second-order the energy becomes, neglecting the first-order correction,

$$E_{n\mathbf{k}}=E_{n0}+\frac{\hbar^2\mathbf{k}^2}{2m}+\frac{\hbar^2}{m^2}\sum_{m\neq n}\frac{|\mathbf{k}.\langle m0|\,\mathbf{p}\,|n0\rangle|^2}{E_{n0}-E_{m0}}. \tag{1.74}$$

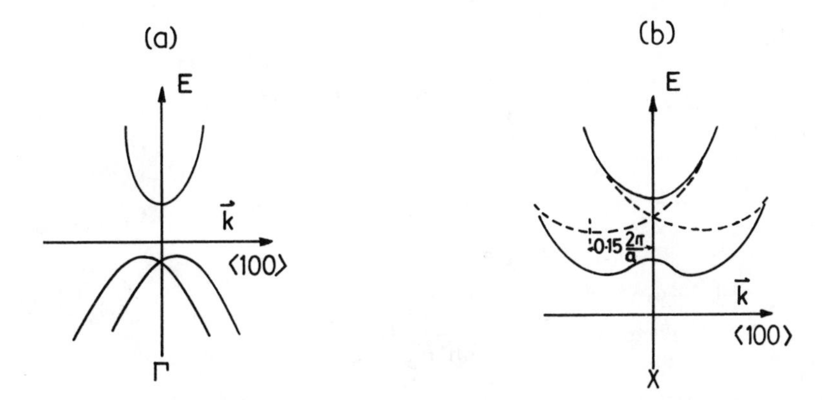

FIG. 1.18. Shifts in extrema caused by lack of inversion symmetry: (a) top of valence band; (b) camel's back structure in the conduction band near X (the broken curves are for silicon).

We can express this result in terms of an effective mass m^* as follows:

$$E_{n\mathbf{k}} = E_{n0} + \sum_{i,j} \frac{\hbar^2}{2m^*_{ij}} k_i k_j \tag{1.75}$$

where

$$\frac{m}{m^*_{ij}} = \delta_{ij} + \frac{2}{m} \sum_{m \neq n} \frac{\langle n0|\, p_i \,|m0\rangle\langle m0|\, p_j \,|n0\rangle}{E_{n0} - E_{m0}} \tag{1.76}$$

and clearly m^* is a second-order tensor. Once again we have an example in eqn (1.76) of the repulsion of bands. A narrow gap between the conduction and valence bands produces small effective masses.

Degenerate perturbation theory is necessary for the valence band. The result near $\mathbf{k}=0$ at the Γ point, considering only the conduction and valence bands, and neglecting terms linear in $\mathbf{k}$ is (Kane 1957)

conduction band

$$E_{c\mathbf{k}} = E_{g\Gamma} + \frac{\hbar^2 k^2}{2m^*_c}$$

$$\frac{m}{m^*_c} = 1 + \frac{2p^2_{cv}}{m}\frac{1}{3}\left(\frac{2}{E_{g\Gamma}} + \frac{1}{E_{g\Gamma}+\Delta_0}\right) \tag{1.77}$$

heavy hole

$$E_{v_1\mathbf{k}} = \frac{\hbar^2 k^2}{2m^*_h} \qquad \frac{m}{m^*_h} = 1 \tag{1.78}$$

light hole

$$E_{v_2\mathbf{k}} = -\frac{\hbar^2 k^2}{2m_l^*} \qquad \frac{m}{m_l^*} = \frac{4p_{cv}^2}{3mE_{g\Gamma}} - 1 \tag{1.79}$$

split-off

$$\begin{aligned} E_{v_3\mathbf{k}} &= -\Delta_0 - \frac{\hbar^2 k^2}{2m_{so}^*} \\ \frac{m}{m_{so}^*} &= \frac{2p_{cv}^2}{3m(E_{g\Gamma}+\Delta_0)} - 1 \end{aligned} \tag{1.80}$$

where p_{cv} is the momentum matrix element between the conduction band $|s\rangle$ and the valence band $|p\rangle$. For germanium $2p_{cv}^2/m = 22{\cdot}5$ eV and for GaAs $2p_{cv}^2/m = 21{\cdot}5$ eV (Phillips 1973). When the effect of more remote bands is included, the effective mass of the heavy-hole band becomes negative, as in the case of the other two. The two valence bands, which are degenerate at $\mathbf{k} = 0$, have the form of warped spheres (Fig. 1.19):

$$E_{v_{1,2}\mathbf{k}} = -\frac{\hbar^2}{2m}\{Ak^2 \pm (B^2k^4 + C^2[k_x^2k_y^2 + k_y^2k_z^2 + k_z^2k_x^2])^{1/2}\}. \tag{1.81}$$

For silicon $A = 4{\cdot}0$, $B = 1{\cdot}1$, and $C = 4{\cdot}1$ which corresponds to a heavy-hole mass of about $0{\cdot}49m$ and a light-hole mass of about $0{\cdot}16m$. For germanium $A = 13{\cdot}1$, $B = 8{\cdot}3$, and $C = 12{\cdot}5$ which corresponds to a heavy-hole mass of about $0{\cdot}28m$ and a light-hole mass of about $0{\cdot}044m$.

The split-off band is spherical, as is the $|s\rangle$ like conduction band at Γ. However, away from $\mathbf{k} = 0$ the bands become non-parabolic and take the

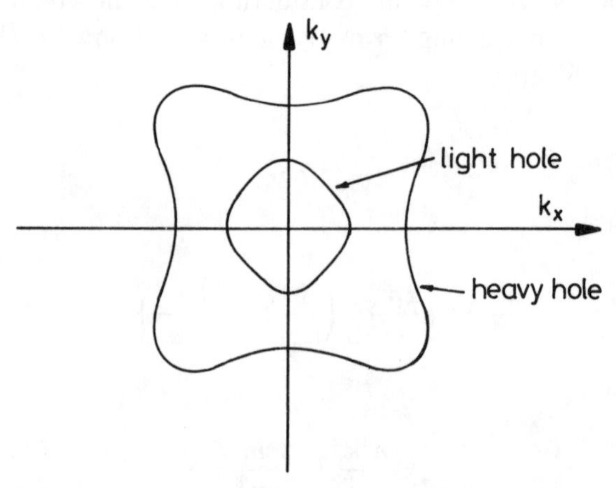

FIG. 1.19. Constant-energy surfaces for valence bands in the (x, y) plane.

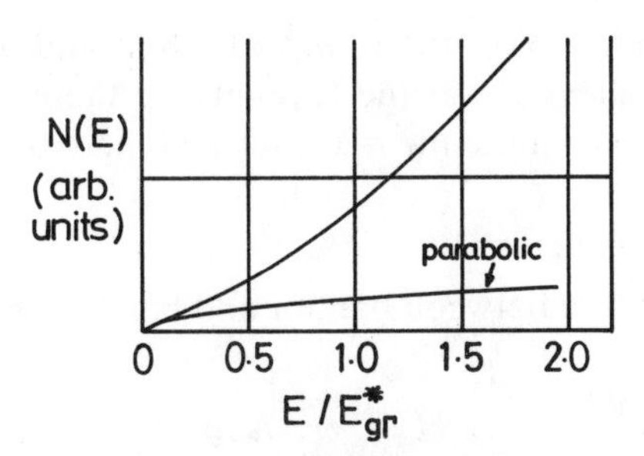

FIG. 1.20. The non-parabolicity of the conduction band of InSb. $N(E)$ is the density of states and $E^*_{g\Gamma}$ is the effective energy gap equal to 0.23 eV at 0 K. (From Ehrenreich 1957.)

form (Fig. 1.20) (for Δ_0 not too close to E_g)

$$\frac{\hbar^2 k^2}{2m^*} = \gamma(E_{\mathbf{k}}) \approx E_{\mathbf{k}}(1 + \alpha E_{\mathbf{k}} + \beta E_{\mathbf{k}}^2) \qquad (1.82)$$

$$\alpha = \frac{1}{E_{g\Gamma}}\left(1 - \frac{m^*}{m}\right)^2$$
$$\beta = -\frac{2}{E_{g\Gamma}^2}\frac{m^*}{m}\left(1 - \frac{m^*}{m}\right) \qquad (1.83)$$

The conduction-band valleys at L and X are prolate spheroidal, each with the general form near the extremum

$$\frac{\hbar^2 k_{x'}^2}{2m_l^*} + \frac{\hbar^2(k_{y'}^2 + k_{z'}^2)}{2m_t^*} \approx E_{\mathbf{k}} \qquad (1.84)$$

where **k** is measured from the minimum, x' is along the principal direction of the prolate spheroid, and y' and z' are at right-angles. Silicon has six equivalent valleys at points along the ⟨100⟩ directions distant from the X

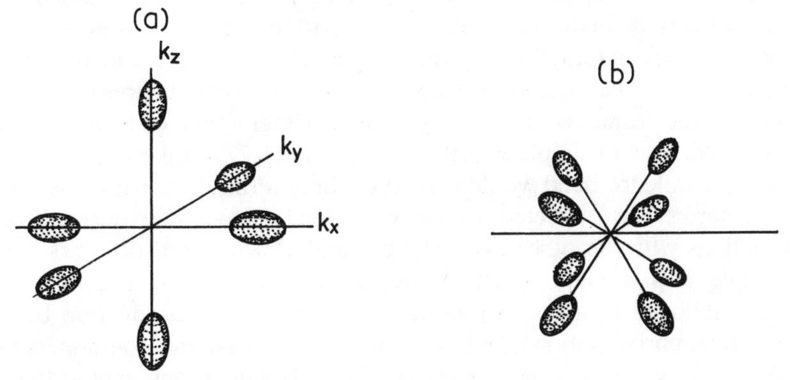

FIG. 1.21. Spheriodal conduction-band balleys: (a) silicon-like valleys; (b) germanium-like valleys.

points by $0{\cdot}15(2\pi/a)$. For these $m_l^* = 0{\cdot}98m$ and $m_t^* = 0{\cdot}19m$. Germanium's lowest valleys are at the L points, so there are four equivalent valleys in this case. For these $m_l^* = 1{\cdot}66m$ and $m_t^* = 0{\cdot}082m$ (Fig. 1.21).

1.11.1. Oscillator strengths

In a radiative transition between band n and band m the perturbation is

$$H_\nu = -\frac{e}{m}\mathbf{A}.\mathbf{p} \tag{1.85}$$

where **A** is the vector potential of the electromagnetic wave. Consequently, the matrix elements which determine the transition rate contain the momentum matrix element $\mathbf{p}_{mn}$ connecting the two bands at $k = 0$:

$$\mathbf{p}_{mn} = \langle m0|\, \mathbf{p}\, |n0\rangle \tag{1.86}$$

which are exactly the quantities that appear in $\mathbf{k}.\mathbf{p}$ theory and determine the curvature of band extrema. The strength of the transition is usually described by a dimensionless quantity known as the oscillator strength which is defined as

$$f_{mn} = \frac{2\,|\mathbf{p}_{mn}|^2}{m\hbar\omega_\nu} \tag{1.87}$$

where $\hbar\omega_\nu = E_{m0} - E_{n0}$. Equation (1.76) shows that oscillator strengths obey the following sum rule for spherical bands:

$$\sum_{m \neq n} f_{mn} = 1 - m/m_n^* \tag{1.88}$$

The optical transition of most interest is that from valence band to conduction band. Limiting our discussion to direct transitions at the centre of the zone, we observe that since nearby bands produce the largest effects in many cases the major part of the sum in eqn (1.88) comes from the interaction between the conduction band and the three valence bands. The sum for the conduction band consists approximately of only three terms, each deriving from its interaction with one of the valence bands as in simple $\mathbf{k}.\mathbf{p}$ theory (*cf.* eqn (1.77)). Likewise, the sum for a given valence band would consist of three terms, but in this case two of the terms are expected to be very small since they derive from interactions with the other two valence bands (such interactions are zero in simple $\mathbf{k}.\mathbf{p}$ theory). Furthermore, a valence band is usually even further removed from more remote bands than is the conduction band, and so the approximation of replacing the sum by its major components is usually more valid for a valence band. These considerations suggest that a reasonably good approximation for the oscillator strength of a direct interband transition is

$$f_{\mathrm{cv}} \approx 1 + m/m_{\mathrm{v}}^* \tag{1.89}$$

where $m_n^* = -m_{\mathrm{v}}^*$ and m_{v}^* is the hole effective mass for the given valence band. Going back to eqn (1.87) we see that this is equivalent to taking

$$\frac{|\mathbf{p}_{\mathrm{cv}}|^2}{2m} = \frac{E_{\mathrm{go}}}{4}\left(1 + \frac{m}{m_{\mathrm{v}}^*}\right). \tag{1.90}$$

This approximation is applied in the theory of radiative transitions (Chapter 5) and in the theory of the Auger effect (Chapter 6).

1.12. Temperature dependence of energy gaps

Direct and indirect band gaps in semiconductors become smaller as the temperature increases. This effect is caused by lattice vibrations acting in four different ways:

(1) anharmonicity which produces thermal expansion and therefore changes of energy gap through the latter's dependence on lattice constant;
(2) smearing out of the periodic potential as measured by the so-called Debye–Waller factor in neutron and X-ray scattering;
(3) mutual repulsion of intraband electronic states through increased electron–phonon coupling in second-order perturbation—the so-called Fan terms;
(4) Fan terms for interband coupling.

As Heine and Van Vechten (1976) have pointed out, all of these effects can conveniently be taken into account by relating to a single parameter, namely the change in frequency of the phonons, following the original discussion of Brooks (1955).

The statistical mechanics of the problem is as follows. The probability of an electron occupying a state s is given by the Fermi factor

$$f(E_{\mathrm{s}}) = \frac{1}{1 + \exp\{(E_{\mathrm{s}} - F)/k_{\mathrm{B}}T\}} \tag{1.91}$$

where E_{s} is the Helmholtz free energy of the crystal when one electron occupies the state s and is given by

$$E_{\mathrm{s}} = -k_{\mathrm{B}}T \ln\left\{\sum_j \exp(-E_{\mathrm{s}j}/k_{\mathrm{B}}T)\right\}. \tag{1.92}$$

$E_{\mathrm{s}j}$ is the energy of the electron when it occupies the state s while the rest of the crystal is in state j. Thus E_{s} is a thermal average energy. If the system is at constant pressure we replace E_s by E_s' where

$$E_s' = E_s - V_s\left(\frac{\partial E_s}{\partial V_s}\right)_T. \tag{1.93}$$

V_s is the volume of the crystal and E_s' is a Gibbs free energy. This incorporates the effect of thermal expansion.

Now E_{sj} will consist of a purely electronic term E_{es}, a vibrational term E_{pj}, and an electron–phonon interaction E_{epsj}:

$$E_{sj} = E_{es} + E_{pj} + E_{epsj}. \tag{1.94}$$

If $\mathbf{q}$ is the phonon wavevector and $\omega(\mathbf{q})$ is the angular frequency, then

$$E_{pj} = \sum_{\mathbf{q},b} \hbar\omega(\mathbf{q})\{n_j(\omega) + \tfrac{1}{2}\} \tag{1.95}$$

where the sum is over all $\mathbf{q}$ and all branches, and $n_j(\omega)$ is the number of quanta excited, i.e. 0, 1, 2, etc.

If the electron–phonon interaction is a small perturbation

$$E_{epsj} = \sum_{\mathbf{q},b} c_s(\mathbf{q})\{n_j(\omega) + \tfrac{1}{2}\} \tag{1.96}$$

where $c(\mathbf{q})$ is the interaction energy. Thus

$$E_{sj} = E_{es} + \sum_{\mathbf{q},b} \hbar\omega_s(\mathbf{q})\{n_j(\omega_s) + \tfrac{1}{2}\} \tag{1.97}$$

where $\omega_s(q)$ is the frequency including the electron–phonon interaction. Noting that

$$\begin{aligned}
&\sum_j \exp\left[-\sum_{\mathbf{q},b} \frac{\hbar\omega_s(\mathbf{q})}{k_B T}\{n_j(\omega_s) + \tfrac{1}{2}\}\right] \\
&= \prod_{\mathbf{q},b} \exp\left(-\frac{\hbar\omega_s(\mathbf{q})}{2k_B T}\right)\left\{1 + \exp\left(-\frac{\hbar\omega_s(\mathbf{q})}{k_B T}\right) + \exp\left(-\frac{2\hbar\omega_s(\mathbf{q})}{k_B T}\right) + \ldots\right\} \\
&= \prod_{\mathbf{q},b} (2\sinh\{\hbar\omega_s(\mathbf{q})/2k_B T\}]^{-1}
\end{aligned} \tag{1.98}$$

we obtain, by substituting eqn (1.97) into eqn (1.92),

$$E_s = E_{es} + k_B T \sum_{\mathbf{q},b} \ln[2\sinh\{\hbar\omega_s(\mathbf{q})/2k_B T\}]. \tag{1.99}$$

Thus the band gap E_g varies with temperature according to

$$E_g = E_{eg} + k_B T \sum_{\mathbf{q},b} \ln \frac{\sinh\{\hbar\omega_u(\mathbf{q})/2k_B T\}}{\sinh(\hbar\omega_l(\mathbf{q})/2k_B T)} \tag{1.100}$$

where $\omega_u(\mathbf{q})$ is the frequency when an electron occupies the upper state and $\omega_l(\mathbf{q})$ that when an electron occupies the lower state. Since $\omega_u(\mathbf{q}) < \omega_l(\mathbf{q})$, because exciting an electron weakens the atomic binding and hence lowers the elastic restoring force, the logarithm is negative and so the gaps reduce as the temperature increases. Modes with already weak restoring force constants, such as the transverse acoustic (TA) modes, are particularly affected by electron–hole generation, and the shift in frequency of these modes provides the greatest contribution. At high temperatures.

$$E_g \approx E_{eg} - k_B T \sum_{\mathbf{q},b} \ln \frac{\omega_l(\mathbf{q})}{\omega_u(\mathbf{q})} \qquad k_B T > \hbar\omega(\mathbf{q}). \tag{1.101}$$

Observed variations obey a linear law of the form $E_g(T) = E_g(0) - \beta T$, $\beta \approx 4 \times 10^{-4}$ eV K^{-1}, at high temperatures (Fig. 1.22). Since the contribution from thermal expansion is usually small, this implies that the sum in eqn (1.101) is about 5. If one bond were broken per electron–hole pair and only N TA modes were involved, we would expect $\omega_l^2 = \omega_u^2(1 - 1/N)$ or $\omega_l = \omega_u(1 - 1/2N)$, whence the sum would be only 0.5. The observation that it is 5 suggests that the equivalent of 10 bonds are disrupted per electron–hole pair.

The dependence of energy gaps on the lattice constant was discussed in Section 1.10. It is interesting to see how little that dependence determines the temperature variation, though it is the principal component in determining the variation with hydrostatic pressure. The latter can readily be measured and the corresponding temperature variation deduced from it using the relation

$$\left(\frac{\partial E_g}{\partial T}\right)_P = \left(\frac{\partial V}{\partial T}\right)_P \left(\frac{\partial P}{\partial V}\right)_T \left(\frac{\partial E_g}{\partial P}\right)_T. \tag{1.102}$$

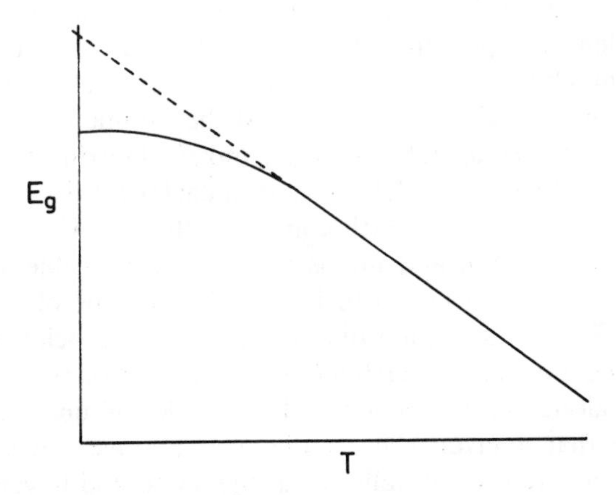

FIG. 1.22. Typical variation of the energy gap with temperature.

Typical pressure dependences are of the form $E_g = E_{g0} + \alpha P$ with $\alpha \approx 7 \times 10^{-6}$ eV cm^2 kg^{-1} (though from the discussion in Section 1.10 it is clear that α can vary widely and even be negative). Corresponding temperature dependences are usually less than 10^{-4} eV K^{-1}.

1.13. Deformation potentials

In general the application of mechanical stress alters the band structure by shifting energies and, where it destroys symmetry, by removing degeneracies. We have already mentioned the effect of hydrostatic pressure in the previous section. Usually the effect is regarded as not changing band curvature, and therefore not changing effective masses, but merely as shifting the energy states of interest. In semiconductors the states of interest are usually those near a band extremum, and the shift in energy of the band edge per unit elastic strain is called the deformation potential Ξ.

If $\mathbf{u}$ is the displacement of a unit cell, the strain tensor is defined by

$$S_{ij} = \frac{1}{2}\left(\frac{\partial u_i}{\partial x_j} + \frac{\partial u_j}{\partial x_i}\right) = S_{ji} \tag{1.103}$$

and the change in energy of a given non-degenerate band edge can be defined by

$$\Delta E = \sum_{ij} \Xi_{ij} S_{ij} \tag{1.104}$$

where Ξ is the deformation potential tensor. Since the strain tensor is symmetric there exist six independent components, and hence there exist six independent components of the deformation potential. For cubic crystals this number reduces to three independent deformation potential components denoted Ξ_d, Ξ_u, and Ξ_p, and the symmetry of the Γ, L, and X valleys reduces this further to two, Ξ_d and Ξ_u. Herring and Vogt (1956) gave a table in which it was defined that in each case the components are related to a system of axes which coincide with the principal axes of the given spheroidal valleys, and this is reproduced in Table 1.7. The constants Ξ_d and Ξ_u have values which are characteristic of the conduction band valley. Ξ_d relates to pure dilation and Ξ_u is associated with a pure shear involving a uniaxial stretch along the major axis plus a symmetrical compression along the minor axis. The Γ valley is unaffected by shear strains. Note that a given shear strain can remove the degeneracy of equivalent conduction band valleys, raising some and lowering others.

TABLE 1.7
Deformation constants

	Valley		
	Γ	L	X
Ξ_{11}	$\Xi_{d\Gamma}$	$\Xi_{dL}+\frac{1}{3}\Xi_{uL}$	$\Xi_{dX}+\Xi_{uX}$
Ξ_{22}	$\Xi_{d\Gamma}$	$\Xi_{dL}+\frac{1}{3}\Xi_{uL}$	Ξ_{dX}
Ξ_{33}	$\Xi_{d\Gamma}$	$\Xi_{dL}+\frac{1}{3}\Xi_{uL}$	Ξ_{dX}
Ξ_{23}	0	$\frac{1}{3}\Xi_{uL}$	0
Ξ_{31}	0	$\frac{1}{3}\Xi_{uL}$	0
Ξ_{12}	0	$\frac{1}{3}\Xi_{uL}$	0

In addition to acoustic strains of the type considered above in which the whole unit cell is deformed, another type known as optical strain exists in materials where each unit cell contains two atoms and is associated with the contrary displacement of the two atoms. Thus, for group IV materials, if $\mathbf{u}_1$ is the displacement of one atom, $\mathbf{u}_2=-\mathbf{u}_1$ is the displacement of the other. If the atoms are separated by a distance d at equilibrium the optical strain can be defined as a vector:

$$S_{op}=2\mathbf{u}_1/d. \tag{1.105}$$

We can in general associate a change of energy with this strain. Usually the change of energy is associated directly with the displacement rather than the strain:

$$\Delta E=\mathbf{D}.\mathbf{u}_1 \tag{1.106}$$

where $\mathbf{D}$ is an optical deformation constant. Optical strains are like acoustic shear strains and affect L valleys but not Γ or X valleys. Harrison (1956) has shown that for L valleys $\mathbf{D}$ is directed along the major axis.

In the case of the degenerate valence bands, uniaxial strains tend to remove the degeneracy. The situation is generally more complicated, and four deformation potentials a, b, d and d_o, where a, b, and d are associated with acoustic strains and d_o with optical strains, are required.

In eqn (1.81) we gave the form of the energy dependence of the wavevector for the heavy- and light-hole bands. When acoustic strains are present the energy takes the rather complicated form (Pikus and Bir 1959)

$$E_{v_{1,2}\mathbf{k}}=-\frac{\hbar^2}{2m}\{Ak^2+a\,\mathrm{Tr}\,\mathbf{S}\pm(E_{kk}^2+E_{ks}^2+E_{ss}^2)^{1/2}\} \tag{1.107}$$

where

$$\mathrm{Tr}\,\mathbf{S} = \sum_i S_{ii} \tag{1.108}$$

$$E_{\mathrm{kk}}^2 = B^2k^4 + C^2(k_x^2k_y^2 + k_y^2k_z^2 + k_z^2k_x^2) \tag{1.109}$$

$$E_{\mathrm{ks}}^2 = Bb\left\{3\left(\sum_i S_{ii}k_i^2\right) - k^2\,\mathrm{Tr}\,\mathbf{S}\right\}$$
$$+2(C^2+3B^2)^{1/2}d(S_{xy}k_xk_y + S_{yz}k_yk_z + S_{zx}k_zk_x) \tag{1.110}$$

$$E_{\mathrm{ss}}^2 = \frac{b^2}{2}\{(S_{xx}-S_{yy})^2 + (S_{yy}-S_{zz})^2 + (S_{zz}-S_{xx})^2\} + d^2(S_{xy}^2 + S_{yz}^2 + S_{zx}^2). \tag{1.111}$$

The degeneracy at $k=0$ is removed when E_{ss} is non-zero, e.g. for a pure shear. Pure dilation produced by hydrostatic pressure does not remove the degeneracy but merely shifts the band edge by an amount determined by the deformation potential a. The response of the gaps between the valence-band and the conduction-band valleys at L and X to hydrostatic pressure is determined by the gap deformation potentials as follows:

$$\begin{array}{ll} \text{valence to } \Gamma \text{ valley} & \Xi_{\mathrm{d}\Gamma} + a \\ \text{valence to L valley} & \Xi_{\mathrm{dL}} + \frac{1}{3}\Xi_{\mathrm{uL}} + a \\ \text{valence to X valley} & \Xi_{\mathrm{dX}} + \frac{1}{3}\Xi_{\mathrm{uX}} + a \end{array} \tag{1.112}$$

Thus valence- and conduction-band deformation potentials always combine and cannot be determined separately by a measurement of the energy gap. Note that the hydrostatic pressure can generally be used to alter the energy gap between the Γ valley and the L and X minima in the conduction band. It is worth pointing out that the sign convention for valence-band deformation potentials is that positive means that a positive strain increases the hole energy, and so a positive a means that a positive strain lowers the valence-band edge.

The shift of band edges with hydrostatic pressure can readily be understood from the standpoint of the Phillips and Van Vechten model. Thus for a homopolar energy gap $E_{\mathrm{h}} \propto d^{-s}$ there is an associated deformation energy shift per unit strain of

$$\Xi_{\mathrm{dh}} = -sE_{\mathrm{h}} \tag{1.113}$$

The optical deformation potential d_{o} for the valence band is usually defined as an energy rather than as an energy per unit displacement as is the case for electrons. However, it is useful to retain the latter description since it is more directly related to the physical mechanism. If this is done with optical phonon scattering in mind we can relate a scalar magnitude

TABLE 1.8
Deformation potentials for some semiconductors

	Ξ_d (eV)	Ξ_u (eV)	a (eV)	b (eV)	d (eV)	d_o (eV)	D_o (eV cm^{-1})
Si	−6·0	7·8		2·1	3·1		(10^9 intervalley)[b]
		9·2		2·5	5·3		
Ge	−9·1	15·9	2·6	−2·4	−4·1	6·4	$7\times10^8\langle111\rangle$
	−12·3	19·3	3·9[a]	−2·7[a]	−4·7[a]		
AlSb	+1·8						
	+2·2	6·2		−1·4	−4·3		
GaSb		20		−2	−4·6		
		22·6		−3·3	−8·4		
GaAs	+7·0	+7·4	−8·7	−1·8	−4·6		($\approx10^9$ intervalley)[b]
			−9·2	−2·0	−6·0		
GaP		6·2		−1·3	−4·0		
InSb	+4·5		−88	−0·2	−4·6		
	+16·2			−2·1	−5		
InP		21		−1·6	−4·4		

The data are the uppermost and lowermost values reported by Neuberger (1971). Ξ_d and Ξ_u refer to the lowest conduction band valley.
[a] Data from Lawaetz 1967.
[b] See Chapter 3, Sections 3.4.1 and 3.4.2.

D_o to d_o as follows:

$$D_0^2 = \frac{3}{2}\frac{d_0^2}{a_o^2} \tag{1.114}$$

where a_o is the lattice constant. (The factor $\frac{3}{2}$ comes out of averaging the effect of optical strains over all directions.) This relation can then be used to describe the interaction between holes and optical phonons.

The effect of strain on the electronic energy considered above is short range and is confined to the unit cell subjected to deformation. In polar materials long-range electric fields are produced by some strains, and these fields also change the energy of the electron. Electric fields produced by acoustic strain are described by the piezoelectric third-order tensor (Nye 1957). Optical strains produce directly a dipole moment, which can couple strongly to light, and is the property giving rise to the designation 'optical'. These polar effects will be discussed further in the section on lattice scattering of electrons. The deformation potentials for some semiconductors are given in Table 1.8.

1.14. Alloys

When alloys of two semiconducting compounds AB and CD are made with the formula $(AB)_x(CD)_{1-x}$ one might expect that the band structure of the alloy could be deduced by linear extrapolation from the band structures of the pure compounds, but in fact this turns out to be a poor approximation. A band gap E_g is generally observed to have the form

$$E_g = a + bx + cx^2 \tag{1.115}$$

where c is the non-linear coefficient which is usually called the bowing parameter. The latter is significant and cannot be neglected. (Fig. 1.23.)

Its origin is twofold. One part of c arises from the dependence of energy gaps on the lattice constant as described in Section 1.10. If, instead of regarding E_g as a linear function of composition, its components such as the lattice constant and the homopolar and ionic energy gaps are taken to vary linearly with composition, then a bowing parameter c_i emerges naturally as Van Vechten and Bergstresser (1970) have shown. In addition to c_i there exists a component c_e associated with disorder. Disorder produces potential fluctuations which will be proportional to the difference in electronegativities of the two component cations or anions. As such, these fluctuations will be associated with the difference of ionic energy gaps of the two compounds. Potential fluctuations will scatter electrons and, to second order, will mix band states which, as we saw in the case of electron–phonon mixing in Section 1.12, drives states at the bottom of a conduction-band valley down and drives states at the top of the valence band up, thus reducing the band gap. The curve E_g versus x therefore sags downwards as a result of disorder, corresponding to an increase in the bowing parameter.

Disorder also appears to be responsible for deviations from the expected behaviour of the conduction-band effective mass and the valence-band spin–orbit splitting as well as the direct band gap in III–V alloys. Potential fluctuations mix conduction and valence band wavefunctions, and therefore the degree of spin–orbit splitting at Γ is reduced and the effective mass also bows downward less markedly than a simple Kane **k.p** model would suggest.

Quantitatively, the reduction in energy caused by disorder in a ternary alloy $MF_{1-x}G_x$ is of the form

$$\Delta E = \frac{x(1-x)C_{FG}^2}{A} \tag{1.116}$$

where C_{FG} is the Phillips electronegativity difference between F and G, and A is a bandwidth parameter equal to 1 eV (Berolo, Woolley, and Van Vechten (1973)). C_{FG} has been computed for several ternary alloys by Van Vechten and Bergstresser (1970) (these values are significantly different from ΔC obtained from Table 1.4).

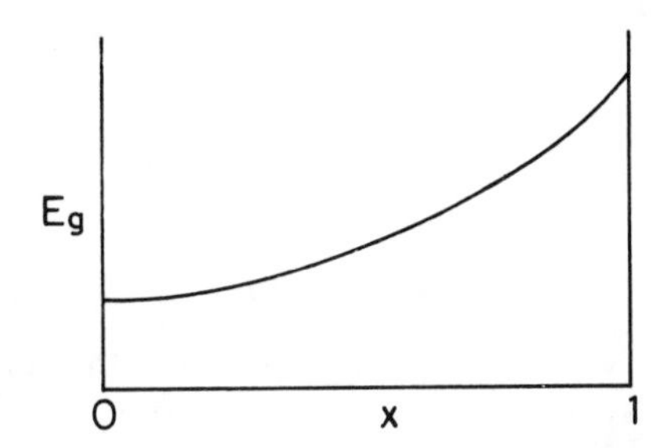

FIG. 1.23. Bowing effect in the energy-gap dependence on composition.

References

BASSANI, F. (1966). *Semicond. Semimet.* **1,** 21.
BEROLO, O., WOOLLEY, J. C., and VAN VECHTEN, J. A. (1973). *Phys. Rev.* **B8,** 3794.
BOUCKAERT, L. P., SMOLUCHOWSKI, R., and WIGNER, E. P. (1936). *Phys. Rev.* **50,** 58.
BROOKS, H. (1955). *Adv. Electr.* **7,** 85.
CHELIKOWSKY, J. R. and COHEN, M. L. (1976). *Phys. Rev.* **B14,** 556.
EHRENREICH, H. (1957). *J. Phys. Chem. Solids.* **2,** 131.
HARRISON, W. (1956). *Phys. Rev.* **104,** 1281.
HEINE, V. and VAN VECHTEN, J. (1976). *Phys. Rev.* **B13,** 1622.
HERRING, C. and VOGT, E. (1956). *Phys. Rev.* **101,** 944.
KANE, E. O. (1957). *J. Phys. Chem. Solids.* **1,** 249.
____ (1966). *Semicond. Semimet.* **1,** 75.
KITTEL, C. (1963). *Quantum theory of solids.* John Wiley, New York.
LAWAETZ, P. (1967). Ph.D. Thesis, Technical University of Denmark, Lyngby.
NEUBERGER, M. (1971). *Handbook of Electronic Materials.* TFI/Plenum, New York.
NYE, J. F. (1957). *Physical Properties of Crystals.* Clarendon Press, Oxford.
PAULING, L. (1960). *The Nature of the Chemical Bond.* Cornell University Press, Ithaca, N.Y.
PHILLIPS, J. C. (1968). *Phys. Rev. Lett.* **20,** 550.
____ (1973). *Bonds and Bands in Semiconductors.* Academic Press, New York.
PIKUS, G. E. and BIR, G. L. (1959). *Sov. Phys.–Solid State* **1,** 1502.
VAN VECHTEN, J. (1969). *Phys. Rev.* **182,** 891; **187,** 1007.
____ and BERGSTRESSER, M. (1970). *Phys. Rev.* **1B,** 3351.

2. Energy levels

2.1. The effective-mass approximation

WHEN there is a perturbing potential present, either internally associated with crystal defects or externally applied, the time-independent one-electron Schrödinger equation becomes

$$(H_0 + V_p)\psi(\mathbf{r}) = E_p\psi(\mathbf{r}) \tag{2.1}$$

where H_0 is the unperturbed Hamiltonian and V_p is the perturbation potential energy which is assumed to satisfy $V_p \ll H_0$ and is in general a function of space. Following standard perturbation theory we form a wave packet out of the unperturbed eigenfunctions, choosing for the latter the Bloch functions $\phi_{n\mathbf{k}}(\mathbf{r})$:

$$\psi(\mathbf{r}) = V^{-1/2}\sum_{n,\mathbf{k}} c_{n\mathbf{k}}\phi_{n\mathbf{k}}(\mathbf{r}) \tag{2.2}$$

where the sum is over all bands, labelled n, and all $\mathbf{k}$. We obtain

$$c_{m\mathbf{k}'}E_{m\mathbf{k}} + \sum_{n,\mathbf{k}} c_{n\mathbf{k}}\langle m\mathbf{k}'|\, V_p\,|n\mathbf{k}\rangle = c_{m\mathbf{k}'}E_p \tag{2.3}$$

We now multiply this equation by $\exp(i\mathbf{k}'.\mathbf{r})$, sum over $\mathbf{k}'$, and define an effective-mass wavefunction for the band m as follows:

$$F_m(\mathbf{r}) = V^{-1/2}\sum_{\mathbf{k}'} c_{m\mathbf{k}'}\exp(i\mathbf{k}'.\mathbf{r}). \tag{2.4}$$

Equation (2.3) may then be cast in the form

$$(H_0^* + V_p^*)F_m(\mathbf{r}) = \mathbf{E}_pF_m(\mathbf{r}) \tag{2.5}$$

where, for a simple band,

$$H_0^* = \frac{\mathbf{p}^2}{2m^*} \qquad H_0^*F_m(\mathbf{r}) = \frac{\hbar^2k^2}{2m^*}F_m(\mathbf{r}) \tag{2.6}$$

and

$$V_p^*F_m(\mathbf{r}) = V^{-1/2}\sum_{\mathbf{k}'}\sum_{n,\mathbf{k}} c_{n\mathbf{k}}\langle m\mathbf{k}'|\, V_p\,|n\mathbf{k}\rangle\exp(i\mathbf{k}'.\mathbf{r}). \tag{2.7}$$

Note that all energies are measured from the band edge.

What we have done is to subsume the effect of the background periodic potential into the effective Hamiltonian H_0^*, so that the electron can be considered to be a particle of mass m^* subject to the perturbation V_p^* rather than a particle of mass m subject to a perturbation V_p. So far no approximation has been used. This procedure becomes useful only if we may take $V_p^* F_m(\mathbf{r}) \approx V_p F_m(\mathbf{r})$ and this may be done if V_p is slowly varying.

If we represent V_p by the Fourier expansion

$$V_p = \sum_{\mathbf{q}} V(\mathbf{q}) \exp(i\mathbf{q}\,.\,\mathbf{r}), \tag{2.8}$$

then

$$\langle m\mathbf{k}'|\, V_p\, |n\mathbf{k}\rangle = V^{-1} \sum_{\mathbf{q}} V(\mathbf{q}) \int u^*_{m\mathbf{k}'} u_{n\mathbf{k}} \exp\{i(\mathbf{k}+\mathbf{q}-\mathbf{k}')\,.\,\mathbf{r}\}\, d\mathbf{r} \tag{2.9}$$

where the integral is over the whole volume of the crystal. Because $u_{n\mathbf{k}}$ is the same function in each unit cell we can convert the integral to one over a unit cell and a sum over the unit cells. By putting

$$\mathbf{r} = \mathbf{r}_0 + \mathbf{R}, \tag{2.10}$$

where $\mathbf{r}_0$ is measured from an origin within a unit cell and $\mathbf{R}$ is the position vector of the unit cell, we obtain

$$\langle m\mathbf{k}'|\, V_p\, |n\mathbf{k}\rangle = \frac{V_0}{V} \sum_{\mathbf{q}} V(\mathbf{q}) \sum_{\mathbf{R}} \exp\{i(\mathbf{k}+\mathbf{q}-\mathbf{k}')\,.\,\mathbf{R}\} \int u^*_{m\mathbf{k}'} u_{n\mathbf{k}} \times \exp\{i(\mathbf{k}+\mathbf{q}-\mathbf{k}')\,.\,\mathbf{r}_0\}\, d\mathbf{r}_0 \tag{2.11}$$

The integral in eqn (2.11) is the same for each cell regardless, and consequently the sum is infinitesimal unless

$$\mathbf{k}+\mathbf{q}-\mathbf{k}' = \mathbf{K} \tag{2.12}$$

where $\mathbf{K}$ is a reciprocal-lattice vector, whence the sum is N where N is the number of unit cells. Since $u_{\mathbf{k}+\mathbf{q}-\mathbf{K}} \exp(i\mathbf{K}\cdot\mathbf{r}_0) = u_{\mathbf{k}+\mathbf{q}}$, eqn (2.11) becomes

$$\langle m\mathbf{k}'|\, V_p\, |n\mathbf{k}\rangle = \sum_{\mathbf{q}} V(\mathbf{q}) \delta_{\mathbf{k}+\mathbf{q}-\mathbf{k}',\mathbf{K}} \int u^*_{m,\mathbf{k}+\mathbf{q}} u_{n\mathbf{k}}\, d\mathbf{r}_0. \tag{2.13}$$

If $u_{n\mathbf{k}}$ varies only slightly over the effective range of $\mathbf{q}$ we can approximate the integral as follows

$$\int u^*_{m\mathbf{k}+\mathbf{q}} u_{n\mathbf{k}}\, d\mathbf{r}_0 \approx \int u^*_{m\mathbf{k}} u_{n\mathbf{k}}\, d\mathbf{r}_0 = \delta_{mn} \tag{2.14}$$

In this approximation, only states in one band are affected by the perturbation. Thus

$$V_p^* F_m(\mathbf{r}) = V^{-1/2} \sum_{\mathbf{k}'} \sum_{n\mathbf{k}} c_{n\mathbf{k}} \sum_{\mathbf{q}} V(\mathbf{q})\, \delta_{\mathbf{k}+\mathbf{q}-\mathbf{k}',\mathbf{K}}\, \delta_{mn} \exp(i\mathbf{k}'\,.\,\mathbf{r}). \tag{2.15}$$

In the case of a simple band when the potential is slowly varying, the effective values of $\mathbf{k}$, $\mathbf{q}$, and $\mathbf{k}'$ are all small, and $\mathbf{K}=0$. Thus

$$V_p^* F_m(\mathbf{r}) = V^{-1/2} \sum_{\mathbf{k}} c_{m\mathbf{k}} \sum_{\mathbf{q}} V(\mathbf{q}) \exp\{i(\mathbf{k}+\mathbf{q}).\mathbf{r}\} = V_p F_m(\mathbf{r}). \quad (2.16)$$

The equation

$$(H_0^* + V_p)F(\mathbf{r}) = E_p F(\mathbf{r}) \quad (2.17)$$

is known as the effective-mass equation. In this approximation the perturbed wavefunction (eqn 2.2) is taken to be of the form

$$\psi(\mathbf{r}) = u_{m0} F_m(\mathbf{r}). \quad (2.18)$$

When there are several equivalent valleys $\mathbf{K}$ need not be zero. The potential couples states in different valleys and therefore the total envelope wavefunction must consist of a linear combination of single-valley envelope functions. Thus for L equivalent valleys at $\mathbf{K}_i$

$$c_{n\mathbf{k}} \to \sum_{i=1}^{L} c_{n\mathbf{k}}^i \exp(i\mathbf{K}_i.\mathbf{r}) \quad (2.19)$$

and

$$F_m(\mathbf{r}) = \sum_{i=1}^{L} \alpha_i F_m^i(\mathbf{r}) \exp(i\mathbf{K}_i.\mathbf{r}) \quad (2.20)$$

where the α_i are constants determined by symmetry and $F_m^i(\mathbf{r})$ is the localized function associated with valley i. The unperturbed effective-mass Hamiltonian for a given spheroidal valley is now

$$H_{0i}^* = \frac{p_{\perp i}^2}{2m_{\perp i}^*} + \frac{p_{\parallel i}^2}{2m_{\parallel i}^*} \quad (2.21)$$

where $p_\perp$ and $p_\parallel$ are the components of the momentum perpendicular and parallel to the major axis and $m_\perp^*$ and $m_\parallel^*$ are the corresponding effective masses. The many-valley effective-mass Schrödinger equation is therefore

$$\sum_{i=1}^{L} \alpha_i \exp(i\mathbf{K}_i.\mathbf{r})(H_{0i}^* + V_p - E_p)F_m^i(\mathbf{r}) = 0. \quad (2.22)$$

In the case of a band with two sets of equivalent valleys eqn (2.20) generalizes to

$$F_m(\mathbf{r}) = A_1 \sum_{i=1}^{L_1} \alpha_i \exp(i\mathbf{K}_i.\mathbf{r}) F_m^i(\mathbf{r}) + A_2 \sum_{j=1}^{L_2} \alpha_j \exp(i\mathbf{K}_j.\mathbf{r}) F_m^j(\mathbf{r}) \quad (2.23)$$

and two coupled equations like eqn (2.22) have to be solved.

When there are several degenerate or nearly degenerate bands at an extremum, as is the case of the valence band, the treatment is inevitably more complicated (Kittel and Mitchell 1954; Luttinger and Kohn 1955).

For the valence band the wavefunction is a linear combination of $u_jF_j(\mathbf{r})$, and the effective mass equation takes the form

$$\sum_{j=1}^{3}\sum_{\alpha\beta} D_{ij}^{\alpha\beta}\frac{\partial}{\partial x_\alpha}\frac{\partial}{\partial x_\beta}F_j(\mathbf{r}) + (\Delta_0\delta_{j3} + V_\mathrm{p} - E_\mathrm{p})F_j(\mathbf{r}) = 0 \qquad (2.24)$$

where $D_{ij}^{\alpha\beta}$ are parameters of the band structure, Δ_0 is the spin-orbit splitting (assumed small), and $\delta_{j3} = 1$ for $j = 3$ (split-off band) and zero otherwise.

The effective-mass equation is useful for describing shallow localized impurity states, for describing the scattering of electrons by defects, other electrons, and lattice vibrations, and for describing the motion of electrons in weak applied fields. However, before exploiting this approach we must turn to the more general features of electron dynamics.

2.2. Electron dynamics†

In the previous section it was shown that a time-independent potential perturbation induced a static wave packet, made up of waves from one band, which was associated with the eigenvalue E_p. If $E_\mathrm{p} > 0$ we can choose a travelling wave packet consisting of waves centred about a particular wavevector $\mathbf{k}$ and describe the effect of the perturbation on the motion of this wave packet. Thus we can take

$$\psi(\mathbf{r}, t) = u_{m\mathbf{k}}F_{m\mathbf{k}}(\mathbf{r}, t) \qquad (2.25)$$

$$F_{m\mathbf{k}}(\mathbf{r}, t) = V^{-1/2}\sum_{\mathbf{k}'} c_{m\mathbf{k}'}\exp\{\mathrm{i}(\mathbf{k}'\cdot\mathbf{r} - \omega' t)\} \qquad (2.26)$$

where the $c_{m\mathbf{k}'}$ are appreciable only near $\mathbf{k}' = \mathbf{k}$. The group velocity is, quite generally,

$$\mathbf{v}(\mathbf{k}) = \nabla_\mathbf{k}\omega(\mathbf{k}) = \frac{1}{\hbar}\nabla_\mathbf{k}E(\mathbf{k}). \qquad (2.27)$$

If the potential perturbation is of the form $-\mathscr{F}\cdot\mathbf{r}$, where $\mathscr{F}$ is a force, then we can conceive of the perturbation inducing transitions of the electron between states within the same band such that the energy of the electron changes with time as follows:

$$\frac{\mathrm{d}E_\mathrm{p}}{\mathrm{d}t} = \mathscr{F}\cdot\mathbf{v}(\mathbf{k}) = \frac{1}{\hbar}\mathscr{F}\cdot\nabla_\mathbf{k}E(\mathbf{k}). \qquad (2.28)$$

Since $E(\mathbf{k})$ is a static function of $\mathbf{k}$, a change of energy must be associated with a change of $\mathbf{k}$. Thus

$$\frac{\mathrm{d}E_\mathrm{p}}{\mathrm{d}t} = \frac{\mathrm{d}\mathbf{k}}{\mathrm{d}t}\cdot\nabla_\mathbf{k}E(\mathbf{k}), \qquad (2.29)$$

† See also Section 11.5.

whence

$$\hbar \frac{\mathrm{d}\mathbf{k}}{\mathrm{d}t} = \mathscr{F} \tag{2.30}$$

which is Newton's law of motion if $\hbar\mathbf{k}$ is interpreted as the momentum of the packet moving through the crystal. This quantity $\hbar\mathbf{k}$ is referred to as crystal momentum, and, as we discovered in the discussion of the $\mathbf{k.p}$ approximation in Section 1.11, it can be assumed to be much smaller than the total momentum, the latter being principally associated with motion within a unit cell as determined by the cell-periodic part of the Bloch function.

The ratio of crystal momentum and total momentum is measured by the ratio of effective mass and free-electron mass. Thus, if $\mathbf{v}(\mathbf{k})$ is the group velocity, then the total momentum is, by definition,

$$\mathbf{p} = m\mathbf{v}(\mathbf{k}) \tag{2.31}$$

whereas the crystal momentum in a parabolic band is

$$\hbar\mathbf{k} = m^*\mathbf{v}(\mathbf{k}). \tag{2.32}$$

Equation (2.30) describes how an electron in a band responds to a force. It may be derived more rigorously by using time-dependent perturbation and regarding the potential perturbation as inducing transitions between band states. Thus, within the effective-mass approximation, we can take

$$F(\mathbf{r}, t) = V^{-1/2} \sum_{\mathbf{k}} c_{\mathbf{k}}(t) \exp\{\mathrm{i}(\mathbf{k.r} - \omega t)\}. \tag{2.33}$$

We can easily obtain the following expression for the initial rate by substituting into the time-dependent effective-mass equation and putting $c_{\mathbf{k}} = 1$ and $c_{\mathbf{k}'} = \delta_{\mathbf{kk}'}$ at $t = 0$:

$$\mathrm{i}\hbar \frac{\mathrm{d}c_{\mathbf{k}}(t)}{\mathrm{d}t} = \langle \mathbf{k}' | \, V_{\mathrm{p}} \, | \mathbf{k} \rangle = -\mathscr{F}.\langle \mathbf{k}' | \, \mathbf{r} \, | \mathbf{k} \rangle \tag{2.34}$$

Now

$$\begin{aligned} \langle \mathbf{k}' | \, r \, | \mathbf{k} \rangle &= V^{-1} \int \mathbf{r} \exp\{\mathrm{i}(\mathbf{k} - \mathbf{k}').\mathbf{r}\} \, \mathrm{d}\mathbf{r} \\ &= V^{-1} \int_{-X}^{X} \int_{-Y}^{Y} \int_{-Z}^{Z} \mathbf{r} \exp(-\mathrm{i}\,\Delta\mathbf{k}.\mathbf{r}) \, \mathrm{d}x \, \mathrm{d}y \, \mathrm{d}z \end{aligned} \tag{2.35}$$

where $\Delta\mathbf{k} = \mathbf{k}' - \mathbf{k}$ is the increase in the wavevector and the crystal has been assumed to be a rectanguloid of dimensions $2X \times 2Y \times 2Z$. Consider the x component

$$\langle \mathbf{k}'|\, x\, |\mathbf{k}\rangle = -V^{-1} 8\left(\frac{X \cos \Delta k_x X}{\mathrm{i}\,\Delta k_x} - \frac{\sin \Delta k_x X}{\mathrm{i}\,\Delta k_x^2}\right) \frac{\sin \Delta k_y Y}{\Delta k_y} \frac{\sin \Delta k_z Z}{\Delta k_z}. \tag{2.36}$$

This is large when Δk_x is small but not zero and when $\Delta k_y = \Delta k_z = 0$. The minimum value of Δk_x which is non-zero is given by the periodic boundary conditions (which also ensure the orthogonality of the waves):

$$\Delta k_{x0} X = \pi \tag{2.37}$$

whence

$$\langle \mathbf{k}'|\, x\, |\mathbf{k}\rangle = -\frac{\mathrm{i}}{\Delta k_{x0}}\, \delta_{\Delta k_y,0}\, \delta_{\Delta k_z,0}. \tag{2.38}$$

Thus

$$\hbar \Delta \mathbf{k} \frac{\mathrm{d}c_{\mathbf{k}}(t)}{\mathrm{d}t} = \mathscr{F} \tag{2.39}$$

or

$$\hbar \frac{\mathrm{d}\mathbf{k}}{\mathrm{d}t} = \mathscr{F}. \tag{2.40}$$

It is worth noting that the expression 'force equals rate of change of crystal momentum' does not contain the effective mass explicitly, and it is therefore particularly useful for considering the motion of holes since it is easy to avoid ambiguities connected with the negative effective mass at the top of the valence band (Fig. 2.1). Electrons always move through **k**-space in the direction of the force.

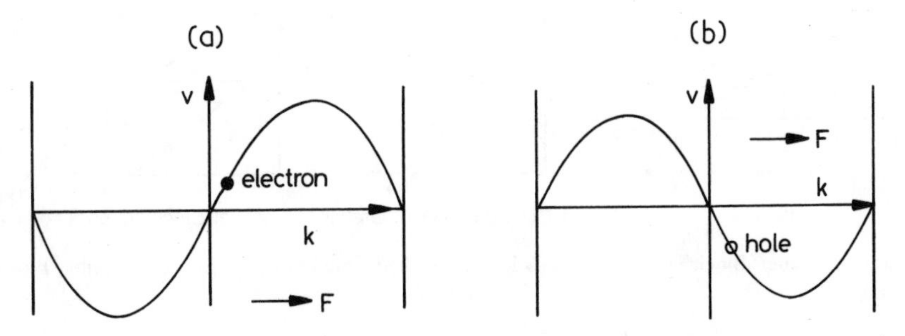

FIG. 2.1. Group velocity: (a) simple conduction band; (b) simple valence band.

2.3. Zener–Bloch oscillations†

If we can neglect the possibility of an electron making interband transitions in the presence of an applied electric field, the law governing the rate of change of crystal momentum (eqn (2.30)) shows that the electron transits through **k**-space at a rate equal to the strength of the force. In a uniform electric field $\mathscr{E}$ this rate is a constant and given by

$$\hbar\frac{\mathrm{d}\mathbf{k}}{\mathrm{d}t} = \bar{e}\mathscr{E} \tag{2.41}$$

where $\bar{e}$ is the elementary charge. (The bar over the e will be taken throughout to denote that the sign of the charge is contained within the symbol, so that we avoid any trouble connected with the conventional assignation of current direction and electron force applied to the negatively charged electron by writing the equations as if the electron were positively charged and eventually substituting for electrons $\bar{e} = -1.6 \times 10^{-19}$C). In the absence of collisions the electron transits the Brillouin zone, is ultimately reflected at the zone boundary, and transits the zone once again to be once again reflected at the zone boundary. Thus the unimpeded motion of an electron in a band under the influence of a constant field is oscillatory in **k**-space (Fig. 2.1) and therefore oscillatory in ordinary space. These oscillations have been termed Zener oscillations (Zener 1934) or alternatively Bloch oscillations, and we shall refer to them as Zener–Bloch oscillations.

Suppose the field $\mathscr{E}$ to be directed along a reciprocal-lattice vector **K** defining the reflecting zone boundary. In one oscillation the electron traverses K. If $K = 2\pi/a$, where a is a dimension of the unit cell, then the angular frequency is given by

$$\omega_z = e\mathscr{E}a/\hbar. \tag{2.42}$$

Since $a \approx 3$ Å, the angular frequency for a field of $2 \times 10^5\,\mathrm{V\,cm^{-1}}$ is about $10^{13}\,\mathrm{s^{-1}}$. The oscillation is confined in space and is centred on a particular unit cell. In such a situation the energy levels in the band are modified by the perturbing potential $\bar{e}\mathscr{E}\cdot\mathbf{r}$ (Fig. 2.2). The situation produces states

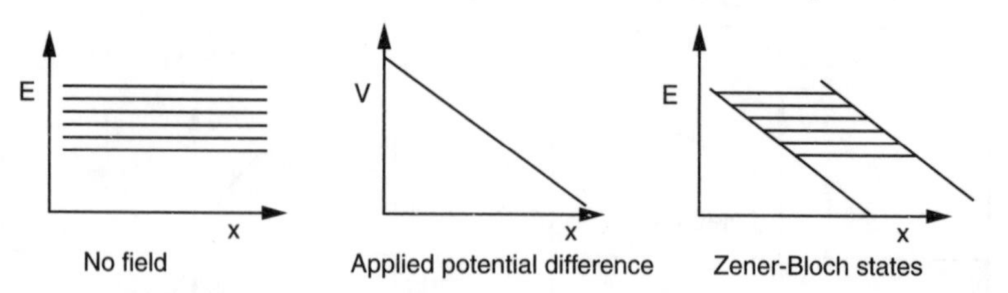

FIG. 2.2. Zener–Bloch states: (a) no field; (b) applied potential differences; (c) Zener–Bloch states.

† See also Section 11.5.

separated in energy by the amount $e\mathscr{E}a$, and the step-like variation of energy along a band edge gives rise to its description as a Stark ladder in an allusion to the Stark effect in atomic physics. Clearly, the spatial amplitude L_z of the oscillation is determined by the bandwidth E_b:

$$L_z = E_b/2e\mathscr{E}. \tag{2.43}$$

Since there is one state per unit cell the number of states in the band remains unchanged, but they become uniformly spaced in energy.

The wavefunction of an electron in a Zener–Bloch state is clearly very different from a travelling plane wave since $\mathbf{k}$ is no longer a good quantum number. Regarding the applied potential as a perturbation, we have

$$(H_0^* - \bar{e}\mathscr{E}\cdot\mathbf{r})\psi_z = E_z\psi_z \tag{2.44}$$

$$\psi_z = V^{-1/2}\sum_{\mathbf{k}} c_{\mathbf{k}}\phi_{\mathbf{k}}(\mathbf{r}) \tag{2.45}$$

where the $\phi_{\mathbf{k}}(\mathbf{r})$ are the Bloch functions of the band. Perturbation theory gives

$$c_{\mathbf{k}'} = \sum_{\mathbf{k}} \frac{c_{\mathbf{k}}\langle\mathbf{k}'| - \bar{e}\mathscr{E}\cdot\mathbf{r}|\mathbf{k}\rangle}{E_z - E_{\mathbf{k}'}}. \tag{2.46}$$

The matrix element is most conveniently calculated by noting that

$$\mathbf{r}\exp(\mathrm{i}\mathbf{k}\cdot\mathbf{r}) = -\mathrm{i}\nabla_{\mathbf{k}}\exp(\mathrm{i}\mathbf{k}\cdot\mathbf{r}). \tag{2.47}$$

Converting the sum over $\mathbf{k}$ into an integral according to

$$\sum_{\mathbf{k}} \to \int \mathrm{d}\mathbf{k}\,V/(2\pi)^3 \tag{2.48}$$

we readily obtain by integrating by parts and using the orthogonality properties of the plane waves

$$\sum_{\mathbf{k}} c_{\mathbf{k}}\langle\mathbf{k}'| - \bar{e}\mathscr{E}\cdot\mathbf{r}|\mathbf{k}\rangle = -\mathrm{i}\bar{e}\mathscr{E}\nabla_{\mathbf{k}}c_{\mathbf{k}}\delta_{\mathbf{k}\mathbf{k}'}. \tag{2.49}$$

By substituting in eqn (2.46) we obtain

$$\frac{\mathrm{d}c_{\mathbf{k}}}{\mathrm{d}\mathbf{k}} = \frac{\mathrm{i}(E_z - E_{\mathbf{k}})c_{\mathbf{k}}}{\bar{e}\mathscr{E}} \tag{2.50}$$

whence

$$c_{\mathbf{k}} = c_0\exp\left\{\int\frac{\mathrm{i}(E_z - E_{\mathbf{k}})}{\bar{e}\mathscr{E}}\mathrm{d}\mathbf{k}\right\}. \tag{2.51}$$

To retain the periodicity of the wavefunction $c_{\mathbf{k}}$ must be periodic. If we let

$$E_{\mathbf{k}} = E_0 + E(\mathbf{k}) \tag{2.52}$$

where E_0 is the energy at the band centre, then periodicity demands that

$$E_z - E_0 = -\bar{e}\mathscr{E}\cdot n'\mathbf{a} \tag{2.53}$$

where n' is an integer and $\mathbf{a}$ is a unit-cell vector. Thus the state whose eigenvalue is E_z is spatially centred about the unit cell at $n'\mathbf{a}$, and if we put $n'\mathbf{a} = \mathbf{r}_0$ we obtain

$$c_{\mathbf{k}} = c_0 \exp\left\{-\mathrm{i}\left(\mathbf{k}\cdot\mathbf{r}_0 + \int \frac{E(\mathbf{k})}{\bar{e}\mathscr{E}}\,\mathrm{d}\mathbf{k}\right)\right\}. \tag{2.54}$$

This is as far as one can go without going into the band structure in detail; thus

$$\psi_z = V^{-1/2} c_0 \sum_{\mathbf{k}} u_{\mathbf{k}}(\mathbf{r}) \exp\left\{\mathrm{i}\int \frac{E(\mathbf{k})}{\bar{e}\mathscr{E}}\,\mathrm{d}\mathbf{k} - \mathrm{i}\mathbf{k}\cdot(\mathbf{r}_0 - \mathbf{r})\right\}. \tag{2.55}$$

Let us take a simple model for the band in the direction of $\mathscr{E}$:

$$E(\mathbf{k}) = -\frac{E_{\mathrm{b}}}{2}\cos ka \qquad -\frac{\pi}{a} < k < \frac{\pi}{a} \tag{2.56}$$

(which is the form obtained in a tight-binding approximation) where E_{b} is the bandwidth. Further, let us assume that $u_{\mathbf{k}}(\mathbf{r})$ is not dependent on $\mathbf{k}$. Then

$$\psi_z = V^{-1/2} c_0 u(\mathbf{r}) \sum_{\mathbf{k}} \exp\left\{-\frac{\mathrm{i}E_{\mathrm{b}}}{2e\mathscr{E}a}\sin \mathbf{k}\cdot\mathbf{a} - \mathrm{i}\mathbf{k}\cdot(\mathbf{r}_0 - \mathbf{r})\right\} \tag{2.57}$$

$$= c_0' u(\mathbf{r}) J_n(E_{\mathrm{b}}/2e\mathscr{E}a) \qquad n = (x_0 - x)/a \tag{2.58}$$

where $J_n(z)$ is a Bessel function, n is an integer, and the field is along the x direction. Near $x = x_0$, $J_n(z)$ behaves like a standing wave of vector $\pi/2a$, i.e. the vector midway between the centre and the boundary of the zone. When $|x_0 - x| \gg a$, an asymptotic expansion gives

$$J_n\left(\frac{E_{\mathrm{b}}}{2e\mathscr{E}a}\right) \approx \frac{(-1)^{|x_0 - x|/a}}{(2\pi|x_0 - x_1|/a)^{1/2}}\left(\frac{e_{\mathrm{n}} L_Z}{2|x_0 - x|}\right)^{|x_0 - x|/a} \tag{2.59}$$

where L_Z (eqn (2.43)) is the 'classical' spatial amplitude and e_{n} is the base of Naperian logarithms. Clearly, the wavefunction attenuates very rapidly once $|x_0 - x| > e_{\mathrm{n}} L_Z/2$, but is a maximum at $|x_0 - x| = L_Z/2$ and declines towards $|x_0 - x| \to 0$. Qualitatively the behaviour is like that of a simple harmonic oscillator—the wavefunction is piled up at the extremities which are the classical turning points. Such oscillators have not been observed to date. The cause is not hard to find. In order for a Zener–Bloch oscillation to establish, the condition

$$\omega_z \tau \geqslant 1, \tag{2.60}$$

where τ is the collision time, must be satisfied. Calculations of τ are usually made for the state near the band edges, and values of order 10^{-13} s are typical. As we have just seen, the electron is a Zener–Bloch oscillation that spends most time near one or other band edge, so perhaps to take an effective τ of 10^{-13} s is not unreasonable, in which case a field in excess of 2×10^5 V cm^{-1} is required to satisfy eqn (2.60). In many cases a field of such magnitude produces electrical breakdown.

Another reason why the oscillations are difficult to obtain is that tunnelling between bands becomes increasingly probable towards high fields. Such tunnelling is observed in tunnel diodes and takes place across the forbidden gap. Furthermore, overlap of bands would tend to make interband transitions very probable, and this would virtually rule out the possibility of oscillations. All in all, the observation of Zener–Bloch oscillations is very difficult in practice. They are nevertheless intriguing manifestations of the quantum properties of crystals, implying that the action of a steady electric field on a perfect crystal is primarily to produce an oscillating electric current, although transitions between Zener–Bloch states would eventually produce a net flow of electrons down the potential gradient. However, provided that transitions between bands occur with a time constant long enough to satisfy eqn (2.60), Zener–Bloch oscillations, though transient, ought to play a role in breakdown or tunnelling phenomena for example by replacing plane waves with Zener–Bloch wavefunctions.

2.4. Landau levels

Turning to the case of electronic motion in a magnetic field, we first observe that, unlike electron fields, magnetic fields do not change the energy of the electron since the force they exert is at right angles to the motion. The basic force equation is

$$\hbar\frac{\mathrm{d}\mathbf{k}}{\mathrm{d}t} = \bar{e}(\mathbf{v}\times\mathbf{B}) \tag{2.61}$$

where $\mathbf{B}$ is the magnetic intensity and $\mathbf{v}$ the group velocity. Let $\mathbf{B}$ be in the x direction. Then

$$\hbar\frac{\mathrm{d}k_x}{\mathrm{d}t} = \bar{e}(v_yB_z - v_zB_y) = 0 \tag{2.62}$$

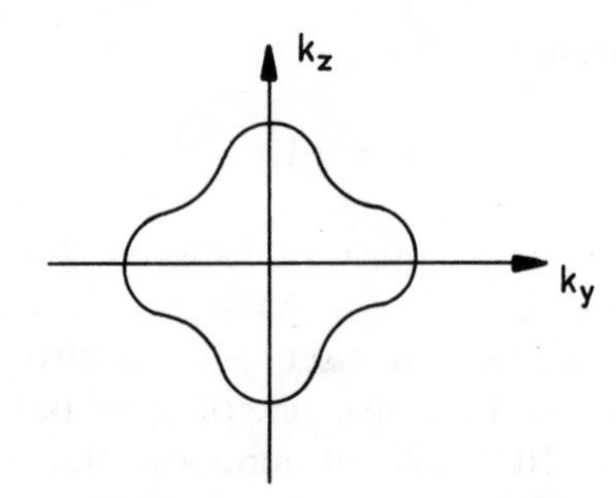

FIG. 2.3. A constant-energy track through **k**-space.

and k_x is a constant of the motion. Also

$$\hbar\frac{\mathrm{d}k_y}{\mathrm{d}t}=\bar{e}v_zB$$
$$\hbar\frac{\mathrm{d}k_z}{\mathrm{d}t}=-\bar{e}v_yB. \tag{2.63}$$

Since $v_z=\hbar^{-1}\,\mathrm{d}E/\mathrm{d}k_z$ and $v_y=\hbar^{-1}\,\mathrm{d}E/\mathrm{d}k_y$, eqn (2.63) is equivalent to

$$\hbar\frac{\mathrm{d}k_y\,\mathrm{d}k_z}{\mathrm{d}t\,\mathrm{d}E}=\bar{e}B. \tag{2.64}$$

The quantity $\mathrm{d}k_y\,\mathrm{d}k_z$ is an elementary area of the electronic orbit in a plane perpendicular to **B**. The electron cannot change energy and so its orbit must lie along a constant energy contour on **k**-space. (Fig. 2.3.) After a period T the orbit is closed and has an area $\mathrm{d}A$. Thus integrating eqn (2.64) gives

$$\hbar^2\frac{\mathrm{d}A}{\mathrm{d}E}=eBT=eB\frac{2\pi}{\omega_c} \tag{2.65}$$

and

$$\omega_c=\frac{2\pi eB}{\hbar^2}\frac{\mathrm{d}E}{\mathrm{d}A} \tag{2.66}$$

where ω_c is the cyclotron frequency. The latter carries information about the rate at which energy surfaces vary with **k** through the quantity $\mathrm{d}E/\mathrm{d}A$. For a simple spherical parabolic band $E=\hbar^2k^2/2m^*$, $A=\pi k^2$, and so $\mathrm{d}E/\mathrm{d}A=\hbar^2/2\pi m^*$, whence

$$\omega_c=eB/m^*. \tag{2.67}$$

Measurement of this frequency is therefore a direct measurement of effective mass. By varying the direction of the magnetic field the curvature of the band can be mapped out. Measurement is effected by observing the absorption of radio frequency or infra-red waves at the resonant frequency.

In the presence of an electromagnetic field defined by vector potential **A** and scalar potential ϕ the Hamiltonian is

$$H = \frac{1}{2m}(\mathbf{p} - \bar{e}\mathbf{A})^2 + \bar{e}\phi \tag{2.68}$$

where

$$\mathscr{E} = -\frac{\partial \mathbf{A}}{\partial t} - \nabla\phi \qquad \mathbf{B} = \nabla \times \mathbf{A} \tag{2.69}$$

In the case where only a uniform magnetic field exists, say in the x direction, a suitable choice for **A** in addition to the periodic potential is

$$A_x = 0 \qquad A_y = -zB/2 \qquad A_z = B_y/2. \tag{2.70}$$

Now

$$\begin{aligned}(\mathbf{p} - \bar{e}\mathbf{A})^2 &= \mathbf{p}^2 - \bar{e}\mathbf{p}.\mathbf{A} - \bar{e}\mathbf{A}.\mathbf{p} + \bar{e}^2\mathbf{A}^2 \\ &= \mathbf{p}^2 + \bar{e}(\mathrm{i}\hbar\nabla.\mathbf{A}) - 2\bar{e}\mathbf{A}.\mathbf{p} + \bar{e}^2\mathbf{A}^2.\end{aligned} \tag{2.71}$$

In the gauge we are considering $\nabla.\mathbf{A} = 0$, and so the perturbing Hamiltonian is

$$H_{\mathrm{B}} = -\frac{\bar{e}}{2m}(2\mathbf{A}.\mathbf{p} - \bar{e}\mathbf{A}^2). \tag{2.72}$$

The perturbation is of a different character from the straightforward potential energy which led to the effective-mass equation since it is a perturbation of the kinetic energy, and so we cannot use the effective-mass formulation. However, we might expect that, provided that the perturbation remains small, the effective-mass equation could be used if the perturbation is changed from eqn (2.72) to the effective Hamiltonian

$$H_{\mathrm{B}}^* = -\frac{\bar{e}}{2m^*}(2\mathbf{A}.\hbar\mathbf{k} - \bar{e}\mathbf{A}^2). \tag{2.73}$$

That this turns out to be the case can be shown by using second-order perturbation theory and retaining terms up to $\mathbf{A}^2$. The general result for the energy to second order is

$$E_{n\mathbf{k}} = E_{n\mathbf{k}0} + \langle n\mathbf{k}|\, H_{\mathrm{B}}\, |n\mathbf{k}\rangle + \sum_{n'\mathbf{k}' \neq n\mathbf{k}} \frac{|\langle n'\mathbf{k}'|\, H_{\mathrm{B}}\, |n\mathbf{k}\rangle|^2}{E_{n\mathbf{k}} - E_{n'\mathbf{k}'}} \tag{2.74}$$

where the eigenfunction has been expanded in terms of all the Bloch functions in the usual way. Now

$$\langle n\mathbf{k}|\, H_{\mathrm{B}}\, |n\mathbf{k}\rangle = -\frac{\bar{e}}{2m}(2\mathbf{A}\,.\,m\mathbf{v}_n(\mathbf{k}) - \bar{e}\mathbf{A}^2)$$

$$= -\bar{e}\mathbf{A}\,.\,\mathbf{v}_n(\mathbf{k}) + \frac{e^2}{2m}\mathbf{A}^2 \qquad (2.75)$$

where $\mathbf{v}_n(\mathbf{k})$ is the group velocity associated with the state $\mathbf{k}$ in the band n. For parabolic bands $\mathbf{v}_n(k) = \hbar\mathbf{k}/m^*$, and so

$$\langle n\mathbf{k}|\, H_{\mathrm{B}}\, |n\mathbf{k}\rangle = -\frac{\bar{e}}{m^*}\mathbf{A}\,.\,\hbar\mathbf{k} + \frac{e^2}{2m}\mathbf{A}^2. \qquad (2.76)$$

Also

$$\sum_{n'\mathbf{k}'} \frac{|\langle n'\mathbf{k}'|\, H_{\mathrm{B}}\, |n\mathbf{k}\rangle|^2}{E_{n\mathbf{k}} - E_{n'\mathbf{k}'}} \approx \frac{e^2}{m^2} \sum_{n'\mathbf{k}'} \frac{|\langle n'\mathbf{k}'|\, \mathbf{A}\,.\,\mathbf{p}|\, n\mathbf{k}\rangle|^2}{E_{n\mathbf{k}} - E_{n'\mathbf{k}'}}$$

$$= \frac{e^2}{m^2} \sum_{n'} \frac{|\mathbf{A}\,.\,\langle n'\mathbf{k}|\, \mathbf{p}\, |n\mathbf{k}\rangle|^2}{E_{n\mathbf{k}} - E_{n'\mathbf{k}}}$$

$$= \frac{e^2}{2m^*}\left(1 - \frac{m^*}{m}\right)\mathbf{A}^2. \qquad (2.77)$$

The last step follows from $\mathbf{k}\,.\,\mathbf{p}$ perturbation theory (eqn (1.76)) for a spherical band near $\mathbf{k} = 0$:

$$\sum_{n'} \frac{|\langle n'\mathbf{k}|\, \mathbf{p}\, |n\mathbf{k}\rangle|^2}{E_{n\mathbf{k}} - E_{n'\mathbf{k}}} = \frac{m^2}{2m^*}\left(1 - \frac{m^*}{m}\right). \qquad (2.78)$$

Substitution in eqn (2.74) gives

$$E_{n\mathbf{k}} = E_{n\mathbf{k}0} - \frac{\bar{e}}{m^*}\mathbf{A}\,.\,\hbar\mathbf{k} + \frac{e^2}{2m^*}\mathbf{A}^2 \qquad (2.79)$$

which justifies eqn (2.73). This result holds good provided that $\mathbf{A}$ varies negligibly over a unit cell.

Thus the effective-mass equation containing the kinetic perturbation produced by the field is

$$\left(\frac{\mathbf{p}^2}{2m^*} - \frac{\bar{e}}{m^*}\mathbf{A}\,.\,\hbar\mathbf{k} + \frac{e^2}{2m^*}\mathbf{A}^2\right)F(\mathbf{r}) = EF(\mathbf{r}) \qquad (2.80)$$

where $F(\mathbf{r})$ is the envelope function. Substitution for $\mathbf{A}$ from eqn (2.70) gives

$$\left\{\frac{\mathbf{p}^2}{2m^*} - \frac{\bar{e}\hbar B}{2m^*}(-zk_y + yk_z) + \frac{e^2}{8m^*}B^2(z^2 + y^2)\right\}F(\mathbf{r}) = EF(\mathbf{r}) \qquad (2.81)$$

or

$$\left[\frac{\mathbf{p}^2}{2m^*}+\frac{1}{2m^*}\{(\tfrac{1}{2}m^*\omega_c z+\hbar k_y)^2+(\tfrac{1}{2}m^*\omega_c y+\hbar k_z)^2\}\right]F(\mathbf{r})$$

$$=\left\{E+\frac{\hbar^2(k_y^2+k_z^2)}{2m^*}\right\}F(\mathbf{r}) \quad (2.82)$$

where $\omega_c = eB/m^*$ is the cyclotron frequency. Apart from the term in $\mathbf{p}^2$, the Hamiltonian in this equation is that for a classical particle in a (y, z) circular trajectory with orbiting frequency ω_c and momentum components $\hbar k_y$ and $\hbar k_z$ (Morse and Feshbach 1953, p. 296). We can take $k_y^2+k_z^2$ as a constant and transform the equation to

$$\left(\frac{\mathbf{p}^2}{2m^*}+\tfrac{1}{2}m^*\omega_c R^2\right)F(\mathbf{r})=E'F(\mathbf{r}) \quad (2.83)$$

where R is the classical radius of the orbit and E' is the energy for a given $k_R=(k_y^2+k_z^2)^{1/2}$. Since k_y and k_z merely specify the position of the centre of the orbit, the states described by this Schrödinger equation are as degenerate as the number of ways of choosing k_y and k_z for constant k_R. The equation describes a freely travelling electron in the x direction and a simple harmonic oscillator in the (y, z) plane. Thus

$$E'=(n+\tfrac{1}{2})\hbar\omega_c+\hbar^2k_x^2/2m^* \quad (2.84)$$

and

$$F(\mathbf{r})=C\exp(\mathrm{i}k_x x)H_n\{R(m^*\omega_c/\hbar)^{1/2}\}\exp(-m^*\omega_c R^2/2\hbar) \quad (2.85)$$

where $H_n(z)$ is a Hermite polynomial and C is a normalizing constant. The energies of the quantized states thus described are known as Landau levels (Fig. 2.4).

The degeneracy of these states is just the number of zero-field $\mathbf{k}$ states between adjacent Landau levels for a given k_x. Thus the number between k_R and $k_R+\mathrm{d}k_R$, i.e. the degeneracy, is

$$g_B=2\pi k_R\,\mathrm{d}k_R\frac{L_yL_z}{4\pi^2}=k_R\,\mathrm{d}k_R\frac{L_yL_z}{2\pi} \quad (2.86)$$

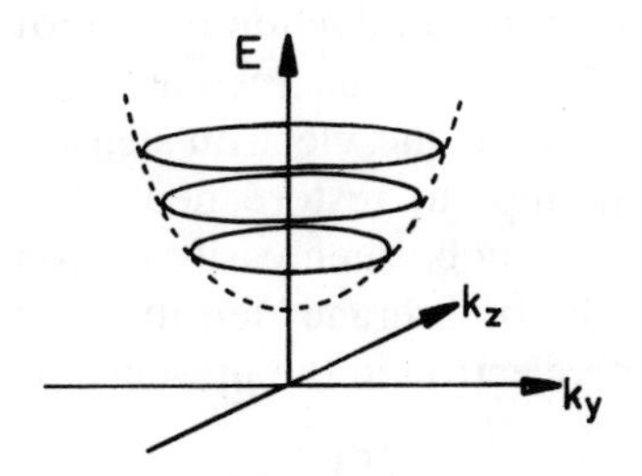

FIG. 2.4. Landau levels.

where L_y and L_z are the dimensions of the periodic crystal in the y and z directions, and the spread of k_R between adjacent Landau levels is given for a parabolic band by

$$\frac{\mathrm{d}(\hbar^2 k_R^2/2m^*)}{\mathrm{d}k_R}\,\mathrm{d}k_R = \hbar\omega_c, \tag{2.87}$$

i.e.

$$\frac{\hbar^2 k_R\,\mathrm{d}k_R}{m^*} = \hbar\omega_c. \tag{2.88}$$

Therefore

$$g_B = eB(L_y L_z/2\pi\hbar). \tag{2.89}$$

Once again, such states are observable only if the collision time τ is such that $\omega_c\tau \gtrsim 1$ or

$$eB\tau/m^* \gtrsim 1. \tag{2.90}$$

Since $e\tau/m^* = \mu$, where μ is the mobility, this condition can be expressed as

$$\mu B \gtrsim 1. \tag{2.91}$$

For $\mu = 1\ m^2\,\mathrm{V}^{-1}\,\mathrm{s}^{-1}$, $B \gtrsim 1\mathrm{T}$. Unlike the case of Zener–Bloch states, Landau 'cyclotron' states can be formed comparatively easily. As mentioned previously, measurement of $\hbar\omega_c$ by observing the frequency of resonant absorption of radiofrequency or infra-red radiation gives m^* directly. In the case of non-spherical energy surfaces the measure is of an average m^* around the orbit.

The effects of magnetic quantization are observed in optical absorption spectra, magnetic freeze-out of carriers at impurities, and the magnetophonon effect in which interaction with lattice vibrational modes is involved (Lax and Mavroides 1960; Button 1970; Harper, Hodby, and Stradling 1973).

2.5. Plasma oscillations

Electrons in the bottom of the conduction band move through a relatively immobile background of positively charged atoms. On average, neutrality prevails, but if fluctuations in the electron density occur strong electric forces appear which attempt to restore neutrality but succeed only in causing oscillations. This may be seen easily by considering the electrons to form a continuum gas of average density n_0 and writing down the equation of motion of an electron in the presence of an electric field $\mathscr{E}$:

$$m^*\frac{\mathrm{d}\mathbf{v}}{\mathrm{d}t} = \bar{e}\mathscr{E} \tag{2.92}$$

where we assume that the fields are such that the effective-mass approximation is valid. Gauss's equation gives

$$\nabla . \mathscr{E} = \bar{e}(n - n_0)/\epsilon \tag{2.93}$$

where ϵ is an appropriate permittivity and $n - n_0$ is the density fluctuation. Finally, if the electron number is conserved, we have the continuity equation

$$\frac{\mathrm{d}n}{\mathrm{d}t} = -\nabla . (\mathbf{v}n). \tag{2.94}$$

Regarding terms like $\mathbf{v}\nabla n$ as second order we readily obtain from these three equations

$$\frac{\mathrm{d}^2 n}{\mathrm{d}t^2} = -\frac{\bar{e}^2 n_0}{\epsilon m^*}(n - n_0) \tag{2.95}$$

corresponding to harmonic oscillations of density with angular frequency given by

$$\omega_\mathrm{p}^2 = \frac{e^2 n_0}{\epsilon m^*} \tag{2.96}$$

where ω_p is the plasma frequency. In this derivation purely elastic forces associated with dilations and contractions of the electron gas, considered as a neutral collection of particles, have been ignored. If they are included solutions exist in the form of travelling plane waves of density fluctuations with the dispersion relation

$$\omega^2(q) = \omega_\mathrm{p}^2 + v^2(\mathbf{q})q^2 \tag{2.97}$$

where $\mathbf{v}(\mathbf{q})$ is the velocity of acoustic waves in the electron gas. Quantization leads to the appearance of plasmons, each of energy $\hbar\omega(\mathbf{q})$ (Bohm and Pines 1953. See Chapter 9 for a fuller discussion).

In order for plasma oscillations to be established the condition

$$\omega_\mathrm{p}\tau \gtrsim 1 \tag{2.98}$$

must be fulfilled, where τ is the collision time. Since τ is about 10^{-13}–10^{-14} s, the condition is satisfied in semiconductors for carrier densities typically in excess of 10^{18} cm^{-3}. In practice such conditions are generally of importance only in narrow-gap materials or in wide-gap materials in very special circumstances such as those associated with intense laser illumination or exceptionally heavy doping.

2.6. Excitons

The minimum energy required to produce an electron–hole pair in pure materials is less than the band-gap energy because of the mutual coulombic attraction which exists between the two particles. The two-particle structure involving an electron and a hole mutually bound together is known as an exciton. Its energy states in semiconductors can be obtained, often very accurately, within the effective-mass approximation.

If we assume spherical bands at $\mathbf{k}=0$, we can write the effective-mass Hamiltonian as follows:

$$H^* = \frac{\mathbf{p}_e^2}{2m_e^*} + \frac{\mathbf{p}_h^2}{2m_h^*} - \frac{e^2/4\pi\epsilon}{|\mathbf{r}_e - \mathbf{r}_h|} = \frac{\mathbf{P}^2}{2(m_e^* + m_h^*)} + \frac{\mathbf{p}^2}{2\mu^*} - \frac{e^2/4\pi\epsilon}{r} \tag{2.99}$$

where m_e^* and m_h^* are the effective masses of the electron and hole, $\mathbf{r}_e$ and $\mathbf{r}_h$ are their position vectors, $\mathbf{P}$ is the momentum of the exciton conjugate to the centre-of-mass coordinate $\mathbf{R}$ where

$$\mathbf{R} = \frac{m_e^* \mathbf{r}_e + m_h^* \mathbf{r}_h}{m_e^* + m_h^*}, \tag{2.100}$$

$\mathbf{r} = \mathbf{r}_e - \mathbf{r}_h$, and μ^* is the reduced effective mass given by

$$\frac{1}{\mu^*} = \frac{1}{m_e^*} + \frac{1}{m_h^*}. \tag{2.101}$$

Clearly the envelope wavefunction must be of the form $F(\mathbf{r})\exp(i\mathbf{k}.\mathbf{R})$, where $F(\mathbf{r})$ obeys the Schrödinger equation

$$\left(\frac{\mathbf{p}^2}{2\mu^*} - \frac{e^2/4\pi\epsilon}{r}\right) F(\mathbf{r}) = EF(\mathbf{r}). \tag{2.102}$$

The total energy is then (Fig. 2.5)

$$E_{\mathbf{k}} = E + \frac{\hbar^2 k^2}{2(m_e^* + m_h^*)} \tag{2.103}$$

Equation (2.102) is the hydrogenic wave equation, and consequently

$$E_n = -\frac{(e^2/4\pi\epsilon)^2}{2(\hbar^2/\mu^*)n^2}, \tag{2.104}$$

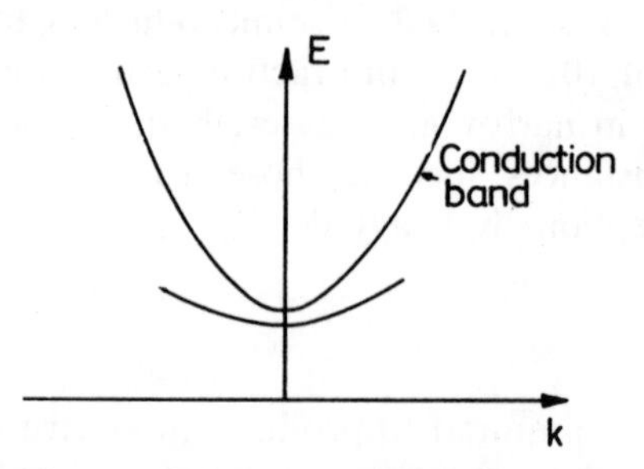

FIG. 2.5. Exciton band. The exciton band is shown juxtaposed to the conduction band. It could equally well be shown with opposite curvature juxtaposed to the valence band.

where n is an integer, and the envelope wavefunction takes the hydrogenic form

$$F(\mathbf{r}) = R_{nl}(\mathbf{r})Y_l^m(\theta, \phi) \qquad (2.105)$$

where $R_{nl}(\mathbf{r})$ is the radial function, $Y_l^m(\theta, \phi)$ is a spherical harmonic, and l and m are the orbital and magnetic quantum numbers respectively. Compared with the hydrogen atom the radius of the ground state is generally large because $\epsilon > \epsilon_0$ and $\mu^* < m$. The effective Bohr radius is given by

$$a_{\mathrm{H}}^* = (m\epsilon/\mu^*\epsilon_0)a_{0\mathrm{H}} \qquad (2.106)$$

where $a_{0\mathrm{H}}$ is the Bohr radius $\hbar^2/(me^2/4\pi\epsilon_0) = 0{\cdot}528$ Å. The factor multiplying $a_{0\mathrm{H}}$ in eqn (2.106) is about 200 in GaAs and the ground state binding energy is about 5 meV.

In the case of spheroidal or ellipsoidal valleys the Hamiltonian must reflect the anisotropy of the effective mass and also any anisotropy of permittivity which may exist (see Section 2.7).

2.7. Hydrogenic impurities

Impurity atoms generally give rise to localized electronic states with energies lying in the forbidden gap. In many cases the potential surrounding the impurity cannot be regarded as slowly varying, and an effective-mass treatment is ruled out. However, there exists a class of impurities for which the effective-mass approximation is reasonably valid, and these impurities give rise to hydrogenic states. This class consists of shallow-level donors and acceptors associated with substitutional impurities whose valency differs from the substituted host atom by unity, plus interstitial lithium which is a small enough atom to occupy interstitial sites and has a loosely bound electron. A substitutional atom with one extra valence electron is a donor whose extra electron is easily removed to the conduction band leaving a fixed positive charge at the impurity site. An atom with one valence electron too little is an acceptor which can easily abstract an electron from the valence band producing a mobile hole and thereby acquiring a negative charge. In each case the mobile particle finds itself moving in a long-range attractive coulombic potential analogous to the hydrogen atom, except that in the immediate vicinity of the origin the potential departs from being a simple coulombic one reduced by the dielectric constant of the host lattice. However, if these core anomalies are neglected the bound states can be described in the same way as exciton states with $\mathbf{k} = 0$ and $\mu^* = m_{\mathrm{e}}^*$ for donors, $\mu^* = m_{\mathrm{h}}^*$ for acceptors. Thus for a spherical band at $\mathbf{k} = 0$ the energy levels are given by

$$E_n = -\frac{R_{\mathrm{H}}^*}{n^2} \qquad R_{\mathrm{H}} = \frac{(e^2/4\pi\epsilon)^2}{2(\hbar^2/m^*)} \qquad (2.107)$$

TABLE 2.1
Hydrogenic factors for spherical conduction bands in sphalerite semiconductors

	m^*/m	ϵ_s/ϵ_0	a_H^* (Å)	R_H (meV)	Polaron α[a]	Polaron R_H^* (meV)
GaAS	0·067	13·18	104	5·25	0·088	5·33
GaSb	0·049	15·69	169	2·71	0·031	2·72
InP	0·080	12·35	81·7	7·14	0·127	7·29
InAs	0·023	14·55	335	1·48	0·063	1·50
InSb	0·014	17·72	670	0·607	0·025	0·610
ZnS	0·3	8·32	14·7	59·0	0·725	66·6
ZnSe	0·16	9·2	30·4	25·7	0·462	27·8
CdTe	0·096	10·6	58·4	11·6	0·298	12.2

[a] Taking ω_{LO} as the zone-edge value.

where R_H^* is the effective Rydberg energy, and the effective Bohr radius is

$$a_H^* = \frac{m\epsilon}{m^*\epsilon_0} a_{0H} \qquad a_{0H} = \frac{\hbar^2/m}{e^2/4\pi\epsilon_0}. \tag{2.108}$$

This result is very good in the case of excited states for which core effects are negligible; the level structure is independent of the chemical nature of the impurity (Table 2.1). The excited states for multiple donors and acceptors can be approximately described by substituting *Ze* for *e* in the above equations, but the ground-state energy is hopelessly underestimated in these cases.

In polar materials there is an ambiguity concerning the permittivity ϵ: which is to be used, the high-frequency or the low-frequency value? The dilemma can be resolved by comparing the characteristic frequency $R^*/\hbar$ of the electron in the ground state, with the frequency ω_{TO} of transversely polarized optical phonons, since it is the excitation of these which provides the component of ionic shielding in the low-frequency permittivity. If $R^*/\hbar \ll \omega_{TO}$ the static permittivity ϵ_s is used; if $R^*/\hbar \gg \omega_{TO}$ the high-frequency permittivity ϵ_∞ is used. In the case of III–V compounds ϵ_s is the better choice.

In weakly polar materials a refinement can be made which takes into account the continuous interaction of the electron with the longitudinal optical modes. The strength of this interaction is determined by the dimensionless coupling constant

$$\alpha = \frac{e^2}{4\pi\hbar}\left(\frac{1}{\epsilon_\infty} - \frac{1}{\epsilon_s}\right)\left(\frac{m^2}{2\hbar\omega_{LO}}\right)^{1/2} \tag{2.109}$$

which is about 0·09 for GaAs. Fröhlich, Pelzer, and Zieman (1950) have shown that the effect is as though the effective mass were raised by the extra inertia of the lattice polarization. The electron is now a 'dressed' particle and is known as a polaron. Its mass is given by

$$m_{\mathrm{p}}^{*}=\frac{1+\alpha/12}{1-\alpha/12}m^{*} \tag{2.110}$$

and the ground-state energy is increased accordingly (Table 2.1) (see Fröhlich 1962).

For an ellipsoidal valley in an anisotropic crystal the effective-mass equation for hydrogenic impurity states must be modified. To obtain the form of the potential we note that with the electric displacement related to the field as follows

$$D_i=\sum_j \epsilon_{ij}\mathscr{E}_j \tag{2.111}$$

and

$$\boldsymbol{\nabla}.\mathbf{D}=\rho \tag{2.112}$$

the potential satisfies

$$\sum_{ij}\epsilon_{ij}\frac{\partial^2\phi}{\partial x_i\,\partial x_j}=-\rho \tag{2.113}$$

or, referred to the principal axes of ϵ,

$$\epsilon_1\frac{\partial^2\phi}{\partial x_1^2}+\epsilon_2\frac{\partial^2\phi}{\partial x_2^2}+\epsilon_3\frac{\partial^2\phi}{\partial x_3^2}=-\rho \tag{2.114}$$

Equation (2.114) is most easily solved by transforming the coordinates according to

$$x_1=\epsilon_1^{1/2}X_1 \quad \text{etc.} \tag{2.115}$$

As a consequence the space-charge density is transformed as follows:

$$\rho=\rho'/(\epsilon_1\epsilon_2\epsilon_3)^{1/2} \tag{2.116}$$

and

$$\begin{aligned}\phi&=-\frac{\bar{e}}{4\pi(\epsilon_1\epsilon_2\epsilon_3)^{1/2}(X_1^2+X_2^2+X_3)^{1/2}}\\&=-\frac{\bar{e}}{4\pi(\epsilon_2\epsilon_3x_1^2+\epsilon_3\epsilon_1x_2^2+\epsilon_1\epsilon_2x_3^2)^{1/2}}.\end{aligned} \tag{2.117}$$

The effective-mass equation for one valley is (assuming that the principal axes of m^* coincide with those of ϵ)

$$\left\{\frac{p_1^2}{2m_1^*}+\frac{p_2^2}{2m_2^*}+\frac{p_3^2}{2m_3^*}-\frac{e^2}{4\pi(\epsilon_2\epsilon_3x_1^2+\epsilon_3\epsilon_1x_2^2+\epsilon_1\epsilon_2x_3^2)^{1/2}}-E\right\}F(x_1,x_2,x_3)=0 \tag{2.118}$$

In the case of silicon, germanium, and the III–V and II–VI sphalerites, which are cubic crystals, there is no anisotropy of the permittivity, and the L and X valleys are spheroidal with $m_1^* = m_2^* = m_\perp^*$ and $m_3^* = m_\parallel^*$. In the case of the II–VI wurzites, which are hexagonal crystals, the symmetry is axial and $\epsilon_1 = \epsilon_2 = \epsilon_\perp$ and $\epsilon_3 = \epsilon_\parallel$, though the anisotropy is small, and at $\mathbf{k} = 0$ the conduction band is slightly spheroidal (for CdS $m_\perp^* = 0{\cdot}19$ and $m_\parallel^* = 0{\cdot}18m$). Equation (2.118) is directly applicable to the wurzite Γ valley.

For L equivalent valleys in cubic crystals we have, from eqns (2.118) and (2.222), the equation

$$\sum_{i=1}^{L} \alpha_i \exp(\mathrm{i}\mathbf{K}_i\,.\mathbf{r}) \left\{ \left(\frac{p_1^2}{2m_\perp^*} + \frac{p_2^2}{2m_\perp^*} + \frac{p_3^2}{2m_\parallel^*} \right)_i - \frac{e^2/4\pi\epsilon}{r} - E \right\} F_i(x_1, x_2, x_3) = 0. \tag{2.119}$$

To zero order, the intervalley mixing can be neglected and the equation can be solved for a single valley. The axial symmetry of the single-valley equation has the effect of splitting the three spin-degenerate p-states of the hydrogenic solution with quantum numbers $m = 0, \pm 1$ into $|\mathrm{p}, 0\rangle$ and $|\mathrm{p}, \pm 1\rangle$. Taking these one-valley states for a given eigenvalue from each valley, we can form different linear combinations compatible with the symmetry of the tetrahedral character of the substitutional impurity site (Table 2.2).

TABLE 2.2
Symmetry types for the tetrahedral point group

Notation	Basis function
A_1	1
	xyz
E	$\begin{cases} 2z^2 - x^2 - y^2 \\ x^2 - y^2 \end{cases}$
T_1	$\begin{cases} x(y^2 - z^2) \\ y(z^2 - x^2) \\ z(x^2 - y^2) \end{cases}$
T_2	$\begin{cases} x \\ y \\ z \end{cases}$
T_2	$\begin{cases} yz \\ zx \\ xy \end{cases}$
T_2	$\begin{cases} x(2x^2 - 3y^2 - 3z^2) \\ y(2y^2 - 3z^2 - 3x^2) \\ z(2z^2 - 3x^2 - 3y^2) \end{cases}$

It turns out that in silicon the six degenerate $|1s\rangle$ states (one for each Δ valley) group into three combinations denoted by the symmetry types A_1, E, and T_2, i.e. a singlet, a doublet, and a triplet (not counting spin). The $|p, 0\rangle$ state behaves likewise, and the $|p, \pm 1\rangle$ state groups into two T_1 and two T_2 combinations, i.e. two double sets of triplets.

The degeneracy of these respective combinations is removed once the intervalley coupling caused by the impurity potential is included. The resultant splitting is most important for the ground state, which divides into A_1, T_2, and E in order of binding energy.

To see how the $|1s\rangle$ state splits into these groupings for X valley donors we note that equivalent valleys lie at k_x and $-k_x$, k_y and $-k_y$, and k_z and $-k_z$. Taking linear combinations of Bloch functions means that symmetric combinations ϕ_x^+, transforming as $\cos kx$ or x^2, and antisymmetric combinations ϕ_x^-, transforming as $\sin kx$ or x, are formed out of opposite valleys. Thus the basic wavefunction for the donor ground state is made up of functions $F_{1sx}\phi_x^\pm$, $F_{1sy}\phi_y^\pm$, and $F_{1sz}\phi_z^\pm$ where F_{1sx} is the envelope for the $|1s\rangle$ state associated with the valley in the x direction. The appropriate combinations are

$$\psi_{1s}(A) = \frac{1}{\sqrt{3}}(F_{1sx}\phi_x^+ + F_{1sy}\phi_y^+ + F_{1sz}\phi_z^+) \tag{2.120}$$

$$\psi_{1s}(T_2) = F_{1sx}\phi_x^- \quad F_{1sy}\phi_y^- \quad F_{1sz}\phi_z^- \tag{2.121}$$

$$\psi_{1s}(E) = \frac{1}{\sqrt{6}}(2F_{1sz}\phi_z^+ - F_{1sx}\phi_x^+ - F_{1sy}\phi_y^+)$$

$$\frac{1}{\sqrt{2}}(F_{1sx}\phi_x^+ - F_{1sy}\phi_y^+) \tag{2.122}$$

In III–V compounds the silicon-like valleys are split at the zone boundary into a lower set of minima with X_1 symmetry and an upper set with X_3 symmetry. The former's Bloch functions pile up more on the anion, the latter's more on the cation. A donor on the anion (group VI impurity) implies that $\psi_{1s}(A)$ and $\psi_{1s}(E)$, which are made up of Bloch functions which have an antinode ϕ^+ at the impurity, are those associated with the X_1 minima. The $\psi_{1s}(T_2)$ consist of Bloch functions with a node at the impurity having X_3 symmetry and therefore do not appear as effective-mass ground states associated with the X_1 minima. When the donor is on the cation site (group IV impurity) states A and E are associated with the X_3 minima, and only the triplet T_2 state is associated with X_1 and is therefore the lowest in energy.

In the case of L valleys in groups IV and III–V semiconductors, the donor ground state is quadruply degenerate and splits in the tetrahedral field into A and T_2, of which the spherically symmetric state A is the lower in energy.

It is worth noting that $\psi_{1s}(E)$ and $\psi_{1s}(T_2)$ both have nodes at the impurity and they are therefore less affected by core effects than $\psi_{1s}(A)$.

Hydrogenic acceptor states are more complicated to derive because of the degeneracy of the valence bands. When the spin-orbit splitting is much larger than the acceptor binding energy the split-off band Γ_7 can be decoupled from quadruply degenerate upper bands Γ_8 otherwise all six bands must be considered together. The full degeneracy must be taken into account in the case of silicon because of its small spin–orbit splitting, but the Γ_8 and Γ_7 bands can be considered separately in the case of germanium. In the latter case the Γ_8 band $|s\rangle$ states have Γ_8 symmetry, $|p\rangle$ states are split into $\Gamma_6+\Gamma_7+2\Gamma_8$, the Γ_7 band $|s\rangle$ states have Γ_7 symmetry, and the $|p\rangle$ states are split into $\Gamma_6+\Gamma_8$. The ground state is therefore Γ_8 which is quadruply degenerate. In silicon the situation is qualitatively the same.

Although excited states are rather well described by hydrogenic effective-mass theory (Pantelides 1978), ground states are not, as Table 2.3 shows. Much of the discrepancy is occasioned not so much by a breakdown of effective-mass theory but by having the wrong impurity potential. A simple coulombic potential with constant permittivity is too

TABLE 2.3
Observed and calculated ground-state A_1 *Energies (meV)*

		Silicon		Germanium		Gallium arsenide	
		Observed	Theory	Observed	Theory	Observed	Theory
Donors	P	45·5	31·2	12·76	9·78	—	
	As	53·7		14·04			
	Sb	42·7		10·19			
Acceptors	B	45	44	10·47	9·73	—	
	Al	68		10·80			
	Ga	71		10·97			
	In	151		11·61			
As site donors	S, Se	—		—		6.10, 5·89	5·72
Ga site donors	Si, Ge	—		—		5·85, 6·08	
As site acceptors	C, Si, Ge	—		—		26.7, 35.2, 41.2	24
Ga site acceptors	Be, Mg, Zn	—		—		30, 30, 31.4	

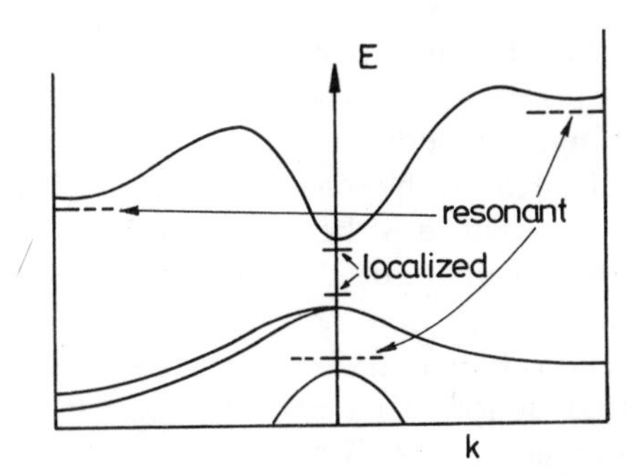

FIG. 2.6. Localized and resonant states.

inaccurate for describing conditions near the core. If more realistic forms are taken, a distinction has to be made between those impurities in the same row of the periodic table as the substitute atom and those from different rows. The former are termed isocoric impurities. Such impurities do not introduce potentials which differ appreciably from what one would expect on the basis of a point-charge model with varying permittivity. Non-isocoric impurities, however, do introduce variations in potential which may be so substantial as to invalidate the effective-mass approach. Thus isocoric impurities may be successfully modelled within effective-mass theory with a point-charge potential with varying permittivity, and the discrepancy between theory and experiment can be reduced in the cases Si:P and Ge:Ga for example. A full account can be found in the review by Pantelides (1978).

There is a consequence of having impurity states associated with individual band extrema which we have yet to mention, namely the appearance of resonant states. Levels associated in the effective-mass approximation with an upper valley of a many-valley conduction band or with a split-off valence band are degenerate with lower energy bands (Fig. 2.6). Consequently a carrier cannot be permanently localized in them and only spends a time τ in the level. The level is therefore broadened by this effect by an amount $\Delta E \approx \hbar/\tau$, which depends on how easily the carrier tunnels through **k**-space and what the density of states is in the band state which is degenerate with the impurity state. Conversely, carriers in the band at this energy may suffer temporary capture. In so far as they disturb the motion of carriers, such states act as resonant scattering centres (see Bassani, Iadonisi, and Prezcori 1974).

2.8. Hydrogen molecule centres

The effective-mass approximation may be applied to more complex centres, the simplest being centres analogous to the hydrogen molecule, either charged or neutral. Donor pairs immediately suggest themselves as possible candidates (as do acceptor pairs), but such complexes are likely to be rare since at high temperatures they are usually ionized and hence unlikely to form pairs against the force of electrostatic repulsion. A more commonly occurring candidate is the bound exciton, and such a complex has been successfully described by the straightforward application of the standard theory for the hydrogen molecule.

Two types of excitons bound at donors are immediately apparent, that bound to an ionized donor D^+eh and that bound to a neutral donor D^+eeh. The first is analogous to H_2^+ and the second to H_2. The analogy is particularly close if the effective mass of the hole is much larger than that of the electron, for then the kinetic energy of the hole can be neglected and the hole acts approximately as a second fixed charge. In this case the ground state energy for D^+eh can be directly related to the hydrogenic energy using exactly the relation between the energies of H_2^+ and H:

$$E(D^+eh) = 1{\cdot}21E(D^+e) \tag{2.123}$$

In general, binding is possible only if $m_e^*/m_h^* < 0{\cdot}4$. Binding to an acceptor is possible only if $m_h^*/m_e^* < 0{\cdot}4$, which is not usually the case. It follows that if $0{\cdot}4 < m_e^*/m_h^* < 2{\cdot}25$ binding to either ionized donor or ionized acceptor is ruled out, however, binding to a neutral donor or to a neutral acceptor is possible whatever the ratio of masses. The energy to separate two hydrogen atoms is 4·48 eV which is $0{\cdot}33R_H$. A general rule for the binding energy E_B of an exciton to a neutral donor or to a neutral acceptor (Haynes 1960) in silicon is

$$E_B \approx 0{\cdot}1E(D^+e) \quad \text{or} \quad 0{\cdot}1E(A^-h). \tag{2.124}$$

Four-body centres which are not hydrogen-molecule-like are donor–acceptor pairs, which commonly occur in compensated material. Information about pairing may be strikingly obtained from the observation of radiation emitted when the electron or the donor recombines directly with the hole on the acceptor. If the separation is much larger than the Bohr radii the centres are virtually independent, and the photon energy $h\nu$ is just

$$h\nu = E_g - E_D - E_A + e^2/4\pi\epsilon R. \tag{2.125}$$

Since R is determined by the position of the lattice sites the resulting spectrum consists of a set of lines, each line being associated with a possible lattice vector (Dean 1973).

A three-body centre may exist which consists of two electrons and a donor (D^+ee) or two holes and an acceptor (A^-hh), which is the analogue of the negatively charged hydrogen ion H^-. The binding energy of the latter is 0·75 eV, which is $0{\cdot}06R_H$, so one might expect the binding energy in a semiconductor to be about $0{\cdot}06R_H$.

2.9. Core effects

Substitutional impurities which have the same valency as the substituted host atom are known as isovalent or, alternatively, iso-electronic impurities. As such they do not necessarily introduce localized states. None are known for iso-electronic impurities in germanium and silicon, but N and Bi replacing P in GaP and O replacing Te in ZnTe do introduce localized levels, though Se in SnTe does not. This variability may be understood in terms of the short-range nature of the potential variation introduced into the crystal by the impurity. Though iso-electronic, the impurity is a different atom, and that difference may be quantified by two quantities, electronegativity and size. Electronegativity difference measures the intrinsic difference in attractiveness to electrons of the impurity and host atoms, and difference in size introduces elastic strain and an associated deformation potential. A net potential $U(\mathbf{r})$ results which is short range compared with a coulombic potential. If $U(\mathbf{r})<0$ the impurity may bind an electron; if $U(\mathbf{r})>0$ it may bind a hole. If $U(\mathbf{r})$ is assumed to be a square well of depth $|U_0|$ and radius r_0 and the effective-mass approximation is applicable, we must have

$$|U_0|\, r_0^2 \geqslant \pi^2\hbar^2/8m^* \tag{2.126}$$

in order to bind an electron or a hole. Thus not all iso-electronic impurities may introduce localized states. Those which do are usually observed via bound excitions and luminescence rather than via the straightforward trapping of an electron or hole.

Core effects are also responsible for the deviations from hydrogenic values of the ground states of ordinary donors and acceptors. The effective-mass approximation can still be used where $U(\mathbf{r})$ is not large. A model potential which is useful both for conceptual purposes and for modelling simple impurity wavefunctions is that shown in Fig. 2.7. The envelope function for the ground state in the core region is of the form

$$F_1(\mathbf{r}) = A_1 j_0(\alpha r) \tag{2.127}$$

where A_1 is a constant, $j_0(\alpha r)$ is the zero-order spherical Bessel function, and

$$\alpha^2 = \frac{2m^*}{\hbar^2}(U_0 - E_\mathrm{T}) \tag{2.128}$$

where E_T is the observed binding energy of the carrier. In the outer region the envelope function is

$$F_2(\mathbf{r}) = A_2 W_{\mu,1/2}(2r/\nu_\mathrm{T} a^*) \tag{2.129}$$

$$\nu_\mathrm{T} = (R_\mathrm{H}^*/E_\mathrm{T})^{1/2} \tag{2.130}$$

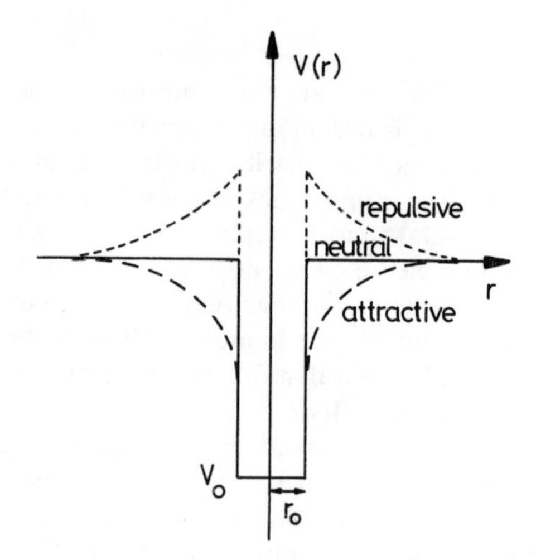

FIG. 2.7. Model potential.

where A_2 is a constant, $W_{\mu,1/2}(z)$ is a Whittaker function (Whittaker and Watson 1946), and $\mu = Z\nu_T$ where Z is the number, including the sign, of the charge on the centre. Usually it is sufficient to take the asymptotic form of the Whittaker function with an appropriate normalizing constant A to represent the whole wavefunction, even that in the core region (Bebb 1969):

$$F(\mathbf{r}) \approx A\left(\frac{2r}{\nu_T a^*}\right)^{\mu-1} \exp\left(-\frac{r}{\nu_T a^*}\right). \tag{2.131}$$

Normalizing to unity gives

$$A = \{4\pi(\nu_T a^*/2)^3 \Gamma(2\mu+1)\}^{-1/2}. \tag{2.132}$$

Though this procedure underestimates the amplitude of the wavefunction outside of the core, it leads to a wavefunction which reduces correctly to the pure hydrogenic result when $\nu_T = 1$ and $Z = 1$. The influence of the core is reflected in the deviation of ν_T from unity, a deviation referred to as the quantum defect. In the case of neutral centres, $\mu = 0$ and eqn (2.131) becomes the wavefunction of the Lucovsky model (1965) used in the calculation of photo-ionization cross-section. The modified Bohr radius is $\nu_T a^*$. This relation quantifies a characteristic feature of the effective-mass theory, namely the deeper the level, the less extended the orbit.

The wavefunction of eqn (2.131) applies in principle to centres with a repulsive coulomb ($Z, \mu < 0$), but no normalization is possible unless $\mu > -\frac{1}{2}$. Physically, this is a reflection of the necessity for the wavefunction to be concentrated centrally around the attractive core for stability. Thus a critical level is defined and only relatively deep levels are stable. However, this is an entirely artificial condition which arises out of the neglect of the core wavefunction. Nevertheless, the necessity for a tight orbit around an attractive core in the case of a centre with long-range coulomb repulsion is a real one. Consequently, it is not possible to neglect details of the core potential, and it follows that in the case of repulsive centres the application of the effective-mass approximation is of dubious validity, even if the level is shallow. An example of the latter is the Au^{2-} centre in germanium which can bind a third electron at a level 0·04 eV below the conduction band.

2.10. Deep-level impurities

In general, core effects in iso-electronic impurities, deep-level single and multiple donors and acceptors, vacancies, and other defects cannot be treated within the effective-mass approximation. The short-range nature of the potential in such centres and the tightness of the orbit of the localized state means that the impurity wavefunction, if expanded in terms of Bloch functions, must consist of contributions from the whole of **k**-space and from many bands. It is then no longer possible to associate a deep level with one or other of the conduction-band or valence-band extrema. To describe impurities of this sort theoretically requires methods akin to band-structure calculations. There is the added complication that lattice distortion around the centre must also be computed when the centre is in its ground state and when it is in the excited state corresponding to one electron in the conduction band. The theoretical task is truly formidable (see Pantelides 1978; Bassani *et al.* 1974). Nor is the experimental definition of a deep-level impurity straightforward. Energies of excitation are commonly measured optically and thermally, but their interpretation is not always obvious. Let us suppose impurity X is known to introduce a localized level in the forbidden gap of some semiconductor. The following observations can be made.

(1) Resonant absorption at a photon energy $h\nu_r$. This is interpreted as a transition between the ground state and an excited state, probably of the same impurity but possibly belonging to a neighbouring impurity. Two well-defined levels separated by $h\nu_r$ are picked out, but there is no indication of where the levels are situated in the forbidden gap. If the lattice distortion changes after excitation, the Franck–Condon principle, which states that because the electron is so light no change in ionic configuration occurs during an optical transition, implies that the excited state is a metastable one and exceeds the stable excited state by the Franck–Condon shift d_{FC} (Fig. 2.8). Thus

$$h\nu_r = E_{ex} + d_{FC} - E_0 \qquad (2.133)$$

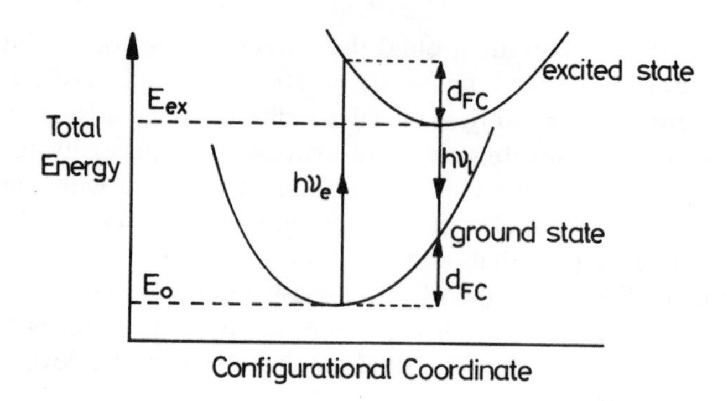

FIG. 2.8. Franck–Condon shift.

and an investigation of thermal broadening may yield d_{FC}. (A fuller account of lattice relaxation is given in Chapter 6; see also Stoneham (1975).)

(2) Edge absorption at a photon energy $h\nu_e$ (i.e. zero absorption up to ν_e and increasing absorption without structure beyond). This is interpreted as a transition of an electron either from the valence band to the empty level or from a full level to the conduction band, to distinguish which usually requires electrical measurements of some sort. Let us suppose that the absorption edge is associated with the excitation of an electron to the conduction band at E_c. Then

$$h\nu_{ec} = E_c + d_{FC} - E_0. \tag{2.134}$$

Sometimes it is possible to obtain the edge corresponding to a transition to the other band, in this case the valence band at energy E_v, and hence

$$h\nu_{ev} = E_0 + d_{FC} - E_v. \tag{2.135}$$

Thus

$$h\nu_{ec} + h\nu_{ev} = E_g + 2d_{FC} \tag{2.136}$$

where E_g is the energy gap of the semiconductor and $2d_{FC}$ is known as the Stokes shift; the sum of the photon energies exceeds the band gap.

(3) Provided that the concentration is of the same order as for absorption, i.e. usually about $10^{16}\ \mathrm{cm}^{-3}$ or greater, it is often possible to observe luminescence, say at $h\nu_l$, corresponding to a transition which is the reverse of optical absorption. Since electrons in a band are on average within an energy of about $k_B T$ of the edge the distinction between resonant luminescence and 'edge' luminescence is not as marked as in the reverse situation pertaining to optical absorption. In either case equations like (2.133) and (2.134) apply but with d_{FC} replaced with $-d_{FC}$.

(4) Measurement of photoconductivity and transport properties can be made when the concentration of impurities is much lower than that required for optical absorption. Although there are many pitfalls to be avoided in the interpretation of results, it is possible to obtain the spectral dependence of the photo-ionization cross-section which gives essentially the same information as obtained from edge absorption. In addition it is sometimes possible to obtain information about the charge carried by the centre from the observation of the capture of carriers. Transport properties such as mobility can yield information about the charge on the centre and about the existence of resonant states.

(5) Measurements of the thermal activation of carriers out of the impurity X are usually carried out via observations of Hall effect and electrical resistivity. What is observed is an exponential dependence of the form $\exp(-E_{th}/k_B T)$ which, if the temperature dependence of the mobility can be disentangled, may be associated with the increase of carrier density with temperature. Consequently the activation energy E_{th} can be related to the impurity centre. Unfortunately there is a dilemma. If the levels are full at absolute zero then $E_{th} = (E_c - E_0)/2$, as is well known from the elementary non-degenerate statistics. However, if the levels are only partially full then $E_{th} = E_c - E_0$. This dilemma may be resolved for instance by appeal to observation of the dependence of photoconductivity on intensity, partially filled levels giving a linear variation and full levels giving a square-root variation. In either case E_{th} is generally different from the optical excitation energy $h\nu_{ec}$, not only because of the Franck–Condon shift but because E_{th} is the activation energy at absolute zero, given a linear dependence of energy on temperature (Section 1.12).

(6) A quite different problem exists concerning the relationship between wavefunction and ionization energy in many-electron centres. Where unpaired electron spins exist the study of electron spin resonance (e.s.r.), i.e. the resonant absorption of microwave radiation in a magnetic field, may yield information about the symmetry of the wavefunction and about its strength on neighbouring lattice sites. More often than not the wavefunction has to be inferred from the premises of some theoretical model. In either situation the wavefunction is usually a one-electron wavefunction associated with an eigenvalue E_i and satisfying, say, a Hartree self-consistent central-field one-electron equation of the form

$$\left\{\frac{\mathbf{p}_i^2}{2m} - \frac{Z(e^2/4\pi\epsilon_0)}{r} + \sum_{i\neq j} \int \mathrm{d}\mathbf{r}_j \psi_j^*(r_j) \frac{e^2/4\pi\epsilon_0}{r_{ij}}\right\}\psi_i(r_i) = E_i\psi_i(r_i) \quad (2.137)$$

where the sum is over the other electrons in the centre and represents their screening of the central charge. The field is self-consistent in that the screening is determined by the wavefunctions of the other electrons, and these wavefunctions are determined by the field. The equation is for one electron since it is assumed that the true wavefunction can be expressed as the product of one-electron wavefunctions. (Correlation of electronic movement is ignored.) The many-electron character of the centre reveals itself in the relation between E_i and the energy E_T to remove an electron for these are not identical because of the phenomenon of electron relaxation. In a one-electron centre $E_i = E_T$ because the field is independent of the occupation of the orbital. In a many-electron centre E_i is only approximately equal to E_T because the field is dependent on screening. As the electron is removed to infinity its screening contribution diminishes and the other electrons move to tighter orbits in which their screening contribution becomes greater. Since electrons are light particles this adjustment will happen very rapidly and the electron being removed will experience a repulsive component. Thus $E_T < E_i$, i.e. the energy to remove an electron will be less than its eigenvalue by an amount associated with this electron relaxation. What this implies for modelling impurity wavefunctions can be seen by referring to the quantum-defect wavefunction (eqn (2.131)). The extent of the wavefunction is determined by the effective Bohr radius a^* and the quantum-defect parameter $\nu_T = (R_H^*/E_T)^{1/2}$ through the product $\nu_T a^*$. With electronic relaxation ν_T should be replaced by $\nu_i = (R_H^*/E_i)^{1/2}$.

(7) A question associated with lattice distortion is whether the impurity introduces local vibrational modes since these may interact strongly with localized electrons. Such modes manifest themselves in experiments on optical absorption and scattering. In order for localized modes to exist at all the impurity mass difference δM from the host must exceed some critical value, rather as a short-range potential well must exceed some critical value in order to bind an electron. Therefore not all impurities possess localized vibrational modes. If $\delta M > 0$, i.e. a heavier impurity, a localized state can be extracted from the optical modes of the crystal but not from the acoustic modes since this condition implies a reduction in frequency and no state can be pulled below a frequency $\omega = 0$. However, if $\delta M < 0$ states can be pulled from the tops of the acoustic band and the optical band. In either case the vibration is of short wavelength with an amplitude that falls off exponentially with distance. The simplest type of localized vibration has spherical symmetry and is referred to as a 'breathing mode'. Although mass difference is the transparent candidate for introducing local modes, it is not the only one; changes in force constants can obviously have the same effect, although these are not easy to predict.

(8) Shifts in localized electronic levels and a reduction in symmetry can be achieved by applying uniaxial stress, an electric field, or a magnetic field, and observing the changes produced in the optical properties of the centre. This approach can help in some cases to interpret the previous observation of electronic states. (It may often merely create a more

TABLE 2.4
Characteristics of deep-level impurities

Property	Effect
(1) Size: radius of ground state wavefunction	Transition cross-sections Interaction with neighbouring defects Solubility
(2) Shape and symmetry: point defect or complex?	As for size
(3) Parity: relative to band-edge Bloch waves.	Transition probabilities
(4) Charge	Electron scattering and capture Defect pairing.
(5) Energy-level structure	Optical absorption; optical and thermal ionization Luminescence; capture and emission through excited states Electron Scattering
(6) Occupation and degeneracy of localized states	Determination of energy levels Electronic relaxation effects
(7) Magnetic moment	E.S.R. Magnetic interaction with other defects.
(8) Coupling to lattice vibrations	Thermal broadening; determination of energy levels. Electron capture rate
(9) Local vibrational mode structure	As for (8); optical absorption and scattering

complex situation to analyse which at best provides confirmatory evidence for assigned models.) A full account can be found in the review by Bassani *et al.* (1974).

The principal characteristics of a defect which require description are listed in Table 2.4. Some impurity levels are noted in Table 2.5. Comprehensive reviews have been written by Milnes (1973), Grimmeiss (1977), and Queisser (1978).

2.11. Scattering states

As well as introducing localized states into the forbidden gap or semi-localized states associated with upper minima, impurities modify the states in the conduction and valence bands. The positive energy solutions of the Schrödinger equation containing the impurity potential describe how an electron moving in a band is affected by the defect. As such they describe the scattering of electrons by the impurity potential. This topic will be taken up in the chapter devoted to impurity scattering.

2.12. Impurity bands

Hydrogenic impurities have spatially extended ground states and can easily reach concentrations at which significant interaction between individual centres occurs and the discrete levels broaden into an impurity

TABLE 2.5
Impurity levels in silicon, germanium, GaAs, and GaP (meV)

Group	Impurity	Silicon		Germanium		GaAs		GaP	
		CB	VB	CB	VB	CB	VB	CB	VB
I(a)	Li	31.0[a]		9·9			230 510		
	Na	31·5[a]							
I(b)	Cu		240[b]	260	40 320		150[b] 400[d]		500[d] 700[d]
	Ag	290	260	100 250	140		110		
	Au	540	350	40 200	50 150				
II(a)	Be						30[c]		50[a]
	Mg	107·5 206·5					30[c]		54[a]
II(b)	Zn	550	310		29 83		24[b]		64[a]
	Cd		100 300		45 160				97[a]
	Hg	360	330		90/160 370				
III	B		45[a]		10·47[a]				
	Al		68[a]		10·80[a]				
	Ga		71[a]		10·97[a]				
	In		151[a]		11·61[a]				
	T1		260		10				
IV	C						26·7[c]		48[a]
	Si					5·85[c]	35·2[c]	83[a]	204[a]
	Ge					6·08[c]	41.2[c]	200[a]	300[a]
	Sn				510			66[a]	
V	N	45·0							
	P	45·5[a]		12·76[a]					
	As	53·7[a]		14·04[a]					
	Sb	42·7[a]		10·19[a]					
	Bi			12					
VI	O	160 380	350			400* 690*		895[a]	
	S	320[d] 590[d]		180[a]		6·10[c]		104[a]	
	Se	300[d] 570[d]				5·89[c]		102[a]	
	Te	140				5.8[d]		89.8[d]	
Transition elements	Cr		890[c]		120		790		1200[d]
	Mn		330[c]				90[c]		410[c]
	Fe		570[c]	270	340		500[d]		750[d]

TABLE 2.5 (*continued*)
Impurity levels in silicon, germanium, GaAs, and GaP (meV)

		Silicon		Germanium		GaAs		GaP	
Group	Impurity	CB	VB	CB	VB	CB	VB	CB	VB
	Co		580[c]		250 310		160[c]		41[c]
	Ni		350[c]	310	230				500[c] 920[c]
	Pt	250	360	200	40				

[a] From Pantelides (1975).
[b] From Partin, Chen, Milnes, and Vassamillet (1979).
[c] From Watts (1977).
[d] Ledebo (private communication)
Other data are from Neuberger (1971).
* May not be O: the 690 meV level is probably associated with a Ga vacancy.
CB, conduction band; VB valence band.

band. Were impurities to be uniformly distributed, an impurity band would exist at all concentrations, however small, and if there were sufficient compensation of donors by uniformly distributed acceptors only a few electrons would occupy the band states and the latter can be described by a one-electron theory. At low concentrations nearest-neighbour overlap will be small and the tight-binding approximation is appropriate.

In general the wavefunction for a periodic potential can be taken in the form of Wannier functions (see eqn (1.32), namely

$$\psi_{n\mathbf{k}}(\mathbf{r}) = \sum_{\mathbf{R}} \phi_n(\mathbf{r}-\mathbf{R})\exp(i\mathbf{k}.\mathbf{R}) \tag{2.138}$$

where $\phi_n(\mathbf{r}-\mathbf{R})$ is a Wannier function centred about the site at $\mathbf{R}$. Neglecting electron–electron interactions and assuming appreciable overlap of Wannier functions only on neighbouring sites we can apply tight-binding theory and obtain for a simple cubic lattice (Fig. 2.9)

$$E = E_0 - 2I_{100}(\cos k_x a + \cos k_y a + \cos k_z a), \tag{2.139}$$

for a face-centred cubic lattice

$$E = E_0 - 4I_{110}(\cos k_x a \cos k_y a + \cos k_y a \cos k_z a + \cos k_z a \cos k_x a), \tag{2.140}$$

and for a body-centred cubic lattice

$$E = E_0 - 8I_{111} \cos k_x a \cos k_y a \cos k_z a \tag{2.141}$$

where a is the lattice constant and I in each case is an overlap integral of

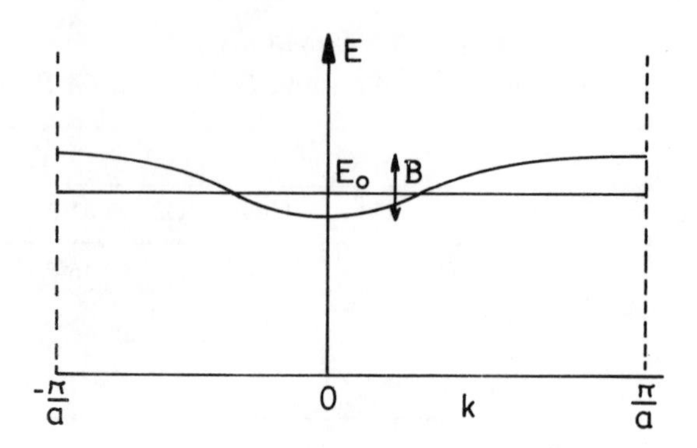

FIG. 2.9. Tight-binding band.

the form

$$I = \int \phi^*(\mathbf{r}-\mathbf{R}) H \phi(\mathbf{r}-\mathbf{R}-\mathbf{a})\, d\mathbf{r} \tag{2.142}$$

where H is the Hamiltonian. The bandwidth B is given generally by

$$B = 2zI \tag{2.143}$$

where z is the co-ordination number (6 for simple cubic, 12 for face-centred cubic, and 8 for body-centred cubic). By taking the Wannier function to be a $|1s\rangle$ hydrogenic function, the Bohr radius a_H^* (eqn (2.108)), and $2 \leqslant a/a_H^* \leqslant 4$ Mott (1972) has shown that

$$I \approx 16 R_H^* \exp(-a/a_H^*). \tag{2.144}$$

The effective mass associated with the impurity band edge ($ka \to 0$) can be found from

$$E = E_0 - zI + \hbar^2 k^2 / 2m_I^*. \tag{2.145}$$

For a simple cubic lattice

$$m_I^* = \hbar^2 / 2Ia^2. \tag{2.146}$$

This model predicts essentially metallic behaviour at any concentration, however low. An electron is free to move through the impurity superlattice with an effective mass m_I^*. Such behaviour has never been observed and there are several good physical reasons for this:

(1) the impurity distribution is non-uniform;
(2) the electron experiences a random potential arising from compensating impurities;
(3) in reality many electrons occupy the band and their mutual interaction cannot be neglected.

The effect of a random set of potential wells was first pointed out by Anderson (1958): it is to produce a localization of the electron. If the potentials vary randomly in the range $U \pm \frac{1}{2}V_0$, localization occurs when the localization parameter P exceeds 2, i.e.

$$P = V_0/B \gtrsim 2. \tag{2.147}$$

Such localization can be understood on the basis of the uncertainty principle. If the macroscopic size of the crystal is ΔL, then the uncertainty in the wave vector is Δk such that $\Delta k\, \Delta L \approx 1$. In terms of energy in the band this translates to

$$\Delta E\, \Delta L \approx Ba. \tag{2.148}$$

The random potential introduces an uncertainty $V_0/2$ into the energy, thus inverting the argument which led to eqn (2.148), and an uncertainty Δx is implied such that

$$(V_0/2)\, \Delta x \approx Ba \tag{2.149}$$

Localization, defined by $\Delta x \approx a$, occurs when $V_0/B \approx 2$ in accordance with eqn (2.147).

Although Anderson localization is predicted specifically for random potential variations, it is likely to be valid for a random spatial distribution of impurities as Mott (1972) has pointed out. Thus point (1) concerning a non-uniform distribution and point (2) concerning random potentials lead to the conclusion that unless the bandwidth exceeds the amplitude of the potential variations the electron remains localized and can migrate through the crystal only by a thermally activated hopping process. This is commonly observed.

However, even when the bandwidth is large enough for substantial delocalization to occur there will always be a few impurity clusters with large V_0 which will produce localization. The band edges will then exhibit density-of-states tails which decay roughly exponentially away from the band. Electrons in such tail states will be localized, whereas above a critical energy E_c they will be mobile. The energy is known as the mobility edge (Fig. 2.10). Such tails are to be expected not only for impurity bands but also for conduction and valence bands.

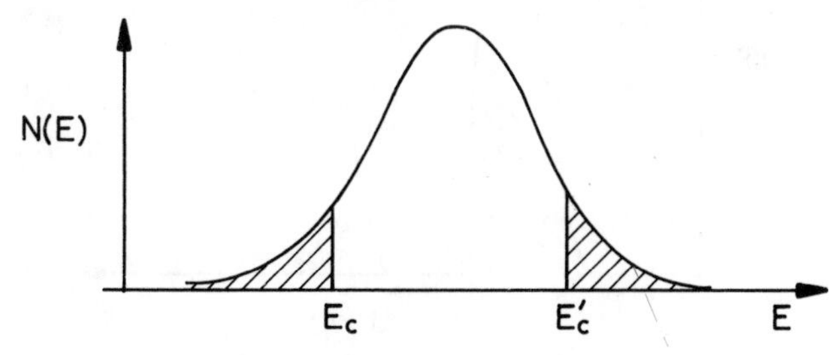

FIG. 2.10. Density of states and mobility edge.

In well-occupied narrow bands the interaction between electrons cannot be ignored. Screening, polarization, and magnetic effects all become important and severely modify the simple tight-binding band structure. The quantitative measure of electron–electron interaction is the energy associated with the repulsion of two electrons of opposite spin on the same atom:

$$U_{ee} = \iint |\phi(r_1)|^2 \, |\phi(r_2)|^2 \, (e^2/4\pi\epsilon r_{12}) \, d\mathbf{r}_1 \, d\mathbf{r}_2. \tag{2.150}$$

For hydrogenic $|1s\rangle$ states (Slater 1963)

$$U_{ee} = \tfrac{5}{4} R_H^*. \tag{2.151}$$

(Since $U_{ee} > R_H$ the H^- ion ought to be unstable, but it is not (see Section 2.8). The electronic motion becomes correlated to reduce U_{ee} just below R_H. This correlation energy is not important in the present context.) With one electron per atom (zero compensation) and $U_{ee} \gg B$, U_{ee} is the energy to activate an electron so that it can migrate through the crystal and is known as the Hubbard gap. In the ground state electrons on adjacent atoms must have opposite spins, and so the system acts as an antiferromagnetic insulator with an energy gap U_{ee} (Fig. 2.11). However, if electrons are removed or added, hole or electron conduction can occur and the system acts like a metal. With a finite band width the Hubbard gap becomes $U_{ee} - B$, and when the concentration of impurities increases to a critical value N_{crit} the gap becomes zero and the behaviour is entirely metallic. This metal–insulator transition occurs when $B = U_{ee}$. From eqns (2.143), (2.144), and (2.151) the critical concentration is given by

$$N_{crit}^{1/3} a_H^* = 0{\cdot}27. \tag{2.152}$$

Detailed discussions of this topic and the closely related one of amorphous semiconductors can be found in the book by Mott and Davis (1971).

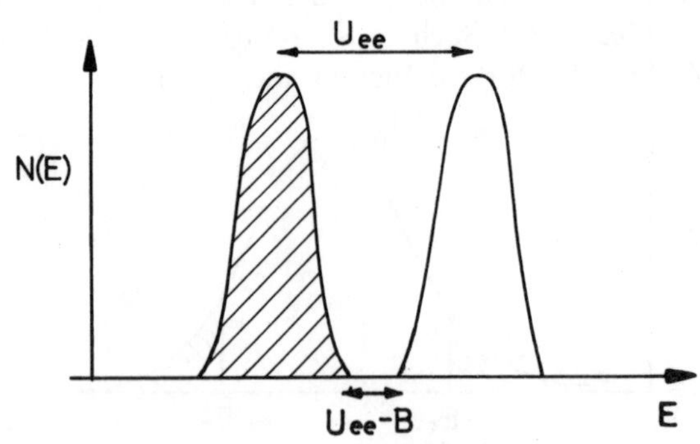

FIG. 2.11. Hubbard bands.

References

ANDERSON, P. W. (1958). *Phys. Rev.* **109,** 1492.
BASSANI, F., IADONISI, G., and PREZCORI, B. (1974). *Rep. Prog. Phys.* **37,** 1099.
BEBB, H. (1969). *Phys. Rev.* **185,** 1116.
BOHM, D. and PINES, D. (1953). *Phys. Rev.* **92,** 609.
BUTTON, K. J. (1970). *Optical properties of solids* (ed. E. D. Haidemenakis), p. 253. Gordon and Breach, London.
DEAN, P. J. (1973). *Prog. Solid. State Chem.* **8,** 1.
FRÖHLICH, H. (1962). *Polarons and excitons* (ed. C. G. Kuper and G. D. Whitfield), p. 1. Oliver and Boyd, Edinburgh.
——, PELZER, H., and ZIEMAN, S. (1950). *Phil. Mag.* **41,** 221.
GRIMMEISS, H. G. (1977). *Annu. Rev. Mater. Sci.* **7,** 341.
HARPER, P. G., HODBY, J. W., and STRADLING, R. (1973). *Rep. Prog. Phys.* **36,** 1.
HAYNES, J. R. (1960). *Phys. Rev. Lett.* **4,** 361.
KITTEL, C. and MITCHELL, A. (1954). *Phys. Rev.* **96,** 1488.
LAX, B. and MAVROIDES, J. G. (1960). *Solid State Phys.* **11,** 261.
LUCOVSKY, G. (1965). *Solid State Commun.* **3,** 299.
LUTTINGER, J. M. and KOHN, W. (1955). *Phys. Rev.* **97,** 869.
MILNES, A. G. (1973). *Deep impurities in semiconductors.* John Wiley, New York.
MORSE, P. M. and FESHBACH, H. (1953). *Methods of theoretical physics.* McGraw-Hill, New York.
MOTT, N. F. (1972). *Phil. Mag.* **26,** 505.
—— and DAVIS, E. A. (1971). *Electronic processes in non-crystalline materials.* Clarendon Press, Oxford.
NEUBERGER, M. (1971). *Handbook of electronic materials,* Vols. 2 and 4. I.F.I. Plenum, New York.
PANTELIDES, S. T. (1975). *Festkorperproblem* **15,** 149.
—— (1978). *Rev. mod. Phys.* **50,** 797.
PARTIN, D. L., CHEN, J. W., MILNES, A. G., and VASSAMILLET, L. F. (1979). *Solid State Electron.* **22,** 455.
QUEISSER, H. J. (1978). *Solid State Electron.* **21,** 1495.
SLATER, J. C. (1963). *Quantum theory of molecules and solids,* Vol. 1. McGraw-Hill, New York.
STONEHAM, A. M. (1975). *Theory of defects in solids.* Clarendon Press, Oxford.
WATTS, R. K. (1977). *Point defects in solids.* John Wiley, New York.
WHITTAKER, E. T. and WATSON, G. N. (1946). *A course in modern analysis* (4th edn). London University Press, London.
ZENER, C. (1934). *Proc. R. Soc. A* **145,** 523.

3. Lattice scattering

3.1. General features

In broad-band semiconductors an electron migrates through the crystal with properties determined principally by the periodic potential associated with the array of ions at the lattice points. Vibrations of the ions about their equilibrium positions produce practically instantaneous changes in the energy of electrons and thus introduce a time-dependent component H_{ep} into the time-independent adiabatic one-electron Schrödinger equation. This interaction is the adiabatic electron–phonon interaction and it is usually weak enough for H_{ep} to be regarded as a small perturbation which in the usual way induces transitions between unperturbed states. The rate is given by the familiar equation of first-order perturbation theory:

$$W = \int \frac{2\pi}{\hbar} |\langle \mathrm{f}|H_{ep}|\mathrm{i}\rangle|^2 \delta(E_{\mathrm{f}} - E_{\mathrm{i}}) \mathrm{d}S_{\mathrm{f}} \tag{3.1}$$

where f and i refer to final and initial states, the integral is over all final states S_{f}, and the time dependence is subsumed in the delta function conserving energy.

Strict conservation of energy is of course an approximation. What first-order perturbation theory gives is not a delta function but a function of energy of the form $\sin\{(E_{\mathrm{f}} - E_{\mathrm{i}})t/\hbar\}/(E_{\mathrm{f}} - E_{\mathrm{i}})\pi$, where t is time, which approaches a delta function when $t \rightarrow \infty$. The difference between this function and a delta function will, however, be unimportant provided the 'duration' t_{d} of the collision is short compared with the time t_{c} between collisions where t_{d} is defined (Paige 1964) by

$$t_{\mathrm{d}} = \frac{\hbar}{|\langle \mathrm{f}|H_{ep}|\mathrm{i}\rangle|^2} \frac{\mathrm{d}}{\mathrm{d}E} |\langle \mathrm{f}|H_{ep}|\mathrm{i}\rangle|^2. \tag{3.2}$$

Usually, where first-order perturbation theory is applicable it is found that $t_{\mathrm{d}} \ll t_{\mathrm{c}}$.†

The interaction energy in its simplest form depends linearly on strain—acoustic strain in the case of acoustic modes and optical strain in the case of optical modes (eqns (1.103) and (1.105)). These strains influence the electron in a band in two distinct ways. In the first way, short-range disturbances of the periodic potential cause practically instantaneous changes in energy, and these changes are the ones quantified by deformation potentials (Section 1.13). Disturbance of the electron's motion by

† A fuller account will be found in Section 11.4.

this effect is referred to as deformation-potential scattering. By its nature it is common to all semiconductors and indeed all solids. In the second way, the distortion of the lattice may destroy local electric neutrality, and produce electrical polarization and associated macroscopic comparatively long-range electric fields to which the electron responds. Disturbance of the electron's motion by this effect is referred to as piezoelectric scattering, associated with acoustic modes, and polar optical mode scattering, associated with optical modes. Such scattering occurs only in polar materials.

The displacement $\mathbf{u}(\mathbf{R})$ associated with a given mode of the unit cell at $\mathbf{R}$ can be expressed in terms of plane waves:

$$\mathbf{u}(\mathbf{R})=\frac{1}{\sqrt{N}}\sum_{\mathbf{q}}\{Q_{\mathbf{q}}\mathbf{a}_{\mathbf{q}}\exp(\mathrm{i}\mathbf{q}.\mathbf{R})+\mathrm{c.c.}\} \tag{3.3}$$

(the time dependence being already incorporated in the delta function conserving energy) where N is the number of unit cells in the periodic crystal, the $Q_{\mathbf{q}}$ are normal coordinates, $\mathbf{a}_{\mathbf{q}}$ is a unit polarization vector, $\mathbf{q}$ is the wavevector, and c.c. stands for complex conjugate. In acoustic waves $\mathbf{u}(\mathbf{R})$ refers to the displacement of the unit cell; in optical waves it refers to the relative displacement of the two atoms in the unit cell. The linear dependence of H_{ep} on strain, either acoustic or optical, means that it is also linearly dependent on the normal coordinates of individual simple harmonic oscillators. This has the well-known consequence that a given mode may only change its phonon occupancy by unity.

The wavefunction Ψ_{i} of the initial state can be expressed as a product of a one-electron wavefunction, i.e. a Bloch function $\psi_{n\mathbf{k}}(\mathbf{r})$, and harmonic oscillator wavefunctions:

$$\Psi_{\mathrm{i}}=\psi_{n\mathbf{k}}(\mathbf{r})\prod_{\mathbf{q},b}\phi_{\mathbf{q},b}(Q) \tag{3.4}$$

where the product is of all modes of a given branch b and of all branches. Also, the final state will consist of a product of a Bloch function $\psi_{n'\mathbf{k}'}(\mathbf{r})$ and oscillator functions $\phi'_{\mathbf{q},b}(Q)$, where the prime denotes that the number $n_{\mathbf{q},b}(\omega_{\mathbf{q},b})$ of quanta excited is different in general from the number excited in the initial state. Because of the linear dependence of H_{ep} on the $Q_{\mathbf{q},b}$, the matrix element will contain factors linear in the normal coordinate. These are as follows:

$$\int\phi'_{\mathbf{q}'',b''}(Q)Q_{\mathbf{q},b}\phi_{\mathbf{q}',b'}(Q)\,\mathrm{d}Q_{\mathbf{q},b}=\left(\frac{\hbar}{2M'\omega_{\mathbf{q},b}}\right)^{1/2}\delta_{\mathbf{q},\mathbf{q}',\mathbf{q}''}\delta_{b,b',b''}$$
$$\times\{n_{\mathbf{q},b}^{1/2}(\omega_{\mathbf{q},b})\delta_{n',n-1}+(n_{\mathbf{q},b}(\omega_{\mathbf{q},b})+1)^{1/2}\delta_{n',n+1}\} \tag{3.5}$$

where $\omega_{\mathbf{q},b}$ is the angular frequency of the mode with wavevector $\mathbf{q}$, M' is the appropriate mass of the oscillator, e.g. the total mass of the unit cell in the case of acoustic modes or in the case of long-wavelength optical modes the reduced mass $\bar{M}$ where

$$1/\bar{M} = 1/M_1 + 1/M_2. \tag{3.6}$$

M_1 and M_2 are the masses of the two atoms. The right-hand side of eqn (3.5) follows from the properties of simple harmonic wavefunctions and depicts the well-known property of single-phonon absorption or emission: a linear dependence of H_{ep} on the normal coordinate entails that the scattering of an electron by a given lattice wave is accompanied by the absorption or the emission of one phonon.

The matrix element also contains an electronic component which consists of a sum of terms, each term being associated with an individual lattice wave having the form

$$\frac{1}{V}\int u^*_{n'\mathbf{k}'}(\mathbf{r})\exp(-\mathrm{i}\mathbf{k}'.\mathbf{r})H_{\mathbf{q},b}(\mathbf{r})\exp(\pm\mathrm{i}\mathbf{q}.\mathbf{R})u_{n\mathbf{k}}(\mathbf{r})\exp(\mathrm{i}\mathbf{k}.\mathbf{r})\,\mathrm{d}\mathbf{r}$$

$$= \frac{1}{V}C_{\mathbf{q},b}I(\mathbf{k},\mathbf{k}')\int \exp\{\mathrm{i}(\mathbf{k}\pm\mathbf{q}-\mathbf{k}').\mathbf{R}\}\mathrm{d}\mathbf{R} \tag{3.7}$$

$$C_{\mathbf{q},b}I(\mathbf{k},\mathbf{k}') = \int_{\text{cell}} \psi^*_{n'\mathbf{k}'}(\mathbf{r})H_{\mathbf{q},b}(\mathbf{r})\psi_{n\mathbf{k}}(\mathbf{r})\,\mathrm{d}\mathbf{r},\; I(\mathbf{k},\mathbf{k}') = \int_{\text{cell}} \psi^*_{n'\mathbf{k}'}(\mathbf{r})\psi_{n\mathbf{k}}(\mathbf{r})\,\mathrm{d}\mathbf{r} \tag{3.8}$$

where the integral over the crystal has been factorized into an integral over a unit cell, which is the same for all unit cells because of the periodic properties of the Bloch functions and of the interaction, and an integral over all the unit cells over the volume V. The interaction constant $H_{\mathbf{q},\mathrm{b}}(\mathbf{r})$ represents the electron-co-ordinate-dependent part of H_{ep} for a given mode which is independent of **R** for lattice-scattering processes. The integral over all the unit cells can be performed immediately, and the electronic part of the matrix element becomes simply

$$C_{\mathbf{q},b}I(\mathbf{k},\mathbf{k}')\delta_{\mathbf{k}\pm\mathbf{q}-\mathbf{k}',\mathbf{K}} \tag{3.9}$$

where **K** is a reciprocal-lattice vector. Scattering events in which $\mathbf{K}=0$ are termed normal processes. In them, crystal momentum is conserved:

$$\mathbf{k}\pm\mathbf{q}-\mathbf{k}'=0. \tag{3.10}$$

If $\mathbf{K}\neq 0$, the scattering event is an umklapp process. In either case only one mode of a given branch can effect the transition for a given initial and final electronic state. Consequently, in a single lattice scattering event only one mode is involved. The minus sign is taken for emission and the plus sign for absorption.

The integral $C_{\mathbf{q},b}I(\mathbf{k},\mathbf{k}')$ provides a final selection rule depending on the symmetries of the interaction and the Bloch functions of initial and final states. For the polar, optical, and piezoelectric interactions the macroscopic fields produced are essentially long-range fields which vary slowly over a unit cell, and thus $H_{\mathbf{q},b}(\mathbf{r})$ is approximately constant over the unit cell. The integral is then zero unless the Bloch functions belong to the same valley. Moreover, because the cell periodic part of the Bloch function does not vary rapidly with $\mathbf{k}$ over those states near the band edge most occupied by electrons (or holes), we can often take $u_{n\mathbf{k}} \approx u_{n,\mathbf{k}\pm\mathbf{q}}$ and consequently obtain $I(\mathbf{k},\mathbf{k}')=1$. This will be valid unless non-parabolicity is marked. Where non-parabolicity is present $I(\mathbf{k},\mathbf{k})$ is less than unity. In the central conduction valley of III–V compounds the cell periodic part of the Bloch function contains an admixture of a $|p\rangle$-like component deriving from the valence band, and this leads to the non-parabolic form

$$\frac{\hbar^2k^2}{2m^*}=E_{\mathbf{k}}(1+\alpha E_{\mathbf{k}}). \tag{3.11}$$

It can be shown that in this case (Fawcett, Boardman, and Swain 1970)

$$I^2(\mathbf{k},\mathbf{k}')=\frac{\{(1+\alpha E_{\mathbf{k}})^{1/2}(1+\alpha E_{\mathbf{k}'})^{1/2}+\alpha(E_{\mathbf{k}}E_{\mathbf{k}'})^{1/2}\cos\theta_{\mathbf{k}}\}^2}{(1+2\alpha E_{\mathbf{k}})(1+2\alpha E_{\mathbf{k}'})}. \tag{3.12}$$

For holes within heavy or light bands (Wiley 1971)

$$I^2(\mathbf{k},\mathbf{k}')=\tfrac{1}{4}(1+3\cos^2\theta_{\mathbf{k}}) \tag{3.13}$$

where $\theta_{\mathbf{k}}$ is the angle between the initial and final state vectors.

Because of the long-range nature of the polar interaction, polar optical and piezoelectric scattering are basically intra-valley scattering events. Both intra-valley and inter-valley scattering are possible through the deformation-potential interaction. In either case we can take $H_{\mathbf{q},b}(\mathbf{r})=C_{\mathbf{q},b}$ for an allowed transition and zero for a transition forbidden by symmetry.

We can summarize these general features in the following formulae for the matrix element and transition rate for an allowed process:

$$|\langle \mathrm{f}|\,H_{\mathbf{q},b}\,|\mathrm{i}\rangle|^2=\frac{\hbar}{2NM'}\frac{C^2_{\mathbf{q},b}I^2(\mathbf{k},\mathbf{k}')}{\omega_{\mathbf{q},b}}(n(\omega_{\mathbf{q},b})+\tfrac{1}{2}\mp\tfrac{1}{2})$$

$$W(\mathbf{k})=\frac{\pi}{NM'}\int\frac{C^2_{\mathbf{q},b}I^2(\mathbf{k},\mathbf{k}')}{\omega_{\mathbf{q},b}}\delta_{\mathbf{k}\pm\mathbf{q}-\mathbf{k}',\mathbf{K}}$$
$$\times(n(\omega_{\mathbf{q},b})+\tfrac{1}{2}\mp\tfrac{1}{2})\delta(E_{\mathbf{k}'}-E_{\mathbf{k}}\mp\hbar\omega_{\mathbf{q},b})\,\mathrm{d}S_{\mathrm{f}} \tag{3.14}$$

where $\hbar\omega_{\mathbf{q},b}$ is the phonon energy and $C^2_{\mathbf{q},b}$ is a coupling parameter which in general depends upon the magnitude and direction of $\mathbf{q}$. The number of final states in an elementary volume of $\mathbf{k}$-space denoted by $\mathrm{d}\mathbf{k}$ is $\mathrm{d}\mathbf{k}\,V/(2\pi)^3$, and so we obtain

$$W(\mathbf{k}) = \frac{V}{8\pi^2 NM'} \int \frac{C_{\mathbf{q},b}^2 I^2(\mathbf{k},\mathbf{k}')}{\omega_{\mathbf{q},b}} \delta_{\mathbf{k}\pm\mathbf{q}-\mathbf{k}',\mathbf{K}}$$

$$\times (n(\omega_{\mathbf{q},b}) + \tfrac{1}{2} \mp \tfrac{1}{2}) \delta(E_{\mathbf{k}'} - E_{\mathbf{k}} \mp \hbar\omega_{\mathbf{q},b})\, \mathrm{d}\mathbf{k}' \quad (3.15)$$

The upper sign in these equations is for absorption and the lower is for emission of a phonon. The integral over $\mathbf{k}'$ is equivalent to an integral over $\mathbf{q}$ (because of eqn (3.10)) and often the latter is more convenient. We cannot proceed further without a more detailed investigation of the form of the interaction and of the consequences of the conservation of energy and momentum.

The above equations describe the rate of scattering of an electron occupying a Bloch state $|\mathbf{k}\rangle$. In many situations we are interested in calculating phenomenological quantities which are properties of an ensemble of electrons rather than of a single electron, and in these cases we must introduce statistical elements and concepts and graft them onto the basic quantum-mechanical results discussed above. For example, we often need to know how the occupation probability $f(\mathbf{k})$ of the state $|\mathbf{k}\rangle$ changes with time because of scattering, and this is given by the so-called collision integral which for non-degenerate statistics is as follows:

$$\frac{\mathrm{d}f(\mathbf{k})}{\mathrm{d}t} = \int \{W(\mathbf{k}'',\mathbf{k})f(\mathbf{k}'') - W(\mathbf{k},\mathbf{k}'')f(\mathbf{k})\}\delta(E_{\mathbf{k}''} - E_{\mathbf{k}} - \hbar\omega_{\mathbf{q},b})\, \mathrm{d}\mathbf{k}'' +$$

$$+ \int \{W(\mathbf{k}',\mathbf{k})f(\mathbf{k}') - W(\mathbf{k},\mathbf{k}')f(\mathbf{k})\}\delta(E_{\mathbf{k}'} - E_{\mathbf{k}} + \hbar\omega_{\mathbf{q},b})\, \mathrm{d}\mathbf{k}' \quad (3.16)$$

where $W(\mathbf{k},\mathbf{k}')$ is the integrand of eqn (3.15) apart from the delta function representing the transition $\mathbf{k}\rightarrow\mathbf{k}'$ and $W(\mathbf{k}',\mathbf{k})$ represents the reverse case. Our primary concern in this chapter and elsewhere is with the elementary quantum processes and so we shall not be concerned much with elements of statistical physics. Our emphasis will be on the response of an individual electron rather than on solving the Boltzmann equation in order to obtain the distribution function of the ensemble.

3.2. Energy and momentum conservation

3.2.1. *Spherical parabolic band*

Consider a normal process in which crystal momentum is conserved in a scattering event within a spherical parabolic band.

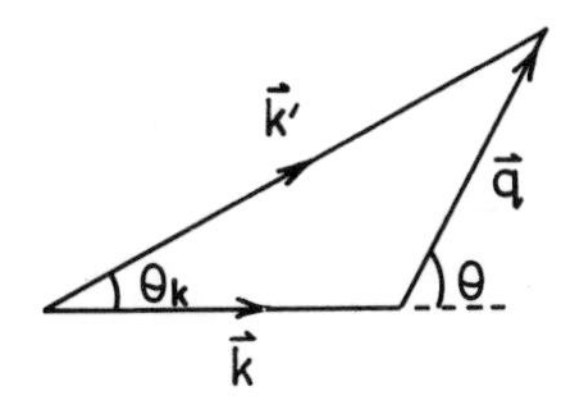

FIG. 3.1. Absorption of a phonon.

3.2.1.1. Absorption. The absorption of a phonon is determined by (Figs. 3.1 and 3.2)

$$k'^2 = k^2 + q^2 + 2kq \cos\theta \tag{3.17}$$

$$\frac{\hbar^2 k'^2}{2m^*} = \frac{\hbar^2 k^2}{2m^*} + \hbar\omega \tag{3.18}$$

whence

$$\cos\theta = -\frac{q}{2k} + \frac{m^*\omega}{\hbar kq}. \tag{3.19}$$

Let us denote the right-hand side of eqn (3.19) by $f(q)$:

$$f(q) = -\frac{q}{2k} + \frac{m^*\omega}{\hbar kq}. \tag{3.20}$$

According to eqn (3.19) energy and momentum conservation imposes the constraint $-1 \leqslant f(q) \leqslant 1$.

Absorption of acoustic phonons. For acoustic phonons $\omega = v_s q$ whence, since $\hbar k = m^* v$,

$$f(q) = -q/2k + v_s/v. \tag{3.21}$$

If $v \geqslant v_s$ the minimum value of q is clearly zero, and the maximum value

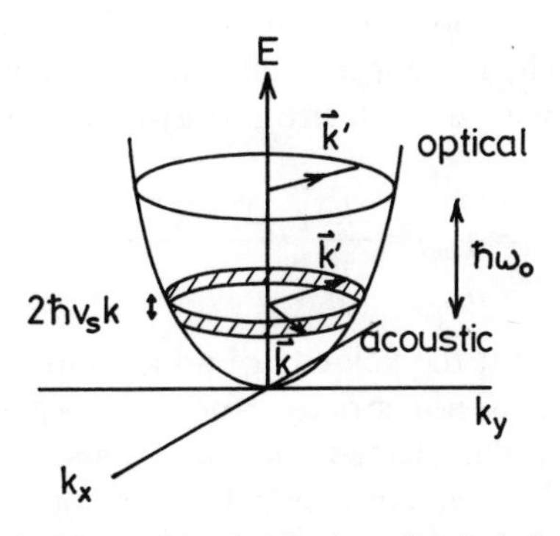

FIG. 3.2. Absorption of acoustic and optical phonons.

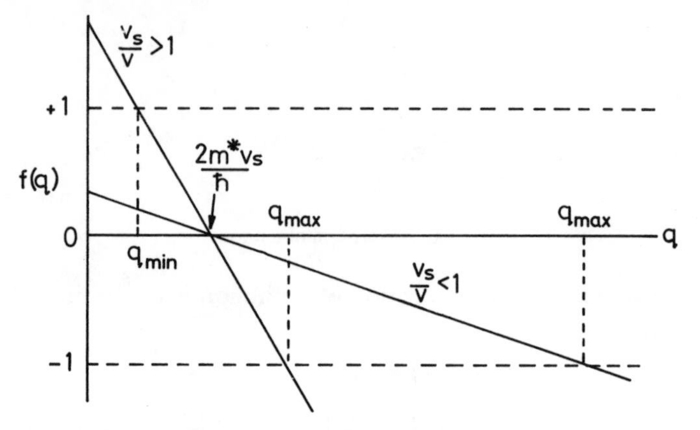

FIG. 3.3. Limitations on q for the absorption of an acoustic phonon.

of q is given by $f(q) = -1$. If $v < v_s$ the minimum value of q is obtained by putting $f(q) = +1$ (Fig. 3.3). In summary, for absorption

$$q_{\min} \leqslant q \leqslant q_{\max} \tag{3.22}$$

$$q_{\min} = \begin{cases} 0 & v \geqslant v_s \text{ corresponding to } \cos\theta = v_s/v \\ 2k(v_s/v - 1) & v < v_s \text{ corresponding to } \theta = 0 \end{cases} \tag{3.23}$$

$$q_{\max} = 2k(v_s/v + 1) \qquad \text{corresponding to } \theta = \pi. \tag{3.24}$$

The average group velocity at room temperature is typically of order 10^7 cm s^{-1}, whereas the velocity of sound is typically of order 10^5 cm s^{-1}; thus most electrons can absorb phonons with wavevectors q satisfying $0 \leqslant q \leqslant 2k$. The range of energy change involved is from zero to $2\hbar v_s k$. With an average k of order 10^7 cm^{-1}, $2\hbar v_s k \approx 1$ meV. This energy is small compared with thermal energies $k_B T \approx 25$ meV, and so scattering by absorption of acoustic phonons is approximately elastic (Fig. 3.2). The direction of $\mathbf{k}$ can change appreciably but its magnitude changes very little. In the limit of very slow electrons only one value of q is possible for all θ:

$$q_{\min} = q_{\max} = \frac{2kv_s}{v} = \frac{2m^* v_s}{\hbar} \qquad v \ll v_s. \tag{3.25}$$

We have assumed that the velocity of an acoustic mode is independent of direction but this is not strictly true (see Appendix, Section 3.9). Usually the variation is neglected and an average velocity is assumed.

Absorption of optical phonons. For Γ point optical phonons $\omega = \omega_0$, where ω_0 is insensitive to the magnitude of q for small q very much less

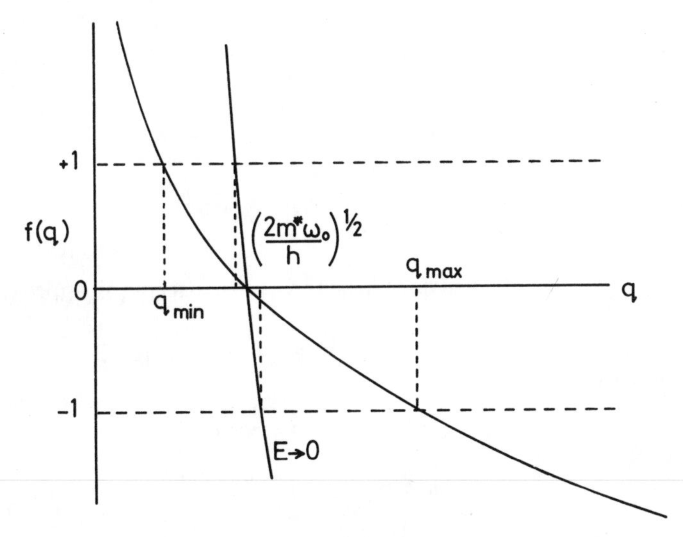

FIG. 3.4. Limitations on q for the absorption of an optical phonon.

than the zone-edge wavevector. In this case (Fig. 3.4)

$$f(q) = -\frac{q}{2k} + \frac{m^*\omega_0}{\hbar kq} \tag{3.26}$$

which has the solution

$$\frac{q}{2k} = \frac{1}{2}\left[-f(q) \pm \left\{f^2(q) + \frac{\hbar\omega_0}{E_k}\right\}^{1/2}\right]. \tag{3.27}$$

Since q is a positive quantity only the plus sign is relevant. Thus with

$$\chi_k^+ = (1 + \hbar\omega_0/E_k)^{1/2} \tag{3.28}$$

$$q_{\min} \leqslant q \leqslant q_{\max} \tag{3.29}$$

$$q_{\min} = k(\chi_k^+ - 1) \qquad \text{corresponding to } \theta = 0 \tag{3.30}$$

$$q_{\max} = k(\chi_k^+ + 1) \qquad \text{corresponding to } \theta = \pi. \tag{3.31}$$

In the limit of small k only one value of q is possible for all θ:

$$q_{\min} = q_{\max} = (2m^*\omega_0/\hbar)^{1/2}. \tag{3.32}$$

Unlike the case of acoustic phonons, optical phonon scattering is markedly inelastic because of the appreciable magnitude of $\hbar\omega_0$ which is typically several tens of millielectronvolts (Fig. 3.3). However, as with acoustic phonon scattering, the modes which may interact are those with wavevectors near the zone centre, i.e. long-wavelength modes. In the case

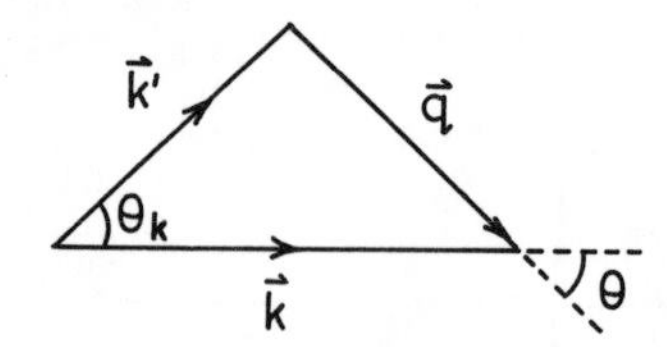

FIG. 3.5. Emission of a phonon.

of very energetic electrons, $E_k \gg \hbar\omega_0$ and the scattering is approximately elastic.

3.2.1.2. Emission. The emission of a phonon (Figs 3.5 and 3.6) is determined by

$$k'^2 = k^2 + q^2 - 2kq \cos\theta \tag{3.33}$$

$$\frac{\hbar^2 k'^2}{2m^*} + \hbar\omega = \frac{\hbar^2 k^2}{2m^*} \tag{3.34}$$

whence

$$\cos\theta = \frac{q}{2k} + \frac{m^*\omega}{\hbar kq} = f(q) \tag{3.35}$$

Emission of acoustic phonons. For acoustic phonons (Fig. 3.7)

$$f(q) = \frac{q}{2k} + \frac{v_s}{v} \tag{3.36}$$

whence it follows from the limitation $-1 \leqslant f(q) \leqslant 1$ that no solution exists for $v < v_s$. An electron must travel with a group velocity in excess of the sound velocity in order to emit a phonon. Provided that $v \geqslant v_s$, the limits

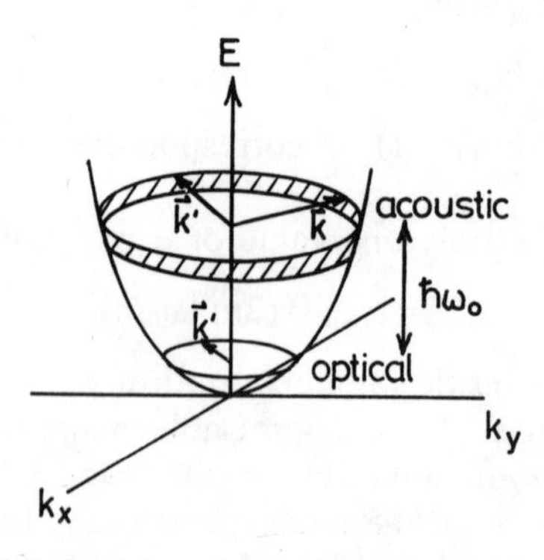

FIG. 3.6. Emission of acoustic and optical phonons.

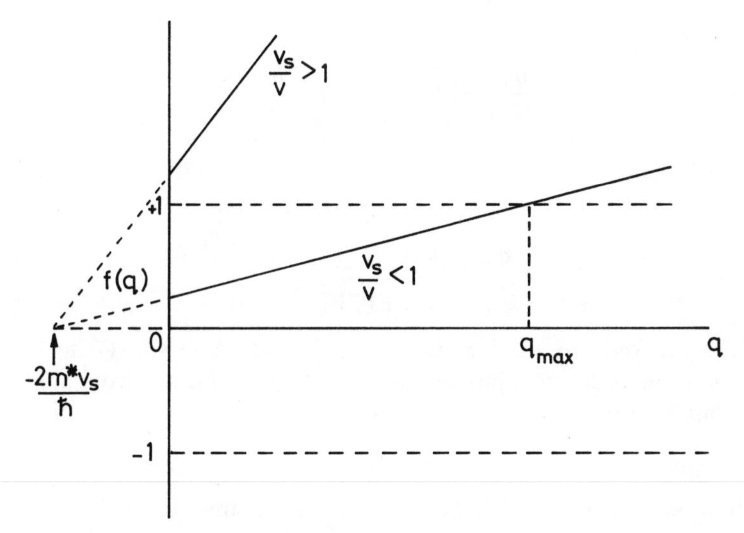

FIG. 3.7. Limitations on q for the emission of an acoustic phonon.

on q are

$$q_{\min} = 0 \qquad \cos\theta = v_s/v \tag{3.37}$$

$$q_{\max} = 2k\left(1 - \frac{v_s}{v}\right) = 2k - \frac{2m^* v_s}{\hbar} \qquad \theta = 0. \tag{3.38}$$

Note that $0 \leq \theta \leq \cos^{-1}(v_s/v)$, i.e. forward emission only.

Emission of optical phonons. For optical phonons (Fig. 3.8)

$$f(q) = \frac{q}{2k} + \frac{m^* \omega_0}{\hbar k q} \tag{3.39}$$

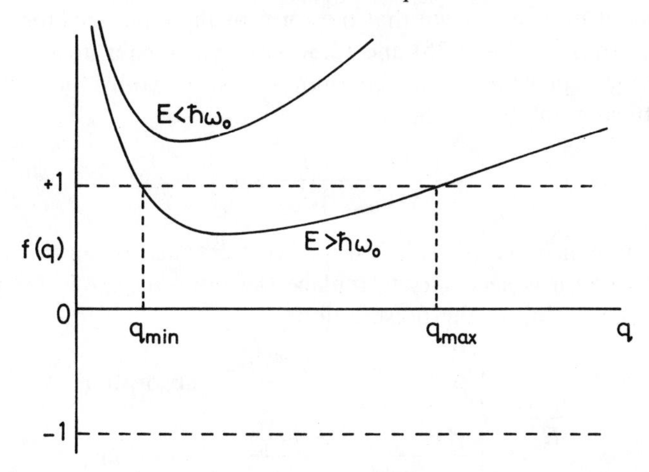

FIG. 3.8. Limitations on q for the emission of an optical phonon.

and

$$\frac{q}{2k}=\frac{1}{2}\left[f(q)\pm\left\{f^2(q)-\frac{\hbar\omega_0}{E_k}\right\}^{1/2}\right]. \tag{3.40}$$

When

$$\chi_k^- = (1-\hbar\omega_0/E_k)^{1/2} \tag{3.41}$$

$$q_{\min}=k(1-\chi_k^-) \qquad \theta=0 \tag{3.42}$$

$$q_{\max}=k(1+\chi_k^-) \qquad \theta=0. \tag{3.43}$$

Once again only forward emission is allowed. Also for χ_k^- to be real, $E_k \geqslant \hbar\omega_0$; in order to emit an optical phonon the electron must have sufficient energy.

3.2.2. Spherical non-parabolic band

In the case of a non-parabolic band we can assume (eqn (1.82))

$$\hbar^2k^2/2m^* = \gamma(E_k). \tag{3.44}$$

For the case of acoustic phonon scattering it is sufficient to relate final and initial states through a truncated Taylor expansion:

$$\gamma'(E_k)\approx\gamma(E_k)+\frac{d\gamma(E_k)}{dE_k}\hbar v_s q. \tag{3.45}$$

The group velocity v is given by

$$v=\frac{\hbar k}{m^*}\left\{\frac{d\gamma(E_k)}{dE_k}\right\}^{-1} \tag{3.46}$$

and hence it is easily shown that the form of the equations for the limits on q (i.e. eqns (3.23)–(3.25) and (3.36)–(3.38)) remain unchanged.

A Taylor expansion cannot be employed successfully for optical phonon scattering unless

$$\frac{\hbar\omega_0}{\gamma(E_k)}\frac{d\gamma(E_k)}{dE_k}\ll 1, \tag{3.47}$$

a condition which is often satisfied for hot electrons. When this condition is not satisfied it is necessary to replace the ratio $\hbar\omega_0/E_k$ in the previous formulae according to the prescription

$$\frac{\hbar\omega_0}{E_k}\rightarrow\begin{cases}\dfrac{\gamma(E_k+\hbar\omega_0)-\gamma(E_k)}{\gamma(E_k)} & \text{absorption}\\[2ex] \dfrac{\gamma(E_k)-\gamma(E_k-\hbar\omega_0)}{\gamma(E_k)} & \text{emission.}\end{cases} \tag{3.48}$$

3.2.3. *Ellipsoidal parabolic bands*

In ellipsoidal bands the energy is no longer simply related to the magnitude of **k**:

$$E_{\mathbf{k}} = \frac{\hbar^2 k_1^2}{2m_1^*} + \frac{\hbar^2 k_2^2}{2m_2^*} + \frac{\hbar^2 k_3^2}{2m_3^*}. \tag{3.49}$$

The trick here is to transform all vector components according to

$$k_1 = \left(\frac{m_1^*}{m_0^*}\right)^{1/2} k_1^* \tag{3.50}$$

where m_0^* is arbitrary (but usually chosen to be one of the m_i^*). Then

$$E_{\mathbf{k}} = \frac{\hbar^2 k^{*2}}{2m_0^*}. \tag{3.51}$$

Unfortunately this means that the acoustic phonon energy $\hbar v_s q$ becomes dependent upon direction in the new vector space. However, since the energy is small it is usually sufficient to perform an average over direction and retain the simple form of the dependence of energy on phonon wavevector:

$$\hbar\omega \approx \hbar\left\langle v_s\left(\frac{m^*}{m_0}\right)^{1/2}\right\rangle q^*. \tag{3.52}$$

The optical phonon energy is not affected by the transformation from ordinary vector space to 'starred' vector space.

When this transformation has been made, the limits are obtained exactly as before but with q and k replaced by q^* and k^*, and v_s replaced by $\langle v_s(m^*/m_0)^{1/2}\rangle$.

3.2.4. *Equivalent valleys*

When the initial state is in valley 1 and the final state is in valley 2, where valleys 1 and 2 are equivalent in energy, momentum conservation demands a short-wavelength phonon (Fig. 3.9). In general an umklapp

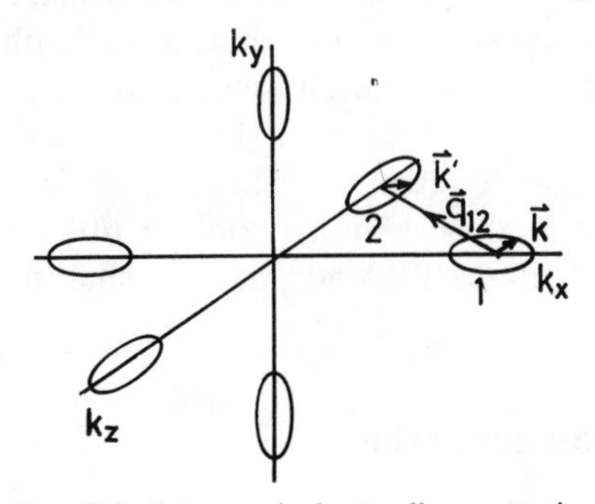

FIG. 3.9. Inter-equivalent-valley scattering.

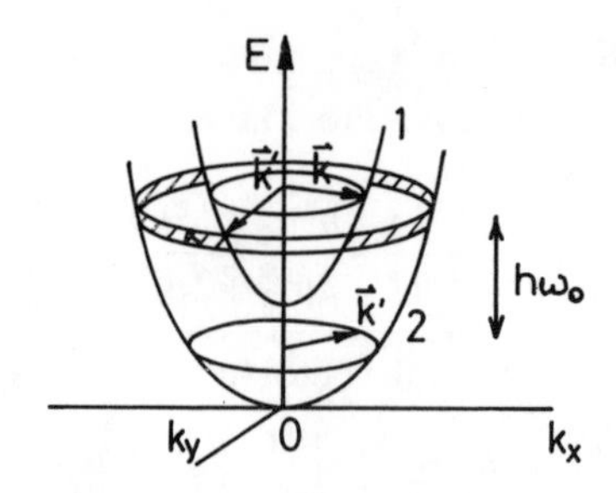

FIG. 3.10. Inter-non-equivalent-valley scattering at $\mathbf{k}=0$.

process may be necessary. Thus if the valley extrema are situated at $\mathbf{k}_1$ and $\mathbf{k}_2$ we require

$$\mathbf{k}_2+\mathbf{k}'=\mathbf{k}_1+\mathbf{k}\pm(\mathbf{q}_{12}+\mathbf{q})+\mathbf{K} \tag{3.53}$$

where $\mathbf{k}'$ and $\mathbf{k}$ are vectors relative to the respective extrema, $\mathbf{q}_{12}$ is a phonon wavevector such that

$$\mathbf{k}_2=\mathbf{k}_1\pm\mathbf{q}_{12}+\mathbf{K} \tag{3.54}$$

and $\mathbf{K}$ is a reciprocal lattice vector. 'Intervalley' phonons have wavevectors whose magnitudes are an appreciable fraction of the Brillouin-zone-boundary vector. Dispersion for these modes, whether acoustic or optical, is such as to allow us to neglect any dependence of frequency on wavevector. Thus, inter-valley scattering, once eqn (3.54) is satisfied, then proceeds exactly like intra-valley optical phonon scattering.

3.2.5. *Non-equivalent valleys*

Scattering from Γ to X, Γ to L, or X to L proceeds as for equivalent-valley scattering in that zone-edge phonons are involved, but the difference of energy between extrema has to be taken into account and also the difference of effective mass.

Non-equivalent valleys at the same point in the zone entail a treatment more akin to intra-valley scattering in that acoustic phonons and optical phonons have to be treated separately since only long-wavelength modes are involved (Fig. 3.10).

In either type of non-equivalent-valley scattering the treatment of energy and momentum conservation is only a little complicated by the energy gap and by the mass difference, and we shall not treat these cases explicitly.

3.3. Acoustic phonon scattering

Energy and momentum conservation restricts intra-valley scattering by acoustic phonons to long-wavelength modes. Such modes cannot change the energy of an electron other than through the elastic strain associated with them. A long-wavelength acoustic *displacement* cannot affect the energy since neighbouring unit cells all move by almost the same amount; only the *differential displacement*, namely the strain, is of importance. The interaction is described in terms of deformation potentials as discussed in Section 1.13. For electrons in a conduction band

$$H_{\text{ep}} = \sum_{ij} \Xi_{ij} S_{ij} \tag{3.55}$$

where Ξ_{ij} are the deformation potentials and S_{ij}, are the strain components:

$$S_{ij} = S_{ji} = \frac{1}{2}\left(\frac{\partial u_i}{\partial x_j} + \frac{\partial u_j}{\partial x_i}\right) \tag{3.56}$$

where $\mathbf{u}$ is the displacement and $\mathbf{R}(x_1 x_2 x_3)$ is the position vector of the unit cell. When $\mathbf{u}$ is expanded in terms of travelling plane waves (eqn (3.3) and Appendix, Section 3.9),

$$S_{ij} = \frac{1}{2\sqrt{N}} \sum_{\mathbf{q}} \{\mathrm{i}Q_{\mathbf{q}}(a_i q_j + a_j q_i) \exp(\mathrm{i}\mathbf{q}\centerdot\mathbf{R}) + \text{c.c.}\} \tag{3.57}$$

where N is the number of unit cells, $\mathbf{a}$ is the unit polarization vector, $\mathbf{q}$ is the wavevector and $Q_{\mathbf{q}}$ is the normal coordinate. Hence

$$H_{\text{ep}} = \frac{1}{2\sqrt{N}} \sum_{\mathbf{q}} \left[\mathrm{i}Q_{\mathbf{q}} \exp(\mathrm{i}\mathbf{q}\centerdot\mathbf{R}) \left\{\sum_{ij} \Xi_{ij}(a_i q_j + a_j q_i)\right\} + \text{c.c.}\right]. \tag{3.58}$$

Table 1.7 (p.39) lists the deformation potentials for Γ, L and X valleys. In the case of intra-valley scattering in a Γ valley, shear strains produce no energy change and consequently

$$H_{\text{ep}} = \frac{1}{\sqrt{N}} \sum_{\mathbf{q}} \{\mathrm{i}Q_{\mathbf{q}} \exp(\mathrm{i}\mathbf{q}\centerdot\mathbf{R}) \Xi_{\text{d}} \mathbf{a}\centerdot\mathbf{q} + \text{c.c.}\} \tag{3.59}$$

where Ξ_{d} is the deformation potential associated with pure dilation. The appearance of the scalar product $\mathbf{a}\centerdot\mathbf{q}$ denotes that only longitudinal modes interact with electrons.†

In the case of L and X valleys it is convenient to consider longitudinal and transverse modes separately. Because of elastic anisotropy the situation is very complicated. Shears may contribute to the energy shift; the effective elastic constants which determine the velocity of the wave vary with direction and so do the polarization vectors (Appendix). Strictly speaking, it is necessary to consider each direction of $\mathbf{q}$ separately in calculating the transition rate, and this was done by Herring and Vogt (1956) in their pioneering paper. However, it is possible to adopt a simpler approach which involves an implicit averaging over the azimuthal

† This is true only for elastically isotropic crystals. In general, LA and TA modes have mixed polarizations.

angle of **q** and regarding elastic anisotropy as small. In this scheme we define an effective deformation potential for longitudinal modes

$$\Xi_L(\theta) = \Xi_d + \Xi_u \cos^2 \theta_q \tag{3.60}$$

and for the sum of both branches of the transverse modes

$$\Xi_T(\theta) = \Xi_u \sin \theta_q \cos \theta_q \tag{3.61}$$

where θ_q is the angle between **q** and the principal axis of the spheroidal valley (Fig. 3.11). Thus, if $\Xi(\theta_q)$ denotes either of the above deformation

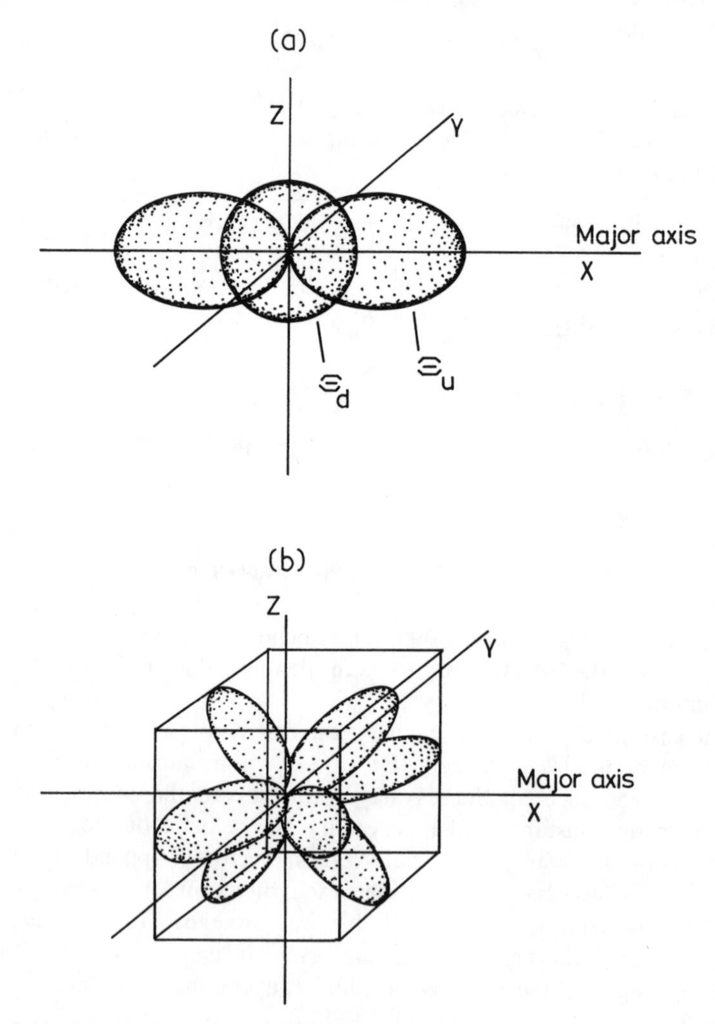

FIG. 3.11. Angular dependence of the deformation potential: (a) longitudinal modes; (b) transverse modes.

potentials (or Ξ_d in the case of a Γ valley) we can write generally

$$H_{ep} = \frac{1}{\sqrt{N}} \sum_{\mathbf{q}} \{iQ_{\mathbf{q}} \exp(i\mathbf{q}\cdot\mathbf{R})\Xi(\theta_q)q + \text{c.c.}\}. \tag{3.62}$$

This means that the coupling coefficient $C_{\mathbf{q}}^2$ in eqn (3.15) is given by

$$C_{\mathbf{q}}^2 = \Xi^2(\theta_q)q^2. \tag{3.63}$$

Having obtained the $\mathbf{q}$ dependence of the coupling strength it is necessary to exhibit the $\mathbf{q}$ dependence of the remaining terms in eqn (3.15), namely $\omega_{\mathbf{q}}$ and $n(\omega_{\mathbf{q}})$. In the spirit of the azimuthal average approximation we can take

$$\omega_{\mathbf{q}} = \bar{v}_s q \tag{3.64}$$

where $\bar{v}_s$ is the velocity of the mode averaged over direction. As regards $n(\omega_{\mathbf{q}})$ there are two simple cases to consider. The most commonly applicable is the case of equipartion:

$$n(\omega_{\mathbf{q}}) = \frac{1}{\exp(\hbar\omega_q/k_B T) - 1} \approx \frac{k_B T}{\hbar\omega_{\mathbf{q}}} \qquad \frac{\hbar\omega_{\mathbf{q}}}{k_B T} \ll 1. \tag{3.65}$$

Since energy and momentum conservation limit q to $2k$, a typical phonon energy is $\hbar\bar{v}_s k$. For electrons obeying non-degenerate statistics at thermal equilibrium and for a parabolic band

$$\hbar^2 k^2/2m^* \approx \tfrac{3}{2}k_B T. \tag{3.66}$$

Thus

$$\frac{\hbar\omega_{\mathbf{q}}}{k_B T} \approx \left(\frac{3m^*\bar{v}_s^2}{k_B T}\right)^{1/2}. \tag{3.67}$$

Typically $3m^*\bar{v}_s^2/k_B \approx 0{\cdot}2$ K, and so equipartition can be assumed to hold for all temperatures above 1 K as regards the scattering of thermal electrons. For hot electrons at low temperatures this assumption may break down, in which case $n(\omega_{\mathbf{q}}) \ll 1$ is usually assumed. We shall consider this situation later, and at present continue to assume that equipartition holds good.

3.3.1. *Spherical band: equipartition*

When $n(\omega_{\mathbf{q}}) \gg 1$ the rates for absorption and emission become almost identical. Substituting from eqns (3.63)–(3.65) into eqn (3.15) gives, for a spherical band, and with $M' = \rho V/N$ where ρ is the mass density and $I(\mathbf{k}, \mathbf{k}') = 1$,

$$W(\mathbf{k}) = \frac{\Xi_d^2 k_B T}{8\pi^2 \hbar \rho \bar{v}_{sL}^2} \int \delta_{\mathbf{k}\pm\mathbf{q}-\mathbf{k}',0}\,\delta(E_{k'} - E_k \mp \hbar\omega_q)\,d\mathbf{k}' \tag{3.68}$$

where $\bar{v}_{sL}$ is the velocity of longitudinal modes. We can replace $\rho\bar{v}_{sL}^2$ by the average elastic constant c_L for longitudinal modes (see Appendix). It is convenient to use the one-to-one correspondence between $\mathbf{k}'$ and $\mathbf{q}$ to convert the integral to one over $\mathbf{q}$. Thus

$$W(\mathbf{k}) = \frac{\Xi_d^2 k_B T}{8\pi^2 \hbar c_L} \int_0^{q_{ZB}} \int_{-1}^{1} \int_0^{2\pi} \delta_{\mathbf{k}\pm\mathbf{q}-\mathbf{k}',0}\,\delta(E_{k'} - E_k \mp \hbar\omega_q) q^2\, dq\, d(\cos\theta)\, d\phi, \tag{3.69}$$

The Kronecker symbol merely reminds us that crystal momentum is conserved. The Dirac delta function is a function of polar angle θ (which we choose to be the angle between $\mathbf{q}$ and $\mathbf{k}$) through the dependence of its argument on the magnitude of $\mathbf{q}$ through $\omega_{\mathbf{q}}$. For a parabolic band

$$\begin{aligned}\delta(E_{k'} - E_k \mp \hbar\omega_{\mathbf{q}}) &= \delta\left(\frac{\hbar^2 k'^2}{2m^*} - \frac{\hbar^2 k^2}{2m^*} \mp \hbar\bar{v}_{sL} q\right)\\ &= \delta\left(\frac{\hbar^2 q^2}{2m^*} \pm \hbar v q \cos\theta \mp \hbar\bar{v}_{sL} q\right)\end{aligned} \tag{3.70}$$

where k' is replaced according to eqns (3.17) or (3.33). (The upper sign is to be taken for absorption, the lower for emission.) Thus, integrating over ϕ and $\cos\theta$ we obtain

$$W(\mathbf{k}) = \frac{\Xi_d^2 k_B T}{4\pi\hbar^2 c_L v} \int_{q_{min}}^{q_{max}} q\, dq \tag{3.71}$$

where the limits on q arise from the delta function, and they are those given in eqns (3.23) and (3.24) for absorption and in eqns (3.37) and (3.38) for emission. Now the thermal average group velocity v_{th} is given by kinetic theory as

$$v_{th} = \left(\frac{8k_B T}{\pi m^*}\right)^{1/2} \tag{3.72}$$

and is typically of order 10^6 cm s^{-1} at temperatures around 1 K. Thus for situations in which equipartition can be assumed, it is usually possible to take $v \gg \bar{v}_{sL}$ and hence take $q_{min} = 0$ and $q_{max} = 2k$. Consequently, for absorption or emission

$$W(\mathbf{k}) = \frac{\Xi_d^2 k_B T k^2}{2\pi\hbar^2 c_L v} = \frac{\Xi_d^2 k_B T (2m^*)^{3/2} E_k^{1/2}}{4\pi\hbar^4 c_L}. \tag{3.73}$$

The density of states of a given spin $N(E_k)$ per unit energy in the band is given by

$$N(E_k) = \frac{(2m^*)^{3/2} E_k^{1/2}}{4\pi^2\hbar^3} \tag{3.74}$$

and thus

$$W(\mathbf{k}) = \frac{\pi \Xi_d^2 k_B T N(E_k)}{\hbar c_L}. \tag{3.75}$$

Returning to eqn (3.68) we see that this result could have been obtained directly by neglecting the phonon energy $\hbar\omega_{\mathbf{q}}$ in the delta function, thus removing the angular dependence of the integrand, for then

$$\frac{1}{(2\pi)^3}\int \delta(E_{k'} - E_k)\,d\mathbf{k}' = \int \delta(E_{k'} - E_k) N(E_{k'})\,dE_{k'} = N(E_k). \tag{3.76}$$

Note that phonon scattering does not flip the spin.

It follows that eqn (3.75) is valid for spherical non-parabolic bands, provided that $N(E_k)$ is the appropriate density of states per unit energy, namely

$$N(E_k) = \frac{(2m^*)^{3/2}\gamma^{1/2}(E_k)}{4\pi^2\hbar^3}\frac{d\gamma(E_k)}{dE_k} \tag{3.77}$$

where $\gamma(E_k)$ is defined in eqns (3.44) and (1.82), and provided that the departure of the overlap integral $I(\mathbf{k}, \mathbf{k}')$ from unity can be neglected.

The total scattering rate is the sum of absorption and emission rates. Since these are nearly equal for equipartition, we have

$$W_{tot}(\mathbf{k}) = \frac{2\pi \Xi_d^2 k_B T N(E_k)}{\hbar c_L} \tag{3.78}$$

which is the final result (see Fig. 3.14).

The free path l between scattering events is given by v/W_{tot}, namely (e.g. from eqns (3.73) and (3.46)

$$l = \frac{\pi \hbar^2 c_L v^2}{\Xi_d^2 k_B T k^2} = \frac{\pi \hbar^4 c_L}{\Xi_d^2 k_B T m^{*2}}\left(\frac{d\gamma(E_k)}{dE_k}\right)^{-2}. \tag{3.79}$$

In a parabolic band l is independent of energy.

3.3.2. Spherical band: zero-point scattering

In the case of hot electrons at low temperature, the assumption of equipartition fails. Instead $n(\omega_{\mathbf{q}}) \ll 1$ for the phonons most involved in the scattering. This means that scattering by absorption or by stimulated emission is negligible, and only emission triggered by zero-point oscillations is important. Thus in place of eqn (3.68) we obtain

$$W(\mathbf{k}) = \frac{\Xi_d^2}{8\pi^2 \rho \bar{v}_{sL}} \int q\,\delta_{\mathbf{k}-\mathbf{q}-\mathbf{k}',0}\,\delta(E_{k'} - E_k + \hbar\omega_{\mathbf{q}})\,d\mathbf{k}'. \tag{3.80}$$

The integration follows the path of eqn (3.69) *et seq.*, and we obtain

$$W(\mathbf{k})=\frac{4\Xi_d^2 m^{*2}\gamma(E_k)}{3\pi\rho\bar{v}_{sL}\hbar^4}\frac{d\gamma(E_k)}{dE_k}. \tag{3.81}$$

For a parabolic band $W(\mathbf{k}) \propto E_k$. The ratio of zero-point rate to equipartition rate is $2m^*v\bar{v}_{sL}/3k_BT$.

3.3.3. Spheroidal parabolic bands

In the case of L or X valleys we have to evaluate, in the case of equipartition,

$$W(\mathbf{k})=\frac{k_BT}{8\pi^2c\hbar}\int \Xi^2(\theta_q)\delta_{\mathbf{k}\pm\mathbf{q}-\mathbf{k}',0}\delta(E_{\mathbf{k}'}-E_{\mathbf{k}}\mp\hbar\omega_{\mathbf{q}})\,d\mathbf{k}' \tag{3.82}$$

for longitudinal modes and for transverse modes, $\Xi(\theta_q)$ being given by eqns (3.60) and (3.61) and c being chosen appropriately. For a given magnitude of $\mathbf{k}'$, $E_{\mathbf{k}'}$ varies with direction in the spheroidal band and thus the delta function is very much dependent on angle. It is therefore convenient to transform the integration to one in 'starred space' (see Section 3.2.3) and over the phonon wavevector. All vectors transform (Fig. 3.12) as

$$k_i \to (m_i^*/m_0^*)^{1/2}k_i^* \tag{3.83}$$

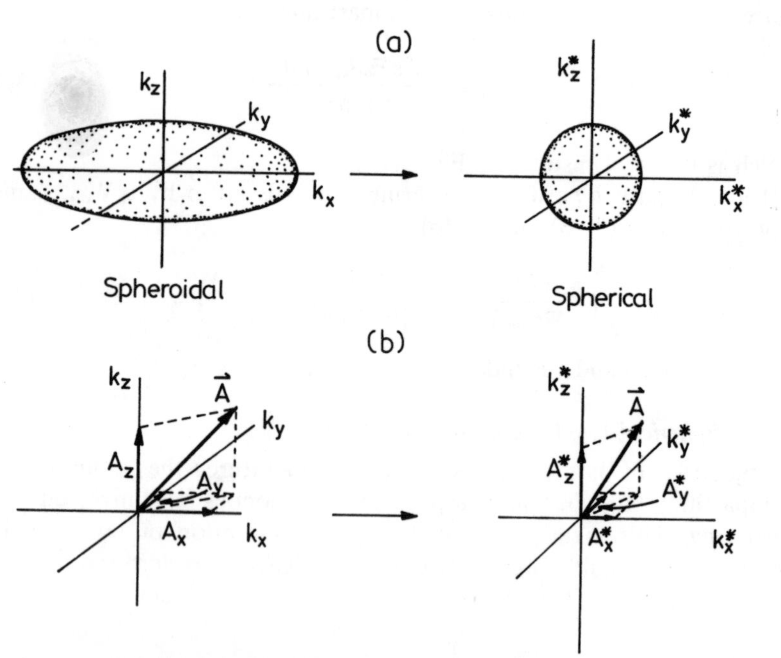

FIG. 3.12. Transformations to starred space: (a) energy surface; (b) vector. $k_x^* = (m_t^*/m_l^*)^{1/2}k_x$; $k_y^* = k_y$; $k_z^* = k_z$.

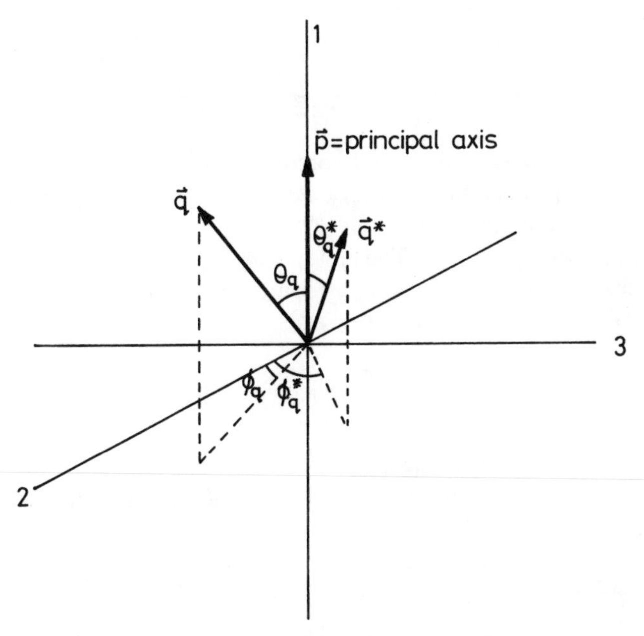

FIG. 3.13. Angle transformations to starred space.

and the energy becomes independent of direction, namely

$$E_{\mathbf{k}} = \hbar^2 k^{*2}/2m_0^*. \tag{3.84}$$

The angle between $\mathbf{q}$ and the principal axis $\mathbf{p}$ transforms as follows (Fig. 3.13):

$$\cos\theta_{\mathbf{q}} = \frac{\mathbf{q}\cdot\mathbf{p}}{\mathbf{q}} = \frac{\sum_i (m_i^*/m_0^*)^{1/2} q_i^*}{\left\{\sum_i (m_i^*/m_0^*) q_i^{*2}\right\}^{1/2}} \tag{3.85}$$

and since

$$q_1^* = q^* \cos\theta_q^* \qquad q_2^* = q^* \sin\theta_q^* \cos\phi_q^* \qquad q_3^* = q^* \sin\theta_q^* \sin\phi_q^* \tag{3.86}$$

then

$$\cos\theta_{\mathbf{q}} = \frac{(m_1^*/m_0^*)^{1/2}\cos\theta_q^* + (m_2^*/m_0^*)^{1/2}\sin\theta_q^*\cos\phi_q^* + (m_3^*/m_0^*)^{1/2}\sin\theta_q^*\sin\phi_q^*}{\{(m_1^*/m_0^*)\cos^2\theta_q^* + (m_2^*/m_0^*)\sin^2\theta_q^*\cos^2\phi_q^* + (m_3^*/m_0^*)\sin^2\theta_q^*\sin^2\phi_q^*\}^{1/2}}. \tag{3.87}$$

Putting $m_1^* = m_l^*$, where m_l^* is the longitudinal component of the effective mass, and $m_2^* = m_3^* = m_t^*$, where m_t^* is the transverse component, we obtain

$$\cos\theta_{\mathbf{q}} = \frac{K_{\mathrm{m}}^{1/2}\cos\theta_q^* + \sin\theta_q^*(\cos\phi_q^* + \sin\phi_q^*)}{\{1+(K_{\mathrm{m}}-1)\cos^2\theta_q^*\}^{1/2}} \tag{3.88}$$

where K_{m} is the mass anisotropy coefficient $m_{\mathrm{l}}^*/m_{\mathrm{t}}^*$.

With this transformation $\Xi^2(\theta_{\mathbf{q}})$ becomes dependent on the azimuthal angle ϕ_{q}^* as well as the polar angle θ_{q}^*, but the delta function is now only weakly dependent on θ_{q}^*. The integrations can be carried out in a straightforward way, first over ϕ_{q}^*, then over $\cos\theta_{\mathrm{q}}^*$, and finally over q^* between the limits 0 and $2k^*$. Integrals of the type

$$I(\mu,\nu) = \int_0^1 x^{\mu}(1+ax^2)^{-\nu/2}\,\mathrm{d}x \tag{3.89}$$

appear, and they are standard.

The final result can be expressed in terms of a longitudinal rate $W_{\mathrm{l}}(E_{\mathbf{k}})$ and a transverse rate $W_{\mathrm{t}}(E_{\mathbf{k}})$:

$$W_{\mathrm{l}}(E_{\mathbf{k}}) = \frac{2\pi\Xi_{\mathrm{l}}^2 k_{\mathrm{B}} T N(E_{\mathbf{k}})}{\hbar c_{\mathrm{L}}} \qquad W_{\mathrm{t}}(E_{\mathbf{k}}) = \frac{2\pi\Xi_{\mathrm{t}}^2 k_{\mathrm{B}} T N(E_{\mathbf{k}})}{\hbar c_{\mathrm{L}}} \tag{3.90}$$

where

$$\Xi_{\mathrm{l}}^2 = \tfrac{3}{4}(\xi_{\mathrm{l}}\Xi_{\mathrm{d}}^2 + \eta_{\mathrm{l}}\Xi_{\mathrm{d}}\Xi_{\mathrm{u}} + \zeta_{\mathrm{l}}\Xi_{\mathrm{u}}^2) \tag{3.91}$$

$$\Xi_{\mathrm{t}}^2 = \tfrac{3}{4}(\xi_{\mathrm{t}}\Xi_{\mathrm{d}}^2 + \eta_{\mathrm{t}}\Xi_{\mathrm{d}}\Xi_{\mathrm{u}} + \zeta_{\mathrm{t}}\Xi_{\mathrm{u}}^2) \tag{3.92}$$

$$N(E_{\mathbf{k}}) = \frac{(2m_{\mathrm{d}}^*)^{3/2}E_{\mathbf{k}}^{1/2}}{4\pi^2\hbar^3} \tag{3.93}$$

$$m_{\mathrm{d}}^* = (m_{\mathrm{l}}^* m_{\mathrm{t}}^{*2})^{1/3}. \tag{3.94}$$

In the above expressions for the deformation potential $\tfrac{3}{4}\xi_{\mathrm{l}}$ and $\tfrac{3}{4}\xi_{\mathrm{t}}$ are equal to unity when elastic anisotropy is neglected, η_{l} and η_{t} depend upon K_{m}, and ζ_{l} and ζ_{t} depend upon K_{m} and the ratio $c_{\mathrm{L}}/c_{\mathrm{T}}$ of the averaged elastic constants.

Explicit expressions have been obtained by Herring and Vogt (1956) for the momentum relaxation rates in germanium and silicon (Table 3.1).

TABLE 3.1

Coefficients for deformation potentials

	ξ_{l}	η_{l}	ζ_{l}	ξ_{t}	η_{t}	ζ_{t}
Si	1·40	2·40	1·62	1·33	1·15	1·07
Ge	1·24	2·32	1·22	1·31	1·61	1·01

After Herring and Vogt 1956.

The distinction between momentum relaxation rate and scattering rate is discussed later, but for the present it is sufficient to note that the magnitude of these rates are within a factor of order unity of each other. For germanium it is found that the ratio of the longitudinal and transverse components is close to unity, whereas for silicon it is about $\frac{3}{2}$ according to Long (1960).

The scattering rate for zero-point scattering can be calculated in the same way. We obtain

$$W_{\mathrm{l,t}}(E_{\mathbf{k}}) = \frac{4\Xi_{\mathrm{l,t}}^2 \mathrm{m}_\mathrm{l}^{*1/2} \mathrm{m}_\mathrm{t}^{*3/2} E_{\mathbf{k}}}{3\pi\rho\bar{v}_\mathrm{s}\hbar^4} \tag{3.95}$$

where

$$\Xi_{\mathrm{l,t}}^2 = 3(\xi_{\mathrm{l,t}}\Xi_\mathrm{d}^2 + \eta_{\mathrm{l,t}}\Xi_\mathrm{d}\Xi_\mathrm{u} + \zeta_{\mathrm{l,t}}\Xi_\mathrm{u}^2), \tag{3.96}$$

Explicit expressions for the momentum relaxation rate of germanium have been obtained by Budd (1964) and Gurevitch (1965):

$$\Xi_\mathrm{l}^2 = 3(1{\cdot}14\Xi_\mathrm{d}^2 + 2{\cdot}19\Xi_\mathrm{d}\Xi_\mathrm{u} + 1{\cdot}13\Xi_\mathrm{u}^2) \tag{3.97}$$

$$\Xi_\mathrm{t}^2 = 3(0{\cdot}665\Xi_\mathrm{d}^2 + 1{\cdot}01\Xi_\mathrm{d}\Xi_\mathrm{u} + 0{\cdot}542\Xi_\mathrm{u}^2). \tag{3.98}$$

These results suggest that the ratio of longitudinal to transverse rates is about $\frac{1}{2}$ in that material.

3.3.4. Momentum and energy relaxation

The rates explicitly dealt with in the foregoing, apart from the immediately preceding expressions, are simple scattering rates. In the description of the response of electrons to fields the more important rates are the momentum relaxation rate and the energy relaxation rate. These are rates which are derived from the scattering rate by weighing the latter by the appropriate change in momentum or energy.

The change of momentum is equal to the change of crystal momentum in the transition from **k** to **k**′:

$$\frac{\mathrm{d}\hbar\mathbf{k}}{\mathrm{d}t} = \int \hbar(\mathbf{k}' - \mathbf{k})\, W(\mathbf{k}, \mathbf{k}')\delta(E_{\mathbf{k}'} - E_{\mathbf{k}} \mp \hbar\omega_{\mathbf{q}})\, \mathrm{d}\mathbf{k}'$$

$$\approx -\hbar\mathbf{k}\int (1 - \cos\theta_{\mathbf{k}})\, W(\mathbf{k}, \mathbf{k}')\, \mathrm{d}\mathbf{k}' \tag{3.99}$$

where $W(\mathbf{k}, \mathbf{k}')\delta(E_{\mathbf{k}'} - E_{\mathbf{k}} \mp \hbar\omega_{\mathbf{q}})$ is the integrand of eqn (3.15), $k' \approx k$ is assumed for acoustic scattering, and $\theta_{\mathbf{k}}$ is the scattering angle. Conservation of energy and momentum entails that

$$q^2 = 2k^2(1 - \cos\theta_{\mathbf{k}}). \tag{3.100}$$

The appropriate weighting factor for the scattering rate is $1-\cos\theta_{\mathbf{k}}$, and this factor must be inserted in the integration of eqn (3.15) to obtain the momentum relaxation rate. From eqn (3.100) it is clear that this is equivalent to multiplying the integrand by $q^2/2k^2$. This does not change the energy dependence of the scattering rate, but it introduces an extra numerical factor of the order of unity. Thus

$$1/\tau_{\mathrm{m}} = W(\mathbf{k}) \qquad \text{equipartition} \tag{3.101}$$

$$1/\tau_{\mathrm{m}} = \tfrac{6}{5}W(\mathbf{k}) \qquad \text{zero point} \tag{3.102}$$

where τ_{m} is the momentum relaxation time and $W(\mathbf{k})$ is the scattering rate. The factor is of order unity because q varies between 0 and $2k$, corresponding to a scattering angle between 0 and 180°. Taking an average scattering angle of 90° implies that $\langle q\rangle = \sqrt{2}k$, which makes the weighting factor unity.

There is usually a net loss of momentum in a collision, but whether a gain or loss of energy occurs depends upon the energy of the electron. The rate of change of energy of the electron is given by

$$\frac{\mathrm{d}E_{\mathbf{k}}}{\mathrm{d}t} = \int \hbar\omega_q W(\mathbf{k},\mathbf{k}'')\delta(E_{\mathbf{k}''}-E_{\mathbf{k}}-\hbar\omega_{\mathbf{q}})\,\mathrm{d}\mathbf{k}'' - \int \hbar\omega_{\mathbf{q}} W(\mathbf{k},\mathbf{k}')\delta(E_{\mathbf{k}'}-E_{\mathbf{k}}+\hbar\omega_{\mathbf{q}})\,\mathrm{d}\mathbf{k}'. \tag{3.103}$$

For the case of a spherical band

$$\frac{\mathrm{d}E_k}{\mathrm{d}t} = \frac{\Xi_{\mathrm{d}}^2(2m^*)^{5/2}}{2\pi\rho\hbar^4}\left\{\frac{2k_{\mathrm{B}}T}{\gamma(E_k)}\frac{\mathrm{d}\gamma(E_k)}{\mathrm{d}E}-1\right\}\gamma^{3/2}(E_k)\frac{\mathrm{d}\gamma(E_k)}{\mathrm{d}E_k} \qquad \text{equipartition} \tag{3.104}$$

$$\frac{\mathrm{d}E_k}{\mathrm{d}t} = -\frac{\Xi_{\mathrm{d}}^2(2m^*)^{5/2}\gamma^{3/2}(E_k)}{2\pi\rho\hbar^4}\frac{\mathrm{d}\gamma(E_k)}{\mathrm{d}E_k} \qquad \text{zero point.} \tag{3.105}$$

In a parabolic band energy is gained when $E_k < 2k_{\mathrm{B}}T$ and lost when $E_k > 2k_{\mathrm{B}}T$. When $E_k \gg 2k_{\mathrm{B}}T$ energy is lost at the zero-point rate. This latter result, which is of importance for hot electrons, can be obtained very simply by observing that 1 in $2n(\omega_q)+1$ collisions results in a loss, and thus

$$\frac{\mathrm{d}E_k}{\mathrm{d}t} = -\frac{1}{2n(\omega_{\mathbf{q}})+1}\hbar\omega_{\mathbf{q}}W(\mathbf{k}). \tag{3.106}$$

Substituting for $n(\omega_{\mathbf{q}})$ and $W(\mathbf{k})$ for equipartition, and taking $\langle\omega_{\mathbf{q}}\rangle = \sqrt{2}v_{\mathrm{s}}k$ gives eqn (3.105). If an energy relaxation time for hot electrons is defined as

$$\frac{dE_k}{dt} = -\frac{E_k}{\tau_E}, \qquad (3.107)$$

then for equipartition,

$$\frac{1}{\tau_E} = \frac{2m^* v_s^2}{k_B T} \frac{1}{\tau_m}. \qquad (3.108)$$

The energy relaxation time in this case is very much longer than the momentum relaxation time since $2m^* v_s^2/k_B T \ll 1$.

The momentum relaxation rate in spheroidal parabolic valleys bears the same relation to the scattering rate as in the case of spherical valleys (eqns (3.101) and (3.102)). Thus, for equipartition eqns (3.90) are valid for momentum relaxation. For zero-point scattering we obtain from eqn (3.95) (see Fig. 3.14)

$$\frac{1}{\tau_m} = \frac{8\Xi_{l,t}^2 m_l^{*1/2} m_t^{*3/2} E_{\mathbf{k}}}{5\pi\rho v_{sL} \hbar^4}. \qquad (3.109)$$

The energy relaxation rate is given by

$$\frac{dE_{\mathbf{k}}}{dt} = \frac{\Xi_0^2 (2m^*)^{5/2}}{2\pi\rho\hbar^4} \left(\frac{2k_B T}{E_{\mathbf{k}}} - 1\right) E_{\mathbf{k}}^{3/2} \quad \text{equipartition} \qquad (3.110)$$

$$\frac{dE_{\mathbf{k}}}{dt} = -\frac{\Xi_0^2 (2m^*)^{5/2} E_{\mathbf{k}}^{3/2}}{2\pi\rho\hbar^4} \quad \text{zero point.} \qquad (3.111)$$

Conwell (1967) has shown that the appropriate average deformation

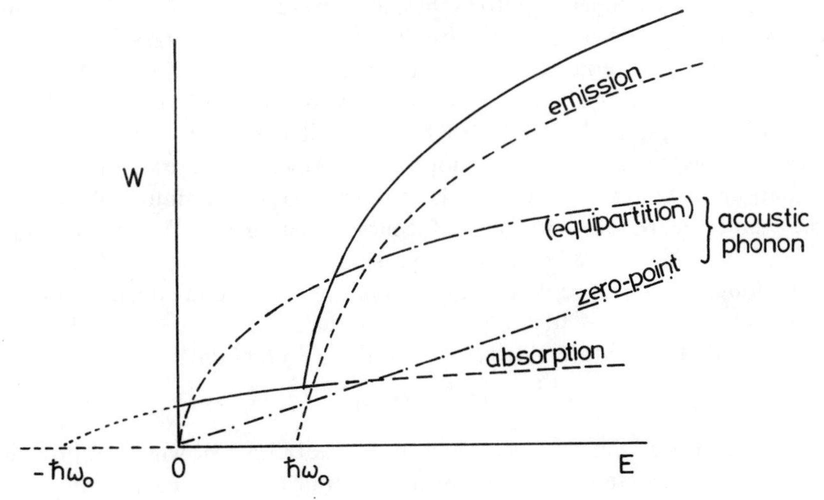

FIG. 3.14. Scattering rate for optical phonon scattering.

potential in this case is

$$\Xi_0^2 = \Xi_d^2 \left\{ \frac{2}{3} + \frac{1}{3} \frac{m_l^*}{m_t^*} \left(\frac{\Xi_u}{\Xi_d} + 1 \right)^2 \right\}. \tag{3.112}$$

The scattering of holes in the degenerate valence bands is somewhat more complex, although no essential new features are present in the problem. Interaction with both longitudinal and transverse modes is allowed, with the strength given by the deformation potentials of Chapter 1, Section 1.13 (Bir and Pikus 1961; Lawaetz 1969, 1971).

3.4. Optical phonon scattering

As in the case of acoustic phonon scattering, energy, and momentum conservation restricts intra-valley scattering by optical phonons to long-wavelength modes. However, unlike the case of acoustic phonons, long-wavelength optical displacement may affect the electronic energy directly. Thus the electron–phonon interaction Hamiltonian takes the form

$$H_{ep} = \mathbf{D}_o' \cdot \mathbf{u} \tag{3.113}$$

where $\mathbf{D}_o'$ is an optical deformation potential constant (*cf.* Chapter 1, Section 1.13) and $\mathbf{u}$ is the relative displacement of the two atoms in the unit cell.

The optical deformation potential constant has been defined in many ways, unfortunately, and that is why we have put a prime on the symbol in eqn (3.113). Equation (3.113) is a natural way of defining the interaction, but although this form has been used frequently following its introduction by Meyer (1958) it has always been accompanied by the assumption that the mass of the oscillator was the total mass of the unit cell and not the reduced mass. (Meyer appeared to adopt a halfway postiion by taking the mass to be that of a single atom. The use of the total mass is explicitly condoned by Conwell (1967, p. 150), and this approach has been universally adopted.) However, the use of the optical displacement calls for the use of the reduced mass, as pointed out in eqn (3.5), and the reduced mass is of course used in the theory of polar interaction with optical modes, as we shall see. If the reduced mass is used along with the accepted magnitude of the interaction constant, which we denote by D_o without the prime, it can be shown to be equivalent to assuming that the interaction in the case of germanium takes the form

$$H_{ep} = \tfrac{1}{2}\mathbf{D}_o \cdot \mathbf{u}. \tag{3.114}$$

This, indeed, was the form assumed by Lawaetz (1967), whose approach was entirely consistent. If we adopt this form of interaction, it means that D_o is not the energy per unit optical displacement but rather the energy per atomic displacement. (In the diamond lattice both atoms have identical masses and have the same displacement.) This has the virtue of making it similar to the acoustic deformation interaction in that the latter is also defined as proportional to the displacement of an individual atom (equal in this case to the displacement of the unit cell). The adoption of eqn (3.114) as the deformation of D_o therefore removes all difficulties in this respect, but only as regards the diamond lattice. In other materials the atoms in the unit cell are not identical and hence eqn (3.114) needs to be generalized. The approach to be adopted here is to define the optical deformation potential constant as the change of energy per unit equivalent acoustic displacement $\mathbf{u}_{ac}$ and to define the latter as the displacement of the unit cell giving the same optical elastic energy as the optical displacement $\mathbf{u}$, namely

$$(M_1+M_2)u_{ac}^2=\bar{M}u^2. \tag{3.115}$$

Thus we take the general optical deformation potential interaction to be

$$H_{ep}=\left(\frac{\bar{M}}{M_1+M_2}\right)^{1/2}\mathbf{D}_o\cdot\mathbf{u} \tag{3.116}$$

An advantage of this approach is that it retains the useful property that the details of the lattice vibration, symbolized by the appropriate oscillator mass M', cancel out of the final result.

In zero-order optical phonon scattering, $\mathbf{D}_o$ is non-zero. This turns out to be the case for intra-valley scattering in L valleys and in the degenerate valence bands at the zone centre, but for Γ_1 and X valleys $\mathbf{D}_o$ is zero (Table 3.2). When zero-order optical phonon processes are forbidden because of symmetry restrictions, it is possible for first-order processes in which the energy of interaction is proportional to the differential displacement, i.e. strain, to occur. In this case deformation potentials analogous to those for acoustic phonons can be introduced, and the calculation of scattering rate follows that for acoustic phonon scattering

TABLE 3.2
Selection rules for zone-centre phonons in intra-valley processes

Valley	Phonons
Γ_1	LA
X_1	LA+TA
L_1	LA+TA+LO+TO
Γ_{15}	LA+TA+LO+TO

After Seitz (1948), Herring and Vogt (1956), Harrison (1956), and Bir and Pikus (1961).

except that the phonon energy cannot be neglected. Usually it is assumed that at room temperature and below first-order optical phonon scattering is negligible compared with acoustic phonon scattering. We consider this type of scattering later.

In the case of the L valley Harrison (1956) has shown that $\mathbf{D}_o$ lies along the principal axis. Thus for a longitudinal optical mode travelling in an arbitrary direction such that its direction cosine with respect to the principal axis is $\alpha_{\mathbf{q}}$,

$$H_{ep} = \left(\frac{\bar{M}}{M_1+M_2}\right)^{1/2} \frac{1}{\sqrt{N}} \sum_{\mathbf{q}} \{Q_{\mathbf{q}} \exp(i\mathbf{q}.\mathbf{R}) + \text{c.c.}\} D_o \alpha_{\mathbf{q}}. \quad (3.117)$$

The coupling coefficient in eqn (3.15) is therefore

$$C_{\mathbf{q}}^2 = \frac{\bar{M}}{M_1+M_2} D_o^2 \alpha_{\mathbf{q}}^2. \quad (3.118)$$

The coupling coefficients for the two transverse modes travelling in the same direction are $D_o^2\beta_{\mathbf{q}}^2$ and $D_o^2\gamma_{\mathbf{q}}^2$, where $\beta_{\mathbf{q}}$ and $\gamma_{\mathbf{q}}$ are the direction cosines of the polarization vectors. The magnitude of the coupling strength is thus the same for all polarizations.

The scattering rate (eqn (3.15)) depends upon the frequency of the mode directly and indirectly through the number of phonons $n(\omega_{\mathbf{q}})$. Strictly speaking, therefore, each mode has to be considered separately, but usually the simplification of regarding all modes as having the same frequency is made. In the diamond lattice there is no difference in frequency at the zone centre, and so a single-frequency approximation for $\mathbf{q} \neq 0$ is a good one. In zinc blende longitudinally polarized optical vibrations produce an electric polarization which increases the restoring force and therefore the frequency over that for transversely polarized modes. If ω_L and ω_T are the longitudinal and transverse angular frequencies, and ϵ and ϵ_∞ are the static and high-frequency permittivities, then

$$(\omega_L/\omega_T)^2 = \epsilon/\epsilon_\infty \quad (3.119)$$

a formula known as the Lyddane–Sachs–Teller relationship. Although $\omega_L \neq \omega_T$ even at the zone centre, the difference between the frequencies is not very large in III–V compounds and it is still a reasonable approximation to adopt a single-frequency approach.

In a single-frequency approach the sum over the three modes allows us to consider the interaction as one with a single mode of frequency ω_o with coupling coefficient

$$C_{\mathbf{q}}^2 = \frac{\bar{M}}{M_1+M_2} D_o^2 \quad (3.120)$$

since $\alpha_{\mathbf{q}}^2 + \beta_{\mathbf{q}}^2 + \gamma_{\mathbf{q}}^2 = 1$. Another approximation, just as irresistible, is obtained by neglecting the dependence of ω_o on q, for then the delta functions conserving energy in eqn (3.15) become independent of direction and the integral over final states can be simply carried out by converting $\mathbf{k}'$ to energy as we did in eqn (3.76). Thus the scattering rate is, with $M' = \bar{M}$ and $I(\mathbf{k}, \mathbf{k}') = 1$,

$$W(\mathbf{k}) = \frac{\pi D_o^2}{\rho \omega_o} [n(\omega_o) N(E_\mathbf{k} + \hbar\omega_o) + \{n(\omega_o) + 1\} N(E_\mathbf{k} - \hbar\omega_o)] \tag{3.121}$$

where for a parabolic band

$$N(E_\mathbf{k}) = \frac{(2m_d^*)^{3/2} E_\mathbf{k}^{1/2}}{4\pi^2 \hbar^3} \tag{3.122}$$

and m_d^* is the density-of-states mass (eqn (3.94) and Fig. 3.14)).

Equation (3.121) is also the expression for the momentum relaxation rate. Multiplying the integrand in eqn (3.15) by $(q/k)\cos\theta$, the fractional change in momentum, does not introduce any new factor because the coupling is independent of direction. When it is summed over all final states this factor gives a contribution of unity. (See discussion p. 117).

The energy relaxation (eqn (3.103)) is also simply obtained in the constant-frequency approximation. The rate of increase of energy of the electron is (Fig. 3.15)

$$\frac{dE_\mathbf{k}}{dt} = \frac{\pi D_o^2 \hbar}{\rho} [n(\omega_o) N(E_\mathbf{k} + \hbar\omega_o) - \{n(\omega_o) + 1\} N(E_\mathbf{k} - \hbar\omega_o)]. \tag{3.123}$$

Although these equations have been derived explicitly for L valleys they are applicable to the degenerate valence bands. The valence band

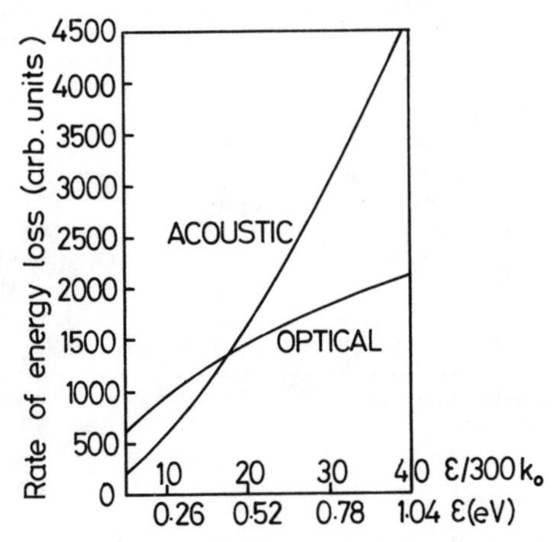

FIG. 3.15. Rate of energy loss to acoustic and optical phonons in n-germanium at 300 K. (After Conwell 1964.)

optical deformation potential d_o replaces D_o in the manner mentioned in Chapter 1 (Section 1.13) and the density-of-states factors must include both light- and heavy-hole bands. However, the overlap factor is no longer unity in this case.

3.4.1. *Inter-valley scattering*

These equations are also valid for inter-valley scattering processes since the conservation of momentum entails that only short-wavelength phonons at or near the zone edge are involved and these have frequencies virtually independent of wavevector, whether they be optical or acoustic modes. Which modes interact with the electrons depends upon the symmetries of the initial and final states (see Table 3.3) and whether an umklapp process is involved (as is the case for silicon). The rate for each mode must now be considered separately. Thus the rate for an inter-valley process between valleys i and j involving a phonon of frequency ω_{ij} with an inter-valley deformation potential constant D_{ij} is

$$W(\mathbf{k}) = \frac{\pi D_{ij}^2}{\rho\omega_{ij}}\{n(\omega_{ij})N(E_{\mathbf{k}} - \Delta E_{ij} + \hbar\omega_{ij}) + (n(\omega_{ij})+1)N(E_{\mathbf{k}} - \Delta E_{ij} - \hbar\omega_{ij})\} \quad (3.124)$$

where ΔE_{ij} is the difference of energy of the valley minima and the appropriate oscillator mass replaces $\bar{M}$ in the interaction eqn (3.116).

TABLE 3.3
Selection rules for phonons in inter-valley scattering processes

Initial Valley	Final Valley	Phonons
X_1	X_1	LO, $M_V > M_{III}$; LA, $M_V < M_{III}$ } zinc blende
Δ_1	Δ_1 (g-scattering opposite valley)	LO[a]
	(f-scattering non-opposite valleys)	LA+TO[a]
L_1	L_1	LO+LA
	X_1	LO, $M_V > M_{III}$; LA, $M_V < M_{III}$ } zinc blende
Γ_1	L_1	LO+LA
L_1	X_1	LO+LA

After Birman, Lax, and Loudon 1966.
[a] Data from Streitwolf 1970.

3.4.2. *First-order processes*

In the case of inter-valley scattering between the Δ_1 mimina in silicon normal processes are ruled out and only umklapp processes are allowed

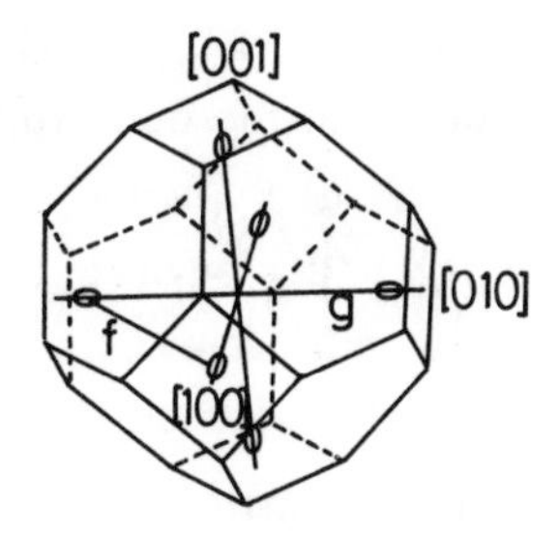

FIG. 3.16. Brillouin zone for silicon showing positions of conduction band valleys. The arrows indicate the two possible types of inter-valley scattering transitions.

(Fig. 3.16). Scattering between opposite valleys, i.e. ⟨100⟩ to ⟨$\bar{1}$00⟩, is denoted a g-process and that between non-opposite valleys, e.g. ⟨100⟩ to ⟨010⟩, an f-process. The reciprocal lattice vector involved in the g-process is $\mathbf{K}_{100}$ and that for an f-process is $\mathbf{K}_{111}$. Momentum conservation then entails a phonon for the g-process with $\mathbf{q}$ along the ⟨100⟩ direction and of magnitude 0·30† of the ⟨100⟩ zone-edge value, and a phonon for an f-process with $\mathbf{q}$ only about 11° off a ⟨100⟩ direction (Σ_1 symmetry) and of exactly the zone-boundary value in that direction. Thus f-processes involve phonons of short wavelength, but the g-process involves a phonon of comparatively long wavelength. Selection rules imply that the latter is a LO phonon (θ_D= 730 K, $\hbar\omega$ = 0·063 eV) and that the f-process phonons are LA and TO (θ_D = 630 K, $\hbar\omega$ = 0·054 eV). Nevertheless, the experimental results in silicon can be fitted satisfactorily only by assuming that an additional scattering mechanism involving a 190 K (0·016 eV) phonon is present. Such phonons exist. They are the LA and TA modes with $\mathbf{q}$ = 0·30 of the zone boundary. However, if they contribute to the scattering they must do so through a first-order process rather than a zero-order process.

First-order process in this context means a process akin to acoustic phonon scattering in which a deformation potential tensor is defined. Thus the coupling coefficient has the general form

$$C_{\mathbf{q}}^2 = \Xi_1^2(\theta_{\mathbf{q}})q^2. \tag{3.125}$$

By taking an isotropic approximation we can write the rate as follows:

$$W(\mathbf{k}) = \frac{\Xi_1^2}{8\pi^2\rho\omega_1}\int q^2\delta_{\mathbf{k}\pm\mathbf{q}-\mathbf{k}',\mathbf{K}}(n(\omega_1)+\tfrac{1}{2}\mp\tfrac{1}{2})\delta(E_{\mathbf{k}'}-E_{\mathbf{k}}\mp\hbar\omega_1)\,\mathrm{d}\mathbf{k}'. \tag{3.126}$$

† This value assumes that the valleys are situated at points 85 per cent towards the zone boundary. If the valleys are at points 83 per cent towards the zone boundary, as has been asserted by Hochberg and Westgate (1970), the value rises to 0·34. Both magnitudes are found in the literature.

Conversion to an integral over **q** and integration over angle, taking account of the $\cos\theta$ dependence in the delta functions, leads to

$$W(\mathbf{k})=\frac{\Xi_1^2}{4\pi\rho\hbar\omega_1 v}\left[n(\omega_1)\int_{q_{min}}^{q_{max}} q^3\,\mathrm{d}q+\{n(\omega_1)+1\}\int_{q_{min}}^{q_{max}} q^3\,\mathrm{d}q\right] \tag{3.127}$$

where the limits to the integrals are those for optical phonons (Section 3.2.1). It follows that for parabolic bands (Ferry 1976)

$$W(\mathbf{k})=\frac{\Xi_1^2(2m^*)^{5/2}E_k}{4\pi\rho\hbar^5\omega_1}\left[n(\omega_1)(E_k+\hbar\omega_1)^{1/2}\left(2+\frac{\hbar\omega_1}{E_k}\right)\right.$$
$$\left.+\{n(\omega_1)+1\}(E_k-\hbar\omega_1)^{1/2}\left(2-\frac{\hbar\omega_1}{E_k}\right)\right]. \tag{3.128}$$

This rate rises with energy more rapidly than in the cases of long-wavelength acoustic mode scattering and short-wavelength modes (Fig. 3.17).

The energy relaxation rate is simply obtained from eqn (3.128):

$$\frac{\mathrm{d}E_k}{\mathrm{d}t}=\frac{\Xi_1^2(2m^*)^{5/2}E_k}{4\pi\rho\hbar^4}\left[n(\omega_1)(E_k+\hbar\omega_1)^{1/2}\left(2+\frac{\hbar\omega_1}{E_k}\right)\right.$$
$$\left.-\{n(\omega_1)+1\}(E_k-\hbar\omega_1)^{1/2}\left(2-\frac{\hbar\omega_1}{E_k}\right)\right]. \tag{3.129}$$

The momentum relaxation rate is obtained from eqn (3.127) by multiplying the first integrand by the factor $q^2/2k^2-\hbar\omega_1/2E_k$ and the second integral by $q^2/2k^2+\hbar\omega_1/2E_k$, each factor being the appropriate value of

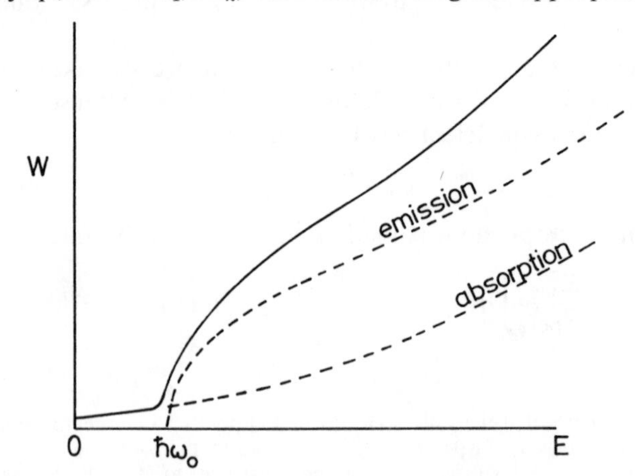

FIG. 3.17. First-order scattering rate for optical phonon scattering.

$(q/k)\cos\theta$. (See discussion p. 117). The result is

$$\frac{1}{\tau_m}=\frac{\Xi_1^2(2m^*)^{5/2}E_k}{4\pi\rho\hbar^5\omega_1}\left[n(\omega_1)(E_k+\hbar\omega_1)^{1/2}\left(\frac{8}{3}+\frac{5}{3}\frac{\hbar\omega_1}{E_k}\right)\right.$$
$$\left.+\{n(\omega_1)+1\}(E_k-\hbar\omega_1)^{1/2}\left(\frac{8}{3}-\frac{5}{3}\frac{\hbar\omega_1}{E_k}\right)\right]. \quad (3.130)$$

(As far as the author is aware expressions for the energy relaxation rate and the momentum relaxation time have not been given before.)

Ferry used $W(\mathbf{k})$ as an approximation for τ_m^{-1} in his analysis of the mobility in silicon and obtained $\Xi_1 = 5{\cdot}6$ eV. Using τ_m^{-1}, we obtain $\Xi_1 = 4{\cdot}7$ eV. The inter-valley coupling constant for each allowed mode turns out to be about 5×10^8 eV cm^{-1} (Ferry 1976; Ridley 1981).

3.5. Polar optical mode scattering

In polar materials the vibrations of oppositely charged atoms give rise to long-range macroscopic electric fields in addition to deformation potentials, and the interaction of the electron with these fields produces additional components of scattering. Polar scattering is indeed the dominant mechanism of scattering in pure III–V and II–VI compounds. First-order polarization occurs in connection with the contrary motion of the two atoms in the primitive unit cell characteristic of the longitudinally polarized optical mode. Second-order polarization in connection with acoustic strain is associated with the piezoelectric effect, and the scattering caused by this effect will be discussed in the next section. The major scattering mechanism in polar materials at room temperature is that associated with longitudinal optical modes, and this has been described by Frohlich (1937) and Callen (1949).

The basic polar interaction energy is

$$H_{ep}=\int\rho(\mathbf{R})\phi(\mathbf{R})\,d\mathbf{R} \quad (3.131)$$

where $\rho(\mathbf{R})$ is the charge density of the electron and $\phi(\mathbf{R})$ is the electric potential associated with polarization in the unit cell at $\mathbf{R}$. If e^* is the magnitude of effective charge on the atoms and V_0 is the volume of a unit cell, the polarization $\mathbf{P}$ is given by

$$\mathbf{P}(\mathbf{R})=e^*\mathbf{u}(\mathbf{R})/V_0 \quad (3.132)$$

where $\mathbf{u}(\mathbf{R})$ is the optical displacement in the unit cell at $\mathbf{R}$. Such a polarization is associated with a bound space-charge density equal to $-\text{div}\,\mathbf{P}$, and so if $\mathbf{u}$ is expanded in terms of plane waves $\text{div}\,\mathbf{P}$ is non-zero for longitudinal modes only.

In order to relate eqns (3.131) and (3.132) we transform the integral in eqn (3.131) as follows:

$$H_{ep} = \int \{\mathrm{div}\,\mathscr{D}(\mathbf{R})\}\phi(\mathbf{R})\,\mathrm{d}\mathbf{R} = -\int \mathscr{D}(\mathbf{R}).\mathrm{grad}\,\phi(\mathbf{R})\,\mathrm{d}\mathbf{R}$$

$$= \int \mathscr{D}(\mathbf{R}).\mathscr{E}(\mathbf{R})\,\mathrm{d}\mathbf{R} \qquad (3.133)$$

where $\mathscr{D}(\mathbf{R})$ is the electric displacement associated with the electron and $\mathscr{E}(\mathbf{R})$ is the electric field associated with the polarization. Since the electric displacement associated with the latter is zero,

$$\mathscr{E}(\mathbf{R}) = -\mathbf{P}(\mathbf{R})/\epsilon_0 \qquad (3.134)$$

and thus

$$H_{ep} = -\frac{1}{\epsilon_0}\int \mathscr{D}(\mathbf{R}).\mathbf{P}(\mathbf{R})\,\mathrm{d}\mathbf{R}. \qquad (3.135)$$

The bare electric displacement at $\mathbf{R}$ associated with an electron at $\mathbf{r}$ is

$$\mathscr{D}(\mathbf{R}) = -\mathrm{grad}\left(\frac{\bar{e}}{4\pi\,|\mathbf{r}-\mathbf{R}|}\right) \qquad (3.136)$$

where $\bar{e}$ is the electronic charge containing its sign. In the presence of screening caused by the motion of charges which is rapid enough to respond to lattice vibrations we can assume a Thomas–Fermi potential and take

$$\mathscr{D}(\mathbf{R}) = -\mathrm{grad}\left\{\frac{\bar{e}}{4\pi\,|\mathbf{r}-\mathbf{R}|}\exp(-q_0\,|\mathbf{r}-\mathbf{R}|)\right\} \qquad (3.137)$$

where q_0 is the reciprocal Debye screening length (see Appendix, Chapter 4 and, more fully, Chapter 9 for a discussion of screening).

By substituting

$$\mathbf{u}(\mathbf{R}) = \frac{1}{\sqrt{N}}\sum_{\mathbf{q}}\{Q_{\mathbf{q}}\mathbf{a}_{\mathbf{q}}\exp(\mathrm{i}\mathbf{q}.\mathbf{R}) + \mathrm{c.c.}\} \qquad (3.138)$$

in eqn (3.132) where $\mathbf{a}_{\mathbf{q}}$ is a unit polarization vector, since $\mathbf{u}(\mathbf{R})$ is a long-wavelength *optical* displacement, we can integrate eqn (3.135) and obtain

$$H_{ep} = -\frac{1}{\sqrt{N}}\frac{\bar{e}e^*}{V_0\epsilon_0}\sum_{\mathbf{q}}\frac{q}{q^2+q_0^2}\{\mathrm{i}Q_{\mathbf{q}}\exp(\mathrm{i}\mathbf{q}.\mathbf{r}) + \mathrm{c.c.}\}. \qquad (3.139)$$

The coupling coefficient of eqn (3.15) is therefore

$$C_{\mathbf{q}}^2 = \left(\frac{\bar{e}e^*}{V_0\epsilon_0}\right)^2\frac{q^2}{(q^2+q_0^2)^2} \qquad (3.140)$$

and the scattering rate is given by

$$W(\mathbf{k}) = \frac{V_0}{8\pi^2 \bar{M}\omega_0}\left(\frac{\bar{e}e^*}{V_0\epsilon_0}\right)^2 \int_0^{q_{ZB}} \int_{-1}^{1} \int_0^{2\mathbf{q}} \frac{q^4}{(q^2+q_0^2)^2}\delta_{\mathbf{k}\pm\mathbf{q}+\mathbf{k}',0}$$

$$\times\,(n(\omega_0)+\tfrac{1}{2}\mp\tfrac{1}{2})\delta(E_{\mathbf{k}'}-E_{\mathbf{k}}\mp\hbar\omega_0)\,\mathrm{d}q\,\mathrm{d}(\cos\theta)\,\mathrm{d}\phi \quad (3.141)$$

where we have put $V = NV_0$. Note that since we are dealing with *optical* displacement we retain the reduced mass $\bar{M}$. Because the coupling coefficient depends upon q and since the magnitude of $\mathbf{q}$ varies with the direction of the final state wavevector, the scattering probability is directionally dependent and this means we must convert the integral over final electronic states into one over phonon modes, as we have done in eqn (3.141), and proceed to integrate following the same method as was used for acoustic phonons (eqns (3.69) *et seq.*).

Integration over the azimuthal angle contributes a factor 2π. Integration over the polar angle is determined by the delta functions. Assuming spherical parabolic bands, we obtain for the total rate

$$W(\mathbf{k}) = \frac{V_0}{4\pi\bar{M}\omega_0}\left(\frac{\bar{e}e^*}{V_0\epsilon_0}\right)^2 \frac{1}{\hbar v}\left[n(\omega_0)\int_{q_{min}}^{q_{max}} \frac{q^3}{(q^2+q_0^2)^2}\,\mathrm{d}q + \{n(\omega_0)+1\}\int_{q_{min}}^{q_{max}} \frac{q^3}{(q^2+q_0^2)^2}\,\mathrm{d}q\right] \quad (3.142)$$

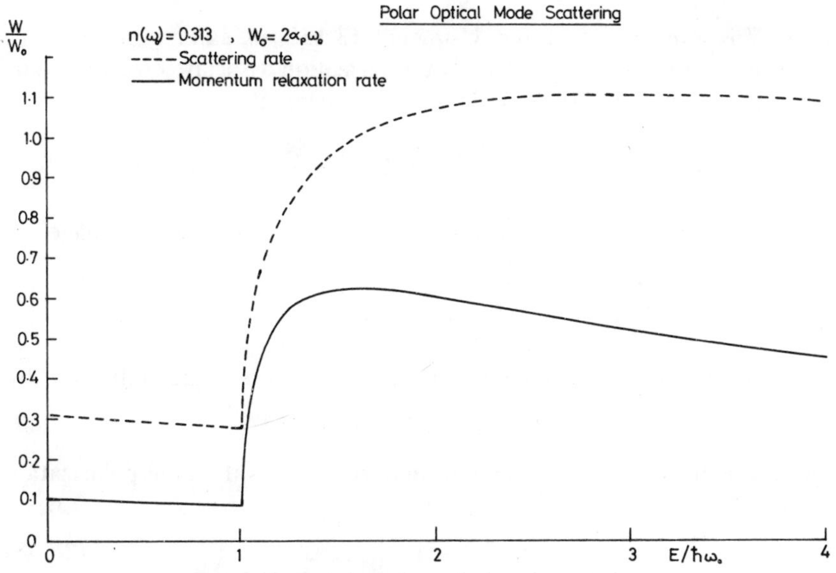

FIG. 3.18. Rate for polar optical mode scattering.

where q_{max} and q_{min}, which are different for the absorption and emission integrals, are given in eqns (3.30) and (3.31) and in eqns (3.42) and (3.43). The integrals are straightforward but involve lengthy expressions. Except for highly doped semiconductors screening is not very important, and we can obtain simpler expressions which are still useful if we neglect screening altogether. Then we obtain for the scattering rate, with $q_0 = 0$,

$$W(\mathbf{k}) = \frac{V_0}{2\pi \bar{M}\hbar\omega_0 v}\left(\frac{\bar{e}e^*}{V_0\epsilon_0}\right)^2\left[n(\omega_0)\sinh^{-1}\left(\frac{E_k}{\hbar\omega_0}\right)^{1/2} + \{n(\omega_0)+1\}\sinh^{-1}\left(\frac{E_k}{\hbar\omega_0}-1\right)^{1/2}\right] \quad (3.143)$$

and it is understood that the second term in the square brackets is zero if $E_k < \hbar\omega_0$ (Fig. 3.18). (See Chapter 9 for a discussion on screening.)

3.5.1. *The effective charge*

The difference between the permittivities at low and high frequencies in polar materials is related to the effective charge on the ions. At high frequencies the contribution to the polarization made by ionic motion vanishes, leaving only the electronic component, whereas in the static case both contributions are present.

The equation of motion which describes a long-wavelength longitudinally polarized optical mode is as follows:

$$\bar{M}\left(\frac{\partial^2 \mathbf{u}}{\partial t^2} + \omega_0^2 \mathbf{u}\right) = \mathcal{F} \quad (3.144)$$

where $\mathcal{F}$ is some applied force. Using eqn (3.132) we can recast this as an equation for the time dependence of polarization associated with the ions in the presence of an applied electric displacement

$$\frac{\bar{M}V_0}{e^*}\left(\frac{\partial^2 \mathbf{P}}{\partial t^2} + \omega_0^2 \mathbf{P}\right) = \frac{e^*\mathcal{D}}{\epsilon_0} \quad (3.145)$$

where ϵ_0 is the permittivity of free space. In the static case we have

$$\mathbf{P} = \frac{e^{*2}\mathcal{D}}{\bar{M}V_0\omega_0^2\epsilon_0}, \quad (3.146)$$

but in general the total polarization $P_{tot}(0)$ is given in the static case by

$$\mathbf{P}_{tot}(0) = \mathcal{D} - \epsilon_0\mathcal{E} = (1-\epsilon_0/\epsilon)\mathcal{D} \quad (3.147)$$

where $\mathcal{E}$ is the electric field, and at high frequencies the total polarization is given by

$$\mathbf{P}_{tot}(\infty) = (1-\epsilon_0/\epsilon_\infty)\mathcal{D} \quad (3.148)$$

where ϵ_∞ is the high-frequency permittivity, Thus the polarization caused by ionic motion is

$$\mathbf{P} = \mathbf{P}_{\text{tot}}(0) - \mathbf{P}_{\text{tot}}(\infty) = \left(\frac{\epsilon_0}{\epsilon_\infty} - \frac{\epsilon_0}{\epsilon}\right)\mathscr{D}. \tag{3.149}$$

Relating eqns (3.149) and (3.146) leads to the expression for the effective charge we require:

$$e^{*2} = \bar{M}V_0\omega_0^2\epsilon_0^2\left(\frac{1}{\epsilon_\infty} - \frac{1}{\epsilon}\right). \tag{3.150}$$

In the case of III–V compounds $\bar{M}V_0\omega_0^2$ is approximately a constant, independent of material, and so a simple expression for the effective charge is

$$\left(\frac{e^*}{e}\right)^2 \approx 3\left(\frac{1}{\kappa_\infty} - \frac{1}{\kappa}\right) \tag{3.151}$$

where κ_∞ and κ are the high frequency and static dielectric constants (Keyes 1962; Hilsum 1975; Ridley 1977).

3.5.2. *Energy and momentum relaxation*

The scattering rate of eqn (3.143) with e^* substituted from eqn (3.150) is

$$W(\mathbf{k}) = \frac{e^2\omega_0}{2\pi\hbar\epsilon_p v}\left[n(\omega_0)\sinh^{-1}\left(\frac{E_k}{\hbar\omega_0}\right)^{1/2} + \{n(\omega_0)+1\}\sinh^{-1}\left(\frac{E_k}{\hbar\omega_0} - 1\right)^{1/2}\right] \tag{3.152}$$

where

$$1/\epsilon_p = 1/\epsilon_\infty - 1/\epsilon. \tag{3.153}$$

The energy relaxation rate can immediately be written down, following eqn (3.103), as follows:

$$\frac{dE_k}{dt} = \frac{e^2\omega_0^2}{2\pi\epsilon_p v}\left[n(\omega_0)\sinh^{-1}\left(\frac{E_k}{\hbar\omega_0}\right)^{1/2} - \{n(\omega_0)+1\}\sinh^{-1}\left(\frac{E_k}{\hbar\omega_0} - 1\right)^{1/2}\right]. \tag{3.154}$$

The momentum relaxation rate is more complicated to obtain. We must return to eqn (3.142) and weigh absorption and emission terms by the respective changes in momentum. An absorption of a phonon travelling at an angle θ to $\mathbf{k}$ contributes a fractional increase of momentum of $(q/k)\cos\theta$ in the direction of $\mathbf{k}$. From energy and momentum conservation $\cos\theta$ is given by eqn (3.19). The fractional increase in momentum for emission is $-(q/k)\cos\theta$, and $\cos\theta$ in this case is given by eqn (3.35). Thus the momentum relaxation rate is given by

$$\frac{1}{\tau_{\rm m}} = \frac{e^2\omega_0}{4\pi\hbar\epsilon_{\rm p} v}\left[n(\omega_0)\int_{q_{\min}}^{q_{\max}} \frac{q^3}{(q^2+q_0^2)^2}\left(\frac{q^2}{2k^2}-\frac{\hbar\omega_0}{2E_k}\right){\rm d}q\right.$$

$$\left. +\{n(\omega_0)+1\}\int_{q_{\min}}^{q_{\max}} \frac{q^3}{(q^2+q_0^2)^2}\left(\frac{q^2}{2k^2}+\frac{\hbar\omega_0}{2E_k}\right){\rm d}q\right]. \tag{3.155}$$

Neglecting q_0 again, we obtain

$$\frac{1}{\tau_{\rm m}} = \frac{e^2\omega_0}{4\pi\hbar\epsilon_{\rm p} v}\left(n(\omega_0)\left(1+\frac{\hbar\omega_0}{E_k}\right)^{1/2} + \{n(\omega_0)+1\}\left(1-\frac{\hbar\omega_0}{E_k}\right)^{1/2}\right.$$

$$\left. +\frac{\hbar\omega_0}{E_k}\left[-n(\omega_0)\sinh^{-1}\left(\frac{E_k}{\hbar\omega_0}\right)^{1/2} + \{n(\omega_0)+1\}\sinh^{-1}\left(\frac{E_k}{\hbar\omega_0}-1\right)^{1/2}\right]\right), \tag{3.156}$$

a result first derived by Callen (1949).†

Taking into account non-parabolicity in a spherical band leads to an energy relaxation rate given by

$$\frac{{\rm d}E_k}{{\rm d}t} = \frac{e^2(2m^*)^{1/2}\omega_0^2}{4\pi\epsilon_{\rm p}\gamma^{1/2}(E_k)} I^2(\mathbf{k},\mathbf{k}')\left[n(\omega_0)\left\{\frac{{\rm d}\gamma(E_k)}{{\rm d}E_k}\right\}_{E_k+\hbar\omega_0}\coth^{-1}\frac{\gamma^{1/2}(E_k+\hbar\omega_0)}{\gamma^{1/2}(E_k)}\right.$$

$$\left. -\{n(\omega_0)+1\}\left\{\frac{{\rm d}\gamma(E_k)}{{\rm d}E_k}\right\}_{E_k-\hbar\omega_0}\tanh^{-1}\left\{\frac{\gamma^{1/2}(E_k-\hbar\omega_0)}{\gamma^{1/2}(E_k)}\right\}\right] \tag{3.157}$$

and a momentum relaxation rate given by

$$\frac{1}{\tau_{\rm m}} = \frac{e^2(2m^*)^{1/2}\omega_0 I^2(\mathbf{k},\mathbf{k}')}{8\pi\epsilon_{\rm p}\hbar\gamma^{1/2}(E_k)}\left[n(\omega_0)\left\{\frac{{\rm d}\gamma(E_k)}{{\rm d}E_k}\right\}_{E_k+\hbar\omega_0}\frac{\gamma^{1/2}(E_k+\hbar\omega_0)}{\gamma^{1/2}(E_k)}\right.$$

$$+\{n(\omega_0)+1\}\left\{\frac{{\rm d}\gamma(E_k)}{{\rm d}E_k}\right\}_{E_k-\hbar\omega_0}\frac{\gamma^{1/2}(E_k-\hbar\omega_0)}{\gamma^{1/2}(E_k)} - n(\omega_0)\left\{\frac{{\rm d}\gamma(E_k)}{{\rm d}E_k}\right\}_{E_k+\hbar\omega_0}$$

$$\times\left\{\frac{\gamma(E_k+\hbar\omega_0)-\gamma(E_k)}{\gamma(E_k)}\right\}\coth^{-1}\left\{\frac{\gamma^{1/2}(E_k+\hbar\omega_0)}{\gamma^{1/2}(E_k)}\right\} + \{n(\omega_0)+1\}$$

$$\left.\times\left\{\frac{{\rm d}\gamma(E_k)}{{\rm d}E_k}\right\}_{E_k-\hbar\omega_0}\left\{\frac{\gamma(E_k)-\gamma(E_k-\hbar\omega_0)}{\gamma(E_k)}\right\}\tanh^{-1}\left\{\frac{\gamma^{1/2}(E_k-\hbar\omega_0)}{\gamma^{1/2}(E_k)}\right\}\right]. \tag{3.158}$$

These expressions were obtained by Conwell and Vassell (1968) (Fig. 3.19).

† When $\hbar\omega/k_{\rm B}T$ is large, it is often more accurate to use solely the absorption rate. See eqn (12.29) and subsequent discussion.

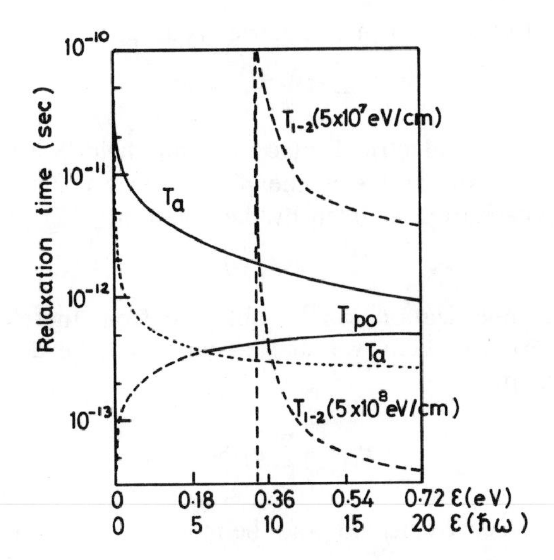

FIG. 3.19. Relaxation times for electron scattering in the (0, 0, 0) valley of GaAs, corrected for non-parabolicity, as a function of energy: τ_a, acoustic relaxation time; τ_{po}, polar optical relaxation time; τ_{1-2}, inter-non-equivalent-valley relaxation time. (After Conwell and Vassell 1968.)

It is worth noting that the energy-independent parameters which compose the term multiplying the curved brackets in eqn (3.156) can be subsumed into a dimensionless coupling constant α_{ep} given by

$$\alpha_{ep} = \frac{e^2/4\pi\epsilon_p}{\hbar}\left(\frac{m^*}{2\hbar\omega_0}\right)^{1/2} \tag{3.159}$$

which is the expression already given in eqn (2.109) in connection with polarons (see also Section 3.7).

3.6. Piezoelectric scattering

In crystals whose lattice lacks inversion symmetry, such as those semiconductors with sphalerite or wurtzite structure (but not those with rocksalt structure), elastic strain may be accompanied by macroscopic electric fields. This piezoelectric effect provides an additional coupling between the electron and acoustic vibrations whose interaction energy is exactly the electrostatic energy derived in the previous section (eqn (3.131)). However, the electric polarization $\mathbf{P}(\mathbf{R})$ is now proportional to acoustic strain and not to optical displacement, and can be obtained from the following phenomenological equation relating electric displacement to electric field and strain in a piezoelectric crystal:

$$\mathscr{D}_i = \sum_j \epsilon_{ij}\mathscr{E}_j + \sum_{k,l} e_{ikl}S_{kl} \tag{3.160}$$

where $\mathscr{D}$ and $\mathscr{E}$ are the electric displacement and field, S is the strain, ϵ is the permittivity tensor, and $\mathbf{e}$ is the piezoelectric constant tensor. By definition the polarization is given by the equation

$$\mathscr{D}_i = \epsilon_0\mathscr{E}_i + P_i \tag{3.161}$$

The bare lattice polarization can be obtained from these equations by putting $\mathscr{D}=0$. For simplicity we shall assume that the dielectric is isotropic. Consequently

$$P_i = \frac{1}{\kappa}\sum_{k,l} e_{ikl}S_{kl} \tag{3.162}$$

where κ is the dielectric constant, and the interaction energy is given by

$$H_{\text{ep}} = -\frac{1}{\epsilon_0}\int \mathscr{D}(\mathbf{R})\mathbf{P}(\mathbf{R})\,\mathrm{d}\mathbf{R} \tag{3.163}$$

where $\mathscr{D}(\mathbf{R})$ is given by eqn (3.137) and $\mathbf{R}$ is the position coordinate of the unit cell. It is important to include the effect of electrical screening, which is why eqn (3.137) rather than eqn (3.136) is used for $\mathscr{D}(\mathbf{R})$.

In the case of zinc blende crystals there is only one piezoelectric constant. In reduced notation $e_{ikl} \to e_{im}$, where m runs from 1 to 6 (Nye 1957). For zinc blende $e_{14} = e_{25} = e_{36}$ and all other components are zero. For wurzite $e_{24} = e_{15}$, $e_{31} = e_{32}$, e_{33} is non-zero, and all the rest are zero. We now follow the calculation for zinc blende. Only the shear strain components give rise to electric fields.

By expanding the displacement in plane waves we obtain

$$P_1 = \sum_{\mathbf{q}} \frac{e_{14}/\kappa}{\sqrt{N}}\{\mathrm{i}(a_2q_3 + a_3q_2)Q_{\mathbf{q}}\exp(\mathrm{i}\mathbf{q}.\mathbf{R}) + \text{c.c.}\} \tag{3.164}$$

where $\mathbf{a}$ is a unit polarization vector and $Q_{\mathbf{q}}$ is the normal coordinate associated with the acoustic mode with wavevector $\mathbf{q}$. The components P_2 and P_3 are of similar form. Performing the integration in eqn (3.163) gives

$$H_{\text{ep}} = -\frac{\bar{e}e_{14}/\epsilon}{\sqrt{N}}\sum_{\mathbf{q}} \frac{q^2}{q^2+q_0^2}\{2\mathrm{i}(a_1\beta\gamma + a_2\gamma\alpha + a_3\alpha\beta)Q_{\mathbf{q}}\exp(\mathrm{i}\mathbf{q}.\mathbf{r}) + \text{c.c.}\} \tag{3.165}$$

where α, β, and γ are the direction cosines with respect to the crystal axis of the direction of propagation of the wave. The coupling constant of eqn

(3.15) for a given mode is therefore

$$C_{\mathbf{q}}^2 = \left(\frac{\bar{e}e_{14}}{\epsilon}\right)^2 \frac{q^4}{(q^2+q_0^2)^2} 4(a_1\beta\gamma + a_2\gamma\alpha + a_3\alpha\beta)^2. \tag{3.166}$$

Piezoelectric coupling has a complicated directional dependence as eqn (3.166) shows. The coupling is zero for modes travelling along a principal crystal axis, whatever the polarization of the mode. Only transversely polarized modes couple when the direction lies in a cube face (e.g. $\alpha \neq 0$, $\beta \neq 0$, $\gamma = 0$). Longitudinally polarized waves couple when travelling along a cube diagonal ($\alpha = \beta = \gamma$), but transverse modes do not because of the cancellation of polarization vector components. Some average over direction is clearly desirable. Meyer and Polder (1953) in their treatment of piezoelectric scattering took averages over ⟨100⟩, ⟨110⟩, and ⟨111⟩ directions of propagation and included in their averaging the anisotropy of the elastic constants since these enter the final result. Harrison (1956) calculated the scattering rates in these directions and then took a weighted average. Hutson (1961) took a spherical average of the piezoelectric constants separately for longitudinal and transverse modes, and this approach was generalized and applied to ellipsoidal bands by Zook (1964). We shall adopt the latter approach and make separate spherical averages of piezoelectric and elastic constants in the case of spherical bands.

For longitudinal waves eqn (3.166) is simply

$$C_{\mathbf{q},1}^2 = \left(\frac{\bar{e}e_{14}}{\epsilon}\right)^2 \frac{q^4}{(q^2+q_0^2)^2} 4(3\alpha\beta\gamma)^2. \tag{3.167}$$

Since the spherical average of $(\alpha\beta\gamma)^2$ can readily be shown to be 1/105, we obtain

$$\langle C_{\mathbf{q},1}^2 \rangle = \frac{12}{35}\left(\frac{\bar{e}e_{14}}{\epsilon}\right)^2 \frac{q^4}{(q^2+q_0^2)^2}. \tag{3.168}$$

For transverse waves, eqn (3.166) becomes

$$C_{\mathbf{q},t}^2 = \left(\frac{\bar{e}e_{14}}{\epsilon}\right)^2 \frac{q^4}{(q^2+q_0^2)^2} 4\{\alpha^2\beta^2 + \beta^2\gamma^2 + \gamma^2\alpha^2 - (3\alpha\beta\gamma)^2\} \tag{3.169}$$

whence

$$\langle C_{\mathbf{q},t}^2 \rangle = \frac{16}{35}\left(\frac{\bar{e}e_{14}}{\epsilon}\right)^2 \frac{q^4}{(q^2+q_0^2)^2}. \tag{3.170}$$

Taking the overlap factor $I(\mathbf{k},\mathbf{k}')$ to be unity and assuming equipartition, we obtain the scattering rate for absorption or emission:

$$W_{l,t}(\mathbf{k}) = \frac{k_B T}{8\pi^2 \rho \bar{v}_{sl,t}^2 \hbar} \int \frac{\langle C_{q\,l,t}^2 \rangle}{q^2} \delta_{\mathbf{k}\pm\mathbf{q}-\mathbf{k}',0}\, \delta(E_{\mathbf{k}'} - E_{\mathbf{k}} \mp \hbar\omega_{\mathbf{q}l,t})\, d\mathbf{k}' \quad (3.171)$$

where $\bar{v}_s^2$ is the square of the spherical average velocity. For longitudinal modes $\rho\bar{v}_{sl}^2 = c_L$ and for transverse modes $\rho\bar{v}_{st}^2 = c_T$, where c_L and c_T are the spherical average elastic constants (see Appendix):

$$c_L = c_{12} + 2c_{44} + \tfrac{3}{5}c^* \quad (3.172)$$

$$c_T = c_{44} + \tfrac{1}{5}c^* \quad (3.173)$$

$$c^* = c_{11} - c_{12} - 2c_{44}. \quad (3.174)$$

The quantity c^* is a measure of elastic anisotropy and is zero for an isotropic crystal. We can lump longitudinal and transverse modes together by defining an average electromechanical coupling coefficient K_{av} such that

$$K_{av}^2 = \frac{e_{14}^2}{\epsilon}\left(\frac{12}{35c_L} + \frac{16}{35c_T}\right). \quad (3.175)$$

Thus

$$W(\mathbf{k}) = \frac{\bar{e}^2 K_{av}^2 k_B T}{8\pi^2 \epsilon\hbar} \int \frac{q^2}{(q^2+q_0^2)^2} \delta_{\mathbf{k}\pm\mathbf{q}-\mathbf{k}',0}\, \delta(E_{\mathbf{k}'} - E_{\mathbf{k}} \mp \hbar\omega_{\mathbf{q}})\, d\mathbf{k}' \quad (3.176)$$

provided that the difference between longitudinal and transverse phonon energies can be neglected. Note that the $\mathbf{q}$ dependence of the integrand is exactly the same as for polar optical mode scattering (eqn 3.142). The integration can be carried out along the lines of eqns (3.68)–(3.73) for acoustic mode scattering. After integrating over direction we obtain for a parabolic spherical band

$$W(\mathbf{k}) = \frac{(e^2/4\pi\epsilon) K_{av}^2 k_B T}{\hbar^2 v} \int_0^{2k} \frac{q^3}{(q^2+q_0^2)^2}\, dq \quad (3.177)$$

where v is the group velocity of the electron. Integrating over q leads to

$$W(\mathbf{k}) = \frac{(e^2/4\pi\epsilon) K_{av}^2 k_B T}{2\hbar^2 v} \left\{ \log\left(1 + \frac{8m^* E_k}{\hbar^2 q_0^2}\right) - \frac{1}{1 + \hbar^2 q_0^2/8m^* E_k} \right\}. \quad (3.178)$$

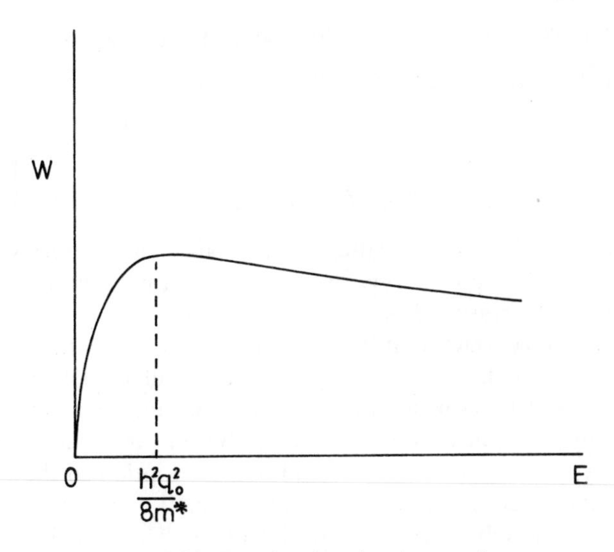

FIG. 3.20. Rate for piezoelectric scattering.

The scattering rate remains finite at small energies only because the reciprocal screening length q_0 is non-zero. By multiplying by 2 to obtain the total scattering rate for absorption and emission we obtain (Fig. 3.20)

$$W(\mathbf{k}) = \frac{(e^2/4\pi\epsilon)K_{\mathrm{av}}^2 k_{\mathrm{B}}T}{\hbar^2 v}\left\{\log\left(1+\frac{8m^*E_k}{\hbar^2 q_0^2}\right) - \frac{1}{1+\hbar^2 q_0^2/8m^*E_k}\right\}. \tag{3.179}$$

(As far as the author is aware, this and the following expressions have not been given before.)

The momentum relaxation rate can easily be obtained, following eqns (3.100) and (3.101), by multiplying the integrand of eqn (3.177) by $q^2/2k^2$. After multiplying by 2 to take account of absorption and emission, we obtain

$$\frac{1}{\tau_{\mathrm{m}}} = \frac{2(e^2/4\pi\epsilon)K_{\mathrm{av}}^2 k_{\mathrm{B}}T}{\hbar^2 v}\left\{1+\frac{1}{1+8m^*E_k/\hbar^2 q_0^2} - \frac{\hbar^2 q_0^2}{4m^*E_k}\log\left(1+\frac{8m^*E_k}{\hbar^2 q_0^2}\right)\right\}. \tag{3.180}$$

The term multiplying the square bracket is identical in form to the rate given by Meyer and Polder (1953) and Hutson (1961) who neglected screening altogether.

The energy relaxation rate can be obtained from the scattering rate following eqn (3.103). Thus, treating longitudinal and transverse modes separately and combining to give a total energy relaxation rate, we obtain

$$\frac{dE_k}{dt}=\frac{4(e^2/4\pi\epsilon)(e_{14}^2/\epsilon)(2m^*)^{3/2}E_k^{1/2}}{5\rho\hbar^2}\left[\frac{k_BT/E_k}{(1+\hbar^2q_0^2/8m^*E_k)^2}-1\right.$$

$$\left.-\frac{1}{1+8m^*E_k/\hbar^2q_0^2}+\frac{\hbar^2q_0^2}{4m^*E_k}\log\left(1+\frac{8m^*E_k}{\hbar^2q_0^2}\right)\right]. \quad (3.181)$$

Like polar optical mode scattering, piezoelectric scattering weakens with increasing electron energy, except at very low energies where the trend reverses because of screening. At low temperatures in pure piezoelectric semiconductors it becomes the dominant scattering mechanism for thermal electrons, but warm electrons will tend to be scattered more via the deformation potential interaction. Since zero-point scattering finds its main application for warm and hot electrons at low temperatures, we have little incentive to extend our description to the case when the assumption of equipartition fails, and we shall not therefore consider this situation although it presents no difficulties. The fact that piezoelectric scattering is important for low energies makes a parabolic approximation a good one.

The foregoing calculation has relied heavily on there being electrical screening. In the absence of screening ($q_0=0$) the scattering rate (eqn (3.179)) diverges logarithmically but momentum and energy relaxation rates remain finite. It is because the momentum relaxation rate remains finite that previous calculations of mobility neglecting screening were successful. Nevertheless, problems arise when screening is very weak. In this respect piezoelectric scattering is similar to charged impurity scattering (see Chapter 4).

Although we have limited discussion explicitly to the case of sphalerite lattices, the approach is readily applicable to wurtzite where a more complicated averaging of the piezoelectric constants is necessary. Thus for longitudinal waves

$$\langle e_l^2\rangle=\tfrac{1}{7}e_{33}^2+\tfrac{4}{35}e_{33}(e_{31}+2e_{15})+\tfrac{8}{105}(e_{31}+2e_{15})^2 \quad (3.182)$$

and for transverse waves

$$\langle e_t^2\rangle=\tfrac{2}{35}(e_{33}-e_{31}-e_{15})^2+\tfrac{16}{105}e_{15}(e_{33}-e_{31}-e_{15})+\tfrac{16}{35}e_{15}^2 \quad (3.183)$$

so that in the isotropic model (Hutson 1961)

$$K_{av}^2=\frac{\langle e_l^2\rangle}{\epsilon c_L}+\frac{\langle e_t^2\rangle}{\epsilon c_T}. \quad (3.184)$$

Values of the piezoelectric coupling constants for some semiconductors are given in Tables 3.4 and 3.5.

TABLE 3.4
Piezoelectric coupling coefficients for III–V compounds

	e_{14} ($C\,m^{-2}$)[a]	κ (static)	ρ ($g\,cm^{-3}$)	$\langle v_s \rangle_{av}$ ($km\,s^{-1}$)	K^2_{av}
GaAs	0·160	13·18	5·36	4·03	0·00252
GaSb	0·126	15·69	5·66	3·35	0·00180
InP	0·035	12·35	4·83	3·85	0·000156
InAs	0·045	14·55	5·71	3·17	0·000274
InSb	0·071	17·72	5·82	2·78	0·000714

[a] From Rode 1970*a*.

TABLE 3.5
Piezoelectric coupling coefficients for II–VI compounds[a]

Sphalerite	e_{14} ($C\,m^{-2}$)	κ	c_L ($10^{10} N\,m^{-2}$)	c_T	K^2_{av}
ZnS	0·17	8·32	12·89	3·60	0·0060
ZnTe	0·027	9·67	8·41	2·48	0·00019
ZnSe	0·045	8·33	10·34	3·29	0·00047
CdTe	0·034	9·65	6·97	1·55	0·00047

Wurtzite	e_{33}	e_{31} ($C\,m^{-2}$)	e_{15}	κ	c_L	c_T	K^2_{av} [b]
CdS	0·49	−0·25	−0·21	9·7	8·8	1·54	0·035
CdSe	0·32	−0·13	−0·15	9·4	7·4	1·72	0·014
ZnO	1·1	−0·16	−0·31	8·2	14·1	2·47	0·074

[a] Values derived from Zook (1964) and Rode (1970*b*).
[b] Values for CdS and ZnO from Hutson (1961); the value for CdSe was calculated from Hutson's formulae.

3.7. Scattering-induced electron mass

The electron–phonon interaction, besides causing transitions between Bloch states in a band, mixes those states and produces energy shifts. The effect can be described using second-order perturbation theory. The perturbed energy of a Bloch state $|\mathbf{k}\rangle$ is as follows

$$E_{\mathbf{k}} = E_{\mathbf{k}0} + \sum_{\mathbf{q}} \left(\frac{M^2_{\mathbf{q}\,\mathrm{em}}}{E_{\mathbf{k}} - E_{\mathbf{k}-\mathbf{q}} - \hbar\omega_{\mathbf{q}}} + \frac{M^2_{\mathbf{q}\,\mathrm{abs}}}{E_{\mathbf{k}} - E_{\mathbf{k}-\mathbf{q}} + \hbar\omega_{\mathbf{q}}} \right) \tag{3.185}$$

where $M_{\mathbf{q}\,\mathrm{em}}$ and $M_{\mathbf{q}\,\mathrm{abs}}$ are the matrix elements for the emission and absorption respectively of a single phonon. Crystal momentum is conserved, and so a state $|\mathbf{k}\rangle$ is coupled via the electron–phonon interaction involving a mode of wavevector $\mathbf{q}$ to a state $|\mathbf{k} \pm \mathbf{q}\rangle$. Since the second-order process returns the electron to its original state, conservation of energy does not impose any condition in which the state is coupled. In the notation of Section 3.1 we have (see eqn (3.14))

$$M_{\mathbf{q}}^2 = \frac{\hbar}{2NM'} \frac{C_{\mathbf{q}}^2 I^2(\mathbf{k}, \mathbf{k}')}{\omega_{\mathbf{q}}} \begin{Bmatrix} n(\omega_{\mathbf{q}})+1 \\ n(\omega_{\mathbf{q}}) \end{Bmatrix}. \tag{3.186}$$

Thus energy shifts are induced which are in general temperature dependent. Such temperature effects have already been mentioned in Chapter 1, Section 1.12. These shifts are also dependent on electron energy and, as such, they can be described in terms of an addition to the effective mass of the carrier. Scattering-induced mass of this sort is associated with the electron being 'dressed' continually with a cloud of virtual phonons. The lattice in the vicinity of the electron is distorted, and the movement of the electron entails the movement of this lattice distortion with a consequential increase in inertia.

We shall estimate this effect at the absolute zero of temperature only. The results will serve to provide a comparison of the intrinsic strengths of the four interactions we have been considering, namely acoustic phonon, optical phonon, polar mode, and piezoelectric. The calculation with polar modes, leading to the concept of the polaron, is perhaps most familiar. In the cases of the other types of interaction we arrive at similar concepts, which we might name as the acouston, the opticon, and the piezon.

At $T=0$ only spontaneous emission is possible; thus we need only $M_{\mathbf{q}\,\mathrm{em}}^2$ for $n(\omega_{\mathbf{q}})=0$. For simplicity, we shall take the overlap factor $I(\mathbf{k}, \mathbf{k}')$ to be unity and take spherical averages of $C_{\mathbf{q}}$ and $\omega_{\mathbf{q}}$. The squared matrix elements are given in Table 3.6.

The sum in eqn (3.185) can be converted to integral form by taking a spherical approximation for the Brillouin zone:

$$E_{\mathbf{k}} = E_{\mathbf{k}_0} - \frac{1}{E_{\mathbf{k}_0}} \int_0^{q_D} \int_{-1}^{+1} \int_0^{2\pi} \frac{M_{\mathbf{q}\,\mathrm{em}}^2 q^2 \,\mathrm{d}q \,\mathrm{d}(\cos\theta) \,\mathrm{d}\phi}{(q/k)^2 - 2(q/k)\cos\theta + \hbar\omega_{\mathbf{q}}/E_{\mathbf{k}_0}} \frac{V}{8\pi^3} \tag{3.187}$$

where the limit $q_D = (6\pi^2 N/V)^{1/3}$ is chosen such that the number of modes in the sphere is equal to N. In eqn (3.187) we have used the conservation of momentum and the parabolic approximation to transform the denominator into a function of q. Since the density of phonon modes is high towards large q, it is convenient to simplify by expanding the reciprocal of

TABLE 3.6

Zero-point isotropic electron–phonon squared matrix elements $M_{\mathbf{q}\,\mathrm{em}}^2$

Acoustic	Optical	Polar	Piezoelectric
$\dfrac{\Xi^2 \hbar q}{2NM\bar{v}_s}$	$\dfrac{D_0^2 \hbar}{2NM\omega_o}$	$\dfrac{e^2 \hbar\omega_o}{2NV_0 \epsilon_p q^2}$	$\dfrac{e^2 \langle e_{14}^2 \rangle_{av} \hbar}{2NM\epsilon^2 \bar{v}_s q}$

the denominator in powers of $\cos\theta$ and to retain terms up to and including the quadratic. Thus, after integrating over the solid angle we obtain

$$E_{\mathbf{k}} = E_{\mathbf{k}_0} - \frac{V}{2\pi^2 E_{\mathbf{k}_0}} \int_0^{q_D} \frac{M^2_{\mathbf{q}\,\mathrm{em}} q^2}{(q/k)^2 + \hbar\omega_q/E_{\mathbf{k}_0}} \left[1 + \frac{4(q/k)^2}{3\{(q/k)^2 + \hbar\omega_{\mathbf{q}}/E_{\mathbf{k}_0}\}^2}\right] dq \tag{3.188}$$

This takes the form

$$E_{\mathbf{k}} = \Delta E_{\mathrm{ep}} + E_{\mathbf{k}_0}(1 - \alpha_{\mathrm{ep}}/6) \tag{3.189}$$

where ΔE_{ep} is an energy shift and α_{ep} is a dimensionless coupling strength which causes an increase in effective mass:

$$\frac{1}{m^*} = \frac{1}{m_0^*}\left(1 - \frac{\alpha_{\mathrm{ep}}}{6}\right) \tag{3.190}$$

The results are given in Table 3.7.

Since ΔE_{ep} merely shifts all the energy states it is not of much interest. The quantities of most interest are the coupling strengths. For perturbation theory to be applicable α_{ep} must be less than unity. Taking typical values for the deformation potential parameters ($\Xi \approx 10$ eV, $D_o \approx 5 \times 10^8$ eV cm^{-1}, $m^* \approx 0{\cdot}2m$) and polar parameters typical of III–V compounds, we obtain $\alpha_{\mathrm{ep}}(\mathrm{ac}) \approx 0{\cdot}03$, $\alpha_{\mathrm{ep}}(\mathrm{op}) \approx 0{\cdot}03$, $\alpha_{\mathrm{ep}}(\mathrm{pol}) \approx 0{\cdot}1$, and $\alpha_{\mathrm{ep}}(\mathrm{piez}) \approx 0{\cdot}1$. These values are small enough to justify the use of perturbation theory in the electron–phonon interaction. For highly polar materials α_{ep} exceeds unity. In these materials the electron is virtually trapped in the potential hole it has dug for itself and perturbation theory is no longer valid. In non-polar and weakly polar semiconductors the total coupling constant lies roughly between 0·1 and 0·3, which corresponds to a contribution to the electron effective mass of a few per cent.

TABLE 3.7
Energy shifts and electron–phonon coupling strengths

	Acoustic	Optical	Polar	Piezoelectric
ΔE_{ep}	$-\frac{\Xi^2 m^* q_D}{4\pi^2\hbar\rho\bar{v}_s}$	$-\frac{D_0^2 m^* q_D}{2\pi^2\hbar\rho\omega_0}$	$-\frac{e^2(m^*\hbar\omega_0)^{1/2}}{2^{5/2}\pi\epsilon_p\hbar}$	$-\frac{e^2\langle e_{14}^2\rangle m^*}{2\pi^2\epsilon^2\hbar\rho\bar{v}_s}\log\left(\frac{\hbar q_D}{2m^*\bar{v}_s}\right)$
α_{ep}	$\frac{8\Xi^2 m^{*2}}{\pi^2\hbar^3\rho\bar{v}_s}\log\left(\frac{\hbar q_D}{2m^*\bar{v}_s}\right)$	$\frac{D_0^2(2m^*)^{3/2}}{\pi\hbar\rho(\hbar\omega_0)^{3/2}}$	$\frac{e^2 m^{*1/2}}{2^{5/2}\pi\epsilon_p\hbar(\hbar\omega_0)^{1/2}}$	$\frac{e^2 K_{\mathrm{av}}^2}{\pi^2\epsilon\hbar\bar{v}_s}$

Note: $q_D = (6\pi^2 N/V)^{1/3}$.

3.8. Mobilities

The scattering rate for an electron with a well-defined energy is scarcely ever directly observable. In almost all situations in practice, what is within the reach of experiment are phenomenological quantities such as mobility, diffusion coefficient, and the various galvanomagnetic coefficients which are properties of a population of electrons rather than of a single electron and which are only indirectly related to the scattering rate. To infer a scattering rate from one of these phenomenological quantities requires a knowledge of how the electrons are distributed in energy. In the case where there are weak fields this distribution can be taken to be substantially that at equilibrium, but slightly perturbed, and the Boltzmann equation can be solved in a straightforward way to obtain the required relation. Where fields are strong, as occurs in hot-electron experiments, the distribution function is quite difficult to compute. Such experiments are useful for providing information about energy relaxation and about scattering processes at high energies. It would take us too far from our discussion of basic processes to go into such cases, and we shall limit our account to the low-field case and further limit it to a summary of low-field mobilities.

The basic statistical relation between mobility μ and the momentum relaxation time τ_m is given by elementary transport theory in the case of a

TABLE 3.8
Momentum relaxation times in a simple band

Scattering mechanism	Momentum relaxation times τ_m
Acoustic phonon (equipartition)	$\dfrac{\pi\hbar^4 c_L}{2^{1/2}\Xi^2 m^{*3/2} k_B T E_k^{1/2}}$
Zero-order optical phonon	$\dfrac{2^{1/2}\pi\hbar^3\omega_0\rho}{D_0^2 m^{*3/2}}[n(\omega_0)(E_k+\hbar\omega_0)^{1/2}+\{n(\omega_0)+1\}(E_k-\hbar\omega_0)^{1/2}]^{-1}$
First-order optical phonon	$\dfrac{\pi\hbar^5\omega_1\rho}{2^{1/2}\Xi_1^2 m^{*5/2}E_k}\Big[n(\omega_1)(E_k+\hbar\omega_1)^{1/2}\Big(\dfrac{8}{3}+\dfrac{5}{3}\dfrac{\hbar\omega_1}{E_k}\Big)$ $+\{n(\omega_1)+1\}(E_k-\hbar\omega_1)^{1/2}\Big(\dfrac{8}{3}-\dfrac{5}{3}\dfrac{\hbar\omega_1}{E_k}\Big)\Big]^{-1}$
Polar optical phonon	$\dfrac{2^{5/2}\pi\hbar\epsilon_p E_k^{1/2}}{e^2\omega_0 m^{*1/2}}\Big(n(\omega_0)\Big(1+\dfrac{\hbar\omega_0}{E_k}\Big)^{1/2}+\{n(\omega_0)+1\}\Big(1-\dfrac{\hbar\omega_0}{E_k}\Big)^{1/2}$ $+\dfrac{\hbar\omega_0}{E_k}\Big[\{n(\omega_0)+1\}\sinh^{-1}\Big(\dfrac{E_k}{\hbar\omega_0}-1\Big)^{1/2}-n(\omega_0)\sinh^{-1}\Big(\dfrac{E_k}{\hbar\omega_0}\Big)^{1/2}\Big]\Big)^{-1}$
Piezoelectric (equipartition)	$\dfrac{2^{3/2}\pi\hbar^2\epsilon E_k^{1/2}}{K_{av}^2 e^2 m^{*1/2}k_B T}\Big\{1+\dfrac{1}{1+8m^*E_k/\hbar^2q_0^2}-\dfrac{\hbar^2q_0^2}{4m^*E_k}\log\Big(1+\dfrac{8m^*E_k}{\hbar^2q_0^2}\Big)\Big\}^{-1}$

spherical band:

$$\mu = \frac{\bar{e}}{m^*}\frac{\langle E_k \tau_m \rangle}{\langle E_k \rangle} \tag{3.191}$$

where, for thermal equilibrium non-degenerate statistics and parabolic bands,

$$\frac{\langle E_k \tau_m \rangle}{\langle E_k \rangle} = \frac{\int_0^\infty E_k^{3/2} \exp(-E_k/k_B T)\tau_m \, dE_k}{\int_0^\infty E_k^{3/2} \exp(-E_k/k_B T) \, dE_k}. \tag{3.192}$$

For an ellipsoidal band three tensor components of mobility occur:

$$\mu_i = \frac{\bar{e}}{m_i^*}\frac{\langle E_k \tau_{mi} \rangle}{\langle E_k \rangle} \tag{3.193}$$

with $\langle E_k \tau_{mi} \rangle$ defined as in eqn (3.192). In Table 3.8 we summarize the

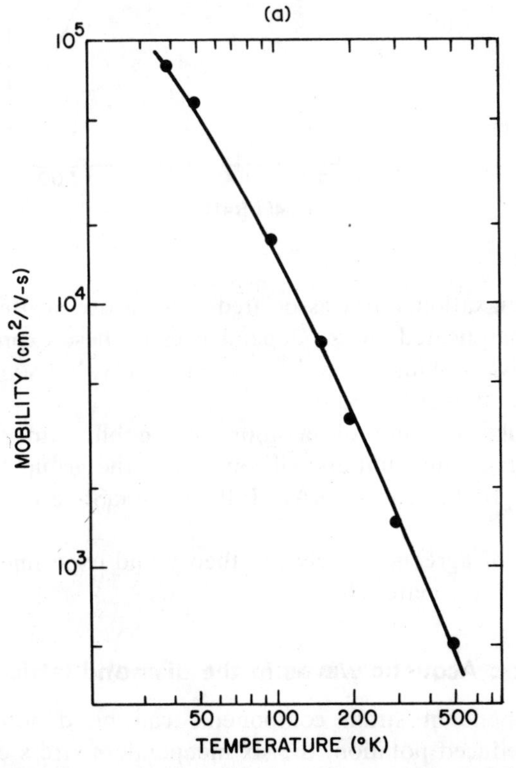

FIG. 3.21. Mobility of electrons in silicon. (a) The solid curve is the experimental result; the points are the theoretical results for mixed acoustic, inter-valley zero-order (630 K), and first-order (190 K) phonon scattering. The experimental curve is from Long (1960). (b) Individual contributions to scattering. (After Ferry 1976.)

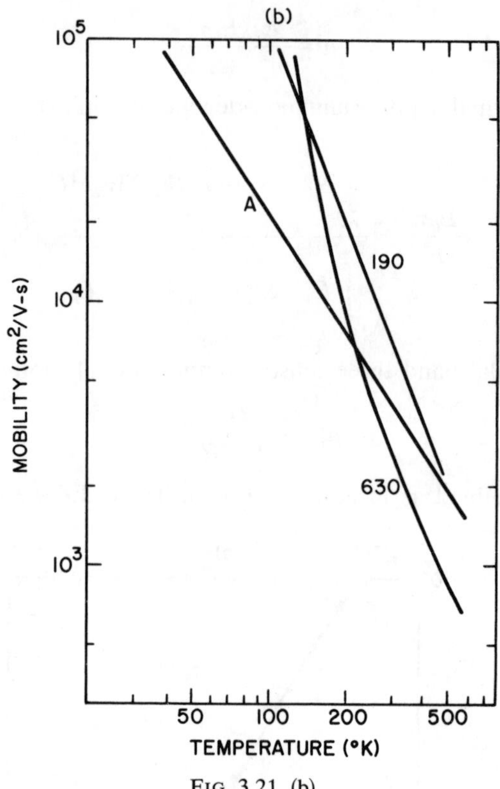

FIG. 3.21. (b)

momentum relaxation times associated with lattice scattering in simple bands. The complicated energy dependences in these expressions lead to some impressive looking integrals for the mobility, but some of them can be reduced to standard functions.

Some calculations and observations of mobility in the Group IV semiconductors germanium and silicon and in the group III–V semiconductors GaP, InSb, GaAs, InAs, InP, and GaSb are shown in Figs. 3.21–3.28.

The degree of agreement between theory and experiment is generally satisfactory in pure materials.

3.9. Appendix: Acoustic waves in the diamond lattice

The six independent strain components can be denoted by S_i ($i = 1, \ldots, 6$) in reduced notation, the six independent stress components by T_j ($j = 1, \ldots, 6$). Cubic symmetry restricts the number of elastic stiffness constants to three, namely c_{11} ($= c_{22} = c_{33}$), c_{12} ($= c_{ij}$, $i \neq j$, $i, j = 1, 2, 3$)

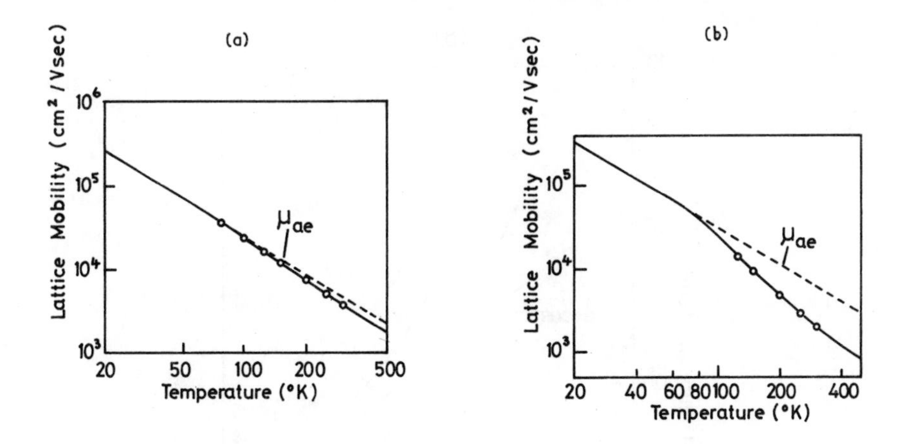

FIG. 3.22. Mobility in germanium: (a) electrons; (b) holes. The points are experimental data from Morin (1954). The lines are theoretical. μ_{ac}, acoustic phonon contribution. (After Conwell 1959.)

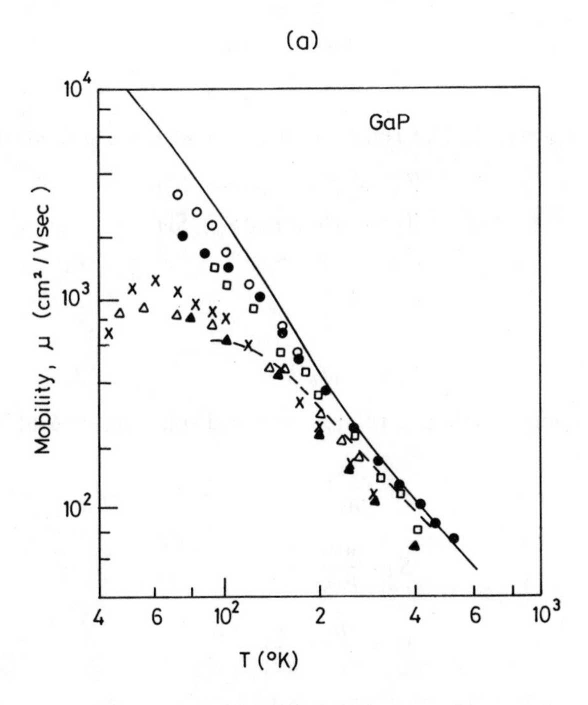

FIG. 3.23. Electron mobility in GaP. (a) Solid curve theory; points experimental results. (b) Individual contributions. (After Rode 1972.)

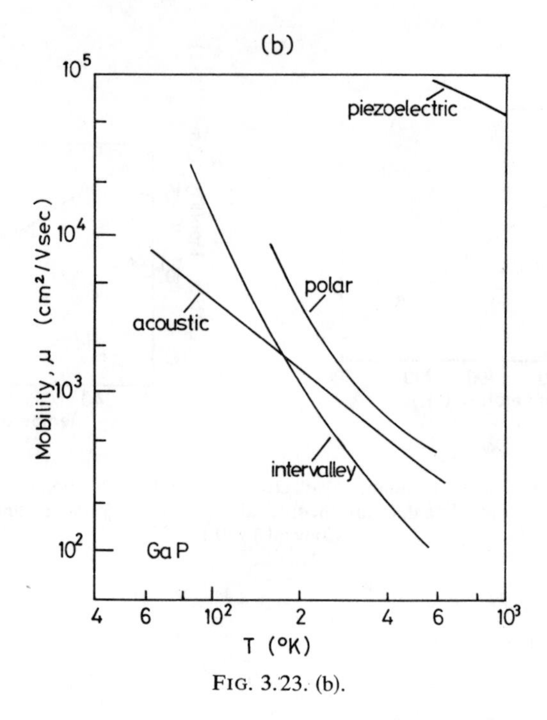

FIG. 3.23. (b).

and c_{44} $(=c_{55}=c_{66})$. The relations between stresses and strains are

$$\begin{aligned} T_1 &= c_{11}S_1+c_{12}(S_2+S_3) \\ T_2 &= c_{11}S_2+c_{12}(S_3+S_1) \\ T_3 &= c_{11}S_3+c_{12}(S_1+S_2) \\ T_4 &= c_{44}S_4 \\ T_5 &= c_{44}S_5 \\ T_6 &= c_{44}S_6. \end{aligned} \tag{3.194}$$

The strain components are related to the displacement $\mathbf{u}$ at $\mathbf{R}$ by

$$\begin{aligned} S_i &= \frac{\partial u_i}{\partial x_i} \qquad i=1,2,3 \\ S_4 &= \frac{\partial u_2}{\partial x_3}+\frac{\partial u_3}{\partial x_2} \\ S_5 &= \frac{\partial u_3}{\partial x_1}+\frac{\partial u_1}{\partial x_3} \\ S_6 &= \frac{\partial u_1}{\partial x_2}+\frac{\partial u_2}{\partial x_1}. \end{aligned} \tag{3.195}$$

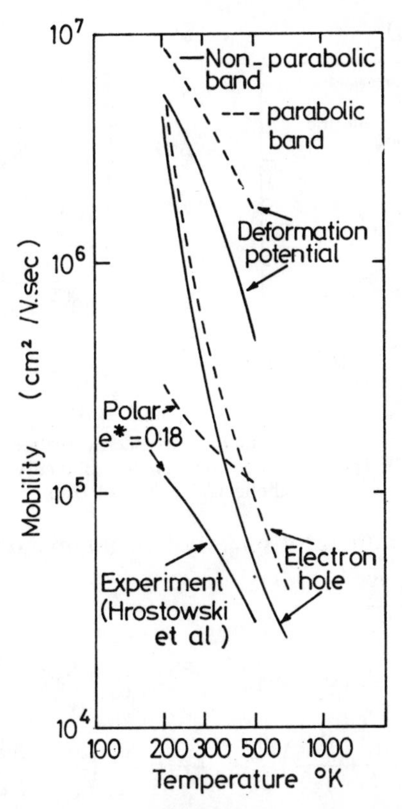

FIG. 3.24. Electron mobility in InSb: solid curve, non-parabolic band; broken curve, parabolic band; dotted curve, experimental data of Hrostowski, Morin, Geballe, and Wheatley (1955). (After Ehrenreich 1957.)

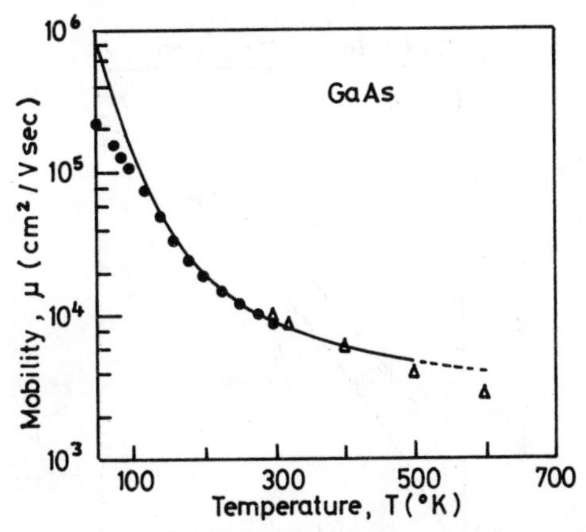

FIG. 3.25. Electron mobility in GaAs: solid curve, theory; points, experimental data. (After Rode 1970*a*.)

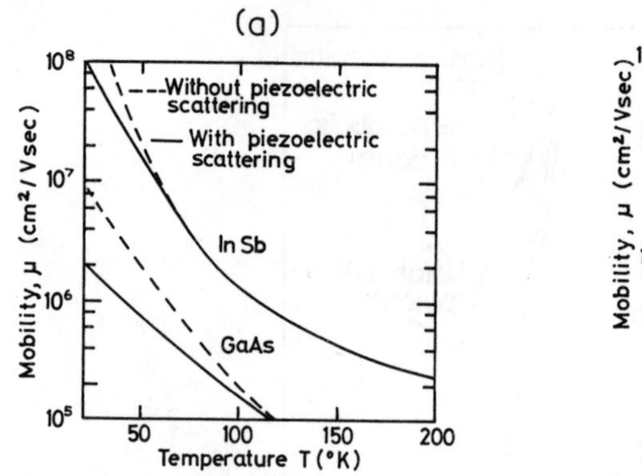

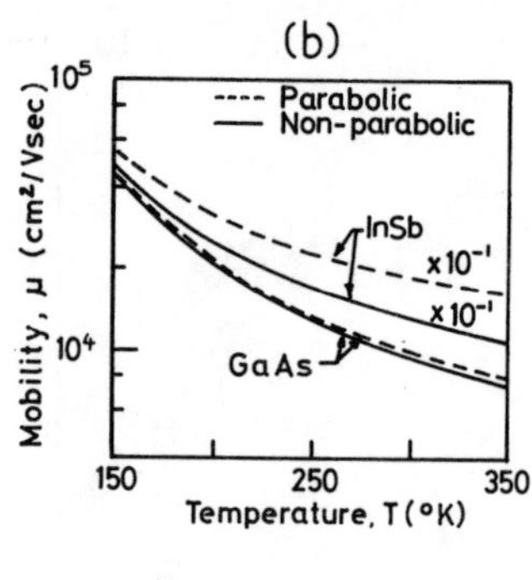

FIG. 3.26. Comparison of effects on electron mobility in InSb and GaAs. (a) The role of piezoelectric scattering: solid curve, without piezoelectric scattering; broken curve, with piezoelectric scattering. (b) The role of non-parabolicity: solid curve, non-parabolic; broken curve, parabolic. (After Rode 1970*a*.)

The equation of motion for one crystallographic axis is

$$\begin{aligned}\rho\frac{\partial^2 u_1}{\partial t^2}&=\frac{\partial T_1}{\partial x_1}+\frac{\partial T_6}{\partial x_2}+\frac{\partial T_5}{\partial x_3}\\&=c_{11}\frac{\partial S_1}{\partial x_1}+c_{12}\frac{\partial}{\partial x_1}(S_2+S_3)+c_{44}\frac{\partial S_6}{\partial x_2}+c_{44}\frac{\partial S_5}{\partial x_3}\\&=c_{44}\left(\frac{\partial^2 u_1}{\partial x_1^2}+\frac{\partial^2 u_1}{\partial x_2^2}+\frac{\partial^2 u_1}{\partial x_3^2}\right)+(c_{12}+c_{44})\frac{\partial}{\partial x_1}\left(\frac{\partial u_1}{\partial x_1}+\frac{\partial u_2}{\partial x_2}+\frac{\partial u_3}{\partial x_3}\right)\\&\quad+(c_{11}-c_{12}-2c_{44})\frac{\partial^2 u_1}{\partial x_1^2}\end{aligned} \tag{3.196}$$

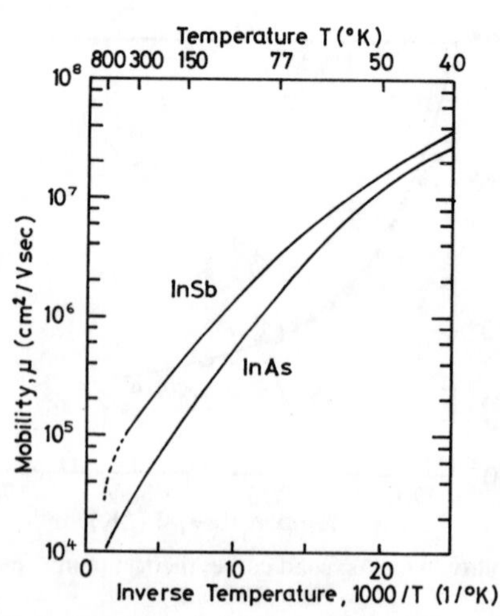

FIG. 3.27. Electron mobility in InSb and GaAs. (After Rode 1970*a*.)

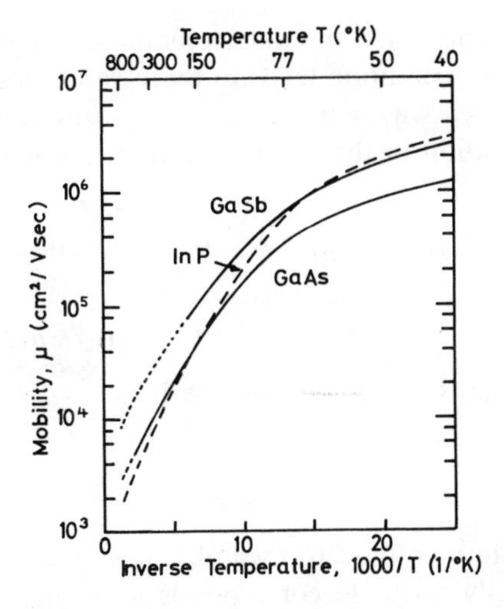

FIG. 3.28. Electron mobility in GaAs, GaSb, and InP. (After Rode 1970*a*.)

where ρ is the density. The general equation of motion is therefore

$$\rho\frac{\partial^2\mathbf{u}}{\partial t^2}=c_{44}\nabla^2\mathbf{u}+(c_{12}+c_{44})\nabla(\nabla.\mathbf{u})+c^*\left(\mathbf{i}_1\frac{\partial^2u_1}{\partial x_1^2}+\mathbf{i}_2\frac{\partial^2u_2}{\partial x_2^2}+\mathbf{i}_3\frac{\partial^2u_3}{\partial x_3^2}\right) \tag{3.197}$$

where $\mathbf{i}_1$, $\mathbf{i}_2$, and $\mathbf{i}_3$ are unit vectors along the axes and $c^*=c_{11}-c_{12}-2c_{44}$ is a measure of the anisotropy of the solid. By substituting the wave solution

$$\mathbf{u}=\mathbf{a}\exp\{\mathrm{i}(\mathbf{q}.\mathbf{R}-\omega t)\} \tag{3.198}$$

we obtain

$$\rho\omega^2\mathbf{a}=c_{44}q^2\mathbf{a}+(c_{12}+c_{44})\mathbf{q}(\mathbf{q}.\mathbf{a})+c^*(\mathbf{i}_1a_1q_1^2+\mathbf{i}_2a_2q_2^2+\mathbf{i}_3a_3q_3^2). \tag{3.199}$$

For acoustic waves the velocity v_s of the wave is given by

$$\omega=v_sq. \tag{3.200}$$

Equation (3.199) can then be written as three simultaneous equations from which we can obtain the values of $\mathbf{a}$ and v_s for any direction of propagation. The equations are, in terms of the direction cosines (α, β, γ) for $\mathbf{q}$,

$$\begin{aligned}a_1\{c_{44}+\alpha^2(c_{12}+c_{44}+c^*)-\rho v_s^2\}+a_2(c_{12}+c_{44})\alpha\beta+a_3(c_{12}+c_{44})\alpha\gamma=0\\a_1(c_{12}+c_{44})\alpha\beta+a_2\{c_{44}+\beta^2(c_{12}+c_{44}+c^*)-\rho v_s^2\}+a_3(c_{12}+c_{44})\beta\gamma=0\\a_1(c_{12}+c_{44})\alpha\gamma+a_2(c_{12}+c_{44})\beta\gamma+a_3\{c_{44}+\gamma^2(c_{12}+c_{44}+c^*)-\rho v_s^2\}=0\end{aligned} \tag{3.201}$$

There are solutions provided the determinant of the coefficients of a_1, a_2, and a_3 vanishes. This condition leads to three values for v_s^2. Each value of v_s^2 corresponds to two waves travelling in opposite directions.

Consider propagation in the $\langle 100\rangle$ direction, i.e. $\alpha=1$, $\beta=\gamma=0$. The equation for v_s^2 is

$$\begin{vmatrix} c_{12}+2c_{44}+c^*-\rho v_s^2 & 0 & 0 \\ 0 & c_{44}-\rho v_s^2 & 0 \\ 0 & 0 & c_{44}-\rho v_s^2 \end{vmatrix} = 0 \tag{3.202}$$

Labelling the solutions v_1^2, v_2^2, and v_3^2 we have

$$\begin{aligned} \rho v_1^2 &= c_{12}+2c_{44}+c^* \\ \rho v_2^2 &= \rho v_3^2 = c_{44}. \end{aligned} \tag{3.203}$$

If we put v_1^2 back into eqn (3.201) we obtain $a_2=a_3=0$ and a_1 undetermined. Consequently this case corresponds to a pure longitudinal wave. Performing the same operation with v_2^2 and v_3^2 we see that these solutions correspond to pure transverse waves.

For propagation along the $\langle 110\rangle$ direction we again obtain pure longitudinal and transverse waves. Labelling the solutions v_1^2, v_2^2, and v_3^2 we have

$$\begin{aligned} \rho v_1^2 &= c_{12}+2c_{44}+\tfrac{1}{2}c^* && \langle 110\rangle \\ \rho v_2^2 &= c_{44} && \langle 001\rangle \\ \rho v_3^2 &= c_{44} && \langle 1\bar{1}0\rangle \end{aligned} \tag{3.204}$$

where the directions are the directions of polarization.

The corresponding values for the $\langle 111\rangle$ direction, in which we again obtain pure longitudinal and transverse waves, are

$$\begin{aligned} \rho v_1^2 &= c_{12}+2c_{44}+\tfrac{1}{3}c^* && \langle 111\rangle \\ \rho v_2^2 &= c_{44}+\tfrac{1}{3}c^* && \langle \bar{1}10\rangle \\ \rho v_3^2 &= c_{44}+\tfrac{1}{3}c^* && \langle 11\bar{2}\rangle. \end{aligned} \tag{3.205}$$

Generally, for other directions of propagation each mode is a mixture of longitudinal and transverse components.

For the purpose of calculating scattering rates the angular dependence of v_{sL} and v_{sT} can be neglected and the following average quantities for the elastic constants can be used to define $\langle v_{sL}\rangle=\langle c_L/\rho\rangle^{1/2}$ and $\langle v_{sT}\rangle=(c_T/\rho)^{1/2}$, namely

$$\begin{aligned} c_L &= c_{12}+2c_{44}+\tfrac{3}{5}c^* \\ c_T &= c_{44}\frac{(c_{44}+\tfrac{1}{3}c^*)}{c_{44}+\tfrac{1}{8}c^*} \approx c_{44}+\tfrac{1}{5}c^* \end{aligned} \tag{3.206}$$

References

Bir, G. L. and Pikus, G. E. (1961). *Sov. Phys.–Solid State,* **2,** 2039.
Birman, J., Lax, M., and Loudon, R. (1966). *Phys. Rev.* **145,** 620.
Budd, H. F. (1964). *Phys. Rev. A* **134,** 1281.
Callen, H. (1949). *Phys. Rev.* **76,** 1394.
Conwell, E. M. (1959). *Sylvania Technol.* **12,** 30.
___ (1964). *Phys. Rev. A* **135,** 1138.
___ (1967). *High field transport in semiconductors.* Academic Press, New York.
___ and Vassell, M. O. (1968). *Phys. Rev.* **166,** 797.
Ehrenreich, H. (1957). *J. Phys. Chem. Solids* **2,** 131.
Fawcett, W., Boardman, D. A., and Swain, S. (1970). *J. Phys. Chem. Solids* **31,** 1963.
Ferry, D. K. (1976). *Phys. Rev. B* **14,** 1605.
Frohlich, H. (1937). *Proc. Roy. Soc.* A **160,** 230.
Gurevitch, Yu. A. (1965). *Sov. Phys.–Solid State* **6,** 1661.
Harrison, W. (1956). *Phys. Rev.* **104,** 1281.
Herring, C. and Vogt, E. (1956). *Phys. Rev.* **101,** 944.
Hilsum, C. (1975). *J. Phys. C* **8,** L578,
Hochberg, A. K. and Westgate, C. R. (1970). *J. Phys. Chem. Solids* **31,** 2317.
Hrostowski, H. J., Morin, F. J., Geballe, T. H., and Wheatley, G. H. (1955). *Phys. Rev.* **100,** 1672.
Hutson, A. R. (1961). *J. appl. Phys.* **32,** 2287.
Keyes, R. W. (1962). *J. appl. Phys.* **30,** 3371.
Lawaetz, P. (1967). Ph.D. Thesis, Technical University of Denmark, Lyngby.
___ (1969). *Phys. Rev.* **183,** 730.
___ (1971). *Phys. Rev. B* **4,** 3460.
Long, D. (1960). *Phys. Rev.* **120,** 2024.
Meyer, H. J. G. (1958). *Phys. Rev.* **112,** 298.
___ and Polder, D. (1953). *Physica* **19,** 255.
Morin, F. J. (1954). *Phys. Rev.* **93,** 62.
Nye, J. F. (1957). *Physical properties of crystals.* Clarendon Press, Oxford.
Paige, E. G. S. (1964). *Prog. Semicond.* **8,** 1.
Ridley, B. K. (1977). *J. appl. Phys.* **48,** 754.
___ (1981). *Solid State Electron* **24,** 147.
Rode, D. L. (1970*a*). *Phys. Rev. B* **2,** 1012.
___ (1970*b*). *Phys. Rev. B* **2,** 4036.
___ (1972). *Phys. Status Solidi B* **53,** 245.
Seitz, F. (1948). *Phys. Rev.* **73,** 549.
Streitwolf, H. W. (1970). *Phys. Status Solidi* **37,** K47.
Wiley, J. D. (1971). *Phys. Rev. B* **4,** 2485.
Zook, J. D. (1964). *Phys. Rev.* A **136,** 869.

4. Impurity scattering

4.1. General features

THE scattering of electrons by the change in potential introduced by an impurity into the lattice differs from lattice scattering in a number of ways. In our simple idealized picture of a semiconductor an electron state and a vibrational state both spread uniformly over the crystal volume. The interaction between them takes place in all unit cells and there is no picture of them being in any degree localized in the theory that we have outlined in the previous chapter. When an impurity is introduced into the lattice its interaction with the electron possesses a more local character, and scattering will occur with appreciable probability only in the vicinity of the impurity site. The same is true of other defects such as vacancies, and we shall use the term impurity to cover all such defects provided that, like an impurity atom, they occupy only one isolated lattice site. This local character means that we must envisage the incoming electron as a wavepacket, at least partially localized in the vicinity of the scattering centre. This aspect of localization means, among other things, that it is possible to speak of a scattering cross-section for the collision as well as a rate. Such a geometric attribute cannot usefully be invented for scattering by phonons.

However, the scattering of a Bloch-wave electron by an impurity is never a truly localized event. The scattering potential of an individual impurity has infinite range and, moreover, there are always many impurity sites present in the crystal. Strictly speaking, we ought to consider the electron as being continually scattered; in other words we must consider the electron to have dynamic properties different from those of the Bloch wave particle—different effective mass for instance. If the impurities were regularly arrayed in a superlattice, we could describe a superband structure with energy gaps at the superzone boundaries and associated negative-mass behaviour—a conduction band structure associated with the impurity band of bound electrons (Chapter 2, Section 2.12). A random array of impurities also possesses a superband structure, albeit with blurred gaps and edges. Whatever the impurity distribution, non-local characteristics of this sort will be present in addition to close-encounter scattering events and a full description requires quantum transport theory.

The analogue in classical physics is the motion of a body in a random array of field sources, such as a star moving and gravitationally interacting with other stars or an electrical particle moving in a plasma. The variation of gravitational or electric fields which the particle experiences is described by the Holtsmark distribution (Holtsmark 1919; Chandresekhar 1943), and in many situations it is possible to describe the motion of the particle satisfactorily in terms of the tail of the distribution which corresponds to high fields. Since this tail is very largely determined by two-body collisions involving relatively short-range interactions we are led to the division of all collisions into two categories: infrequent strong two-body interactions and very frequent weak many-body interactions. In a simple quantum-mechanical picture a time-average of the very frequent weak collisions determines a set of time-independent states, and transitions between these states are induced by infrequent strong two-body collisions. We shall adopt this model for the electron in the conduction band of a semiconductor, and we shall adopt the further simplification that the impurity density is so low that the time-independent states are satisfactorily described by Bloch waves. As we shall see, a rigorous distinction between two-body and three-or-more-body collisions is necessary for dealing with coulombic centres.

The close approach of an electron to an impurity site often implies large perturbations of the electron's motion which renders the usual approach based upon first-order time-dependent perturbation theory rather inaccurate. Consequently it is generally necessary to solve the time-independent Schrödinger equation and find solutions of the form of a Bloch plane wave plus a scattered wave, using the concept of the phase shifts of partial waves (e.g. Schiff 1955). The simpler Born approximation can only be adopted for fast electrons. A criterion based on some characteristic radius r_T associated with the impurity centre is that for the Born approximation to be applicable we must have $k^2r_T^2 \gg 1$. When $k^2r_T^2 \ll 1$ the method of phase shifts produces rather simple results. For coulombic centres, for example, $r_T = a_H^*$ where a_H^* is the effective Bohr radius, whereas for neutral centres r_T is not likely to be much greater than the order of a unit cell dimension. Since impurity scattering is usually strongest at low temperatures where the electrons have low thermal energy, it is not usually practical to make the Born approximation.

It is useful to have a quantitative idea of the magnitude of the electron wavevector. In a parabolic band

$$k = \frac{(2m^*E_k)^{1/2}}{\hbar} = 5{\cdot}123 \times 10^9 \left(\frac{m^*}{m}\right)^{1/2} E_k^{1/2} m^{-1} \tag{4.1}$$

with E_k in eV. Thus for $m^*/m = 0{\cdot}1$ and $E_k = 0{\cdot}01$ eV we obtain $k = 1{\cdot}620 \times 10^8\ \mathrm{m}^{-1}$ which corresponds to a wavelength of 388 Å. For an impurity core of radius 10 Å, $(kr_T)^2 \approx 0{\cdot}026$, and so the phase-shift method is the only alternative for neutral centres and for calculating the effects of the core in a charged centre. However, the coulombic component of scattering by a charged centre can be described by either the Born approximation or the phase-shift method.

The general solution of the Schrödinger equation describing positive energy states in a spherically symmetrical potential is (e.g. Schiff 1955)

$$\psi(r)=\frac{A}{kr}\sum_{l=0}^{\infty}\{F_l(kr)\cos\delta_l+G_l(kr)\sin\delta_l\}P_l(\cos\theta) \tag{4.2}$$

where A is a normalizing constant, l the angular momentum quantum number, $F_l(kr)$ is the regular and $G_l(kr)$ the irregular radial wavefunction, $P_l(\cos\theta)$ is the Legendre spherical harmonic, and δ_l is the phase shift determined by fitting $\psi(r)$ to the internal solution in the core $r\leqslant r_T$ in both value and slope. This form encompasses both neutral and charged centres.

The asymptotic form of eqn (4.2) is, for a non-coulombic potential,

$$\psi(r)\underset{r\to\infty}{\longrightarrow}A\left\{\exp(\mathrm{i}\mathbf{k}.\mathbf{r})+\frac{f(\theta)}{r}\exp(\mathrm{i}kr)\right\} \tag{4.3}$$

whence it follows that the differential cross-section for scattering into the solid angle defined by the polar angle θ is given by

$$\sigma(\theta)=|f(\theta)|^2. \tag{4.4}$$

Since

$$\exp(\mathrm{i}\mathbf{k}.\mathbf{r})=\sum_{l=0}^{\infty}(2l+1)\mathrm{i}^l j_l(kr)P_l(\cos\theta) \tag{4.5}$$

where $j_l(kr)$ is the regular spherical Bessel function, and since $F_l(kr)$ and $G_l(kr)$ can be expressed in terms of regular and irregular spherical Bessel functions, it can be shown that

$$f(\theta)=\frac{1}{k}\sum_{l=0}^{\infty}(2l+1)\exp(i\delta_l)\sin\delta_l P_l(\cos\theta). \tag{4.6}$$

A coulomb potential presents well-known difficulties since its effect is felt even in the asymptotic solution, which never becomes a plane wave. Nor can the incident wave be other than an asymptotic coulomb wave in form. Given an incident coulomb wave one obtains

$$f_c(\theta)=-\frac{\mu(E_k)}{2k\sin^2(\theta/2)}\exp(\mathrm{i}[\mu(E_k)\log\{\sin^2(\theta/2)\}+\pi+2\eta_0]) \tag{4.7}$$

$$\eta_0=\arg\Gamma\{1-\mathrm{i}\mu(E_k)\}\qquad \mu(E_k)=Z(R_H^*/E_k)^{1/2} \tag{4.8}$$

where R_H^* is the effective Rydberg energy, whence

$$\sigma(\theta)=\frac{\mu^2}{4k^2\sin^4(\theta/2)} \tag{4.9}$$

which diverges as $\theta \to 0$. This problem will be discussed in detail in the next section.

When the coulomb potential does not continue inwards to $r = 0$, and this is the case in a semiconductor where core deviations always occur, the scattering wave becomes a mixture of coulomb and core scattering, and one obtains

$$f(\theta) = f_c(\theta) + \frac{1}{k}\sum_{l=0}^{\infty} (2l+1)\exp\{i(\delta_l + 2\eta_l)\}\sin \delta_l P_l(\cos\theta) \quad (4.10)$$

$$\eta_l = \arg \Gamma(l+1-i\mu) \quad (4.11)$$

where $f_c(\theta)$ is the pure coulomb term given in eqn (4.7), and the sum is the contribution from the core which is eqn (4.6) with an extra phase shift $2\eta_l$. In this case the functions $F_l(kr)$ and $G_l(kr)$ in eqn (4.2) are the regular and irregular coulomb wavefunctions (Abramowitz and Stegun 1972). Equation (4.10) with eqn (4.4) is a useful general form for describing two-body impurity scattering by either charged or neutral centres to which we shall have recourse when we consider central-cell modifications of charged-impurity scattering in Section 4.4.

When the constant-energy surfaces of the band containing the scattering particle are not spherical the theory of impurity scattering becomes rather difficult. In what follows we shall generally assume that spherical energy surfaces prevail. Where the mass is anisotropic it is possible to use the fact that small-angle scattering dominates charged-impurity scattering to define diagonalized tensor components of the scattering rate (Ham 1955), and a similar result can be obtained for neutral-impurity scattering, this time by virtue of its isotropy (Brooks 1955), with the effective Bohr radius given by the theory of shallow impurity levels (Kohn 1957).

Inter-valley scattering may conceivably occur by a capture–emission process involving effective-mass impurity states in which an electron is captured from one valley and emitted into another (Weinreich, Sanders, and White 1959). Otherwise, the large change in crystal momentum involved makes direct inter-valley scattering by impurities an extremely weak process. In what follows we shall always assume that impurity scattering is intra-valley scattering.

4.2. Charged-impurity scattering

A purely coulombic potential, i.e. one which is inversely proportional to r, distorts a plane electron wave at all distances, and consequently the scattering cross-section is effectively infinite. There are four ways to overcome this problem and these are associated with the following names and terms: (1) Conwell–Weisskopf, (2) Brooks–Herring, (3) uncertainty broadening and (4) third-body exclusion.

4.2.1. Conwell–Weisskopf approximation

The approach adopted by Conwell and Weisskopf (1950) was based on the classical picture of Rutherford scattering in that the electron was regarded as a reasonably well-localized wave packet whose path could be described by a classical orbit with a well-defined impact parameter b. From classical theory the relation between impact parameter and scattering angle θ is (Fig. 4.1)

$$b = \frac{\mu}{k}\cot\left(\frac{\theta}{2}\right) \tag{4.12}$$

where k is the electron wavevector and μ is defined in eqn (4.8). The Conwell–Weisskopf approximation consists of limiting the impact parameter to half the average separation distance of the impurities:

$$b_{\max} = \tfrac{1}{2}N_{\mathrm{I}}^{-1/3} \tag{4.13}$$

where N_{I} is the impurity concentration. Though this is an arbitrary prescription, it is in the spirit of regarding scattering as essentially a two-body process: in this case the electron and the nearest impurity centre.

As a consequence of eqn (4.12) a minimum angle of scattering is defined:

$$\theta_{\min} = 2\cot^{-1}\left(\frac{k}{2\mu N_{\mathrm{I}}^{1/3}}\right). \tag{4.14}$$

The total cross-section is then

$$\sigma = 2\pi\int_{\theta_{\min}}^{\pi} \sigma(\theta)\sin\theta\,\mathrm{d}\theta. \tag{4.15}$$

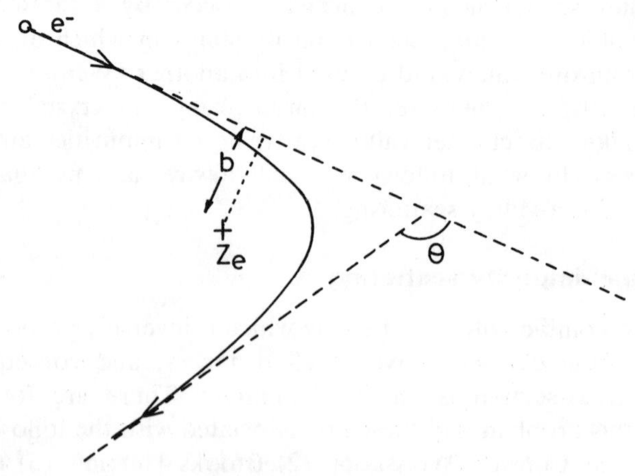

FIG. 4.1. Rutherford scattering.

Substituting $\sigma(\theta)$ from eqn (4.9) we obtain, not surprisingly,

$$\sigma = \pi b_{\max}^2. \tag{4.16}$$

Usually, we are interested in momentum and energy relaxation. Since the mass of the impurity embedded in the lattice greatly exceeds that of the electron, the collisions are very close to being elastic. There is therefore no energy relaxation associated with impurity scattering. A momentum relaxation cross-section can be defined as follows:

$$\begin{aligned}\sigma_m &= 2\pi \int_{\theta_{\min}}^{\pi} \sigma(\theta)(1-\cos\theta)\sin\theta \, d\theta \\ &= -4\pi\left(\frac{\mu}{k}\right)^2 \log\left\{\sin\left(\frac{\theta_{\min}}{2}\right)\right\} \qquad (4.17) \\ &= 2\pi\left(\frac{\mu}{k}\right)^2 \log\left\{1 + b_{\max}^2\left(\frac{k}{\mu}\right)^2\right\} \qquad (4.18)\end{aligned}$$

where $\mu = Z(R_{\mathrm{H}}^*/E_k)^{1/2}$.

This cross-section remains finite for small energies of incidence, and approaches $2\pi b_{\max}^2$ as k approaches zero. In this limit $\sigma_m = 2\sigma$. However, in most applications $b_{\max}^2(k/\mu)^2 \gg 1$, and the cross-section decreases rapidly with increasing energy. Writing σ_m as an explicit function of electron energy, we obtain

$$\sigma_m = \frac{\pi}{2}\left(\frac{Ze^2}{4\pi\epsilon E_k}\right)^2 \log\left\{1 + \left(\frac{4\pi\epsilon E_k}{N_{\mathrm{I}}^{1/3} Ze^2}\right)^2\right\}. \tag{4.19}$$

4.2.2. Brooks–Herring approach

Since electrons in the bands of semiconductors, and even in localized states given sufficient time, may move in response to electric fields, screening of the coulomb field may occur and this has the effect of causing the potential experienced by an electron in the band to drop off more rapidly with r. The simplest form which contains the effect of screening is the potential

$$V(\mathbf{r}) = \frac{Ze}{4\pi\epsilon\,|\mathbf{r}-\mathbf{R}|}\exp(-q_0\,|\mathbf{r}-\mathbf{R}|) \tag{4.20}$$

where q_0 is the reciprocal screening length (see Appendix). (See Debye and Hückel (1923) for electrolytes, Mott (1936) for metals, and Dingle (1955) and Mansfield (1956) for semiconductors.)

The effect of screening on scattering can be calculated simply in the Born approximation. The scattering rate is given by

$$W(\mathbf{k}) = \int \frac{2\pi}{\hbar} |\langle \mathbf{k}'|\, \bar{e}V(\mathbf{r})\, |\mathbf{k}\rangle|^2 \, \delta(E_{k'} - E_k)\, d\mathbf{k}' \frac{V}{8\pi^3} \tag{4.21}$$

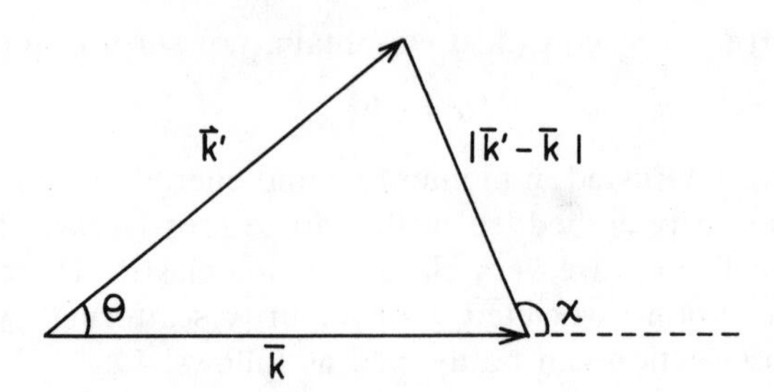

FIG. 4.2. Conservation of crystal momentum.

where **k** and **k**′ are the initial and final state wavevectors and V is the volume of the cavity. The matrix element is, taking the origin of coordinates at the impurity centre,

$$\langle \mathbf{k}' |\bar{e}V(\mathbf{r})| \mathbf{k}\rangle = \frac{1}{V}\int \exp(-i\mathbf{k}'.\mathbf{r})\frac{Ze^2}{4\pi\epsilon r}\exp(-q_0 r)\exp(i\mathbf{k}.\mathbf{r})\,\mathrm{d}\mathbf{r} \qquad (4.22)$$

$$= \frac{Ze^2/\epsilon V)}{|\mathbf{k}'-\mathbf{k}|^2+q_0^2}.$$

Since conservation of energy ensures that $\mathbf{k}'=\mathbf{k}$, we have (Fig. 4.2)

$$|\mathbf{k}'-\mathbf{k}|^2 = 2k^2(1-\cos\theta) = 4k^2\sin^2(\theta/2). \qquad (4.23)$$

The scattering rate is thus

$$W(\mathbf{k}) = \frac{2\pi Z^2 e^4}{\hbar\epsilon^2 V^2}\int\frac{\delta(E_{k'}-E_k)}{\{4k^2\sin^2(\theta/2)+q_0^2\}^2}k'^2\,\mathrm{d}k'\sin\theta\,\mathrm{d}\theta\,\mathrm{d}\phi\frac{V}{8\pi^3}. \qquad (4.24)$$

Replacing $k'^2\,\mathrm{d}k'/8\pi^3$ by $\{N(E_{k'})/4\pi\}\,\mathrm{d}E_{k'}$, where $N(E_{k'})$ is the density of final states per unit energy, we obtain the scattering rate per unit solid angle as follows:

$$W(\theta) = \frac{Z^2e^4N(E_k)}{2\hbar\epsilon^2 V}\frac{1}{\{4k^2\sin^2(\theta/2)+q_0^2\}^2}. \qquad (4.25)$$

The relation between rate and cross-section is simply

$$W(\theta) = \frac{v}{V}\sigma(\theta) \qquad (4.26)$$

where v is the group velocity of the electron. Consequently

$$\sigma(\theta) = \frac{Z^2e^4N(E_k)}{32\hbar\epsilon^2 vk^4}\frac{1}{\{\sin^2(\theta/2)+(q_0/2k)^2\}^2} \qquad (4.27)$$

$$= \frac{\mu^2}{4k^2}\frac{1}{\{\sin^2(\theta/2)+(q_0/2k)^2\}^2}. \qquad (4.28)$$

(We are assuming that the band is parabolic.) The divergence at $\theta = 0$ has been removed by the screening.

The total cross-section is now

$$\sigma = \frac{4\pi(\mu/q_0)^2}{1+(q_0/2k)^2} \tag{4.29}$$

and the momentum relaxation cross-section is

$$\sigma_{\mathrm{m}} = 2\pi\left(\frac{\mu}{k}\right)^2\left[\log\left\{1+\left(\frac{2k}{q_0}\right)^2\right\} - \frac{1}{1+(q_0/2k)^2}\right] \tag{4.30}$$

or, as a function of energy (Brooks 1951),

$$\sigma_m = \frac{\pi}{2}\left(\frac{Ze^2}{4\pi\epsilon E_k}\right)^2\left\{\log\left(1+\frac{8m^*E_k}{\hbar^2 q_0^2}\right) - \frac{1}{1+\hbar^2 q_0^2/8m^*E_k}\right\}. \tag{4.31}$$

Though this formulation solves the problem of infinite cross-sections in a less arbitrary way than the Conwell–Weisskopf approximation, it does so not by going to the heart of the matter but by mixing in a separate and distinct physical process, namely electrical screening. There is no necessity in a real situation for screening to be present or sufficiently strong to limit the cross-section to plausible magnitudes. Once screening is too weak to limit the cross-section to the geometric value determined by the average separation of impurities (which is the Conwell–Weisskopf cross-section), it is not possible to consider the scattering process as a simple two-body collision and the whole approach breaks down.

However, where screening is appreciable the Brooks–Herring result is useful, and there have been several treatments of the screening problem which illuminate the screening process and introduce more sophisticated models (e.g. Takimoto 1959; Hall 1962; Moore 1967; Falicov and Cuevas 1967; Csavinsky 1976). The Thomas–Fermi potential, on which the Brooks–Herring treatment is based, is calculated in a self-consistent way (see Appendix), but its validity in a scattering event relies on the assumption that the approaching electron does not perturb the field by its presence. During the scattering event an electron can repel screening electrons and attract screening holes, and so the amount of screening can vary throughout the process. Such effects can occur only if the screening charges can interact rapidly enough. The characteristic response time for screening is the dielectric relaxation time $\tau_\epsilon = \epsilon\rho$, where ϵ is the permittivity and ρ the resistivity. Modifications to the screened potential will occur only if

$$W^2\tau_\epsilon^2 \ll 1 \tag{4.32}$$

where W is the scattering rate. In such a situation screening becomes a

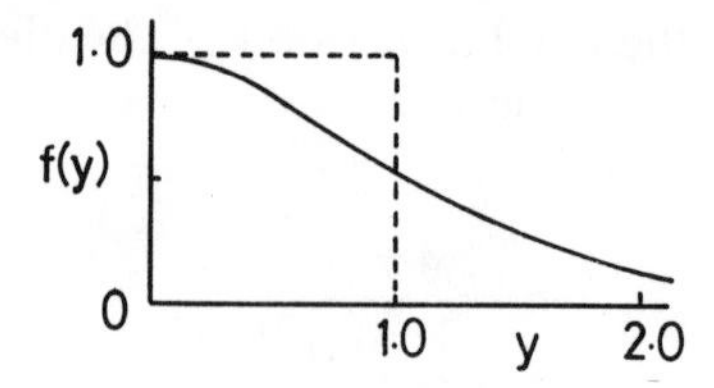

FIG. 4.3. Screening function. The broken line represents a step-like approximation. (After Takimoto 1959.)

complicated many-body process involving electron correlation effects and plasma collective motion. Takimoto (1959) modified the Thomas–Fermi potential by including correlation, and obtained essentially the same result as that which he obtained using the many-body treatment of Nakajima (1954) and that which he obtained using the collective motion approach of Bohm and Pines (1951, 1952, 1953). For non-degenerate semiconductors Takimoto showed that q_0^2 should be replaced by $q_0^2F(y)$, where $F(y) \leqslant 1$ and is given by (Fig. 4.3):

$$F(y) = \frac{1}{y\sqrt{\pi}} \int_0^\infty x \exp(-x^2) \log\left|\frac{x+y}{x-y}\right| \mathrm{d}x \tag{4.33}$$

$$y = \frac{\hbar\,|\mathbf{k}' - \mathbf{k}|}{2(2m^* k_{\mathrm{B}} T)^{1/2}}. \tag{4.34}$$

McIrvine (1960) showed that when $F(y)$ is small (y large) the corresponding potential has spatial oscillations of the form

$$V(\mathbf{r}) = \frac{Ze}{4\pi\epsilon r} \exp(-\alpha r) \cos \alpha r \tag{4.35}$$

where

$$\alpha = \frac{\hbar q_0^2}{2(2m^* k_{\mathrm{B}} T)^{1/2}}. \tag{4.36}$$

Takimoto's result lies near that of the Brooks–Herring approximation at low to moderate carrier concentrations ($n \leqslant 10^{17}$ cm^{-3}) but falls between the predictions of the Brooks–Herring and Conwell–Weisskopf approximations towards higher concentrations (see also Hall 1962).

Another problem connected with scattering by a screened potential is the failure of the Born approximation towards low incident energies. Blatt (1957) performed a numerical phase-shift analysis and showed that the Born approximation overestimated the cross-section at low energies and that differences between negatively and positively charged centres became discernible. We shall pursue the phase-shift approach in Section 4.4.

4.2.3. Uncertainty broadening

In quantum mechanics energy levels are never precisely defined. The transition between two states involving photons, as is well known, takes place over a range of photon energies given by the natural line width Lorentzian

$$F(E) = \frac{1}{\pi} \frac{\hbar\Gamma}{(E - E_0)^2 + \hbar^2\Gamma^2} \tag{4.37}$$

where E is the emitted energy, E_0 is the energy separation of the two states, and Γ is the transition rate. The uncertainty in the energy of the transition is $\hbar\Gamma$ which corresponds to a lifetime of Γ^{-1}. Since

$$\int F(E)\,\mathrm{d}E = \frac{1}{\pi} \tan^{-1}\left(\frac{E - E_0}{\hbar\Gamma}\right) \tag{4.38}$$

it is permissible when Γ is very small to make the replacement

$$F(E) \xrightarrow[\Gamma \to 0]{} \delta(E - E_0) \tag{4.39}$$

and in fact this is usually done in scattering theory. Fujita, Ko, and Chi (1976) have made the useful point that uncertainty broadening removes the divergence in pure coulombic scattering as regards momentum transfer. (That uncertainty broadening also softens the transition between the emission of phonons being forbidden to being allowed was noted by Morgan (1963).) Thus an extension of the theory of charged impurity scattering to include uncertainty broadening removes the necessity for screening to be present, at any rate for momentum transfer.

The scattering rate is the expression of eqn (4.21) with the energy-conserving delta function replaced by $F(E)$, namely

$$W(\mathbf{k}) = \int \frac{2\,|\langle \mathbf{k}'\,|eV(\mathbf{r})|\,\mathbf{k}\rangle|^2\,\Gamma}{(E_{k'} - E_k)^2 + \hbar^2\Gamma^2} \frac{V}{8\pi^3}\,\mathrm{d}\mathbf{k}'. \tag{4.40}$$

For a pure coulombic potential

$$W(\mathbf{k}) = 2\left(\frac{Ze^2}{\epsilon V}\right)^2 \frac{V}{8\pi^3} \int \frac{\Gamma}{\{(E_{k'} - E_k)^2 + \hbar^2\Gamma^2\}\,|\mathbf{k}' - \mathbf{k}|^4}\,\mathrm{d}\mathbf{k}' \tag{4.41}$$

which still diverges for small-angle scattering. The total momentum relaxation rate for N_I impurity centres per unit volume is

$$\frac{1}{\tau_\mathrm{m}} = 2\left(\frac{Ze^2}{\epsilon}\right)^2 \frac{N_\mathrm{I}}{8\pi^3} \int \frac{\Gamma(1 - \cos\theta)}{\{(E_{k'} - E_k)^2 + \hbar^2\Gamma^2\}\,|\mathbf{k}' - \mathbf{k}|^4}\,\mathrm{d}\mathbf{k}'. \tag{4.42}$$

Γ can be identified with $1/\tau_m$, and the term $E_{k'} - E_k$ in the denominator can be neglected since the integrand still peaks sharply at $k' = k$. We obtain, after changing variables to q, χ, and ϕ, where (see Fig. 4.2)

$$q=|\mathbf{k}'-\mathbf{k}| \qquad \cos\theta=\frac{k+q\cos\chi}{(k^2+q^2+2kq\cos\chi)^{1/2}}, \tag{4.43}$$

and allowing q to range from 0 to ∞,

$$\frac{1}{\tau_{\mathrm{m}}^2}=\left(\frac{Ze^2}{\epsilon}\right)^2\frac{N_{\mathrm{I}}}{4\pi^3\hbar^2}\int_0^{\infty}\int_{-1}^{1}\int_0^{2\pi}\frac{1}{q^2}\left\{1-\frac{k+q\cos\chi}{(k^2+q^2+2kq\cos\chi)^{1/2}}\right\}\mathrm{d}q\,\mathrm{d}(\cos\chi)\,\mathrm{d}\phi$$

$$=\left(\frac{Ze^2}{\epsilon}\right)^2\frac{N_{\mathrm{I}}}{\pi^2\hbar^2}\left\{\int_0^{k}\frac{\mathrm{d}q}{3k^2}+\int_k^{\infty}\frac{1}{q^2}\left(1-\frac{2}{3}\frac{k}{q}\right)\mathrm{d}q\right\}$$

$$=\left(\frac{Ze^2}{\epsilon}\right)^2\frac{N_{\mathrm{I}}}{\pi^2\hbar^2k}. \tag{4.44}$$

Thus

$$\frac{1}{\tau_{\mathrm{m}}}=\frac{Ze^2N_{\mathrm{I}}^{1/2}}{\pi\epsilon\hbar^{1/2}(2m^*E_{\mathrm{k}})^{1/4}}. \tag{4.45}$$

This expression was first given by Fujita *et al.* (1976). For comparison, the momentum relaxation rate in the Conwell–Weisskopf approximation is $v\sigma_{\mathrm{m}}N_{\mathrm{I}}$, i.e.

$$\frac{1}{\tau_{\mathrm{m}}}=\left(\frac{Ze^2}{4\pi\epsilon}\right)^2\frac{\pi N_{\mathrm{I}}}{(2m^*E_{\mathrm{k}}^3)^{1/2}}\log\left\{1+\left(\frac{4\pi\epsilon E_{\mathrm{k}}}{N_{\mathrm{I}}^{1/3}Ze^2}\right)^2\right\}. \tag{4.46}$$

The rate with uncertainty broadening (eqn (4.45)) exceeds that predicted by the Conwell–Weisskopf formula and differs significantly in its dependence on energy and impurity concentration.

The fact that the uncertainty broadening rate exceeds the Conwell–Weisskopf rate means that the effective cross-section is so large that, as in the case of the Brooks–Herring approach with weak screening, the simple two-body collision process envisaged by the theory breaks down. Moreover, the inclusion of uncertainty broadening removes the divergence only for momentum relaxation—the scattering rate itself is still infinite. Thus, this approach, although self-consistent, does not solve the basic problem associated with the coulomb potential.

4.2.4. *Statistical screening*

A reconciliation of the Conwell–Weisskopf and Brooks–Herring approaches can be effected by the simple device of weighting the scattering event with the probability of its being a truly two-body nearest-scatterer process (Ridley 1977). If a consistent one-centre scattering approximation is to be adopted, as has been the case in all treatments of this problem in semiconductors, then it is logically necessary to exclude the possibility of there being a second scattering centre closer to the particle being scattered. If only the closest impurity scatters and the interaction with all others averages to zero, such a procedure is called for and entailed by the demands of consistency.

The weighting factor we require can be obtained by using the idea of a classical trajectory for which the impact parameter is b (Fig. 4.1). If p denotes the probability of there being no second scattering centre for which the impact parameter would lie between b and $b+db$, then

$$p = 1 - 2\pi N_I ab \, db \tag{4.47}$$

where N_I is the density of impurities and a is their average separation distance, and this follows because $2\pi N_I ab \, db$ is the probability that such a centre exists. If $P(b)$ is the probability that no scattering centre exists with an impact parameter less than b, and if $P(b+db)$ is the probability that no scattering centre exists with an impact parameter less than $b+db$, it follows that

$$P(b+db) = P(b)p. \tag{4.48}$$

Consequently

$$P(b) = C \exp(-\pi N_I ab^2) \tag{4.49}$$

and since $P(0)=1$ it follows that $C=1$, and therefore

$$P(b) = \exp(-\pi N_I ab^2). \tag{4.50}$$

The impact parameter is related to the differential cross-section $\sigma(\theta)$ as follows:

$$\sigma(\theta) \, d\Omega = -2\pi b \, db \tag{4.51}$$

where θ is the scattering angle, $d\Omega$ is the element of solid angle, and $b=0$ when $\theta=\pi$. Thus the probability $P(b)$ of there being no third body can be related to the two-body cross-section. The latter must then be weighted by $P(b)$ in order to include in the overall scattering process only two-body processes. $P(b)$ can be referred to as the third-body exclusion factor or, in an obvious analogy, the third-body screening factor. This exclusion factor has to be applied to all two-body scattering processes in which the bodies are localized. It is applicable to all impurity scattering in principle, by neutral as well as by charged centres.

Its application to charged-impurity scattering removes the divergence inherent in coulomb scattering simply by building into the theory the decreasing probability of there being two-body scattering events with scattering angles tending to zero. From the Brooks–Herring expression for the differential cross-section (eqn (4.28)) and from eqn (4.51) we have

$$b^2 = \frac{(\mu/k)^2 \cos^2(\theta/2)}{\{\sin^2(\theta/2) + (q_0/2k)^2\}\{1 + (q_0/2k)^2\}} \tag{4.52}$$

whence the weighted differential cross-section becomes

$$\sigma(\theta) = \frac{(\mu/2k)^2}{\{\sin^2(\theta/2) + (q_0/2k)^2\}^2} \exp\left\{-\frac{\pi N_I a(\mu/k)^2 \cos^2(\theta/2)}{\{\sin^2(\theta/2) + (q_0/2k)^2\}\{1 + (q_0/2k)^2\}}\right\}. \tag{4.53}$$

Even when screening is weak this cross-section no longer diverges at small angles. Thus the total cross-section is

$$\sigma = 2\pi \int_0^\pi \sigma(\theta) \sin\theta \, d\theta = \frac{1}{N_I a}\left[1 - \exp\left\{-\frac{4\pi N_I a(\mu/q_0)^2}{1 + (q_0/2k)^2}\right\}\right] \tag{4.54}$$

which tends to $(N_I a)^{-1}$ in the limit of weak screening ($q_0 \to 0$). This is exactly the Conwell–Weisskopf result provided that we take the average separation to be given by

$$a = 4N_I^{-1/3}/\pi. \tag{4.55}$$

(According to Chandresekhar (1943) the average separation should be taken to be $(2\pi N_I)^{-1/3}$; see Appendix 4.10.) The Brooks–Herring result (eqn (4.29)) is obtained in the limit $N_I \to 0$ (Fig. 4.4).

The momentum relaxation cross-section is given by

$$\sigma_m = 2\pi\left(\frac{\mu}{k}\right)^2 L(E_k) \tag{4.56}$$

where

$$L(E_k) = e^z \left(E_1(z) - E_1(y)\right) - \frac{1}{y}\left[1 - \exp\left\{-\frac{y}{1 + (q_0/2k)^2}\right\}\right] \tag{4.57}$$

$$z = \frac{\pi N_I a(\mu/k)^2}{1 + (q_0/2k)^2} \qquad y = 4\pi N_I a\left(\frac{\mu}{q_0}\right)^2.$$

Here $E_1(x)$ is the exponential integral:

$$E_1(x) = \int_x^\infty \frac{e^{-t}}{t} \, dt. \tag{4.58}$$

In the limit $N_I \to 0$, $L(E_k)$ goes to the Brooks–Herring value:

$$L(E_k) \to \log\left\{1 + \left(\frac{2k}{q_0}\right)^2\right\} - \frac{1}{1 + (q_0/2k)^2}. \tag{4.59}$$

In the limit of weak screening

$$L(E_k) \to \exp\left\{\pi N_I a\left(\frac{\mu}{k}\right)^2\right\} E_1\left\{\pi N_I a\left(\frac{\mu}{k}\right)^2\right\} \tag{4.60}$$

which replaces the logarithm in the Conwell–Weisskopf expression (eqn (4.18)) and gives the same cross-section in the limit of small electron energy.

This theory of statistical screening (so-called) successfully reconciles the two approaches and provides a criterion for deciding when screening is decisive and when it is not. If we define a dimensionless quantity η such that

$$\eta = 16N_I^{2/3}\left(\frac{\mu}{q_0}\right)^2 = \left(\frac{4ZN_I^{1/3}}{q_0}\right)^2 \frac{R_H^*}{E_k} \tag{4.61}$$

it is evident that the following criteria apply for the use of the two formulae (Fig. 4.4):

$$\begin{array}{ll} \eta \ll 1 & \text{Brooks–Herring} \\ \eta \gg 1 & \text{Conwell–Weisskopf.} \end{array} \tag{4.62}$$

Although statistical screening is useful in making an analytic bridge between the two most useful formulae, it is only another step forward in providing a clear conceptual picture of the physical reality of charged-impurity scattering which is essentially a many-body phenomenon. Quantum transport theories which explicitly treat such many-body phenomena as coherent scattering from pairs of centres and dressing effects of impurities have tended to be too opaque to have found general application in describing semiconductor phenomena, and they have predicted corrections whose magnitudes are not much larger than those introduced

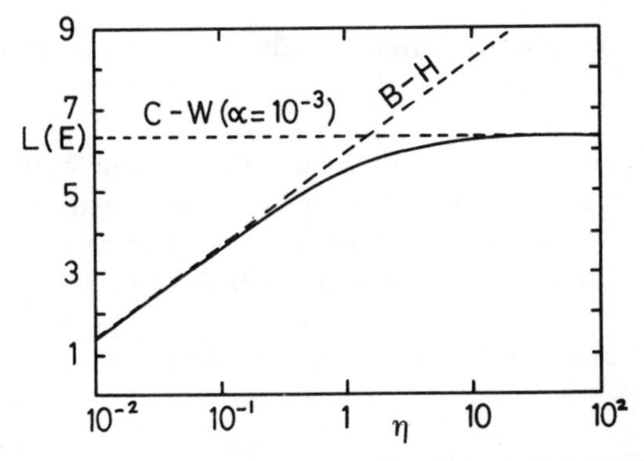

FIG. 4.4. Cross-over from Brooks–Herring (B–H) to Conwell–Weisskopf (C–W) scattering with decreased screening based on third-body exclusion. $L(E)$ is given by eqn (4.57) and η by eqn (4.61). The parameter defining C–W scattering is $\pi N_I a(\mu/k)^2$. (After Ridley 1977.)

by uncertainties in the assessments of impurity concentration and the strength of accompanying scattering mechanisms and have been consequently regarded with reservation. However, even when such theories convincingly and clearly describe physically distinguishable elements beyond the screened two-body interaction, the Conwell–Weisskopf and Brooks–Herring formulae and their reconciliation through third-body exclusion continue to provide a useful quantitative conceptual framework for the experimentalist.

4.3. Neutral-impurity scattering

4.3.1. Hydrogenic models

Non-ionized shallow donors and acceptors at low temperatures provide an important class of neutral impurities. It is not surprising that scattering by these centres was the first to be treated theoretically, especially as these impurities resemble neutral hydrogen in their electronic structure. Pearson and Bardeen (1949) were the first to point out that the cross-sections of neutral hydogenic impurities and of ionized centres are comparable for slow electrons. The theory of slow collisions with neutral hydrogen was developed by Massey and Moisewitch (1950) and their results were applied to the analogous situation in semiconductors by Erginsoy (1950).

Since slow electrons are involved the basic approach has been via the phase-shift method. For very slow electrons such that $k^2 a_H^{*2} \ll 1$, where a_H^* is the effective Bohr radius, only the zero-order phase shift ($l=0$) is appreciable and the scattering is spherically symmetrical. From eqns (4.4) and (4.6) we have

$$\sigma(\theta) = \sin^2\delta_0/k^2 \tag{4.63}$$

and since there is spherical symmetry the total cross-section is

$$\sigma = 4\pi \sin^2\delta_0/k^2 \tag{4.64}$$

where δ_0 is the $l=0$ phase shift. The calculation of δ_0 is quite difficult since it must include the effect of the static field of hydrogen and also the dynamic effects of electron exchange and the polarization of the atom by the incident electron. All these components were included in the calculation of Massey and Moisewitch, and Erginsoy pointed out that for energies such that $ka_H^* \leqslant 0{\cdot}5$ their results could be expressed in the simple analytic form

$$\sigma = 20a_H^*/k. \tag{4.65}$$

Even at the upper end of the range of applicable energies ($ka_H^* = 0{\cdot}5$) this cross-section is about ten times the cross-sectional area of the Bohr orbit, i.e. typically of the order $10^{-12}\,\text{cm}^2$. Note that because of the spherical nature of the scattering this cross-section is also that for describing momentum transfer.

Erginsoy's formula has always had rivals. Mattis and Sinha (1970), applied the results of an elaborate calculation for hydrogen and found a cross-section that was a factor of 2 smaller than Erginsoy's. However,

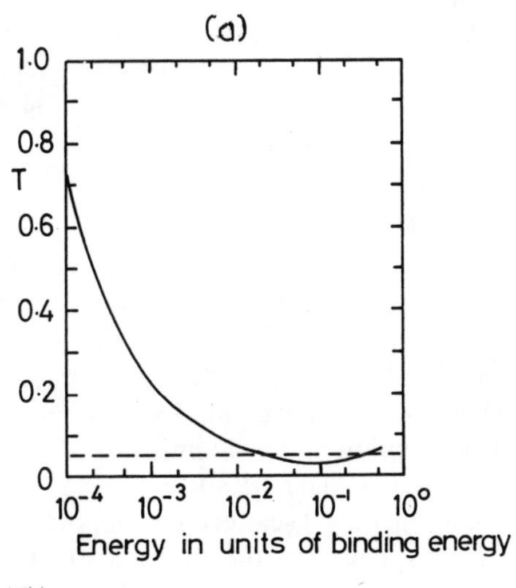

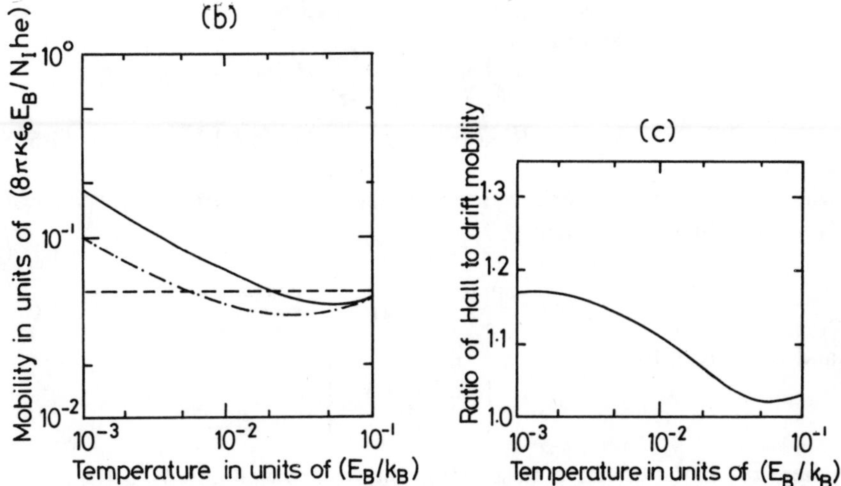

FIG. 4.5. (a) Relaxation time versus energy for neutral-impurity scattering based on hydrogenic models: solid curve, McGill and Baron 1975; broken line, Erginsoy 1950. Time is in units of $8\pi\epsilon R_H^* m^*/N_I\hbar e^2$. (b) Mobility versus temperature: solid curve, McGill and Baron 1975; broken line, Erginsoy 1950; chain line, Sclar 1956. (c) Ratio of Hall to drift mobility versus temperature. (After McGill and Baron 1975.)

quite different dependences on energy were predicted by Anselm (1953) and Sclar (1956) who based their approaches on scattering by a spherically symmetrical square-well potential. Sclar's approach illuminated the role of the weakly bound negative-ion state, and we shall discuss this in Section 4.3.2. More recently, McGill and Baron (1975), basing their approach on the results of the polarized-orbital calculation of Temkin and Lamkin (1961), obtained numerical results more akin to those predicted by Sclar than by Erginsoy (Fig. 4.5).

Evidently the problem of scattering by hydrogenic impurities is still an open one, though a measure of agreement as to magnitude does exist within a factor of 2 or so between the various models for energies between 0·01 and 1·00 times the ionization energy.

4.3.2. *Square-well models*

Scattering from spherically symmetrical square-well potentials can be treated analytically, and the results ought to be applicable to deep-level impurities as well as to core scattering by charged impurities. Their application to scattering by neutral hydrogenic impurities has already been mentioned. Thus the square-well model for neutral impurity scattering offers a useful description which illuminates a great many practical situations (El-Ghanem and Ridley, 1980).

We shall assume that only $l=0$ scattering is important and compute the phase shift of the $l=0$ wave from the internal $(0 \leq r \leq r_T)$ and external $(r_T \leq r)$ wavefunctions $\psi_{int}(r)$ and $\psi_{ext.}(r)$. For a square-well these wavefunctions are as follows:

$$\begin{aligned} \psi_{int.}(r) &= Aj_0(\alpha r) \\ \psi_{ext}(r) &= B\{j_0(kr)\cos\delta_0 - n_0(kr)\sin\delta_0\} \end{aligned} \tag{4.66}$$

where $j_0(\rho)$ is the spherical Bessel function and $n_0(\rho)$ is the spherical Neumann function, namely

$$j_0(\rho) = \frac{\sin\rho}{\rho} \qquad n_0(\rho) = -\frac{\cos\rho}{\rho} \tag{4.67}$$

and α is given by

$$\alpha = \left\{\frac{2m_T^*(E+V_0)}{\hbar^2}\right\}^{1/2} \tag{4.68}$$

where m_T^* is the effective mass within the square well and V_0 is the depth of the well. Fitting the slope and the value at $r=r_T$ leads to

$$\sin\delta_0 = \frac{k\cot(kr_T) - \alpha\cot(\alpha r_T)}{\{k^2+\alpha^2\cot^2(\alpha r_T)\}^{1/2}\{1+\cot^2(kr_T)\}^{1/2}}. \tag{4.69}$$

In the limit of small energies such that $\cot kr_T \approx (kr_T)^{-1}$ we obtain

$$\sin \delta_0 \approx \frac{k\{1-\alpha r_T \cot(\alpha r_T)\}}{\{k^2+\alpha^2 \cot^2(\alpha r_T)\}^{1/2}}. \tag{4.70}$$

All the characteristics of the potential are contained in the terms α and r_T. Phase shifts for partial waves with angular momentum $(l \neq 0)$ are much smaller when $k^2 r_T^2 \ll 1$ except in those cases where resonances exist. We shall assume in what follows that resonances associated with $l > 0$ do not exist and continue to explore only scattering with zero angular momentum.

Three principal cases can be identified. The first is defined by the condition

$$|\alpha r_T \cot(\alpha r_T)| \gg 1 \tag{4.71}$$

whence (since $k^2 r_T^2 \ll 1$)

$$\sin \delta_0 \approx \pm k r_T \tag{4.72}$$

and

$$\sigma \approx 4\pi r_T^2. \tag{4.73}$$

This case corresponds to scattering by a sphere, and is often referred to as hard-sphere scattering. Since r_T for neutral centres will be of order of the unit cell dimension, i.e. two or three ångströms, the cross-section will not much exceed 10^{-14} cm^2, which is small compared with other electron scattering cross-sections. Consequently, there exists a class of neutral impurities, for which the condition of eqn (4.71) is satisfied, which do not scatter electrons appreciably.

The second case is the special one defined by

$$\alpha r_T \cot(\alpha r_T) = 1 \tag{4.74}$$

which gives a zero cross-section. This corresponds to the Ramsauer–Townsend effect for slow electrons and defines a second class of neutral impurities which do not scatter electrons appreciably.

The third case is one in which

$$|\alpha r_T \cot(\alpha r_T)| \lesssim k r_T (\ll 1) \tag{4.75}$$

whence

$$\sin \delta_0 \approx \frac{k}{(k^2+\alpha^2 \cot^2(\alpha r_T))^{1/2}} \qquad \sigma \approx \frac{4\pi}{k^2+\alpha^2 \cot^2(\alpha r_T)}. \tag{4.76}$$

The cross-section can be particularly large in this circumstance. This case is referred to as resonant scattering because it depends upon $\alpha r_T \cot(\alpha r_T) \approx 0$ which is the condition for the appearance of a bound state. The neutral impurity potential is such that it can actually bind an electron to form a netatively charged ion, or it can nearly do so. The first alternative forms the basis of the Sclar formula, and the second forms the basis of resonant scattering analogous to Breit–Wigner scattering in nuclear physics.

4.3.3. *Sclar's formula*

Let us suppose that the square-well potential of the neutral impurity can bind an electron into a state E_T below the conduction band. This means that (Schiff 1955)

$$\alpha_T \cot(\alpha_T r_T) = -\beta_T \tag{4.77}$$

where

$$\alpha_T = \left\{\frac{2m_T^*(V_0 - E_T)}{\hbar^2}\right\}^{1/2} \qquad \beta = \left(\frac{2m_T^* E_T}{\hbar^2}\right)^{1/2}. \tag{4.78}$$

If E_T is assumed to be very small compared with the depth of the well, we can replace α in eqn (4.76) by α_T without much error. Moreover, if E_T is small, it is plausible to adopt an effective-mass approximation and replace m_T^* by the conduction band mass. We then arrive at Sclar's result:

$$\sin \delta_0 = \frac{E_k^{1/2}}{(E_k + E_T)^{1/2}} \tag{4.79}$$

$$\sigma = \frac{4\pi(\hbar^2/2m^*)}{E_k + E_T}. \tag{4.80}$$

Sclar applied this to the hydrogenic centre and took E_T as the observed binding energy for a second electron on hydrogen reduced by a factor $(m^*/m)(\epsilon/\epsilon_0)^{-2}$ (the factor scaling the Rydberg energy):

$$E_T = 0{\cdot}75\,\frac{m^*}{m}\left(\frac{\epsilon_0}{\epsilon}\right)^2 \text{eV} \tag{4.81}$$

which is approximately 5×10^{-4} eV in germanium and 2×10^{-3} eV in silicon. This formula gives a cross-section similar in magnitude to that of Erginsoy (1950) but different in energy dependence, the latter being more akin to the dependence found by McGill and Baron (1975) (Fig. 4.5).

If we apply Sclar's formula (eqn (4.80)) to any neutral impurity which can capture an electron, it is clear that small scattering cross-sections will be obtained unless E_T is small. Thus neutral impurities which can bind an electron into a deep level will not act as strong scattering centres.

4.3.4. *Resonance scattering*

Returning to eqn (4.76), we see that if the potential is such that at an energy $E_k = E_r$

$$\alpha_r r_T \cot(\alpha_r r_T) = 0 \tag{4.82}$$

we obtain a resonant form of scattering. This will occur when the impurity potential is just not strong enough to bind an electron. Let us put $\Delta E = E_k - E_r$ and expand $\alpha r_T \cot(\alpha r_T)$ about the resonance:

$$\alpha r_T \cot(\alpha r_T) \approx \alpha_r r_T \cot(\alpha_r r_T) + \frac{\Delta E}{2V_0}\{\alpha_r r_T \cot(\alpha_r r_T) - (\alpha_r r_T)^2 \operatorname{cosec}^2(\alpha_r r_T)\}. \tag{4.83}$$

Taking $\alpha_r r_T = \pi/2$ at resonance, we obtain

$$\alpha r_T \cot(\alpha r_T) \approx -\frac{\pi^2 \Delta E}{8 V_0} \tag{4.84}$$

whence

$$\sin \delta_0 = \frac{\Gamma}{(\Delta E^2 + \Gamma^2)^{1/2}} \tag{4.85}$$

$$\sigma = \frac{4\pi}{k^2} \frac{\Gamma^2}{\Delta E^2 + \Gamma^2} \tag{4.86}$$

where

$$\Gamma = \frac{8V_0}{\pi^2} k r_T = (E_0 E_k)^{1/2}. \tag{4.87}$$

This has the form of a resonance with a width Γ which is energy dependent. It is proportional to the density of states per unit energy in a parabolic band, and if $\Gamma^2 = E_0 E_k$, E_0 can be regarded as a characteristic energy for the scattering process.

Resonance scattering will be readily detectable only if the width is not so small that it spans very little of the thermal distribution of electrons in the band, for then only a few electrons will have the right energy (Fig. 4.6). Even so, its effect is likely to be evident in transport properties, mainly in the Hall factor because of its comparatively rapid energy dependence.

In non-degenerate material the resonance energy must lie no more than a few tens of millielectronvolts above the band edge for it to influence thermal electrons. The resonant level in a multi-valley conduction band structure may lie lower than one or more sets of minima. For example, a resonant level 10 meV above the bottom of the conduction band in GaAs would lie well below both the L and X sets of minima. In such a case the impurity potential relative to the upper valleys may be deep enough to produce metastable bound states associated with these upper minima at energies which overlap the lower Γ band. Such states will act as resonant scattering states. (The simplest examples of these are effective-mass states characteristic of particular minima, the higher lying of which are degenerate with band states though such states are not usually associated with neutral centres). Resonant scattering states can thus arise from the many-valley band structure of semiconductors as well

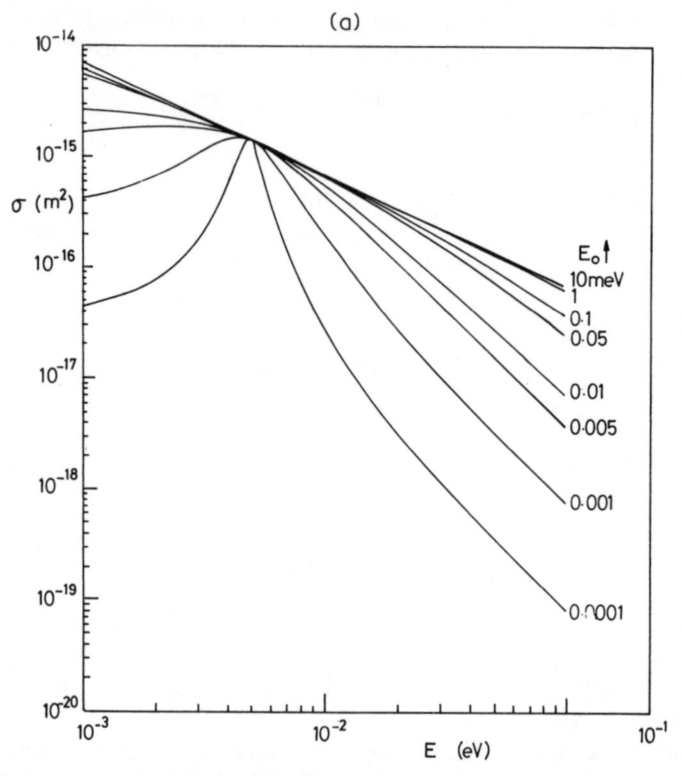

FIG. 4.6. Resonance scattering of electrons in n-type GaAs: (a) cross-section σ_c for Breit–Wigner resonance versus energy and its dependence on the width E_0 of the resonant level ($E_r = 5$ meV); (b) mobility times N_r versus temperature and its dependence on E_0 ($E_r = 5$ meV); (c) Hall scattering factor r_H versus temperature T and its dependence on E_0 ($E_r = 5$ meV) (After El-Ghanem and Ridley 1980.)

as from properties of the impurity potential, and of course they may be associated with partial waves with l greater than zero. However, whatever their origin, they must lie close to the band edge in non-degenerate material to be effective scatterers. Although high-lying resonant states may influence hot electrons, the effect is not likely to be great because the typically large breadth of the electron distribution in the band will provide few carriers at the resonant energy.

In that Sclar's solution can be regarded as an approximate zero-energy resonance, we can conclude that neutral-impurity scattering is important in non-degenerate material only for resonance or quasi-resonance scattering, and then only if the impurity potential provides a resonant level E_r which satisfies the rough criterion

$$-R_H^* \lesssim E_r \lesssim 50\,\text{meV}, \tag{4.88}$$

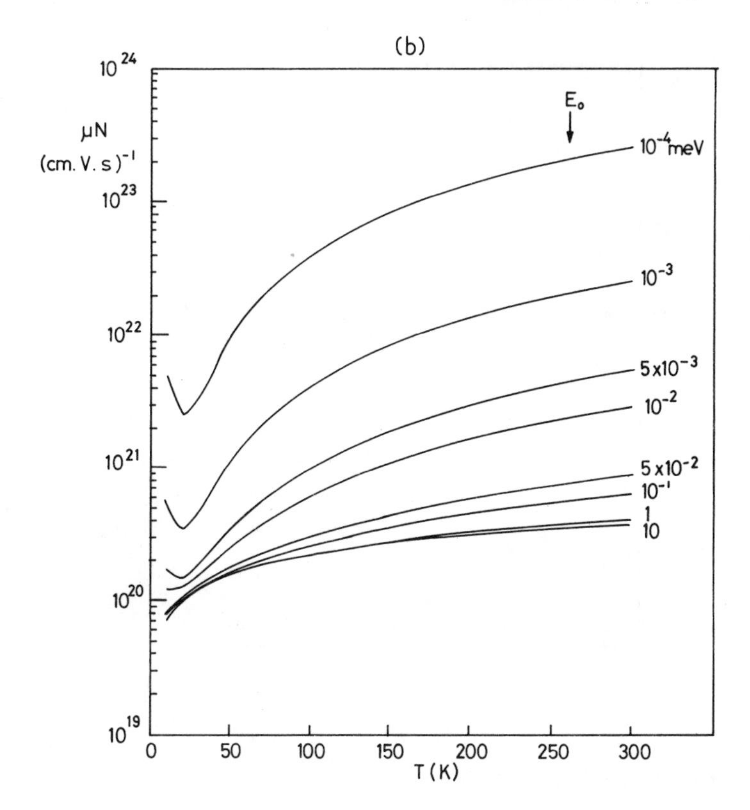

the lower limit being an order of magnitude deeper than the negative hydrogen ion centre (binding energy about $0.06R_H^*$) and the upper limit being roughly the average thermal energy of electrons at room temperature.

4.3.5. *Statistical screening*

Viewing neutral-impurity scattering as a strictly two-body process means multiplying by the statistical screening factor $P(b)$ of eqn (4.50). Since $\sigma(\theta)$ for neutral-impurity scattering is independent of the angle, eqn (4.51) yields

$$b^2 = 2\sigma(\theta)(1+\cos\theta) \tag{4.89}$$

whence

$$P(b) = \exp\{-2\pi N_I a\sigma(\theta)(1+\cos\theta)\}. \tag{4.90}$$

The two-body differential cross-section is now angle dependent with the consequence that the total cross-section now differs from the momentum

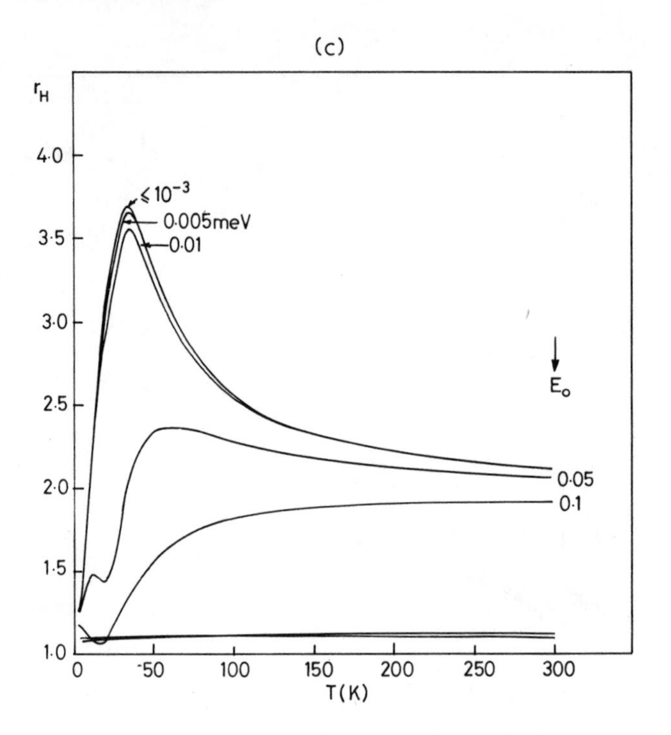

transfer cross-section. Thus

$$\sigma = 4\pi\sigma(\theta)\frac{1-e^{-2y}}{2y} \tag{4.91}$$

where

$$y = 2\pi N_I a\sigma(\theta) \tag{4.92}$$

and for the momentum relaxation cross-section

$$\sigma_m = 4\pi\sigma(\theta)\frac{2y+e^{-2y}-1}{2y^2}. \tag{4.93}$$

For low concentrations and small intrinsic differential cross-sections we have

$$\sigma \approx \sigma_m \approx 4\pi\sigma(\theta) \qquad y \ll 1, \tag{4.94}$$

whereas for high concentrations and large intrinsic differential cross-sections we have

$$\sigma \approx 1/N_I a \qquad \sigma_m = 2/N_I a \tag{4.95}$$

and $a \approx N_I^{-1/3}$. The cross-sections of eqn (4.95) are purely geometric limits independent of the scattering process.

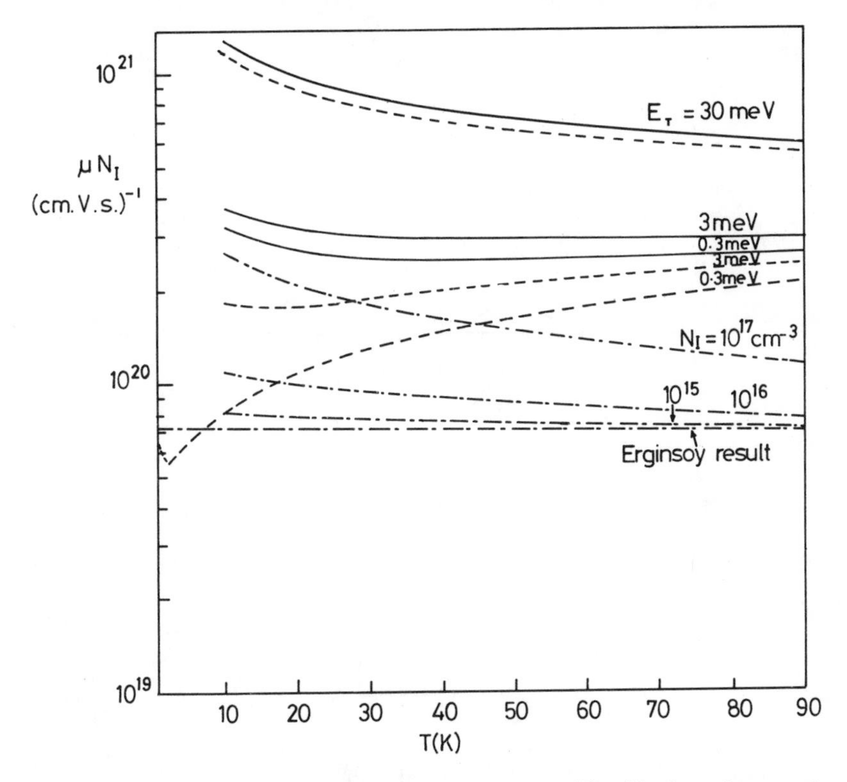

FIG. 4.7. Neutral impurity scattering: μN_I versus temperature T and its dependence on the neutral impurity concentration N_I and the level depth E_T. Broken curve, Sclar theory without third-body screening; solid curve, Sclar theory with third-body screening ($N_I = 10^{17}\ \mathrm{cm}^{-3}$); chain line, Erginsoy theory with third-body screening (N_I values of 10^{17}, 10^{16}, and $10^{15}\ \mathrm{cm}^{-3}$) (After El-Ghanem and Ridley 1980.)

The effect of the third-body exclusion in n-type GaAs is depicted in Fig. 4.7 for both the Sclar and the Erginsoy mobilities. The effect becomes significant at concentrations of $10^{16}\ \mathrm{cm}^{-3}$ and above for neutral hydrogenic models.

4.4. Central-cell contribution to charged-impurity scattering

Neither the screened nor the pure coulomb potential surrounding a charged impurity extends to the core. The permittivity, which is determined by the polarization of the surrounding lattice, does not remain constant with radius, and electronegativity and size differences between the host atoms and the impurity lead to short-range deviations in potential which can be very marked. Core, or central-cell, effects of this sort are responsible for chemical shifts in the ground-state energies of shallow donors and acceptors, and for the existence of deep levels (Section 2.9).

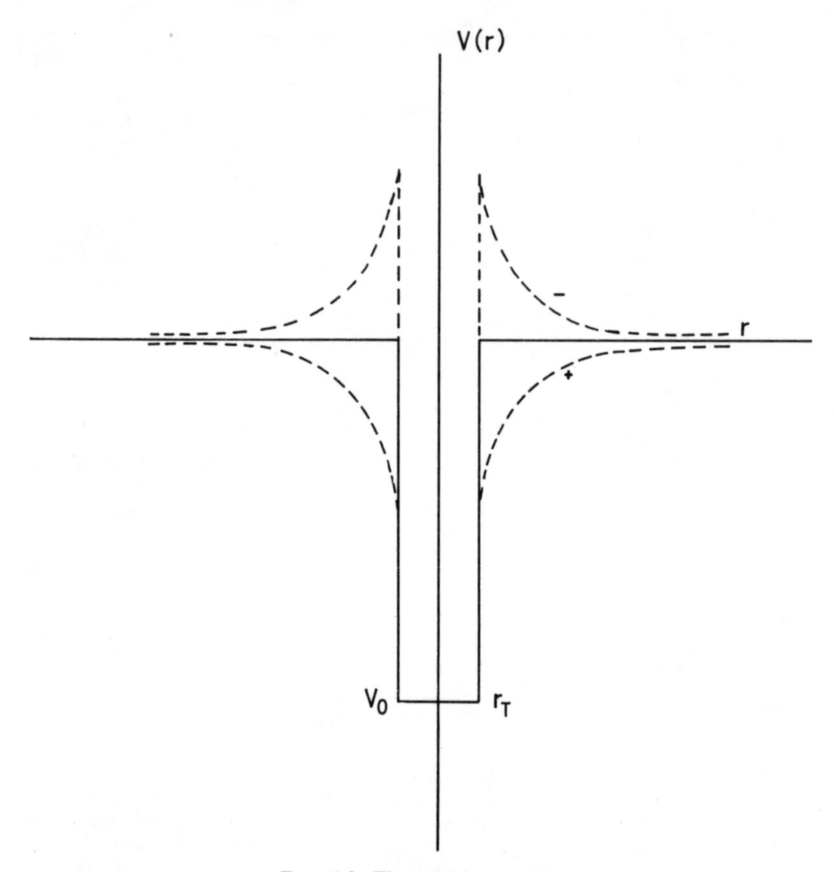

FIG. 4.8. The model potential.

They also introduce additional features into the scattering of electrons by charged impurities.

The simplest model (Fig. 4.8) is to assume the existence of a square-well at $0 \leq r \leq r_T$ which adds to the scattering of a Conwell–Weisskopf impurity. Thus in eqn (4.4) we take

$$f(\theta) = f_c(\theta) + f_N(\theta) \tag{4.96}$$

where $f_c(\theta)$ is the coulomb term (eqn (4.7)) and $f_N(\theta)$ is the 'neutral' term arising from the square-well. Retaining only the $l=0$ contribution we obtain for the differential cross-section

$$\sigma(\theta) = \frac{\mu^2}{4k^2 \sin^4(\theta/2)} + \frac{\sin^2\delta_0}{k^2} + \frac{\mu \sin \delta_0}{k^2 \sin^2(\theta/2)} \cdot \cos\left[\mu \log\left\{\sin^2\left(\frac{\theta}{2}\right)\right\} + \delta_0\right]. \tag{4.97}$$

The first term is the pure coulomb result (eqn (4.9)), the second is the result for a short-range spherically symmetric potential, and the third is an interference term. The momentum transfer cross-section derived from this expression has been given by El-Ghanem and Ridley (1980).

The phase shift for the $l=0$ wave can, as usual, be obtained by matching the internal and external solutions at $r=r_T$. The internal solution is

$$\psi_{\text{int}}(r)=Aj_0(\alpha r) \tag{4.98}$$

where $j_0(\alpha r)$ is the zero-order spherical Bessel function. The external solution is

$$\psi_{\text{ext}}(r)=\frac{B}{kr}\{F_0(kr)\cos\delta_0+G_0(kr)\sin\delta_0\} \tag{4.99}$$

where $F_0(kr)$ is the regular and $G_0(kr)$ is the irregular $l=0$ coulomb wave. When differentiation with respect to r is denoted with a prime we obtain

$$\tan\delta_0=\left(\frac{F_0}{G_0}\right)_{r_T}\frac{(F_0'/F_0)_{r_T}-\alpha\cot\alpha r_T}{\alpha\cot\alpha r_T-(G_0'/G_0)_{r_T}}. \tag{4.100}$$

In the case of a small radius core such that $2|\mu|\gg kr_T$ and $kr_T\to 0$

$$r_T(F_0'/F_0)_{r_T}\approx 1 \qquad r_T(G_0'/G_0)_{r_T}\approx 0 \qquad (F_0/G_0)_{r_T}\approx C_0kr_T \tag{4.101}$$

where the coulomb factor C_0 *is*

$$C_0=\frac{2\pi\mu}{1-e^{-2\pi\mu}}. \tag{4.102}$$

The condition $2|\mu|\gg kr_T$ is also a condition on the energy of the incident electron, namely for a parabolic band

$$E_k\ll 2|Z|\,R_H^*(a_H^*/r_T) \tag{4.103}$$

where R_H^* is the effective Rydberg energy and a_H^* is the effective Bohr radius. Since $r_T\ll a_H^*$ the condition holds for $E_k\lesssim R_H^*$. The phase shift is given by

$$\sin\delta_0=\frac{C_0kr_T\{1-\alpha r_T\cot(\alpha r_T)\}}{[(C_0kr_T)^2\{1-\alpha r_T\cot(\alpha r_T)\}^2+\{\alpha r_T\cot(\alpha r_T)\}^2]^{1/2}}. \tag{4.104}$$

As in the neutral-impurity case, three specially simple cases can be recognized:

(1) hard-sphere core: $|\alpha r_T\cot(\alpha r_T)|\gg 1$

$$\sin\delta_0\approx\pm C_0kr_T \tag{4.105}$$

(2) Ramsauer–Townsend: $\alpha r_T \cot(\alpha r_T) = 1$

$$\sin \delta_0 = 0 \tag{4.106}$$

(3) resonant case: $|\alpha r_T \cot(\alpha r_T)| \ll 1$

$$\sin \delta_0 = \frac{C_0 k r_T}{[\{(C_0 k r_T)^2 + (\alpha r_T \cot(\alpha r_T))\}^2]^{1/2}}. \tag{4.107}$$

Resonance occurs when $\alpha r_T \cot \alpha r_T = 0$.

These expressions are just generalizations of the neutral-impurity results which are recovered by putting $\mu = 0$, whence $C_0 = 1$. As for neutral impurities, core scattering is important only in the third case where a bound state or a nearly bound state exists near the band edge.

Equations (4.97) and 4.107), which describe the effective central-cell contribution, show that negatively charged centres scatter less strongly than positively charged centres. The latter enhance the electron charge density at $r = 0$ by the coulomb factor C_0, which for positive centres and low energies is approximately equal to $2\pi\mu$, and this allows more core scattering to take place. Negatively charged centres repel the incoming electron and diminish its charge density at $r = 0$ by the coulomb factor, which is now approximately $-2\pi\mu \exp(2\pi\mu)$, i.e. very small; thus little core scattering is possible. If core scattering is weak, i.e. $\sin \delta_0 \approx 0$, then the distinction between opposite charges is correspondingly weak. If core scattering is strong, then it will enhance the scattering from positive centres but only weakly affect scattering from negative centres. Negatively charged centres will therefore tend to scatter as pure (or screened) coulombic without any central-cell correction.

Positively charged centres, however, always possess a bound state near the band edge, and therefore the central-cell contribution is potentially important. Provided that the bound state is not too deep it is possible to model the resonant case by using a quantum-defect wavefunction for the localized state:

$$\psi_T(r) = A(r/\nu a_H^*)^{\mu_T - 1} \exp(-r/\nu a_H^*) \tag{4.108}$$

$$\nu = (R_H^*/E_T)^{1/2} \qquad \mu_T = Z\nu$$

where R_H^* is the effective Rydberg energy, a_H^* is the effective Bohr radius and A is a normalizing constant. Following the argument which led to the Sclar formula, we can write the condition for a bound state to be

$$\alpha_T r_T \cot \alpha_T r_T = \mu_T - (r_T/\nu a_H^*) \tag{4.109}$$

$$\alpha_T = \left\{\frac{2m^*(V_0 - E_T)}{\hbar^2}\right\}^{1/2} \tag{4.110}$$

where V_0 is the depth of the well. Equation (4.109) reduced to eqn (4.77) for the neutral case when $\mu = 0$. When $r_T = 0$ there is no central well and the solution of eqn (4.109) is $\mu_T = 1$, i.e. $E_T = Z^2 R_H^*$, which is the pure coulombic solution. Provided $(E_k + E_T)/V_0 \ll 1$, α in eqn (4.104) can be replaced by α_T, whence the phase shift for positively charged centres is given by

$$\sin \delta_0 = \frac{C_0 k r_T \{1 - \mu_T + (r_T/\nu a_H^*)\}}{[(C_0 k r_T)^2 \{1 - \mu_T + (r_T/\nu a_H^*)\}^2 + \{\mu_T - (r_T/\nu a_H^*)\}^2]^{1/2}}. \tag{4.111}$$

Since $C_0 k r_T$ is small by hypothesis, a resonance occurs when $\mu_T \approx r_T/\nu a_H^*$

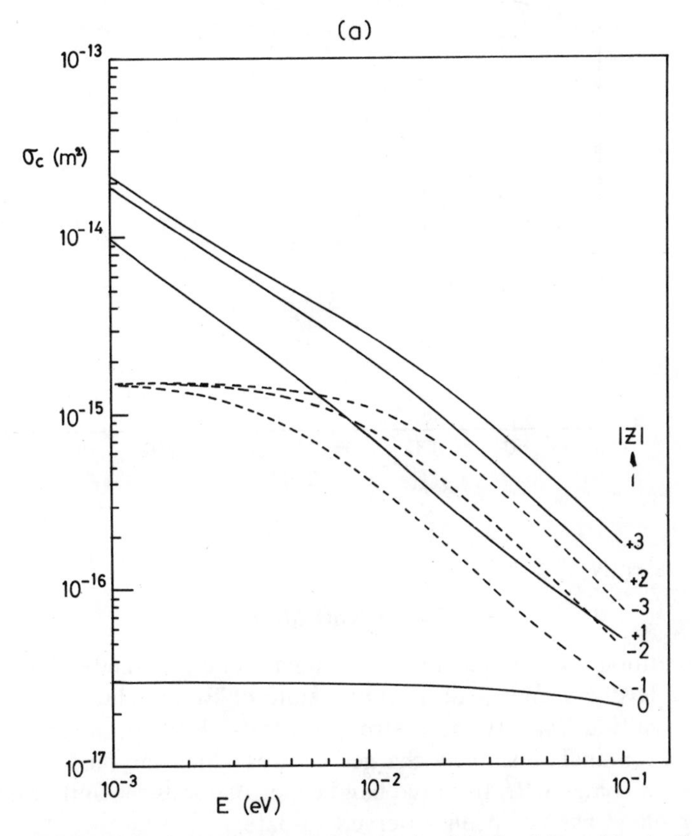

FIG. 4.9. Charged-impurity scattering of electrons by deep-level impurities in GaAs. (a) Scattering cross-section $\sigma_c(E)$ versus energy E and its dependence on the charge at the centre for $\nu = 0{\cdot}1$ and $r_0/a_H^* = 0{\cdot}05$: broken curves, repulsive centre, $Z = -1, -2, -3$; solid curves, attractive centre, $Z = +1, +2, +3$. σ_c for $Z = 0$ is also shown. (b) Mobility times N_I versus temperature and its dependence on the charge Z. (c) Hall scattering factor r_H versus temperature and its dependence on the charge at the centre: broken curves, repulsive centre, $Z = -1, -2, -3$; solid curves, attractive centre, $Z = +1, +2, +3$. (After El-Ghanem and Ridley 1980.)

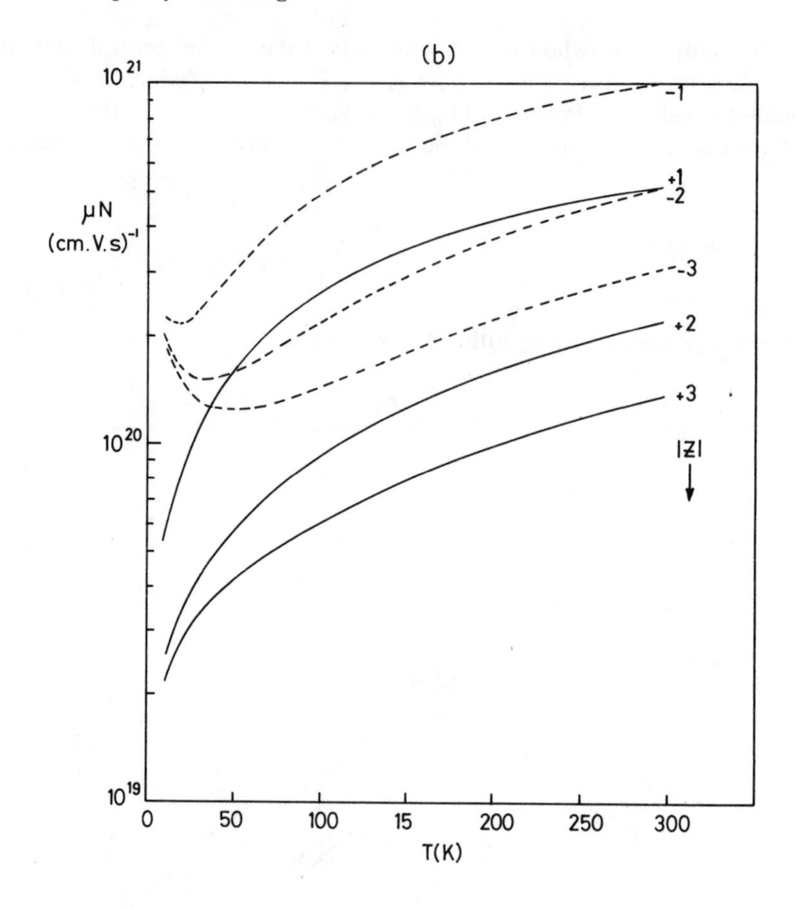

or equivalently

$$E_T = ZR_H^*(a_H^*/r_T). \tag{4.112}$$

This condition corresponds to a bound-state energy exactly equal to the effective Rydberg energy of a ground state of Bohr radius r_T instead of a_H^*. Typically $a_H^*/r_T \approx 10$, so a strong central-cell contribution might be expected for single positively charged centres which can bind an electron into a state some $10R_H^*$ from the band edge. As far as the author is aware such an effect has not been observed to date.

The phase shift in eqn (4.111) reduces to Sclar's result (eqn (4.79)) for neutral centres ($C_0 = 1$, $\mu_T = 0$) with narrow wells ($r_T/va_H^* \ll 1$). The momentum relaxation cross-section for a deep-level charged impurity is shown in Fig. 4.9(a), and the corresponding mobility and Hall factor are shown in Figs. 4.9(b) and 4.9(c). In the case of negatively charged centres there is little contribution from the core. The cross-section approaches

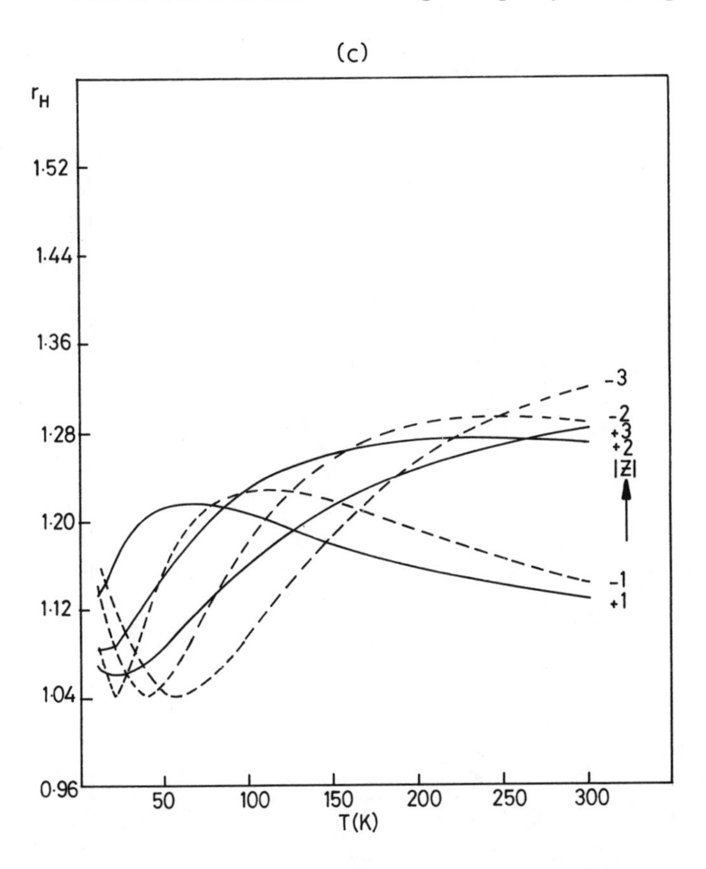

the limiting value set by the cut-off radius a at low energies, whereas for positively charged centres it continues to rise because of enhanced core scattering. The latter effect suggests that third-body exclusion ought to be taken into account at low energies in the case of positively charged impurities, and this would diminish the difference between the two polarities. Towards high energies the effect of core scattering is evident, manifesting itself in the disparity between cross-sections for opposite polarity. In this example the disparity is of moderate proportions. Large differences are expected only for near-resonant core scattering.

Equation (4.111) and eqn (4.97) are useful model expressions which describe in simple analytic form all the principal features of charged- and neutral-impurity scattering. A more sophisticated approach to central-cell corrections using Green's functions has been made by Ralph, Simpson, and Elliott (1975) and applied to the case of silicon and germanium at 300 K (Fig. 4.10).

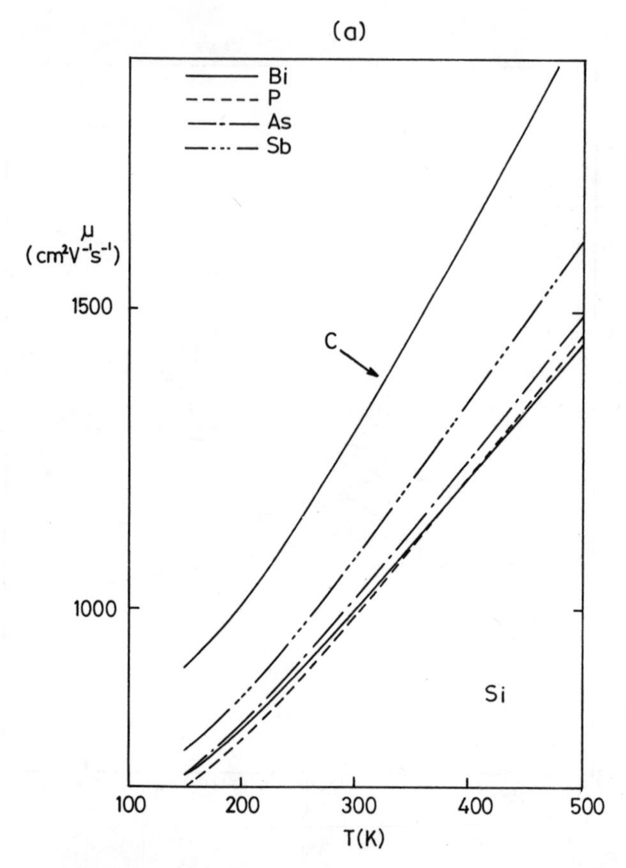

FIG. 4.10. Calculated drift mobility of electrons in (a) silicon and (b) germanium versus temperature for an impurity density of 10^{18} cm^{-3}. The upper solid curve C is for the coulomb potential; the other curves are for the following donor species: ——, bismuth; – – –, phosphorus; –·–, arsenic; –· · ·–, antimony. (After Ralph *et al.* 1975.)

4.5. Dipole scattering

Oppositely charged impurities in compensated (or just plainly impure) semiconductors may associate themselves into pairs and these will scatter electrons like dipoles. In practice one may expect there to be a distribution of dipole moments depending on the separation of the atoms and the charges they bear. If the electric dipole moment is **M** then the scattering potential is

$$V(\mathbf{r}) = \frac{\mathbf{M}.\mathbf{r}}{4\pi\epsilon r^3}. \tag{4.113}$$

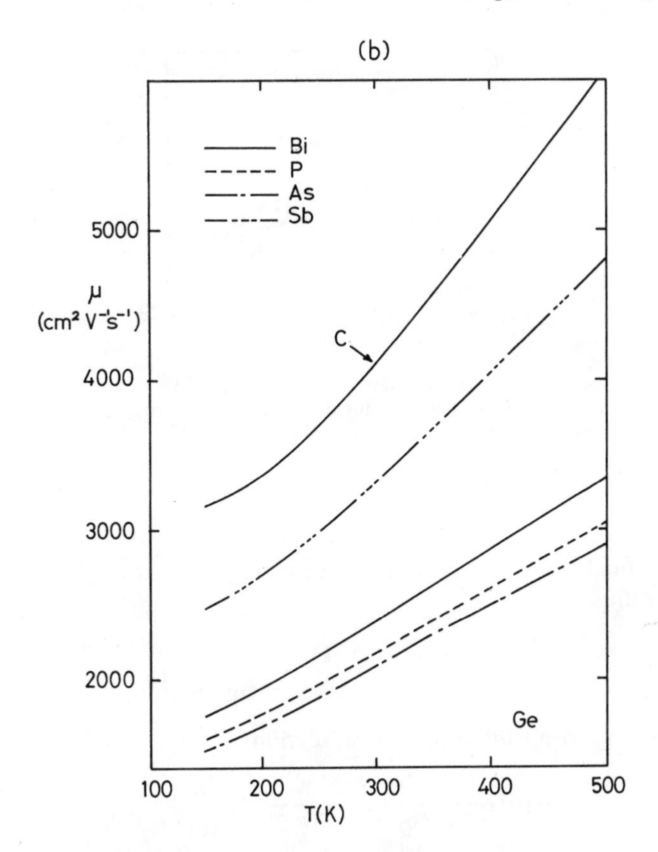

In the Born approximation, the matrix element is

$$\langle \mathbf{k}' | \bar{e}V(\mathbf{r}) | \mathbf{k} \rangle = \frac{1}{V} \int \exp(-\mathrm{i}\mathbf{k}'.\mathbf{r}) \bar{e} \frac{\mathbf{M}.\mathbf{r}}{4\pi\epsilon r^3} \exp(i\mathbf{k}.\mathbf{r})\, \mathrm{d}\mathbf{r} \tag{4.114}$$

$$= \frac{\mathrm{i}\bar{e}\mathbf{M}.(\mathbf{k}'-\mathbf{k})}{\epsilon V |\mathbf{k}'-\mathbf{k}|^2}. \tag{4.115}$$

Conserving energy ensures that $k' = k$ and $|\mathbf{k}'-\mathbf{k}|^2 = 4k^2 \sin^2(\theta/2)$, where θ is the angle through which the electron is scattered. For a random distribution of dipole orientation we can make the replacement

$$|\mathbf{M}.(\mathbf{k}'-\mathbf{k})|^2 = \tfrac{1}{3}\mathbf{M}^2 |\mathbf{k}'-\mathbf{k}|^2. \tag{4.116}$$

The scattering rate is then (eqn (4.21))

$$W(\mathbf{k}) = \int \frac{2\pi}{\hbar} \frac{e^2\mathbf{M}^2\delta(E_{k'}-E_k)}{12\epsilon^2 V^2 k^2 \sin^2(\theta/2)} k'^2\, \mathrm{d}k' \sin\theta\, \mathrm{d}\theta\, \mathrm{d}\phi \frac{V}{8\pi^3}. \tag{4.117}$$

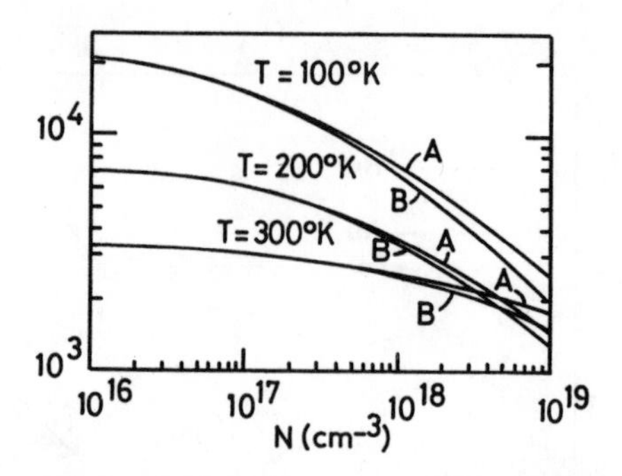

FIG. 4.11. Dipole scattering contribution to the electron drift mobility in compensated germanium: curves A, acoustic phonon plus point charge scattering; curves B, with dipole scattering. Compensation ratio $(N_D - N_A)/(N_D + N_A) = 0{\cdot}01$. (After Stratton 1962.)

When $k'^2\,dk'/8\pi^3$ is replaced by $N(E_{k'})\,dE_{k'}/4\pi$, where $N(E_{k'})$ is the density of final states, we obtain the scattering rate for a single dipole per unit solid angle:

$$W(\theta) = \frac{e^2\mathbf{M}^2 N(E_k)}{24\epsilon^2 V\hbar}\frac{1}{k^2\sin^2(\theta/2)} \tag{4.118}$$

corresponding to a differential cross-section

$$\sigma(\theta) = \frac{e^2\mathbf{M}^2 N(E_k)}{24\epsilon^2\hbar v k^2\sin^2(\theta/2)} = \frac{\mu^2 r_0^2}{3\sin^2(\theta/2)} \tag{4.119}$$

which diverges towards small scattering angles. In this expression $\mu^2 = R_H^*/E_k$ and $\mathbf{M} = \bar{e}r_0$. The ratio of this cross-section to that for coulombic scattering is

$$\frac{\sigma_d(\theta)}{\sigma_c(\theta)} = \tfrac{4}{3}k^2 r_0^2 \sin^2\!\left(\frac{\theta}{2}\right) = \langle(\mathbf{k}' - \mathbf{k})\,.\,\mathbf{r}_0\rangle \tag{4.120}$$

Since r_0 is of order of the nearest-neighbour distance, this ratio is very small in most situations.

The divergence towards small angles disappears when screening and third-body exclusion are taken into account. The latter has not been treated. Screening was taken into account by Stratton (1962) who used a screened potential of the form

$$V(\mathbf{r}) = \frac{\mathbf{M}\,.\,\mathbf{r}}{4\pi\epsilon r^3}(1 + q_0 r)\exp(-q_0 r) \tag{4.121}$$

where q_0 is the reciprocal Debye length. Fortunately, the momentum relaxation cross-section does not diverge, and so we shall limit further discussion of this scattering mechanism to quoting that result. Thus

$$\sigma_m = \int \sigma(\theta)(1-\cos\theta)\,d\Omega = \frac{8\pi\mu^{*2}r_0^2}{3} = \frac{8\pi R_H^* r_0^2}{3E_k}. \quad (4.122)$$

Stratton's calculation of mobility in compensated germanium is shown in Fig. 4.11.

4.6. Electron–hole scattering

A scattering mechanism closely related in essence to charged-impurity scattering is that between an electron in the conduction band and a hole in the valence band. Electron–hole scattering can be a significant factor in determining the resistivity of narrow-gap intrinsic semiconductors, such as InSb at room temperature.

We can approximate the interaction by a screened coulombic force between two point particles and obtain the rate in the Born approximation as in the Brooks–Herring approach (eqns (4.20) and (4.21)). The matrix element which determines the rate of transition when particle 1, wavevector $\mathbf{k}_1$, collides with particle 2, wavevector $\mathbf{k}_2$, and ends the collision with wavevector $\mathbf{k}_1'$ with particle 2 having a wavevector $\mathbf{k}_2'$ (Fig. 4.12) is

$$\langle \mathbf{k}_1'\mathbf{k}_2'|\bar{e}V(\mathbf{r})|\,\mathbf{k}_1\mathbf{k}_2\rangle = I(\mathbf{k}_1,\mathbf{k}_1')I(\mathbf{k}_2,\mathbf{k}_2')\frac{1}{V^2}\cdot$$

$$\times \iint \exp\{-i(\mathbf{k}_1'.\mathbf{r}_1+\mathbf{k}_2'.\mathbf{r}_2)\}\frac{e^2\exp(-q_0|\mathbf{r}_1-\mathbf{r}_2|)}{4\pi\epsilon|\mathbf{r}_1-\mathbf{r}_2|}\exp\{i(\mathbf{k}_1.\mathbf{r}_1+\mathbf{k}_2.\mathbf{r}_2)\}\,d\mathbf{r}_1\,d\mathbf{r}_2 \quad (4.123)$$

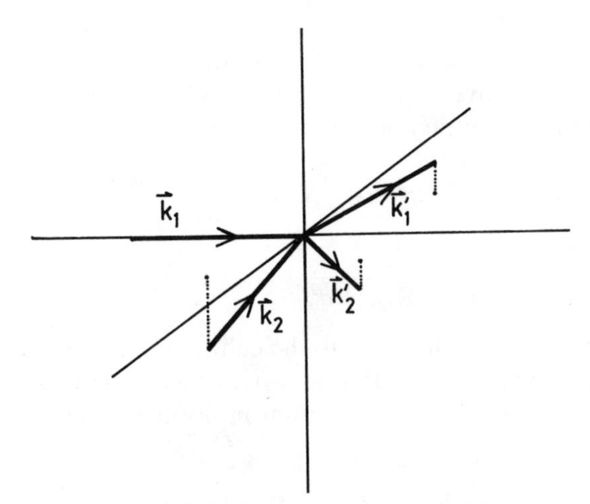

FIG. 4.12. Collision between two particles.

where $I(\mathbf{k}_1, \mathbf{k}_1')$, $I(\mathbf{k}_2, \mathbf{k}_2')$ are the overlap integrals over the unit cell involving the cell-periodic parts of the Bloch functions:

$$I(\mathbf{k}_1, \mathbf{k}_1')I(\mathbf{k}_2, \mathbf{k}_2') = \int_{\text{cell}} u^*_{\mathbf{k}_1'}(\mathbf{r}_1)u_{\mathbf{k}_1}(\mathbf{r}_1)\,\mathrm{d}\mathbf{r}_1 \int_{\text{cell}} u^*_{\mathbf{k}_2'}(\mathbf{r}_2)u_{\mathbf{k}_2}(\mathbf{r}_2)\,\mathrm{d}\mathbf{r}_2. \tag{4.124}$$

Both normal and umklapp processes are theoretically possible, but for semiconductors only the former are normally of interest. For scattering events in which all wavevectors lie close to the band edges the integrals in eqn (4.124) are usually assumed to be unity. In general, they are less than unity.

The interaction depends only on the separation distance of the particles, so it is convenient to transform to a frame of reference in which the centre of mass of the two particles is at rest. The transformation (non-relativistic) to the centre-of-mass frame is effected by converting $\mathbf{k}$ to $\mathbf{K}$ where

$$\mathbf{K} = \mathbf{k} - \mathbf{k}_{\text{cm}} \qquad \mathbf{k}_{\text{cm}} = (\mathbf{k}_1 + \mathbf{k}_2)/2 \tag{4.125}$$

whence

$$\mathbf{k}_1 = \mathbf{k}_{\text{cm}} + \mathbf{K}_{12} \qquad \mathbf{k}_2 = \mathbf{k}_{\text{cm}} - \mathbf{K}_{12} \qquad \mathbf{K}_{12} = \tfrac{1}{2}(\mathbf{k}_1 - \mathbf{k}_2). \tag{4.126}$$

The corresponding transformation of spatial coordinates is

$$\mathbf{R} = \mathbf{r} - \mathbf{r}_{\text{cm}}, \qquad \mathbf{r}_{\text{cm}} = \frac{m_1^*\mathbf{r}_1 + m_2^*\mathbf{r}_2}{m_1^* + m_2^*} \tag{4.127}$$

whence

$$\mathbf{r}_1 = \mathbf{r}_{\text{cm}} + \frac{m_2^*}{m_1^* + m_2^*}\mathbf{r}_{12} \qquad \mathbf{r}_2 = \mathbf{r}_{\text{cm}} - \frac{m_1^*}{m_1^* + m_2^*}\mathbf{r}_{12} \qquad \mathbf{r}_{12} = \mathbf{r}_1 - \mathbf{r}_2. \tag{4.128}$$

The integral in eqn (4.123) splits into a product of two integrals, one over $\mathbf{r}_{\text{cm}}$ and the other over $\mathbf{r}_{12}$. The former gives unity and entails the conservation of momentum, and the latter is

$$\begin{aligned}\langle \mathbf{K}_{12}' |\bar{e}V(\mathbf{r}_{12})| \mathbf{K}_{12}\rangle &= \frac{1}{V}\int \exp(-i\mathbf{K}_{12}'.\mathbf{r}_{12})\,\frac{e^2 \exp(-q_0\mathbf{r}_{12})}{4\pi\epsilon r_{12}}\,\exp(i\mathbf{K}_{12}.\mathbf{r}_{12})\,\mathrm{d}\mathbf{r}_{12} \\ &= \frac{e^2/\epsilon V}{|\mathbf{K}_{12}' - \mathbf{K}_{12}|^2 + q_0^2}.\end{aligned} \tag{4.129}$$

The problem is exactly analogous to the collision of a particle of mass $\bar{m}^*$, equal to the reduced mass of the two particles, with a fixed centre. Conservation of energy and momentum in normal processes entails that

$$\begin{gathered}\mathbf{k}_1 + \mathbf{k}_2 = \mathbf{k}_1' + \mathbf{k}_2' \\ |\mathbf{K}_{12}' - \mathbf{K}_{12}| = 2K_{12}\sin(\theta/2)\end{gathered} \tag{4.130}$$

where θ is the angle between $\mathbf{K}_{12}$ and $\mathbf{K}'_{12}$. The relative momentum relaxation cross-section analogous to eqn (4.31) is therefore, when the overlap integrals are unity,

$$\sigma_m = \frac{\pi}{2}\left(\frac{e^2}{4\pi\epsilon E_{12}}\right)^2\left\{\log\left(1+\frac{8\bar{m}^*E_{12}}{\hbar^2 q_0^2}\right) - \frac{1}{1+\hbar^2 q_0^2/(8\bar{m}^*E_{12})}\right\} \tag{4.131}$$

where

$$E_{12} = \frac{\hbar^2 K_{12}^2}{2\bar{m}^*} = \frac{\hbar^2(\mathbf{k}_1 - \mathbf{k}_{cm})^2}{2m_1^*} + \frac{\hbar^2(\mathbf{k}_2 - \mathbf{k}_{cm})^2}{2m_2^*}. \tag{4.132}$$

All the discussion of Section 4.2 related to small-angle scattering, third-body exclusion, etc. in the case of charged-impurity scattering applies to electron–hole scattering. However, if electron–hole scattering cannot be neglected in a given situation it is likely that electrical screening by free carriers cannot be neglected either, and so eqn (4.131) is a reasonably good formula.

The step from relative to absolute momentum relaxation is usually facilitated in practice by the disparity between electron and hole masses. In many cases the hole mass exceeds the electron mass significantly, and consequently the hole can be considered to be at rest. Equation (4.131) then describes the absolute momentum relaxation cross-section for electrons with $\bar{m}^* = m_e^*$ and $E_{12} = E_k$, where E_k is the electron energy, as in the Brooks–Herring formula (Ehrenreich 1957; see Fig. 3.24). If $m_h^* \not\gg m_e^*$ the conversion from the centre-of-mass frame to the laboratory frame has to be carried out in detail (Schiff 1955; Chapman and Cowling 1958). In most cases, fortunately, $m_h^* \gg m_e^*$ and the problem of electron-hole scattering reduces to that for charged-impurity scattering.

4.7. Electron–electron scattering

Electrons and holes are distinguishable particles. The matrix element in eqn (4.129), which we label M_{12}, implies that particle 1 and particle 2 are distinguishable. In electron–electron collisions the particles are identical and no observable effect would occur if the particle emerging with wavevector $\mathbf{k}'_1$ were exchanged for one emerging with $\mathbf{k}'_2$. In this case the matrix element would be M_{21}, identical to M_{12} except that $\mathbf{K}'_{12}$ would be replaced by $\mathbf{K}'_{21} = (\mathbf{k}'_2 - \mathbf{k}'_1)/2$, whence for normal processes (umklapp processes are usually expected to make insignificant contributions)

$$|\mathbf{K}'_{21} - \mathbf{K}_{12}| = 2K_{12}\cos(\theta/2). \tag{4.133}$$

If the two electrons have identical spins the two processes, which occur with amplitudes M_{12} and M_{21}, interfere and the squared matrix element is $|M_{12} - M_{21}|^2$, whereas if the spins are of opposite sign they do not interfere and the squared matrix element is $M_{12}^2 + M_{21}^2$. The net squared matrix element describing the collision rate of an electron is therefore

$$\begin{aligned} M^2 &= \tfrac{1}{2}(M_{12}^2 + M_{21}^2 + |M_{12} - M_{21}|^2) \\ &= M_{12}^2 + M_{21}^2 - M_{12}M_{21}. \end{aligned} \tag{4.134}$$

The factor $\frac{1}{2}$ arises because in half of the collisions the spins are aligned and in the other half they are opposed. The minus sign arises because the total electron wavefunction changes sign on the interchange of two particles.

Following the argument in the Brooks–Herring approach (Section 4.2.2) we can obtain the net differential cross-section in the centre-of-mass frame (taking the overlap integrals $I(\mathbf{k}, \mathbf{k}')$ to be unity):

$$\sigma(\theta) = \left(\frac{e^2}{16\pi\epsilon E_{12}}\right)^2 \left[\frac{1}{\{\sin^2(\theta/2) + (q_0/2K_{12})^2\}^2} + \frac{1}{\{\cos^2(\theta/2) + (q_0/2K_{12})^2\}^2} - \frac{1}{\{\sin^2(\theta/2) + (q_0/2K_{12})^2\}\{\cos^2(\theta/2) + (q_0/2K_{12})^2\}}\right] \tag{4.135}$$

where θ is the angle between $\mathbf{K}_{12}$ and $\mathbf{K}_{12}'$, $\mathbf{K}_{12} = (\mathbf{k}_1 - \mathbf{k}_2)/2$, and $E_{12} = \hbar^2 K_{12}^2/2\bar{m}^*$ ($\bar{m}^* = m^*/2$ in this case). This is the Mott formula for proton–proton scattering suitably modified by screening (see Messiah 1966, p. 608).

The total cross-section derived from eqn (4.135) is

$$\sigma = \left(\frac{e^2}{16\pi\epsilon E_{12}}\right)^2 8\pi \left[\frac{1}{(q_0/2K_{12})^2\{1 + (q_0/2K_{12})^2\}} + \frac{1}{1 + 2(q_0/2K_{12})^2} \log\left\{\frac{(q_0/2K_{12})^2}{1 + (q_0/2K_{12})^2}\right\}\right]. \tag{4.136}$$

Since we have integrated over all directions, this expression is also the total cross-section in the laboratory frame.

A relative momentum relaxation rate can be calculated by multiplying the first term in eqn (4.135) by $1 - \cos\theta$, the second by $1 + \cos\theta$, and the third by $\{(1 + \cos\theta)(1 - \cos\theta)\}^{1/2}$, and one obtains

$$\frac{1}{\tau_m} = \frac{e^4}{8\pi\epsilon^2 V (2\bar{m}^*)^{1/2} E_{12}^{3/2}} \left[\log\left(1 + \frac{8\bar{m}^* E_{12}}{\hbar^2 q_0^2}\right) - \frac{1}{1 + \hbar^2 q_0^2/8\bar{m}^* E_{12}} - \frac{\pi}{2}\left\{1 - \frac{(\hbar^2 q_0^2/2\bar{m}^* E_{12})(1 + \hbar^2 q_0^2/8\bar{m}^* E_{12})}{1 + (\hbar^2 q_0^2/2\bar{m}^* E_{12})(1 + \hbar^2 q_0^2/8\bar{m}^* E_{12})}\right\}\right]. \tag{4.137}$$

This rate measures the rapidity with which relative momentum attenuates when there is one electron in the cavity of volume V. The total rate is obtained by summing over all electrons, but this requires a knowledge of the distribution of electrons over the initial states. In the case of a fast incident electron such that $\mathbf{K}_{12} \approx \mathbf{k}_1/2$ the energy E_{12} associated with the relative motion is insensitive to the electron distribution, and hence we can simply replace $1/V$ in eqn (4.137) by the electron density n. Moreover, for a stationary target $E_{12} = E_k/2$, where E_k is the energy of the incident electron. Thus for a fast electron eqn (4.137) becomes

$$\frac{1}{\tau_m} = \frac{ne^4}{2^{3/2}\pi\epsilon^2 m^{*1/2} E^{3/2}} L(E_k, q_0) \tag{4.138}$$

where $L(E_k, q_0)$ stands for the square bracket in eqn (4.137).

The total momentum is of course conserved, and so electron–electron collisions cannot relax the momentum gained from external fields or incident fast particles. However, they do randomize momentum and attempt to eliminate relative motion at a rate given approximately by eqn (4.138).

The energy relaxation rate is most easily obtained in the case where particle 2 is stationary and the masses of the two particles are equal. In such a case it is easy to show that the angle between $\mathbf{k}_1'$ and $\mathbf{k}_2'$ is a right angle and

$$k_2' = 2K_{12}\sin(\theta/2) = K_{12}\{2(1-\cos\theta)\}^{1/2}. \tag{4.139}$$

The energy lost by particle 1 is $\hbar^2 k_2'^2/2m^*$ associated with M_{12}^2 and $\hbar^2 k_1'^2/2m^*$ associated with M_{21}^2. The energy relaxation rate is therefore obtained from eqn (4.135) by using exactly the same weighting factors as were used to calculate the relative momentum relaxation rate. The result is, for n electrons per unit volume,

$$\frac{dE_k}{dt} = -\frac{ne^4}{4\pi\epsilon^2(2m^*E_k)^{1/2}} L(E_k, q_0) = -\frac{E_{12}}{\tau_m} = -\frac{E_k/2}{\tau_m} \tag{4.140}$$

which is the result due to Pines (1953).

The physically significant rate for electron–electron scattering is therefore that for relaxing relative momentum. For fast incident electrons it can be calculated from eqn (4.138). If we take $L(E_k, q_0)$ as unity and $E_k = 0{\cdot}1$ eV we obtain, roughly,

$$1/\tau_m \approx 10^{-5} n \tag{4.141}$$

with n in units of cm^{-3}. Thus for τ_m to be a picosecond or less, and therefore competing equally with typical scattering times, the carrier density must be of order 10^{17} cm^3 or more. Typical mean free paths in metals are in the range 400–1000 Å corresponding to scattering times of order 10^{-13} s. In non-degenerate semiconductors electron–electron scattering rates are usually negligible.

Equation (4.138) is based upon the Born approximation for a fast incident electron. As such it overestimates the electron–electron scattering rate in populations at thermal equilibrium by a factor of about 4, as compared with the rate derived numerically on the basis of a phase-shift analysis (Abrahams 1954). Taking $E_k = 0{\cdot}1$ eV rather than $\frac{3}{2}k_BT$ to obtain the estimate of eqn (4.141) was an attempt to compensate for this effect.

4.8. Mobilities

The phenomenological quantity most closely associated with scattering rate is the drift mobility, which is obtained from the momentum relaxation time by averaging over the electron distribution as discussed in

TABLE 4.1

Momentum relaxation times for impurity scattering in non-degenerate material

Scattering mechanism	Momentum relaxation time
Charged impurity scattering	
(1) Conwell–Weisskopf	$\dfrac{2^{9/2}\pi\epsilon^2 m^{*1/2}E_k^{3/2}}{Z^2e^4N_I}\left\{\log\left(1+\dfrac{4\pi\epsilon E_k}{Ze^2N_I^{1/3}}\right)\right\}^{-1}$
(2) Brooks–Herring	$\dfrac{2^{9/2}\pi\epsilon^2 m^{*1/2}E_k^{3/2}}{Z^2e^4N_I}\left\{\log\left(1+\dfrac{8m^*E_k}{\hbar^2q_0^2}\right)-\dfrac{1}{1+\hbar^2q_0^2/8m^*E_k}\right\}^{-1}$
(3) Takimoto	$\dfrac{2^{9/2}\pi\epsilon^2 m^{*1/2}E_k^{3/2}}{Z^2e^4N_I}\left\{\log\left(1+\dfrac{8m^*E_k}{\hbar^2q_0^2F(y)}\right)-\dfrac{1}{1+\hbar^2q_0^2F(y)/8m^*E_k}\right\}^{-1}$
(4) Uncertainty broadening	$\dfrac{2^{1/4}\pi\epsilon\hbar^{1/2}m^{*1/4}E_k^{1/4}}{Ze^2N_I^{1/2}}$
(5) Third-body exclusion	$\dfrac{2^{9/2}\pi\epsilon^2 m^{*1/2}E_k^{3/2}}{Z^2e^4N_I}L^{-1}(E_k)$ ($L(E_k)$ given in eqn (2.57))
(6) Central-cell contribution	See Section 4.4
Neutral impurity scattering	
(1) Erginsoy	$\dfrac{e^2m^{*2}}{80\pi\epsilon\hbar^3N_I}$
(2) Sclar	$\dfrac{m^{*3/2}}{2^{3/2}\pi\hbar^2N_I}\left(\dfrac{E_k+E_T}{E_k^{1/2}}\right)$
(3) Resonance	$\dfrac{m^{*3/2}}{2^{3/2}\pi\hbar^2N_I}\dfrac{(E_r-E_k)^2+E_0E_k}{E_0E_k^{1/2}}$
(4) Third-body exclusion	Multiply above by $\dfrac{2y^2}{2y+e^{-2y}-1}$, $y=2\pi N_I a\sigma(\theta)$
Dipole scattering	$\dfrac{3\times 2^{3/2}\pi\epsilon^2\hbar^2E_k^{1/2}}{e^4m^{*1/2}r_0^2N_{dip}}$

Chapter 3, Section 3.8. The momentum relaxation times are summarized in Table 4.1. Many of these contain logarithmic or more complicated dependences on energy and numerical methods are necessary to evaluate the corresponding mobilities. All refer to non-degenerate semiconductors. In degenerate material, as well as the necessity for using Fermi–Dirac statistics, there are considerable complications associated with electrical screening and other many-body effects which are beyond the scope of this book. Generally speaking, in non-degenerate material agreement between theory and experimentally determined mobility is not as convincing as it is for lattice scattering. This is partly because drift mobility is not easy to measure, and many investigations have depended on a measurement of the Hall mobility which is greater than the drift mobility by the well-known scattering factor r_H. El-Ghanem and Ridley (1980) have recently shown that in the presence of a small amount of resonance scattering r_H can be surprisingly high (Fig. 4.13), and this effect has not been considered in previous work. Other complications such as the

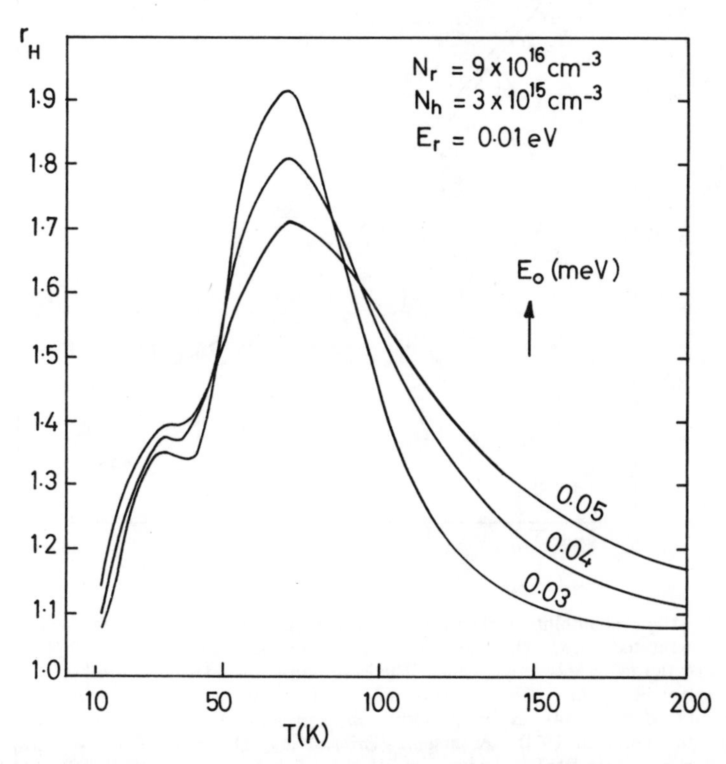

FIG. 4.13. The Hall scattering factor r_H versus temperature for mixed scattering between Erginsoy and Breit–Wigner types of resonance and its dependence on the width E_0 of the resonance level in n-type GaAs: $N_r = 9 \times 10^{16}$ cm^{-3}; $N_n = 3 \times 10^{15}$ cm^{-3}; $E_r = 0{\cdot}01$ eV.

non-uniform distribution of impurities, impurity pairing and complexing, and the effects of band-tailing at low temperatures no doubt also account for the difficulty of achieving the sort of agreement we have in the case of pure lattice scattering.

Figure 4.14, which shows the variation of the Hall mobility with the reciprocal Hall constant for silicon, is a typical representation of the variation of mobility with impurity concentration in semiconductors. Equally typical, this time of the form of variation of mobility with temperature, are the curves of Fig. 4.15 which show the electron drift mobility in GaAs nominally containing oxygen.

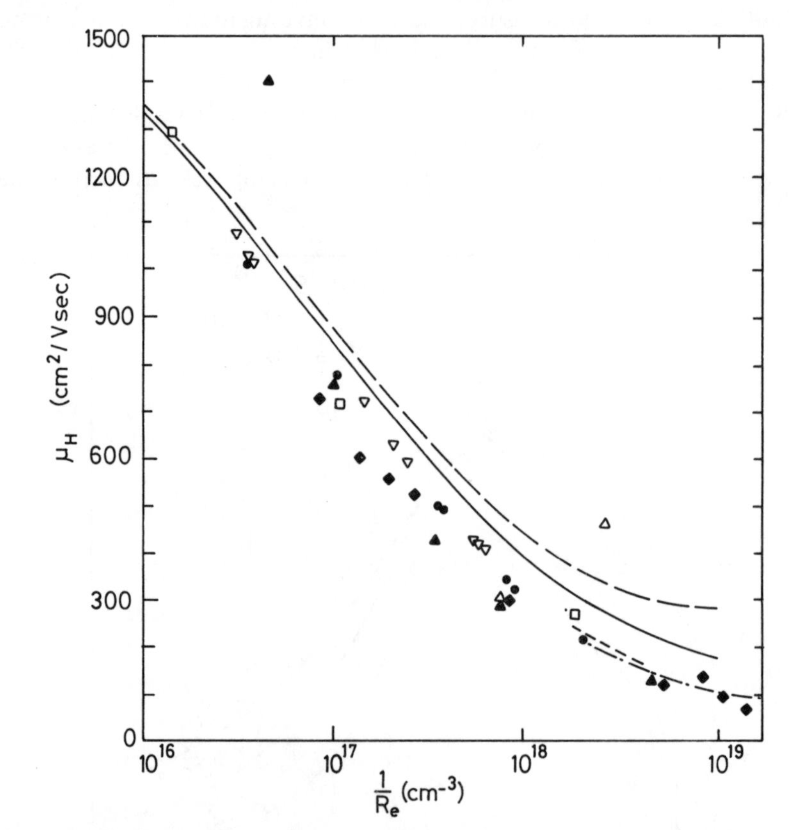

FIG. 4.14. The Hall mobility of electrons at 300 K versus the reciprocal of the Hall coefficient for uncompensated silicon. The upper curve represents the combined effects of phonon and coulomb scattering in anisotropic bands. The lower curve includes the central cell scattering calculated for phosphorus. The experimental points and curves are as follows: □, arsenic (Morin and Maita 1954); ▲, phosphorus (Brinson and Dunstan 1970); △, antimony (Brinson and Dunstan 1970); ◬, arsenic (Brinson and Dunstan 1970); ◆, phosphorus (Gränacher and Czaja 1967); ▽, antimony (Wolfstirn 1960); ⩢, arsenic (Wolfstirn 1960); ●, phosphorus (B. J. Goldsmith and F. Berz, unpublished); broken curve, antimony (Furukawa 1961); chain curve, arsenic (Furukawa 1961). (After Ralph *et al.* 1975.)

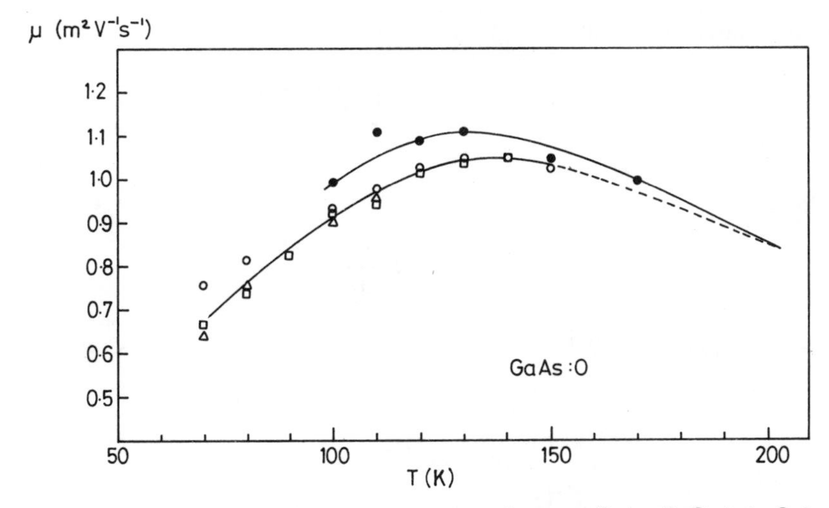

FIG. 4.15. Electron drift mobility versus temperature in n-type GaAs: O: ●, dark; ○, low; △, medium; □, high background illumination with silicon-filtered light. (After Arikan, Hatch, and Ridley 1980.)

4.9. Appendix: Debye screening length

At thermodynamic equilibrium the density of electrons occupying an energy level E_i when the solid is neutral is given by

$$n_i(0) = N_i f(E_i) \tag{4.142}$$

where N_i is the density of levels and $f(E_i)$ is the occupation probability. When an electrical potential $V(\mathbf{r})$ is present the density of occupied levels changes to

$$n_i(V) = N_i f\{E_i + \bar{e}V(\mathbf{r})\} \tag{4.143}$$

where the symbol $\bar{e}$, as usual, contains the sign of the charge. As a result there exists a space charge density

$$\rho_i(\mathbf{r}) = \bar{e}N_i[f\{E_i + \bar{e}V(\mathbf{r})\} - f(E_i)]. \tag{4.144}$$

Poisson's equation describes the spatial variation of potential

$$\nabla^2 V(\mathbf{r}) = -\rho/\epsilon \tag{4.145}$$

$$= -\frac{\bar{e}}{\epsilon}\sum_i N_i[f\{E_i + \bar{e}V(\mathbf{r})\} - f(E_i)]. \tag{4.146}$$

As long as $\bar{e}V(\mathbf{r})$ is small compared with E_i, the right-hand side can be approximated by a truncated Taylor series and we obtain

$$\nabla^2 V(\mathbf{r}) = -\frac{e^2}{\epsilon} V(\mathbf{r}) \sum_i N_i \frac{\mathrm{d}f(E_i)}{\mathrm{d}E_i} \tag{4.147}$$

We can now define a reciprocal screening length q_0 as follows:

$$q_0^2 = -\frac{e^2}{\epsilon} \sum_i N_i \frac{\mathrm{d}f(E_i)}{\mathrm{d}E_i}. \tag{4.148}$$

If the occupation probability of a level with degeneracy g_i is assumed to be

$$f(E_i) = \frac{1}{1 + g_i^{-1} \exp\{(E_i - F)/k_\mathrm{B}T\}}, \tag{4.149}$$

where F is the Fermi level, we can put

$$\frac{\mathrm{d}f(E_i)}{\mathrm{d}E_i} = -\frac{f(E_i)\{1 - f(E_i)\}}{k_\mathrm{B}T} \tag{4.150}$$

whence

$$q_0^2 = \frac{e^2}{\epsilon k_\mathrm{B}T} \sum_i \frac{n_i p_i}{N_i} \tag{4.151}$$

where n_i and p_i are the densities of filled and empty levels respectively in neutral material. Note that only partially filled levels contribute and that the sum includes localized states as well as band states.

In the case of spherical symmetry eqn (4.147) with eqn (4.148) becomes

$$\frac{1}{r} \frac{\mathrm{d}^2}{\mathrm{d}r^2}\{rV(r)\} = q_0^2 V(r) \tag{4.152}$$

the solution of which for $V(r) \to 0$, $r \to \infty$ and $V(r) \to Ze^2/4\pi\epsilon r$, $r \to 0$ is

$$V(r) = \frac{Ze^2}{4\pi\epsilon r} \exp(-q_0 r). \tag{4.153}$$

Two examples of screening are (1) by free electrons in the conduction band, present in non-degenerate density n ($\ll N_\mathrm{c}$), when

$$q_0^2 = \frac{e^2 n}{\epsilon k_\mathrm{B}T} \tag{4.154}$$

and (2) by a combination of free electrons obeying non-degenerate statistics and electrons trapped in a compensated donor level, denisty N_D, when

$$q_0^2 = \frac{e^2}{\epsilon k_B T}\left\{n + \frac{(N_D - N_A - n)(N_A + n)}{N_D}\right\} \tag{4.155}$$

where N_A is the density of acceptors.

4.10. Appendix: Average separation of impurities

Let the probability of there being a nearest neighbour situated between radii r and $r + dr$ be $Q(r)\,dr$. Then if $P(r)$ is the probability of there being no neighbour closer than r it follows that

$$Q(r)\,dr = P(r)4\pi r^2\,drN, \tag{4.156}$$

where N is the average density of impurities. The quantities $Q(r)$ and $P(r)$ are also related, through their definitions, as follows

$$P(r) = 1 - \int_0^r Q(r)\,dr. \tag{4.157}$$

Substituting $P(r)$ from eqn (4.156) and differentiating, we obtain

$$\frac{d}{dr}\left(\frac{Q(r)}{4\pi r^2 N}\right) = -Q(r), \tag{4.158}$$

from which, after integration,

$$Q(r) = 4\pi r^2 N \exp\left(-\frac{4\pi r^3}{3}N\right) \tag{4.159}$$

(with $Q(r) = 0$ when $r = 0$ as boundary condition). The mean distance apart is then

$$\langle r\rangle = \int_0^\infty rQ(r)\,dr = \left(\frac{3}{4\pi N}\right)^{1/3}\Gamma\left(\frac{4}{3}\right). \tag{4.160}$$

The numerical factor multiplying $N^{-1/3}$ is 0·55396, which is close to $(2\pi)^{-1/3}$ (=0·54193), so for convenience we take

$$\langle r\rangle \approx (2\pi N)^{-1/3}. \tag{4.161}$$

4.11. Appendix: Alloy scattering

The band structure of an alloy of the form $A_xB_{1-x}C$ is described using the virtual-crystal approximation, in which an average of the component pseudo-potentials is assumed, corresponding to a uniform distribution of A and B over the cation lattice sites. Fluctuations from uniformity are then seen to give rise to local changes in the potential experienced on average by the electron, and hence to scattering. This perturbing potential can be written:

$$V(\mathbf{r}) = \sum_{\mathbf{q}} V_{\mathbf{q}} \exp(i\mathbf{qr}). \tag{4.162}$$

The Fourier amplitude is taken to be the root mean square deviation from the average energy and the same for all **q**. If V_a and V_b are the potentials associated with cations A and B respectively the average is

$$V_0 = V_a x + V_b(1-x). \tag{4.163}$$

In a region where the fractional occupancy of A is x' the deviation is

$$V' - V_0 = (V_a - V_b)(x' - x), \tag{4.164}$$

and the rms deviation is standard for a binomial distribution, namely:

$$|\langle V' - V_0 \rangle| = |V_a - V_b| \left[\frac{x(1-x)}{N_c}\right]^{1/2}, \tag{4.165}$$

where N_c is the number of cation sites, equal to the number of unit cells. the matrix element for scattering is then

$$\langle \mathbf{k}' \mid H \mid k \rangle = |V_a - V_b| \left[\frac{x(1-x)}{N_c}\right]^{1/2} \delta_{\mathbf{k}\pm\mathbf{q}-\mathbf{k}',0}, \tag{4.166}$$

and the scattering rate becomes

$$W(\mathbf{k}) = \frac{2\pi}{\hbar}(V_a - V_b)^2 \Omega_0 x(1-x) N(E), \tag{4.167}$$

whre $N(E)$ is the density of states for a given spin:

$$N(E) = \frac{(2m^*)^{3/2} E(\mathbf{k})^{1/2}}{4\pi^2\hbar^3} \tag{4.168}$$

and Ω_0 is the unit cell volume. The collisions are elastic.

The interaction potential $V_a - V_b$ has been variously interpreted as the difference of psuedo potentials, the difference of band edge energies in the two component binary compounds, and the difference of electron affinities. However, elastic strain will also add a deformation-potential component, so it is probably prudent to regard $V_a - V_b$ as a quantity to be determined empirically, though this is not easy to do.

Alloy scattering can also be treated as scattering from spherically symmetrical islands of deviant potential. If the spheres are regarded as 'hard' then the energy dependence is entirely determined by the velocity of the particle, which gives the same dependence as eqns (4.167) and (4.168).

References

ABRAHAMS, E. (1954). *Phys. Rev.* **95,** 839.

ABRAMOWITZ, M. and STEGUN, I. A. (1972). *Handbook of mathematical functions.* Dover Publications, New York.

ANSELM, A. I. (1953). *Zh. ekspor. teor. Fiz.* **24,** 85.

Arikan, M. C., Hatch, C. B., and Ridley, B. K. (1980). *J. Phys. C* **13,** 635.
Blatt, F. J. (1957). *J. Phys. Chem. Solids* **1,** 262.
Bohm, D. and Pines, D. (1951). *Phys. Rev.* **82,** 625.
——— ——— (1952). *Phys. Rev.* **85,** 836.
——— ——— (1953). *Phys. Rev.* **92,** 609.
Brinson, M. E. and Dunstan, W. (1970). *J. Phys. C* **3,** 483.
Brooks, H. (1951). *Phys. Rev.* **83,** 879.
——— (1955). *Adv. Electron. Electron Phys.* **7,** 85.
Chandresekhar, S. (1943). *Rev. mod. Phys.* **15,** 1.
Chapman, S. and Cowling, T. G. (1958). *The mathematical theory of non-uniform gases.* Cambridge University Press, Cambridge.
Conwell, E. M. and Weisskopf, V. (1950). *Phys. Rev.* **77,** 388.
Csavinsky, P. (1976). *Phys. Rev. B* **14,** 1649.
Debye, P. and Hückel, E. (1923). *Phys. Z.* **24,** 185, 305.
Dingle, R. B. (1955). *Philos. Mag.* **46,** 831.
Ehrenreich, H. (1957). *J. Phys. Chem. Solids* **2,** 131.
El-Ghanem, H. M. A. and Ridley, B. K. (1960). *J. Phys. C* **13,** 2041.
Erginsoy, C. (1950). *Phys. Rev.* **79,** 1013.
Falicov, L. M. and Cuevas, M. (1967). *Phys. Rev.* **164,** 1025.
Fujita, S., Ko, C. L., and Chi, J. Y. (1976). *J. Phys. Chem. Solids* **37,** 227.
Furukawa, Y. (1961). *J. Phys. Soc. Jpn* **16,** 577.
Gränacher, I. and Czaja, W. (1967). *J. Phys. Chem. Solids* **28,** 231.
Hall, G. L. (1962). *J. Phys. Chem. Solids* **23,** 1147.
Ham, F. S. (1955). *Solid State Phys.* **1,** 127; *Phys. Rev.* **100,** 1251.
Holtsmark, J. (1919) *Annbr Phys. Lpz* **58,** 577.
Kohn, W. (1957) *Solid State Phys.* **5,** 257.
McGill, T. and Baron, R. (1975) *Phys. Rev. B* **11,** 5208.
McIrvine, E. C. (1960). *J. Phys. Soc. Jpn* **15,** 928.
Mansfield, R. (1956). *Proc. Phys. Soc. B* **69,** 76.
Massey, H. S. W. and Moisewitsch, B. L. (1950). *Phys. Rev.* **78,** 180.
Mattis, D. and Sinha, D. (1970). *Ann. Phys. N.Y.* **61,** 214.
Messiah, A. (1966). *Quantum mechanics.* John Wiley, New York.
Moore, E. J. (1967). *Phys. Rev.* **160,** 607, 618.
Morgan, T. N. (1963). *J. Phys. Chem. Solids* **24,** 1657.
Morin, F. J. and Maita, J. P. (1954). *Phys. Rev.* **96,** 28.
Mott, N. F. (1936). *Proc. Camb. Philos. Soc.* **32,** 281.
Nakajima, S. (1954). *Proc. Int. Conf. on Theoretical Physics, Kyoto and Tokyo.*
Pearson, G. L. and Bardeen, J. (1949). *Phys. Rev.* **75,** 865.
Pines, D. (1956). *Rev. mod. Phys.* **28,** 184.
Ralph, H. I., Simpson, G., and Elliot, R. J. (1975) *Phys. Rev. B* **11,** 2948.
Ridley, B. K. (1977). *J. Phys. A* **10,** L79; *J. Phys. C* **10,** 1589.
Schiff, L. (1955). *Quantum mechanics.* McGraw-Hill, New York.
Sclar, N. (1956). *Phys. Rev.* **104,** 1548, 1559.
Stratton, R. (1962). *J. Phys. Chem. Solids* **23,** 1011.
Takimoto, N. (1959). *J. Phys. Soc. Jpn* **14,** 1142.
Temkin, A. and Lamkin, J. C. (1961). *Phys. Rev.* **121,** 788.
Weinreich, G., Sanders, T. M., and White, H. G. (1959). *Phys. Rev.* **114,** 33.
Wolfstirn, K. B. (1960). *J. Phys. Chem. Solids* **16,** 279.

5. Radiative transitions

5.1. Transition rate

The disturbance of an electron by an electromagnetic field with a Lorentz-gauge vector potential **A** and a scalar potential equal to zero can usually be treated as a small perturbation which induces transitions between unperturbed states at a rate given by first-order perturbation theory:

$$W=\int_{\mathrm{f}}\frac{2\pi}{\hbar}|\langle\mathrm{f}|\,H_\nu\,|\mathrm{i}\rangle|^2\,\delta(E_\mathrm{f}-E_\mathrm{i})\,\mathrm{d}S_\mathrm{f} \tag{5.1}$$

where (see eqn (2.72))

$$H_\nu=-\frac{e}{m}\mathbf{A}.\mathbf{p}. \tag{5.2}$$

The relation between the vector potential and the photon number n_ν in a monochromatic wave can be derived as follows. We now take

$$\mathbf{A}=A_0\mathbf{a}\cos(\mathbf{q}_\nu.\mathbf{r}-\omega_\nu t) \tag{5.3}$$

where **a** is a unit vector parallel to **A** and note that **A** is related to the electric field $\mathscr{E}$ and magnetic field $\mathscr{H}$ as follows:

$$\mathscr{E}=-\frac{\partial\mathbf{A}}{\partial t}=-\omega_\nu A_0\mathbf{a}\sin(\mathbf{q}_\nu.\mathbf{r}-\omega_\nu t) \tag{5.4}$$

$$\mathscr{H}=\frac{1}{\mu}\boldsymbol{\nabla}\times\mathbf{A}=-\frac{1}{\mu}A_0(\mathbf{q}_\nu\times\mathbf{a})\sin(\mathbf{q}_\nu.\mathbf{r}-\omega_\nu t) \tag{5.5}$$

where μ is the permeability. We can now derive the Poynting vector **S** thus:

$$\begin{aligned}\mathbf{S}=\mathscr{E}\times\mathscr{H}&=\frac{\omega_\nu A_0^2}{\mu}\{\mathbf{a}\times(\mathbf{q}_\nu\times\mathbf{a})\}\sin^2(\mathbf{q}_\nu.\mathbf{r}-\omega_\nu t)\\&=\frac{\omega_\nu A_0^2}{\mu}\mathbf{q}_\nu\sin^2(\mathbf{q}_\nu.\mathbf{r}-\omega_\nu t)\end{aligned} \tag{5.6}$$

The time-average vector is therefore

$$\langle\mathbf{S}\rangle=\frac{A_0^2}{2\mu}\omega_\nu\mathbf{q}_\nu=\frac{A_0^2q_\nu^2}{2\mu}\mathbf{v}_\nu \tag{5.7}$$

with $\omega_\nu=\mathbf{v}_\nu.\mathbf{q}_\nu$ where $\mathbf{v}_\nu$ is the phase velocity.

Provided that we have a physical situation in which there is little absorption of energy and the group velocity and phase velocity of the wave can be identified with each other, which is the usual case, we can express the energy density $\langle E_\nu \rangle$ of the radiation as follows:

$$\langle E_\nu \rangle = \frac{\langle \mathbf{S} \rangle}{\mathbf{v}_\nu} = \frac{A_0^2 q_\nu^2}{2\mu}. \tag{5.8}$$

In terms of photons the energy density is just $n_\nu \hbar\omega_\nu / V$, where V is the volume of the crystal and n_ν is the number of photons. Furthermore, the magnitude of the velocity is $(\mu\epsilon_\nu)^{-1/2}$ where ϵ_ν is the optical permittivity. Consequently we can write

$$A_0^2 = \frac{2\hbar n_\nu}{V\epsilon_\nu \omega_\nu} \tag{5.9}$$

which is the desired relationship.

When we rewrite eqn (5.3) as

$$\mathbf{A} = \frac{A_0 \mathbf{a}}{2} [\exp\{\mathrm{i}(\mathbf{q}_\nu . \mathbf{r} - \omega_\nu t)\} + \exp\{-\mathrm{i}(\mathbf{q}_\nu . \mathbf{r} - \omega_\nu t)\} \tag{5.10}$$

and investigate the time dependence of the transition probability in the usual way, we find that the second term in eqn (5.10) induces stimulated emission of photons and only the first term induces absorption. Thus for absorption

$$H_{\nu\ \mathrm{abs}} = -\frac{e}{m}\left(\frac{\hbar n_\nu}{2V\epsilon_\nu \omega_\nu}\right)^{1/2} \exp(\mathrm{i}\mathbf{q}_\nu . \mathbf{r})\mathbf{a}.\mathbf{p} \tag{5.11}$$

and for emission, including spontaneous emission,

$$H_{\nu\,\mathrm{em}} = -\frac{e}{m}\left\{\frac{\hbar(n_\nu + 1)}{2V\epsilon_\nu \omega_\nu}\right\}^{1/2} \exp(-\mathrm{i}\mathbf{q}_\nu . \mathbf{r})\mathbf{a}.\mathbf{p} \tag{5.12}$$

where the time dependence is considered to have been absorbed in the delta function of eqn (5.1) which expresses conservation of energy.

It is well known that a transition rate can be defined only if there is a spread of states in energy or, where the transition is between discrete states, there is a finite bandwidth of incident radiation. Usually in semiconductors we are interested in transitions to and from conduction bands, and we shall assume that in the case of absorption a spread of final states always exist (Fig. 5.1).

It is usually possible to ignore the momentum associated with the photon since $\hbar q_\nu$ is small for visible and infrared light. As a consequence the delta function conserving energy can be regarded as independent of direction, and the sum over final states, regarding these as band states of

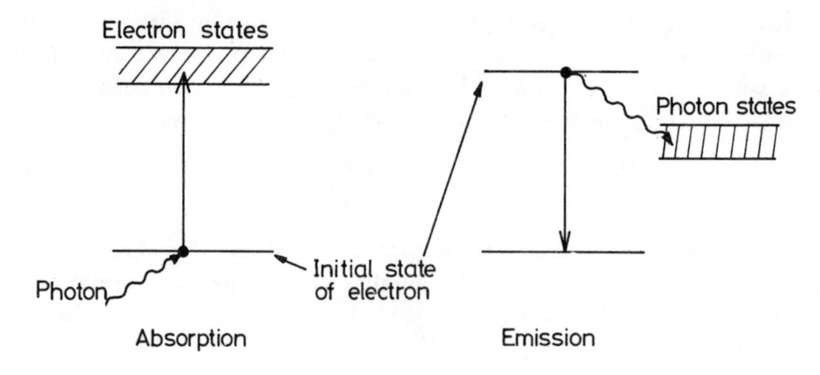

FIG. 5.1. Scheme of radiative transitions: (a) absorption; (b) emission.

given spin with density between E and $E+\mathrm{d}E$ given by $N(E)$, can be replaced by an integral:

$$W=\frac{2\pi}{\hbar}\int|\langle\mathrm{f}|\,H_\nu\,|\mathrm{i}\rangle|^2\,\delta(E_\mathrm{f}-E_\mathrm{i})\,VN(E)\,\mathrm{d}E$$

$$=\frac{2\pi}{\hbar}|\langle\mathrm{f}|\,H_\nu\,|\mathrm{i}\rangle|^2\,VN(E_\mathrm{f}) \tag{5.13}$$

where V is the volume of the periodic crystal. By expressing the rate in terms of the fine-structure constant α, the Rydberg energy R_H and the Bohr radius a_H† we obtain for the photoionization rate,

$$W_\mathrm{abs}=16\pi^2\alpha a_\mathrm{H}^2\left(\frac{R_\mathrm{H}}{\hbar\omega_\nu}\right)\left(\frac{c\epsilon_0}{\epsilon_\nu}\right)n_\nu\frac{|\mathbf{a}\cdot\mathbf{p}_\mathrm{if}|^2}{2m}N(E_\mathrm{f}) \tag{5.14}$$

where, taking $\mathbf{q}_\nu\approx 0$,

$$\mathbf{p}_\mathrm{if}=\langle\mathrm{f}|\,\mathbf{p}\,|\mathrm{i}\rangle. \tag{5.15}$$

In the reverse transition an electron in a definite band state is radiatively captured by a definite final state (Fig. 5.1). In this case a spread of final states does not exist and we must sum over a bandwidth of emitted radiation. If the density of photon states lying between $\hbar\omega_\nu$ and $\hbar\omega_\nu+\mathrm{d}(\hbar\omega_\nu)$ is $\rho_\nu(\hbar\omega_\nu)$, then

$$W_\mathrm{em}=16\pi^2\alpha a_\mathrm{H}^2\left(\frac{R_\mathrm{H}}{\hbar\omega_\nu}\right)\left(\frac{c\epsilon_0}{\epsilon_\nu}\right)(n_\nu+1)\frac{|\mathbf{a}\cdot\mathbf{p}_\mathrm{if}|^2}{2m}\rho_\nu(\hbar\omega_\nu). \tag{5.16}$$

By counting the modes of one type of polarization in our periodic crystal

---

† $$\alpha=\frac{e^2/4\pi\epsilon_0}{\hbar c};\qquad R_\mathrm{H}=\frac{e^2/4\pi\epsilon_0}{2a_\mathrm{H}};\qquad a_\mathrm{H}=\frac{\hbar^2/m}{e^2/4\pi\epsilon_0}$$

cavity we obtain

$$\rho_\nu(\hbar\omega_\nu)\,\mathrm{d}(\hbar\omega_\nu)=\frac{4\pi q_\nu^2\,\mathrm{d}q_\nu}{8\pi^3}=\frac{\omega_\nu^2}{2\pi^2\hbar v_\nu^3}\,\mathrm{d}(\hbar\omega_\nu) \tag{5.17}$$

so that

$$\rho_\nu(\hbar\omega_\nu)=\frac{\omega_\nu^2}{2\pi^2\hbar v_\nu^3} \tag{5.18}$$

where v_ν is the velocity of light in the crystal. Unless light of high intensity is present it is usually permissible to take $n_\nu \ll 1$ in eqn (5.16).

Where absorption of a photon causes a transition between two well-defined states, the rate is obtained by summing over the energy spectrum of the incident radiation. The number of radiation modes of a given polarization with wavevectors in a solid angle $\mathrm{d}\Omega$ and with magnitudes between q_ν and $q_\nu+\mathrm{d}q_\nu$ is

$$\frac{V}{8\pi^3}\,q_\nu^2\,\mathrm{d}q_\nu\,\mathrm{d}\Omega=V\rho_\nu(\hbar\omega_\nu)\frac{\mathrm{d}\Omega}{4\pi}\,\mathrm{d}E \tag{5.19}$$

and so

$$W_{\mathrm{abs}}=16\pi^2\alpha a_{\mathrm{H}}^2\left(\frac{R_{\mathrm{H}}}{\hbar\omega_\nu}\right)\left(\frac{c\epsilon_0}{\epsilon_\nu}\right)\frac{|\mathbf{a}\cdot\mathbf{p}_{\mathrm{if}}|^2}{2m}\,\rho_\nu(\hbar\omega_\nu)\int n_\nu\frac{\mathrm{d}\Omega}{4\pi} \tag{5.20}$$

where $\rho_\nu(h\omega_\nu)\int n_\nu\,\mathrm{d}\Omega/4\pi$ is just the spectral density of photons in the beam with frequency $\omega_\nu=(E_{\mathrm{f}}-E_{\mathrm{i}})/\hbar$.

5.1.1. *Local field correction*

In deriving the foregoing rates in terms of photon number, we have tacitly assumed that it is directly related to the incident intensity of the beam through eqns (5.7) and (5.9). This assumption will be valid provided that the centre absorbing or emitting photons experiences the electromagnetic wave present in the rest of the cavity. However, this will not be so if the centre scatters light. Scattering will occur whenever the polarizability at and near the centre differs from the rest of the medium, and so in general the field experienced in the absorption and emission process differs from the average value. This effect is usually introduced into all the above expressions by means of a rate multiplication factor $(\mathscr{E}_{\mathrm{eff}}/\mathscr{E}_0)^2$ where $\mathscr{E}_{\mathrm{eff}}$ is the effective electric field at the centre and $\mathscr{E}_0$ is the average field. We can expect this factor to be near unity for all transitions in which both the initial and final electronic states are extended appreciably over many unit cells. In contrast, for tightly bound electrons the ratio may be appreciably greater than unity. In the simple case of a centre entirely within a spherical cavity the local field is given by the Lorentz relationship (see

Dexter 1958):

$$\frac{\mathscr{E}_{\text{eff}}}{\mathscr{E}_0} = 1 + \frac{\eta_r^2 - 1}{3} \tag{5.21}$$

where η_r is the refractive index. In semiconductors this ratio would be about 5 and hence the local field correction would amount to a factor of 25! More realistically, a weighting should be introduced to reflect the extent of the initial electron state sampling the average field. For example, if V_T is the volume of the state and V_A is the volume of an atom in the lattice, then one can take

$$\frac{\mathscr{E}_{\text{eff}}}{\mathscr{E}_0} = 1 + \frac{V_A}{V_T} \frac{\eta_r^2 - 1}{3} \tag{5.22}$$

to be a plausible ratio. Even for deep-level states the ratio V_A/V_T is likely to be no greater than 10^{-2}, and hence $(\mathscr{E}_{\text{eff}}/\mathscr{E}_0)^2 \lesssim 2$.

In what follows we shall assume the ratio to be unity and not include it explicitly in the formulae.

5.1.2. Photon drag

In all optical transitions the small momentum of the photon is given up to or contributed by the charged particle which interacts with the light. As we have mentioned this can usually be ignored in calculations, but where mobile carriers are involved the momentum of the photon may manifest itself in charge motion which is detectable by the production of either an electric current or a voltage. This effect is known as photon drag, and it has received considerable attention in connection with the fast detection of laser light (see Gibson and Kimmitt 1980). The size and sign of the effect depends upon the band structure and on the details of conduction processes and their energy dependence. Only in the simplest cases is it found that photon drag behaves classically as the result of radiation pressure. One important conclusion which has emerged from the study of photon drag is that the momentum of a photon in a dielectric medium of refractive index η_r is indeed the Minkowski (1910) expression

$$p_\nu = \eta_r \hbar q_0 = \hbar q_\nu \tag{5.23}$$

where q_0 is the wavevector in free space and q_ν is the wavevector in the dielectric. The wavevector conservation law which appears in optical processes involving electrons in bands is thus equivalent to momentum conservation.

5.2. Photo-ionization and radiative capture cross-sections

Transition rates involving photons depend upon ambient light intensity. To obtain a measure of the radiative processes which is independent of light intensity it is convenient to define a cross-section for the process.

Let us suppose that in our volume V there is a density of centres n_T which can obsorb a photon and lose an electron to a band, and that each centre presents an effective area σ_ν, which we shall call the photo-ionization cross-section, to the photon flux. The number of photons absorbed per second is therefore the photon flux $n_\nu v_\nu/V$ times the total area $n_T V\sigma_\nu$ presented by the centres. Since this rate is the photo-ionization rate for $n_T V$ centres, the rate for one centre is

$$W_{\text{abs}} = n_\nu v_\nu \sigma_\nu / V. \tag{5.24}$$

Thus, from eqn (5.14), we obtain for the photo-ionization cross-section

$$\sigma_\nu = 16\pi^2 \alpha a_H^2 \left(\frac{R_H}{\hbar\omega_\nu}\right) \frac{1}{\eta_r} \frac{|\mathbf{a}.\mathbf{p}_{\text{if}}|^2}{2m} VN(E_f) \tag{5.25}$$

where η_r is the refractive index which is equal to $(\epsilon_\nu/\epsilon_0)^{1/2}$ provided that the crystal is transparent.

In the case of radiative capture we can define a capture cross-section such that

$$W_{\text{em}} = v_{\mathbf{k}}\sigma_{\mathbf{k}}/V \tag{5.26}$$

where $v_{\mathbf{k}}$ is the group velocity associated with the band state $|\mathbf{k}\rangle$. Assuming spontaneous emission only, we obtain

$$\sigma_{\mathbf{k}} = 16\pi^2 \alpha a_H^2 \left(\frac{R_H}{\hbar\omega_\nu}\right)\left(\frac{c}{\eta_r^2 v_{\mathbf{k}}}\right) \frac{|\mathbf{a}.\mathbf{p}_{\text{if}}|^2}{2m} V\rho_\nu(\hbar\omega_\nu). \tag{5.27}$$

In general we have

$$N(E_{\mathbf{k}})\,dE_{\mathbf{k}} = \frac{4\pi k^2\,dk}{8\pi^3} = \frac{k^2}{2\pi^2\hbar v_{\mathbf{k}}}\,dE_{\mathbf{k}} \tag{5.28}$$

and, from eqn (5.18),

$$\rho_\nu(\hbar\omega_\nu) = \frac{q_\nu^2}{2\pi^2\hbar v_\nu} \tag{5.29}$$

whence we obtain the simple relationship

$$\sigma_{\mathbf{k}}/\sigma_\nu = q_\nu^2/k^2. \tag{5.30}$$

For a transition energy of about 0·6 eV, ω_ν is about $10^{15}\ \text{s}^{-1}$ and hence q_ν is roughly $10^5\ \text{cm}^{-1}$. States near the band edge are associated with values of k around $10^7\ \text{cm}^{-1}$, and so the ratio q_ν^2/k^2 is of order 10^{-4}. It turns out that typical values of σ_ν range around $10^{-17}\ \text{cm}^2$, and thus we can predict that radiative capture cross-sections are only of order $10^{-21}\ \text{cm}^2$.

It is often useful to work with the capture rate per unit density $c_{\mathbf{k}}$, which is alternatively called the volume capture rate. From eqn (5.26)

$$c_{\mathbf{k}} = W_{\text{em}}V = v_{\mathbf{k}}\sigma_{\mathbf{k}}. \tag{5.31}$$

5.3. Wavefunctions

In order to evaluate the matrix element we need to know the form of the wavefunctions of the initial and final states. In a photo-ionization process the final state is in general a superposition of a Bloch wave of the band and a spherical scattered wave:

$$\psi_{n\mathbf{k}}(\mathbf{r}) = V^{-1/2}\left\{u_{n\mathbf{k}}(\mathbf{r})\exp(\mathrm{i}\mathbf{k}.\mathbf{r}) + \phi_{s\mathbf{k}}(\theta, \phi)\frac{\exp(\mathrm{i}kr)}{r}\right\}\bigg|_{r\to\infty}. \tag{5.32}$$

This is also the appropriate form for electron–hole generation across the gap, since the electron and hole scatter off one another. Where the scattering potential is slowly varying over a unit cell, as it is at long range from a charged centre, the wavefunction contains no appreciable components from other bands; thus

$$\psi_{n\mathbf{k}}(\mathbf{r}) = V^{-1/2}u_{n\mathbf{k}}(\mathbf{r})\left\{\exp(\mathrm{i}\mathbf{k}.\mathbf{r}) + \phi_{\mathbf{k}}(\theta, \phi)\frac{\exp(\mathrm{i}kr)}{r}\right\}\bigg|_{r\to\infty} \tag{5.33}$$

or alternatively

$$\psi_{n\mathbf{k}}(\mathbf{r}) = V^{-1/2}u_{n\mathbf{k}}(\mathbf{r})S_{\mathbf{k}}(\mathbf{r}) \tag{5.34}$$

where $S_{\mathbf{k}}(\mathbf{r})$ is an envelope function associated with $\mathbf{k}$, e.g. a coulomb wavefunction. Near an impurity centre the potential may vary rapidly and there could be strong core effects. Such core effects would influence the wavefunction in the band only if the scattering were strong—a tautological remark meant to emphasize that modifications of the Bloch function are associated with scattering strength. Since cores are short range by nature they present a small collision cross-section to electrons unless there exists a resonance. In the absence of resonant scattering the envelope function is determined primarily by coulomb scattering. Where resonances exist the effect of the core is felt far from the centre, and it adds to any coulombic effects. However, whether resonances exist or not, the short-range effect of the core is a difficult problem to solve and in working out matrix elements implicit reliance is usually placed on the greater weight thrown on the region of space outside of the core by the three-dimensional integration, and the wavefunction outside the core is considered to be an adequate form to take for the purposes of computing the transition rate. We intend to follow this approximation. Thus we take the form given in eqn (5.34) where, in general, $S_{\mathbf{k}}(\mathbf{r})$ is determined by coulomb and resonant scattering.

For the case of spherical scattering

$$S_{\mathbf{k}}(\mathbf{r}) = (kr)^{-1}\sum_{l=0}^{\infty}(2l+1)\exp(\mathrm{i}\phi_l)\{F_l(kr)\cos\delta_l + G_l(kr)\sin\delta_l\}P_l(\cos\theta) \tag{5.35}$$

where ϕ_l is a phase factor, $F_l(kr)$ is the regular and $G_l(kr)$ the irregular coulomb function, δ_l is the phase shift of the lth wave induced by the core, and $P_l(\cos\theta)$ is a Legendre polynomial of order l. Strictly, $S_{\mathbf{k}}(\mathbf{r})$ never achieves the form of a pure plane wave at large distances because of the long-range character of the coulomb field, but this is highly artificial. Many other influences enter at large distances and randomize the phase of the wave, so at sufficiently large distances from the scattering centre it is reasonable to assume that $S_{\mathbf{k}}(\mathbf{r}) \rightarrow \exp(i\mathbf{k}.\mathbf{r})$.

The principal spatial domain of the matrix element is governed by the overlap of the initial and final wavefunctions. If the domain is large, as it is in the case of electron–hole pair production or even in the case of excitation from an effective-mass centre with its large ground-state orbit, it is a reasonable approximation over most of the final energy range to neglect scattering effects and take

$$S_{\mathbf{k}}(\mathbf{r}) = \exp(i\mathbf{k}.\mathbf{r}). \tag{5.36}$$

However, if the domain is small, as it is in the case of a deep-level impurity with its tightly bound ground state, it is simplest, if somewhat crude, to take the form of the wavefunction at $r=0$. This implies selecting only the $l=0$ term in eqn (5.35) ($kr \ll 1$):

$$S_{\mathbf{k}}(\mathbf{r}) \approx (kr)^{-1}\{F_0(kr)\cos\delta_0 + G_0(kr)\sin\delta_0\}. \tag{5.37}$$

If $2|\mu| \gg kr$,

$$S_{\mathbf{k}}(\mathbf{r}) \approx C_0^{1/2}\left(\cos\delta_0 + \frac{1}{C_0 kr}\sin\delta_0\right) \tag{5.38}$$

where

$$C_0 = \frac{2\pi\mu}{1-\exp(-2\pi\mu)} \tag{5.39}$$

and

$$\mu = Z(R_{\mathrm{H}}/E_{\mathbf{k}})^{1/2}. \tag{5.40}$$

If $2|\mu| \not\gg kr$, $|\mu|$ is so small that the centre acts as a neutral centre ($C_0 = 1$ in eqn (5.38)) $S_{\mathbf{k}}(\mathbf{r})$ denotes the enhancement of the electron wavefunction at positively charged ($Z>0$) sites, or the attenuation at negatively charged ($Z<0$) sites. For neutral sites the coulomb factor C_0 is unity.

Whether or not one takes scattering into account therefore depends upon the extent of the domain in which overlap occurs. However, it also depends upon electron energy in the final state. In the absence of core resonance at energies above zero, scattering is appreciable only for low-energy electrons, specifically for electrons with $kr<1$. If the effective radius of the overlap is r_{T} then scattering effects will be important when the final state wavevector obeys $0 < kr_T < 1$. Thus even in the case of electron–hole production or the photo-ionization of a hydrogenic state, scattering effects will be important very close to threshold. Indeed, coulomb scattering affects the final wavefunction when $ka_H^* \lesssim 1$, where a_H^* is the effective Bohr radius, equivalent to $E_k \lesssim R_H^*$ where R_H^* is the effective Rydberg energy. If the core is considered to scatter resonantly, the phase shift can be written in the form

$$\tan \delta_0 = \frac{\Gamma(E_{\mathbf{k}})}{E_{\mathbf{k}} - E_r} \tag{5.41}$$

where E_r is an energy for resonance (which may be negative corresponding to a bound state) and $\Gamma(E_{\mathbf{k}})$ is the bandwidth of the resonance. For non-resonant scattering $\delta_0 \approx 0$. Equation (5.41) is a useful form for all core scattering.

In general, the wavefunction of a localized state can be expressed in terms of a sum of Bloch functions $\phi_{p\mathbf{k}}(\mathbf{r})$:

$$\psi_T(\mathbf{r}) = \sum_{p,\mathbf{k}} c_{p\mathbf{k}} \phi_{p\mathbf{k}}(\mathbf{r}) \tag{5.42}$$

where the sum is over all bands and all k in the first Brillouin zone. Shallow states can be regarded as being composed of Bloch functions drawn from the nearest band only. Thus

$$\psi_T(\mathbf{r}) \approx V^{1/2} \sum_{\mathbf{k}} c_{\mathbf{k}} u_{\mathbf{k}}(\mathbf{r}) \exp(i\mathbf{k} \cdot \mathbf{r}). \tag{5.43}$$

However, if the level is deep we may expect many bands and many **k** states to contribute. Consequently the general expansion in terms of Bloch functions must be retained. If we consider the effective radius of the state to be r_T, inside which we have a sum of cell-periodic Bloch functions, the spread of k involved is roughly

$$0 < k \lesssim r_T^{-1} \tag{5.44}$$

and so for these states

$$c_{p\mathbf{k}} \approx \left(\frac{V_T}{V}\right)^{1/2} c_{p0} \tag{5.45}$$

where V_T is the volume of the state. Thus an approximate 'billiard-ball' model wavefunction for a deep-level ground state can be written

$$\psi_T(\mathbf{r}) \approx \left(\frac{V_T}{V}\right)^{1/2} \sum_p \sum_{\mathbf{k}=0}^{r_T^{-1}} c_{p0} u_{p\mathbf{k}}(\mathbf{r}). \tag{5.46}$$

In this way we separate the effect of band mixing produced by the impurity potential which determines the coefficients c_{p0} from the effect of the spatial extent of the state as measured by the effective volume V_T.

We can now proceed to calculate the momentum matrix elements and the rates and cross-sections for the following transitions:

(a) valence band to conduction band;
(b) shallow acceptor to conduction band;
(c) shallow donor to conduction band;
(d) deep level to conduction band.

5.4. Direct interband transitions

Since both the initial and the final states are extended it will be sufficient to ignore scattering corrections for $E_{\mathbf{k}} \gg R_{\mathrm{H}}^{*}$ and take Bloch functions. Thus for the conduction band state

$$\psi_{\mathrm{c}\mathbf{k}_{\mathrm{c}}}(\mathbf{r}) = V^{-1/2} u_{\mathrm{c}\mathbf{k}_{\mathrm{c}}}(\mathbf{r}) \exp(\mathrm{i}\mathbf{k}_{\mathrm{c}} \cdot \mathbf{r}) \tag{5.47}$$

and for the valence band state

$$\psi_{\mathrm{v}\mathbf{k}_{\mathrm{v}}}(\mathbf{r}) = V^{-1/2} u_{\mathrm{v}\mathbf{k}_{\mathrm{v}}}(\mathbf{r}) \exp(\mathrm{i}\mathbf{k}_{\mathrm{v}} \cdot \mathbf{r}) \tag{5.48}$$

where the subscripts c and v refer to conduction and valence states. Since $\mathbf{p}$ is an operator with odd parity and the cell-periodic components of the Bloch functions are $|\mathrm{s}\rangle$ like and $|\mathrm{p}\rangle$ like, the momentum matrix element is non-zero and largely independent of k near $k = 0$. The sum over unit cells in the matrix element as usual leads to the conservation of crystal momentum, i.e. (neglecting the photon momentum) (Fig. 5.2)

$$\mathbf{k}_{\mathrm{c}} = \mathbf{k}_{\mathrm{v}}. \tag{5.49}$$

Thus there is no spread of final states, but we can still use eqn (5.13) with $VN(E_{\mathrm{f}})$ replaced by $\mathrm{d}E^{-1}$ and eventually sum over initial states. Also $\mathbf{p}_{\mathrm{if}} = \mathbf{p}_{\mathrm{cv}}$ where $\mathbf{p}_{\mathrm{cv}}$ is the momentum matrix element between conduction and valence band at the zone centre (see $\mathbf{k} \cdot \mathbf{p}$ approximation). A sum over initial states with wavevector magnitude k_{v} gives the total photo-ionization cross-section from a given energy, corresponding to the expression in eqn (5.25). Since $\mathbf{p}_{\mathrm{cv}}$ is, to a first approximation, independent of

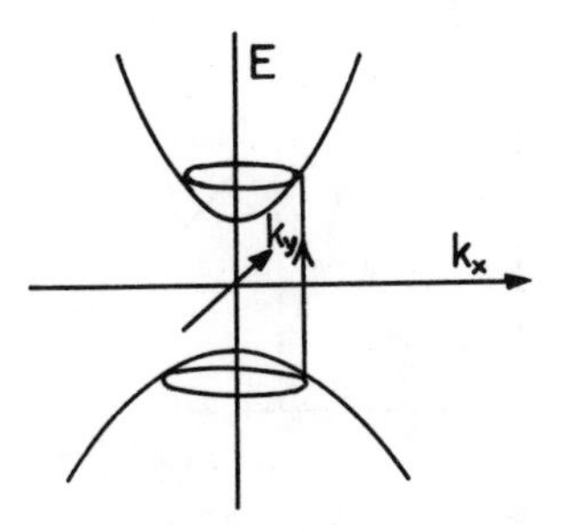

FIG. 5.2. Vertical transitions between bands.

energy, the rate is simply proportional to the density of states determined by the exciton reduced mass.

If the bands are parabolic $N(E_f)$ is proportional to $(\hbar\omega_\nu - E_{cv})^{1/2}$. If we put $x = \hbar\omega_\nu / E_{cv}$ we obtain the spectral dependence of the photo-ionization cross-section of the form

$$\sigma_\nu \propto (x-1)^{1/2}/x \tag{5.50}$$

which maximizes at $x = 2$. Long before this energy is reached, the bands become highly non-parabolic, and the true maximum will be largely determined by points in the zone where the density of states maximizes. This transition is termed 'allowed' because $\mathbf{p}_{cv}$ is non-zero at threshold.

Near threshold we must include the effects of coulomb scattering. Effectively, that means enhancing the matrix element by the coulomb factor C_0 for the exciton. By putting $Z = +1$ in eqn (5.40) and taking $E_\mathbf{k} < E_{ex}$, where E_{ex} is the binding energy of the exciton, we can write

$$C_{0\,ex} \approx 2\pi (E_{ex}/E_\mathbf{k})^{1/2} \tag{5.51}$$

whence

$$\sigma_\nu \propto 1/x. \tag{5.52}$$

The enhancement of the wavefunction caused by the coulombic attraction between electron and hole exactly balances the diminution of the density of states towards zero energy. There is thus a sudden onset of photo-ionization at threshold (Fig. 5.3). Figure 5.4 shows the effect of excitonic absorption in GaAs.

The expression for the photo-ionization cross-section for direct inter-band transitions not too near threshold is, for unpolarized light,

$$\sigma_\nu = \tfrac{8}{3}\alpha a_H^2 \left(\frac{R_H}{\hbar\omega_\nu}\right) \frac{1}{\eta_r} \frac{p_{cv}^2}{2m} \left(\frac{2m_r^*}{\hbar^2}\right)^{3/2} E_\mathbf{k}^{1/2} V \tag{5.53}$$

where we have used

$$|\mathbf{a} \cdot \mathbf{p}_{cv}|^2 = \tfrac{1}{3} p_{cv}^2 \tag{5.54}$$

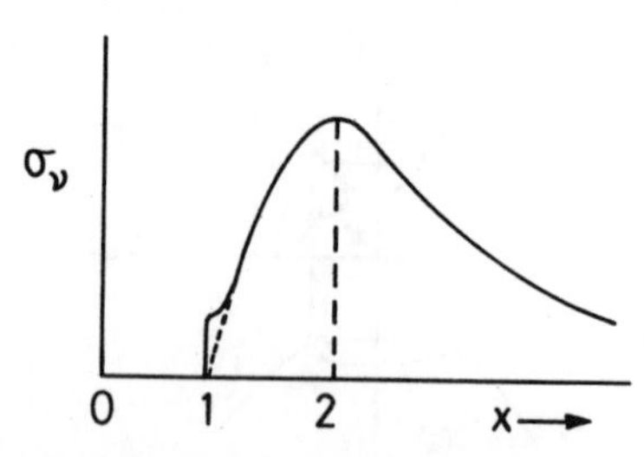

FIG. 5.3. Photo-ionization cross-section for direct-gap electron–hole pair production. The dotted portion of the curve represents the plane-wave final state.

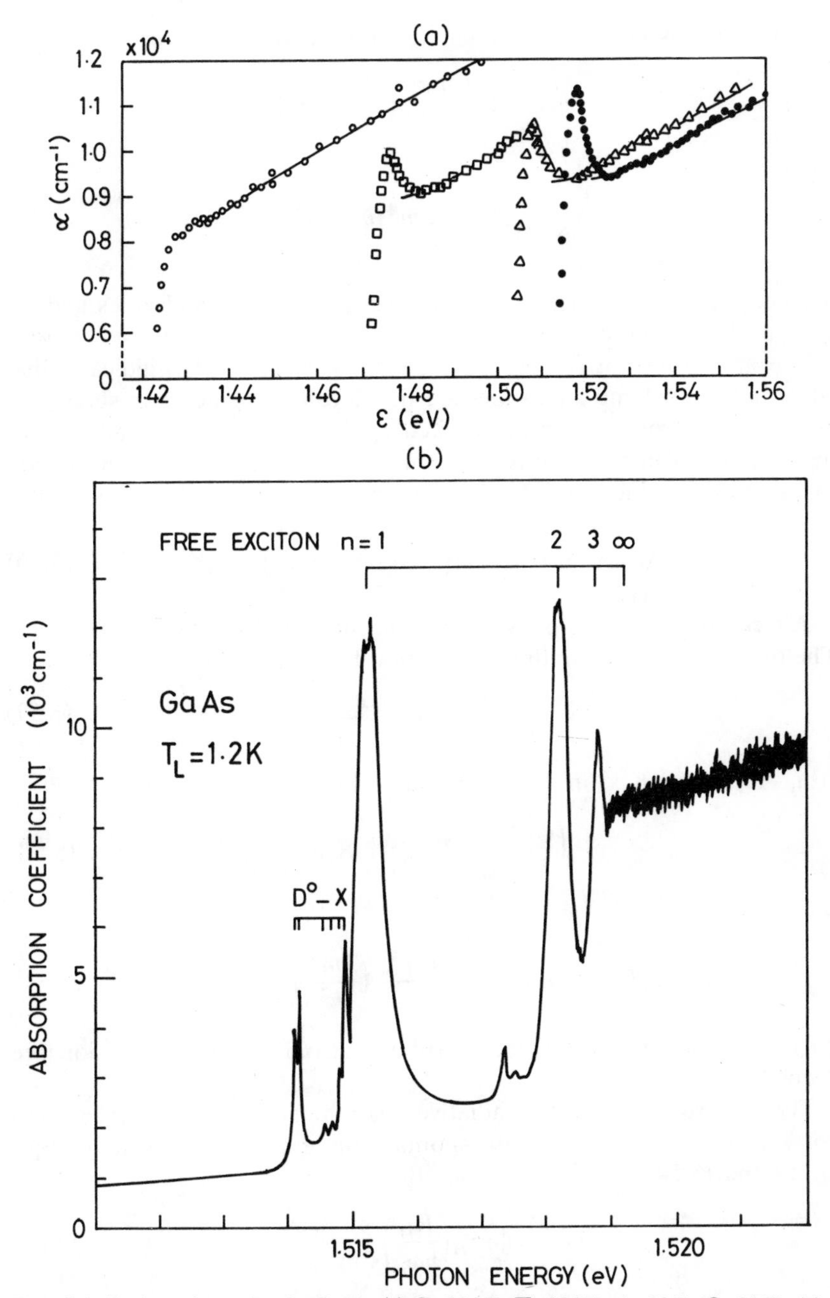

FIG. 5.4. Exciton absorption in GaAs: (a) ○, 294 K; □, 186 K; △, 90 K; ●, 21 K. After Sturge 1962.) (b) Fine detail in epitaxial material at 1·2 K. The $n = 1$ peak is limited by luminescence. Also shown are impurity lines (D_0X) from excitons bound to some 10^{15} cm^{-3} donors (After Weisbuch 1977).

and we have assumed that the bands are parabolic such that

$$\hbar\omega_\nu = E_{\mathrm{g}} + \frac{\hbar k^2}{2}\left(\frac{1}{m_{\mathrm{e}}^*} + \frac{1}{m_{\mathrm{h}}^*}\right)$$

whence

$$N_{\mathrm{cv}}(E_{\mathbf{k}}) = \frac{2(2m_{\mathrm{r}}^*/\hbar^2)^{3/2}E_{\mathbf{k}}^{1/2}}{4\pi^2} \tag{5.55}$$

whence m_{r}^* is the reduced effective mass of electrons and holes and we have included spin degeneracy.

The appearance of the arbitary volume V in eqn (5.53) indicates that when both initial and final states are non-localized plane-wave states, the concept of a cross-section has limited application. The physically meaningful quantity in this case is the photo-ionization rate, and so using eqn (5.24) (or returning to eqn (5.14)), we have

$$W_{\mathrm{abs}} = \tfrac{8}{3}\alpha a_{\mathrm{H}}^2\left(\frac{R_{\mathrm{H}}}{\hbar\omega_\nu}\right)\left(\frac{cn_\nu}{\eta_{\mathrm{r}}^2}\right)\frac{p_{\mathrm{cv}}^2}{2m}\left(\frac{2m_{\mathrm{r}}^*}{\hbar^2}\right)^{3/2}E_{\mathbf{k}}^{1/2}. \tag{5.56}$$

Alternatively, we can convert the rate into an absorption constant K_ν. The one-dimensional continuity equation for photons is

$$\frac{\partial n_\nu}{\partial t} = \left(\frac{\partial n_\nu}{\partial t}\right)_{\mathrm{vol}} - \frac{\partial(v_\nu n_\nu)}{\partial x}. \tag{5.57}$$

If $n_\nu = n_{\nu0}\exp(-K_\nu x)$ in the steady state, then

$$\left(\frac{\partial n_\nu}{\partial t}\right)_{\mathrm{vol}} = W_{\mathrm{abs}} = -K_\nu v_\nu n_\nu \tag{5.58}$$

whence

$$K_\nu = \tfrac{8}{3}\alpha a_{\mathrm{H}}^2\left(\frac{R_{\mathrm{H}}}{\hbar\omega_\nu}\right)\frac{1}{\eta_{\mathrm{r}}}\frac{p_{\mathrm{cv}}^2}{2m}\left(\frac{2m_{\mathrm{r}}^*}{\hbar^2}\right)^{3/2}E_{\mathbf{k}}^{1/2}. \tag{5.59}$$

(Strictly, we should subtract the contribution from stimulated emission (see Section 5.12).)

The reverse process is the radiative recombination of an electron and a hole. The rate associated with spontaneous emission is given by eqns (5.16) and (5.54):

$$W_{\mathrm{em}} = \tfrac{8}{3}\alpha a_{\mathrm{H}}^2\left(\frac{R_{\mathrm{H}}}{\hbar\omega_\nu}\right)\frac{p_{\mathrm{cv}}^2}{2m}\frac{\eta_{\mathrm{r}}\omega_\nu^2}{\hbar c^2}. \tag{5.60}$$

As regards magnitudes, $\alpha a_{\mathrm{H}}^2 = 2{\cdot}044\times10^{-19}\,\mathrm{cm}^2$, $R_{\mathrm{H}}/\hbar\omega_\nu$ is approximately 10 when $\hbar\omega_\nu$ is about 1 eV, $p_{\mathrm{cv}}^2/2m$ is $E_{\mathrm{g}}/2$, and $N(E_{\mathbf{k}})$ is typically $10^{19}\,\mathrm{eV}^{-1}\,\mathrm{cm}^{-3}$ at $E_{\mathbf{k}} \approx 0{\cdot}01$ eV. The refractive index is about 4, and so from eqn (5.56) the rate of photo-ionization per photon is about $2\times 10^{12}\,\mathrm{s}^{-1}$, corresponding via eqn (5.59) to an absorption constant of $4\times 10^{2}\,\mathrm{cm}^{-1}$. The rate and absorption constant are very large as a consequence of the high density of states and strong matrix element. The capture rate per hole from eqn (5.60) is about $2\times 10^{7}\,\mathrm{s}^{-1}$. This is a substantial rate, but it assumes that the final state is unoccupied. When the statistical average rate is computed, W_{em} has to be weighted with the probability of a hole occupying the state. Statistical factors of this sort will generally be ignored for the present. For the purposes of comparison a volume capture rate associated with the capture of an electron by a valence band state can be defined by dividing W_{em} by the effective density of states N_{v} in the valence band which is typically of order $10^{18}\,\mathrm{cm}^{-3}$. Thus $c_{\mathrm{n}}\approx 10^{-11}\,\mathrm{cm}^{3}\,\mathrm{s}^{-1}$ for interband radiative capture.

5.4.1. Excitonic absorption

Sharp absorption peaks on the long-wavelength side of the functional absorption edge are observed at low temperatures (Fig. 5.4) and are associated with the formation of excitons. In the effective-mass approximation the final state wavefunction (Chapter 2, Section 2.6) for the electron is

$$\psi_{\mathrm{ex}}(\mathbf{r})=\frac{1}{V^{1/2}}u_{\mathrm{c0}}(\mathbf{r})F_{\mathrm{ex}}(\mathbf{r})\exp(\mathrm{i}\mathbf{k}_{\mathrm{ex}}\cdot\mathbf{R}) \tag{5.61}$$

where for simple bands the ground-state $|\mathrm{s}\rangle$ function is given by

$$F_{\mathrm{ex}}(\mathbf{r})=(\pi a_{\mathrm{ex}}^{*3})^{-1/2}\exp(-r_{\mathrm{eh}}/a_{\mathrm{ex}}^{*}) \tag{5.62}$$

where a_{ex}^{*} is the effective Bohr radius of the exciton, $\mathbf{R}$ is the centre-of-mass co-ordinate and $\mathbf{r}_{\mathrm{eh}}$ is the separation of the electron and the hole. The energy of the exciton is

$$E=E_{\mathrm{ex}}+\frac{\hbar^{2}k_{\mathrm{ex}}^{2}}{2(m_{\mathrm{e}}^{*}+m_{\mathrm{h}}^{*})}\,. \tag{5.63}$$

It is necessary that $k_{\mathrm{ex}}\approx 0$ (neglecting the photon's momentum), and that the electron and hole be created on the same atom. The probability of this, using eqn (5.62), is $V_0/\pi a_{\mathrm{ex}}^{*3}$, where V_0 is the volume of a unit cell so that, from eqn (5.25),

$$\sigma_{\mathrm{v}}=16\pi^{2}\alpha a_{\mathrm{H}}^{2}\left(\frac{R_{\mathrm{H}}}{\hbar\omega}\right)\frac{1}{\eta_{\mathrm{r}}}\frac{|\mathbf{a}\cdot\mathbf{p}_{\mathrm{cv}}|^{2}}{2m}\frac{V_{0}}{\pi a_{\mathrm{ex}}^{*3}}\cdot VN(E_{\mathrm{f}}) \tag{5.64}$$

where

$$V_{0}VN(E_{\mathrm{f}})\approx V\delta(E_{\mathrm{f}}-E_{\mathrm{ex}}) \tag{5.65}$$

giving an infinitely sharp absorption line. In reality the line is broadened by collisions and the delta function replaced by

$$F(E_p) = \frac{1}{\pi} \frac{\hbar\Gamma}{(E_f - E_{ex})^2 + \hbar^2\Gamma^2}, \tag{5.66}$$

where Γ is the scattering rate. The exciton is created with zero spin and so to accommodate the angular momentum of the photon it ends up in the triplet state. Radiative annihilation entails the same conditions.

A full treatment of excitonic absorption has been given by Elliot (1957) in which forbidden as well as allowed transitions are considered, and the influence of anisotropic masses has been discussed by Altarelli and Lipari (1976).

5.5. Photo-deionization of a hydrogenic acceptor

The excitation to the conduction band of an electron from the valence band in the vicinity of a hydrogenic acceptor is exactly equivalent to the excitation of a hole from the conduction band to one of the localized effective-mass states of the acceptor centre. In the process the negative charge on the acceptor is lost, and so the process is one of deionization.

As in Section 5.4 the conduction band state is assumed to be a Bloch function but, unlike the previous case, there is no coulombic scattering to be taken into account since the acceptor is neutral when the electron is in the conduction band. (Really, we are assuming that the acceptor remains neutral for times which are long compared with the period associated with the transition under consideration. This assumption would break down if thermal ionization were too rapid.)

The localized state is described by the effective-mass wavefunction

$$\psi_T(\mathbf{r}) = u_v(\mathbf{r})F_T(\mathbf{r}) \tag{5.67}$$

where $u_v(\mathbf{r})$ is the cell-periodic part of the valence-band Bloch function and $F_T(\mathbf{r})$ is the envelope function. If there is no resonant scattering of the electron by the neutral acceptor, the momentum matrix element is, in the long-wavelength limit,

$$\mathbf{p}_{if} = \mathbf{p}_{cv} V^{-1/2} \int \exp(-i\mathbf{k}.\mathbf{r})F_T(\mathbf{r})\,d\mathbf{r} \tag{5.68}$$

where $\mathbf{p}_{cv}$, as before, is the interband momentum matrix element which is assumed to be insensitive to $\mathbf{k}$ for the final states of interest. For the ground state we shall ignore the degeneracy of the valence band and take the simple-band form

$$F_T(\mathbf{r}) = (\pi a_H^{*3})^{-1/2} \exp(-r/a_H^*) \tag{5.69}$$

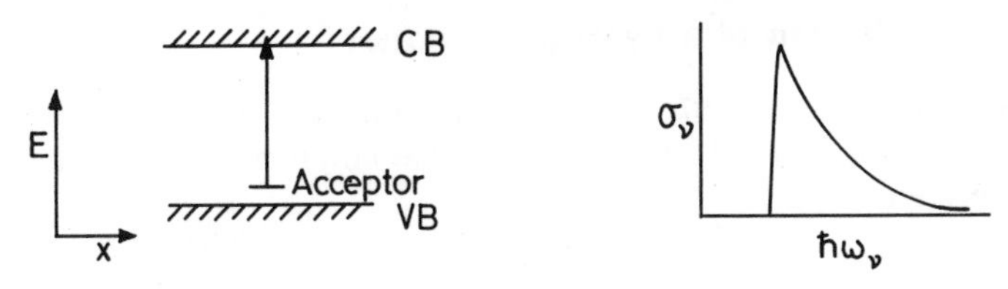

FIG. 5.5. Photo-deionization of an acceptor.

whence

$$\mathbf{p}_{\mathrm{if}} = \mathbf{p}_{\mathrm{cv}} \frac{8}{i} \left(\frac{\pi a_{\mathrm{H}}^{*3}}{V} \right)^{1/2} \frac{1}{(1+k^2 a_{\mathrm{H}}^{*2})^2}. \tag{5.70}$$

In the case of a parabolic conduction band and unpolarized light we obtain for the photo-deionization cross-section from eqn (5.25)

$$\sigma_\nu = \tfrac{256}{3} \alpha a_{\mathrm{H}}^2 \left(\frac{R_{\mathrm{H}}}{\hbar\omega_\nu} \right) \frac{1}{\eta_{\mathrm{r}}} \frac{p_{\mathrm{cv}}^2}{2m} \frac{\pi a_{\mathrm{H}}^{*3}}{\{1+(E_{\mathbf{k}}/R_{\mathrm{H}}^*)\}^4} \left(\frac{2m^*}{\hbar^2} \right)^{3/2} E_{\mathbf{k}}^{1/2}. \tag{5.71}$$

This function peaks sharply at $E_{\mathbf{k}} = R_{\mathrm{H}}^*/7$ (Fig. 5.5). The magnitude is determined by the effective volume of the state, characterized by a_{H}^{*3}.

The reverse process—radiative trapping of an electron by a neutral acceptor—is described by the cross-section $\sigma_{\mathbf{k}}$ related to σ_ν via eqn (5.30) with q_ν given by

$$q_\nu \approx \frac{E_{\mathrm{cv}} - R_{\mathrm{H}}^*}{\hbar(c/\eta_{\mathrm{r}})} \tag{5.72}$$

where R_{H}^* is the hydrogenic ground-state binding energy for the valence band. Thus

$$\sigma_{\mathbf{k}} = \tfrac{256}{3} \alpha a_{\mathrm{H}}^2 \left(\frac{R_{\mathrm{H}}}{\hbar\omega_\nu} \right) \frac{1}{\eta_{\mathrm{r}}} \frac{p_{\mathrm{cv}}^2}{2m} \frac{\pi a_{\mathrm{H}}^{*3}}{\{1+(E_{\mathbf{k}}/R_{\mathrm{H}}^*)\}^4} \frac{(2m^*/\hbar^2)^{1/2}}{\hbar^2(c/\eta_{\mathrm{r}})^2} \frac{(E_{\mathrm{cv}} - R_{\mathrm{H}}^*)^2}{E_{\mathbf{k}}^{1/2}} \tag{5.73}$$

or, in terms of volume capture rate (eqn (5.31)),

$$c_{\mathbf{k}} = \tfrac{256}{3} \alpha a_{\mathrm{H}}^2 \left(\frac{R_{\mathrm{H}}}{\hbar\omega_\nu} \right) \frac{1}{\eta_{\mathrm{r}}} \frac{p_{\mathrm{cv}}^2}{2m} \frac{\pi a_{\mathrm{H}}^{*3}}{\{1+(E_{\mathbf{k}}/R_{\mathrm{H}}^*)\}^4} \frac{2(E_{\mathrm{cv}} - R_{\mathrm{H}}^*)^2}{\hbar^3(c/\eta_{\mathrm{r}})^2}. \tag{5.74}$$

The capture cross-section becomes very large as $E_{\mathbf{k}}$ approaches zero, but since the velocity of the electron approaches zero the capture rate remains finite as eqn (5.74) shows (see also Dumke 1963).

Near the maximum σ_ν is typically of order 10^{-15} cm^2 and, since q_ν is about 10^5 cm^{-1}, the radiative capture cross-section is only some 10^2 times smaller, i.e. 10^{-17} cm^2, and $c_{\mathbf{k}}$ is about 10^{-10} cm^3 s^{-1}. The latter is of the same order of magnitude as that associated with direct interband transitions.

5.6. Photo-ionization of a hydrogenic donor

In the photo-excitation of an electron from the ground state of a hydrogenic donor to a scattering state in the conduction band the initial state wavefunction is

$$\psi_T(\mathbf{r}) = u_{co}(\mathbf{r})F_T(\mathbf{r}) \tag{5.75}$$

where $F_T(\mathbf{r})$ is the radial hydrogenic function (eqn (5.69)) and the final state wavefunction is a Bloch function except near the threshold. In the latter case coulombic scattering enhances the wavefunction by the coulomb factor. Let us first take the final state to be adequately described by a Bloch function.

In the long-wavelength limit the momentum matrix element is given by

$$\mathbf{p}_{if} = V^{-1/2}\int u^*_{c\mathbf{k}}(\mathbf{r})\exp(-i\mathbf{k}.\mathbf{r})\mathbf{p}\{u_{co}(\mathbf{r})F_T(\mathbf{r})\}\,d\mathbf{r} \tag{5.76}$$

or equivalently

$$\mathbf{p}^*_{if} = V^{-1/2}\int u^*_{co}(\mathbf{r})F^*_T(\mathbf{r})\mathbf{p}^*\{u_{c\mathbf{k}}(\mathbf{r})\exp(i\mathbf{k}.\mathbf{r})\}\,d\mathbf{r}. \tag{5.77}$$

In the spirit of the effective-mass approximation we can neglect the variation of the envelope function over a unit cell and, provided that $\mathbf{k}$ for the final state is small compared with a reciprocal lattice vector, the variation of the wave factor over a unit cell can also be neglected. The cell-periodic part of the Bloch function varies slightly with $\mathbf{k}$ according to $\mathbf{k}.\mathbf{p}$ perturbation theory as functions from other bands become mixed with $u_{co}(\mathbf{r})$, but to a good approximation this variation can be neglected if it is not too large. Thus

$$\mathbf{p}^*_{if} \approx \mathbf{p}^*_{c\mathbf{k}} V^{-1/2}\int F^*_T(\mathbf{r})\exp(i\mathbf{k}.\mathbf{r})\,d\mathbf{r} \tag{5.78}$$

where

$$\mathbf{p}^*_{c\mathbf{k}} = V^{-1}\int u^*_{c\mathbf{k}}(\mathbf{r})\exp(-i\mathbf{k}.\mathbf{r})\mathbf{p}^*\{u_{c\mathbf{k}}(\mathbf{r})\exp(i\mathbf{k}.\mathbf{r})\}\,d\mathbf{r}.$$

Thus $\mathbf{p}^*_{c\mathbf{k}}$ is the expectation value for the momentum of a Bloch state in the conduction band. This is simply given by

$$\mathbf{p}^*_{c\mathbf{k}} = m\mathbf{v}_\mathbf{k} \tag{5.79}$$

where $\mathbf{v}_\mathbf{k}$ is the group velocity, and in this formulation it includes all the effects of $\mathbf{k}.\mathbf{p}$ mixing. This approximation is equivalent to that applied in the case of photo-deionization of an acceptor as a comparison of eqns (5.78) and (5.68) shows: $\mathbf{p}_{cv}$ has been replaced by $\mathbf{p}_{c\mathbf{k}}$. However, unlike that case $\mathbf{p}_{c\mathbf{k}}$ is zero when $\mathbf{k}=0$, whereas $\mathbf{p}_{cv}$ is independent of $\mathbf{k}$. The inter-band momentum is non-zero and optical transitions are allowed, but in the present case the intra-band momentum is zero at $\mathbf{k}=0$ and the optical transition at threshold is forbidden. In that respect the photo-ionization of a donor differs significantly from the photo-ionization of a free hydrogen atom, which does have an allowed transition at $\mathbf{k}=0$ even when the final state is taken to be a plane wave. Calculating the transition momentum to a higher order of approximation adds terms of order m^*/m and restores this property to the hydrogenic donor, but since m^*/m is typically of order 0·1 we can neglect this effect in most cases.

Therefore after integration eqn (5.78) becomes

$$\mathbf{p}_{\text{if}}^* = m\mathbf{v}_{\mathbf{k}} 8i\left(\frac{\pi a_{\text{H}}^{*3}}{V}\right)^{1/2} \frac{1}{(1+k^2 a_{\text{H}}^{*2})^2} \tag{5.80}$$

and hence the photo-ionization cross-section from eqn (5.25) for un-polarized light and a parabolic band is

$$\sigma_\nu = \tfrac{256}{3}\alpha a_{\text{H}}^2 \left(\frac{R_{\text{H}}}{\hbar\omega_\nu}\right) \frac{1}{\eta_{\text{r}}} \frac{m}{m^*} \frac{\pi a_{\text{H}}^{*3}}{\{1+(E_{\mathbf{k}}/R_{\text{H}}^*)\}^4} \left(\frac{2m^*}{\hbar^2}\right)^{3/2} E_{\mathbf{k}}^{3/2}. \tag{5.81}$$

The magnitude is determined by the effective volume of the state which is proportional to a_{H}^{*3}. By putting $E_{\mathbf{k}} = \hbar\omega_\nu - R_{\text{H}}^*$ we can easily show that σ_ν peaks at $\hbar\omega_\nu = 10R_{\text{H}}^*/7$.

Near threshold we must include the coulomb factor:

$$C_0 \approx 2\pi (R_{\text{H}}^*/E_{\mathbf{k}})^{1/2}. \tag{5.82}$$

Multiplying the right-hand side of eqn (5.81) by C_0 shifts the peak to $\hbar\omega_\nu = 10R_{\text{H}}^*/9$ which is only $R_{\text{H}}^*/9$ from threshold. Thus the inclusion of C_0 is justified for estimating the position of the peak (Fig. 5.6).

According to eqn (5.30) the radiative capture cross-section is given by σ_ν multiplied by $(q_\nu/k)^2$:

$$\sigma_{\mathbf{k}} = \sigma_\nu \frac{(E_{\mathbf{k}} + R_{\text{H}}^*)^2}{2m^*(c/\eta_{\text{r}})^2 E_{\mathbf{k}}}. \tag{5.83}$$

Because of the small photon energies involved q_ν is small. Roughly, $2m^*(c/\eta_{\text{r}})^2 \approx 10^4$ eV, so with $E_{\mathbf{k}} \approx R_{\text{H}}^*/10$ and $R_{\text{H}}^* \approx 0{\cdot}01$ eV the factor is

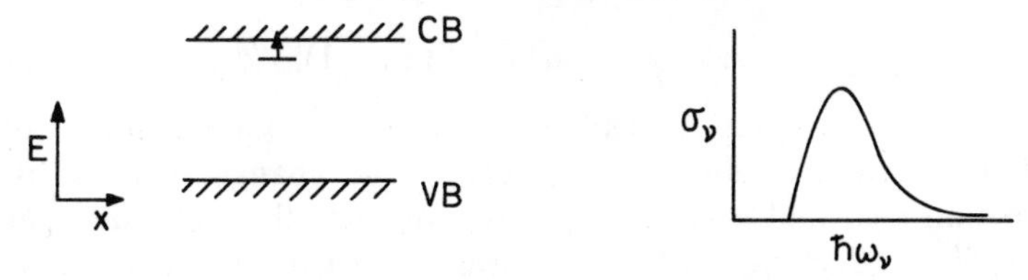

FIG. 5.6. Photo-ionization of a donor.

10^{-5}. Since σ_ν near the maximum is of order $10^{-15}\,\mathrm{cm}^3$, we obtain a capture cross-section of order $10^{-20}\,\mathrm{cm}^2$. This is quite negligible in comparison with non-radiative processes.

It is of interest to compare the strength of this transition, which is forbidden at threshold, with that associated with de-ionization of an acceptor, which is allowed. The ratio of the interaction energies is

$$\frac{mv_{\mathbf{k}}/2}{p_{\mathrm{cv}}^2/2m}=\frac{(m/m_{\mathrm{c}}^*)E_{\mathbf{k}}}{p_{\mathrm{cv}}^2/2m}.$$

From $\mathbf{k.p}$ theory (eqn (1.90)) for simple bands

$$\frac{p_{\mathrm{cv}}^2}{2m}=\frac{E_{\mathrm{g}}}{4}\left(1+\frac{m}{m_{\mathrm{v}}^*}\right). \tag{5.84}$$

Hence

$$\frac{m}{m_{\mathrm{c}}^*}\frac{E_{\mathbf{k}}}{(p_{\mathrm{cv}}^2/2m)}\cong\frac{4E_{\mathbf{k}}}{E_{\mathrm{cv}}(m_{\mathrm{c}}^*/m+m_{\mathrm{c}}^*/m_{\mathrm{v}}^*)}\cdot \tag{5.85}$$

Thus forbidden transitions have smaller intrinsic strengths than those of allowed transitions, but the difference is not large except very near threshold ($E_{\mathbf{k}}\approx 0$). This small difference is because the $\mathbf{k.p}$ interaction has already mixed $|\mathrm{p}\rangle$-like states into the conduction band.

5.7. Photo-ionization of quantum-defect impurities

When core effects increase the depth of a donor state, or are the reason for binding into a shallow state associated with a neutral centre, we have to deal with quantum-defect centres as distinct from hydrogenic centres. We still assume the effective-mass approximation and we take the quantum-defect form of the ground-state wavefunction (see Section 2.9) (Bebb and Chapman 1971):

$$\psi_{\mathrm{T}}(\mathbf{r})=u_{\mathrm{c0}}(\mathbf{r})F_{\mathrm{T}}(\mathbf{r}) \tag{5.86}$$

$$F_{\mathrm{T}}(\mathbf{r})=A\left(\frac{2r}{\nu_{\mathrm{T}}a_{\mathrm{H}}^*}\right)^{\mu-1}\exp\left(-\frac{r}{\nu_{\mathrm{T}}a_{\mathrm{H}}^*}\right) \tag{5.87}$$

$$\nu_{\mathrm{T}}=(R_{\mathrm{H}}^*/E_{\mathrm{T}})^{1/2}\qquad \mu=Z\nu_{\mathrm{T}} \tag{5.88}$$

$$A=\{4\pi(\nu_{\mathrm{T}}a_{\mathrm{H}}^*/2)^3\Gamma(2\mu+1)\}^{-1/2}. \tag{5.89}$$

The normalizing constant A has been obtained by normalizing to unity, even though this procedure, by neglecting the change in form of the wavefunction near and in the core, underestimates the amplitude outside the core. The effect in most cases turns out to be less than 10% for positively charged or neutral centres, but for negatively charged centres

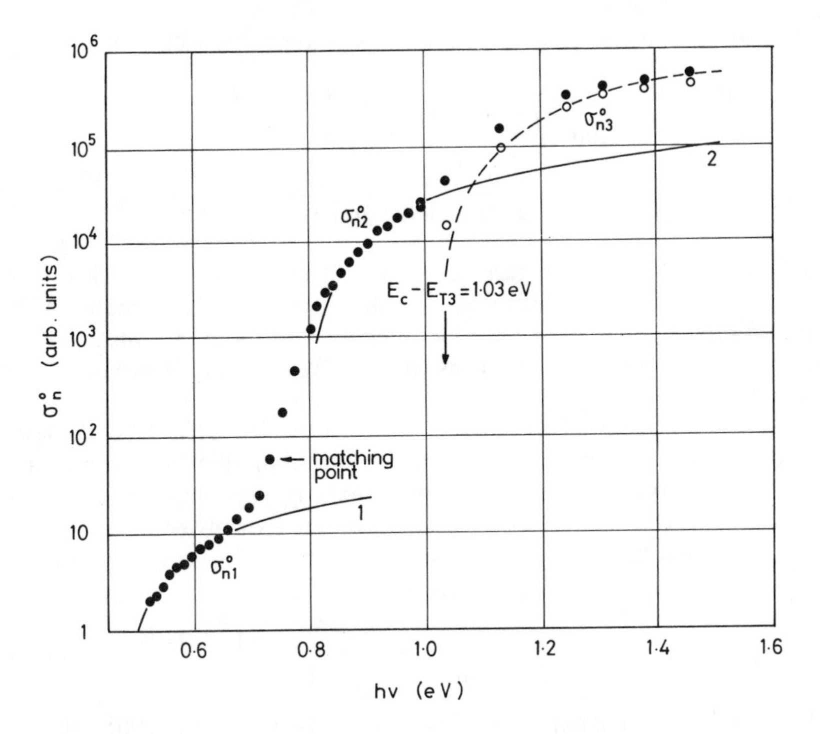

FIG. 5.7. Photo-ionization cross-sections associated with the deep levels in GaAs containing oxygen at 100 K. The solid curves are theoretical and are based on the Lucovsky model. (After Grimmeiss and Ledebo 1975.)

($Z<0$) the error is too large to be neglected. We therefore restrict attention for the moment to relatively shallow levels associated with neutral or positively charged centres. When $Z=1$, $\mu=\nu_T$, and when further $\nu_T=1$, the quantum-defect wavefunction transforms to the hydrogenic form. When $Z=0$, and consequently $\mu=0$, the quantum-defect wavefunction transforms to that assumed by Lucovsky (1965). The latter model has been commonly used to describe experimental results even for deep-level impurities (Fig. 5.7).

The momentum matrix element with the final state described by a Bloch function is

$$\mathbf{p}_{\mathrm{if}} = m\mathbf{v}_k 2^{\mu+1}\mathrm{i}\left\{\frac{2\pi(\nu_T a_H^*)^3\Gamma^2(\mu+1)}{V\Gamma(2\mu+1)}\right\}^{1/2}\frac{\sin\{(\mu+1)\tan^{-1}k\nu_T a_H^*\}}{k\nu_T a_H^*(1+k^2\nu_T^2 a_H^{*2})^{(\mu+1)/2}} \tag{5.90}$$

which correctly reduces to eqn (5.80) when $\mu=\nu_T=1$. For unpolarized light and a parabolic band the photo-ionization cross-section becomes

$$\sigma_\nu = \frac{16\times 2^{2\mu}}{3}\alpha a_{\rm H}^{2}\left(\frac{R_{\rm H}}{\hbar\omega_\nu}\right)\frac{1}{\eta_{\rm r}}\frac{m}{m^*}\frac{2\pi(\nu_{\rm T}a_{\rm H}^*)^3\Gamma^2(\mu+1)}{\Gamma(2\mu+1)}$$
$$\times\frac{\sin\{(\mu+1)\tan^{-1}(E_{\mathbf{k}}/E_{\rm T})^{1/2}\}}{(E_{\mathbf{k}}/E_{\rm T})\{1+(E_{\mathbf{k}}/E_{\rm T})\}^{\mu+1}}\left(\frac{2m^*}{\hbar^2}\right)^{3/2}E_{\mathbf{k}}^{3/2}. \quad (5.91)$$

The capture cross-section can be obtained from σ_ν in the usual way by using eqn (5.30). Near threshold, and this is particularly germane to the capture process, the right-hand side of eqn (5.91) should be multiplied by the coulomb factor if the centre is charged. We are tacitly assuming that core-resonant scattering is absent.

The peak of σ_ν for a hydrogenic centre is at $\hbar\omega_\nu \approx 10R_{\rm H}^*/9$. As the level deepens the maximum moves further away from threshold and in the limit reaches $\hbar\omega_\nu = 2E_{\rm T}$. This is most exactly seen by noting that μ diminishes as the level deepens, and the difference between charged and neutral centres becomes smaller. If μ is neglected relative to unity we obtain the cross-section for a neutral centre:

$$\sigma_\nu = \tfrac{16}{3}\alpha a_{\rm H}^{2}\left(\frac{R_{\rm H}}{\hbar\omega_\nu}\right)\frac{1}{\eta_{\rm r}}\frac{m}{m^*}\frac{2\pi(\nu_{\rm T}a_{\rm H}^*)^3(2m^*/\hbar^2)^{3/2}E_{\mathbf{k}}^{3/2}}{\{1+(E_{\mathbf{k}}/E_{\rm T})\}^2}. \quad (5.92)$$

This is essentially the expression first derived by Lucovsky (1965), though he meant it to apply to deep-level centres. The maximum of the response for a neutral centre, whatever its depth, is at $\hbar\omega_\nu = 2E_{\rm T}$.

The quantity $(\nu_{\rm T}a_{\rm H}^*)^3$ measures the effective volume of the localized state. The effective-mass approximation invariably relates tighter orbits with deeper levels, and since the overlap integral involving plane-wave states peaks near $k\nu_{\rm T}a_{\rm H}^* = 1$ the maximum is thrown towards excitation into shorter wavelength states as the level deepens (Fig. 5.8).

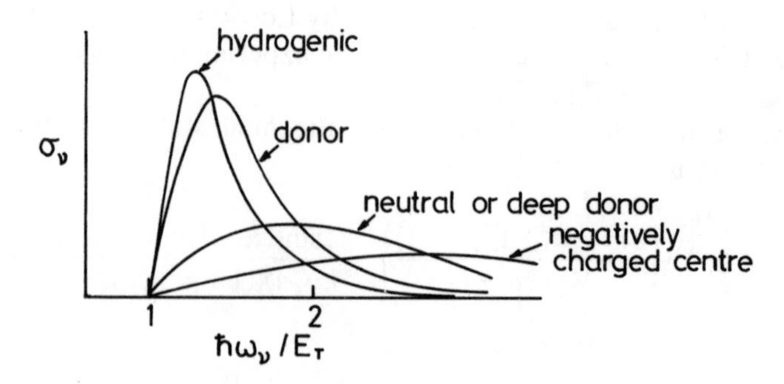

FIG. 5.8. Photo-ionization of quantum-defect centres.

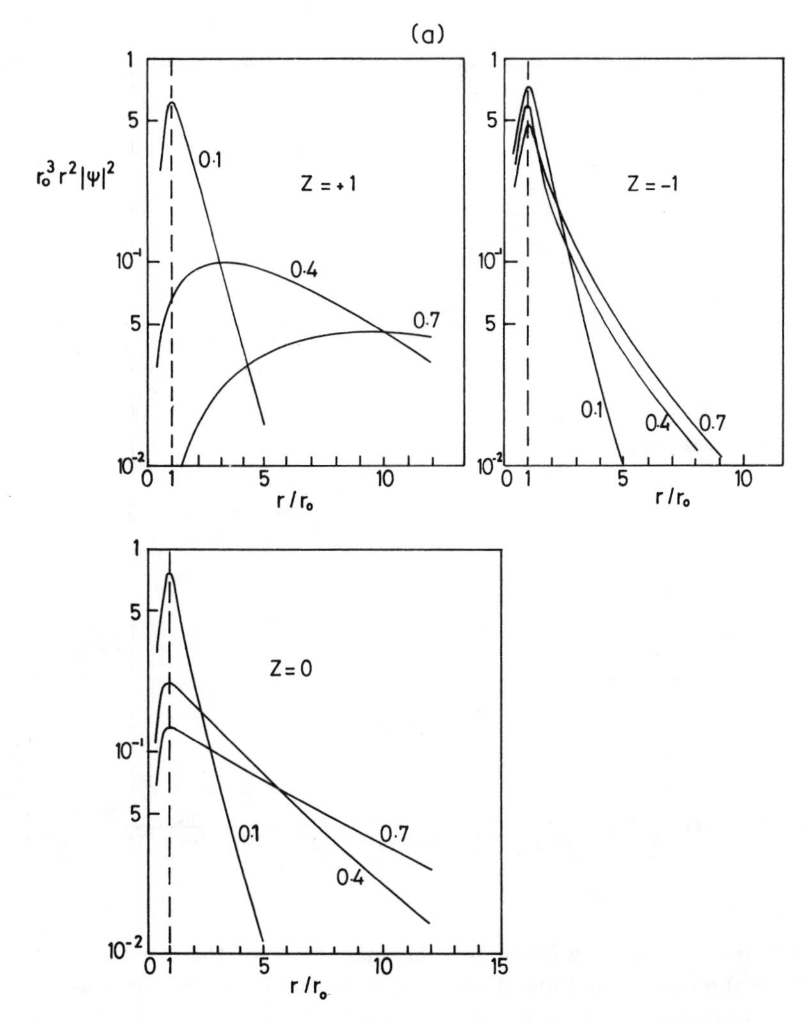

FIG. 5.9. (a) Electron charge density near an impurity centre. The numbers on the curves are the values of ν; r_0, radius of square well. (b) Ratio of electron charge density inside the core to that outside: $r_0/a^* = 0{\cdot}05$; $\nu = (R^*/E_T)^{1/2}$.

The case of photo-ionization of centres with multiple negative charges, i.e. centres whose photo-excited electron finds itself in a long-range repulsive potential, cannot be satisfactorily treated by assuming a quantum-defect wavefunction normalized to unity and neglecting the form at the core. However, if the level is shallow, an effective-mass approach using a model potential can be used along the lines described in Section 2.9 and as discussed for example by Amato and Ridley (1980). When $\mu > -\frac{1}{2}$ the spectral dependence is given reasonably well by the

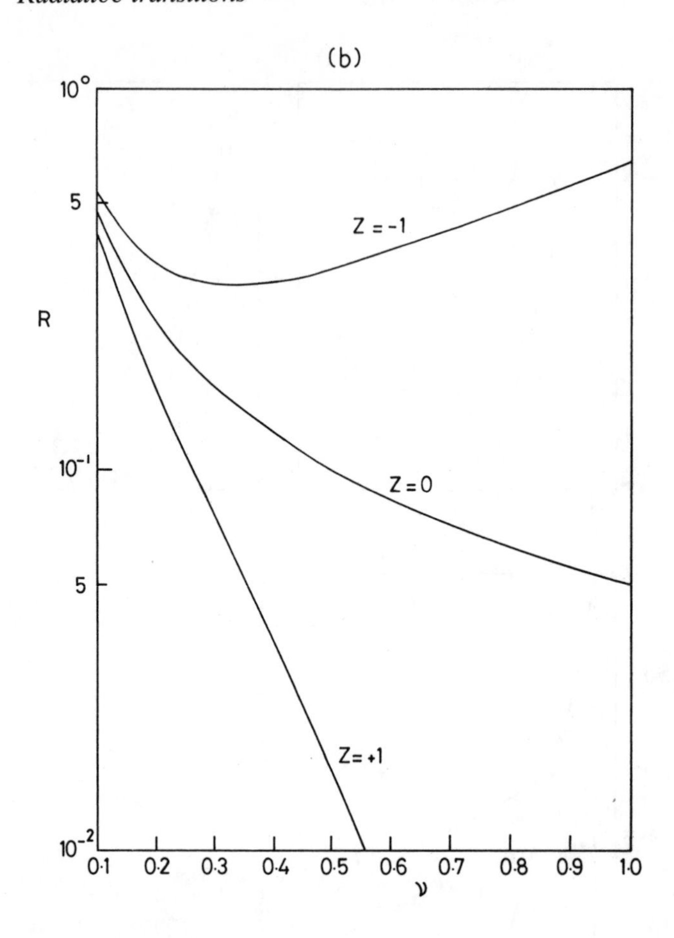

quantum-defect wavefunction but the magnitude of the cross-section is not, for the reason mentioned previously. Normalization of the quantum-defect wavefunction to unity is not possible when $\mu \leqslant -\frac{1}{2}$, and consequently the core cannot be neglected for shallow levels. Figure 5.9 shows the electron charge density for a model potential and the ratio of the charge in the core to that in the tail for various charges. This ratio is small in the case of neutral and positively charged centres with shallow levels, but it is always significant in the case of a negatively charged centre.

Shallow negatively charged centres therefore form a special subclass of shallow-level impurities. To treat the photo-ionization of such centres within the effective-mass approximation it is necessary to adopt models for both core and tail, fit them appropriately, and normalize the total wavefunction. When this is carried out the spectral dependence turns out to have a maximum *beyond* $\hbar\omega_\nu = 2E_T$ with a trend towards $\hbar\omega_\nu \gg 2E_T$ as

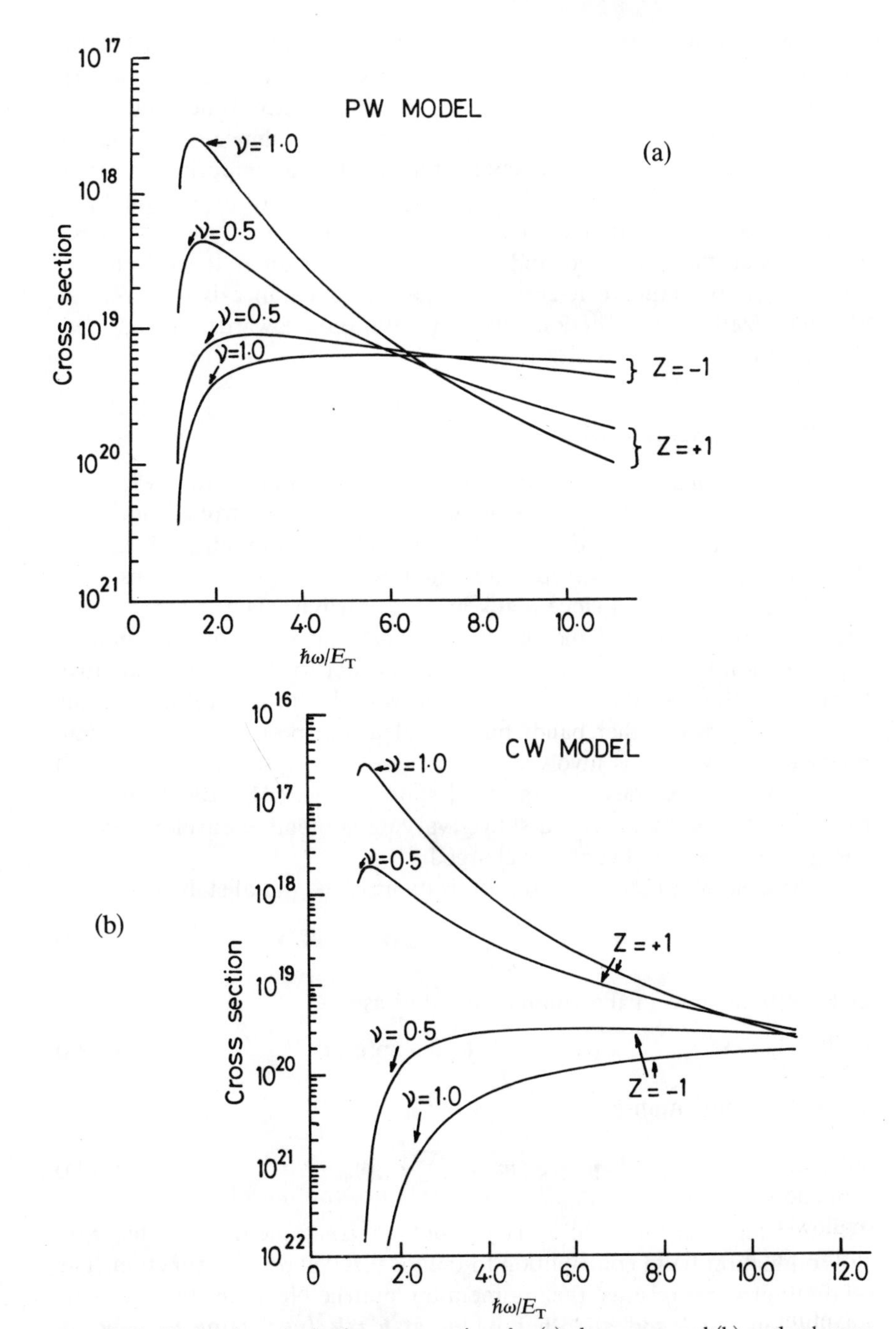

FIG. 5.10. Model photo-ionization cross-sections for (a) plane-wave and (b) coulomb wave final states. (After Amato and Ridley 1980.)

the level becomes shallower. The shift of the maximum away from the neutral value is to be expected, as the discussion of deep-level impurities in the next section will show, as the wavefunction concentrates at the core to avoid the repulsive potential. The spectral dependence of the photo-ionization cross-section of shallow negatively charged impurities is therefore expected to be similar to that for deep-level impurities. It should also be noted that the action of the coulomb factor is to lower the cross-section near the threshold and to push the maximum towards higher energies or to eliminate it entirely. The net appearance is to make the photo-ionization cross-section approach saturation towards high energies (Fig. 5.10).

5.8. Photo-ionization of deep-level impurities

The effective-mass approximation breaks down entirely for deep-level impurities. In this situation the impurity potential is strong enough to bind deeply into a ground state whose wavefunction is highly localized. An expansion of this wavefunction in terms of Bloch functions is in general not very useful since many bands contribute. Nevertheless, since the final state is one of the bands, a formal expansion of the impurity wavefunction in Bloch functions is useful in order to clarify the qualitative features of the transition. The final state will also in general involve the Bloch functions of other bands but provided that no long-lived resonant positive-energy state is involved, i.e. if there is no resonant scattering, it will be sufficient to describe the final state by the single Bloch function belonging to the band with a slowly varying coulombic envelope at low energies if the excited centre is charged.

In the case of a neutral centre let us express the initial state as

$$\psi_{\mathrm{T}}(\mathbf{r}) = V^{-1/2} \sum_{n,\mathbf{k}'} c_{\mathbf{nk}'} u_{\mathbf{nk}'}(\mathbf{r}) \exp(\mathrm{i}\mathbf{k}'.\mathbf{r}) \tag{5.93}$$

and the final state in the conduction band as

$$\psi_{\mathbf{k}}(\mathbf{r}) = V^{-1/2} u_{\mathbf{ck}}(\mathbf{r}) \exp(\mathrm{i}\mathbf{k}.\mathbf{r}). \tag{5.94}$$

Then the momentum is

$$\mathbf{p}_{\mathrm{if}} = c_{\mathbf{ck}} m v_{\mathbf{ck}} + \sum_{\mathrm{n}}{}' c_{\mathbf{nk}} \mathbf{p}_{\mathbf{cnk}}. \tag{5.95}$$

The first term on the right is the 'forbidden' component associated with the conduction-band contribution to the localized state wavefunction. The second term represents the momentum matrix elements between the conduction band and all other bands at $\mathbf{k}' = \mathbf{k}$ (neglecting the photon momentum). This sum can be split into two sections, one in which $\mathbf{p}_{\mathbf{cnk}}$ is zero at $\mathbf{k}=0$ and another in which $\mathbf{p}_{\mathrm{cn}\mathbf{k}}$ is non-zero at $\mathbf{k}=0$. We can rewrite eqn (5.95) as

$$\mathbf{p}_{\mathrm{if}} = m \sum_{\mathrm{s}} c_{\mathrm{s}\mathbf{k}} \mathbf{v}_{\mathrm{s}\mathbf{k}} + \sum_{\mathrm{t}} c_{\mathrm{t}\mathbf{k}} \mathbf{p}_{\mathrm{ct}\mathbf{k}} \tag{5.96}$$

where the subscript s denotes forbidden component and the subscript t denotes allowed components. Since bands contribute in inverse proportion to their remoteness in energy, we can adopt a two-band approximation and retain contributions only from the conduction and valence bands:

$$\mathbf{p}_{\mathrm{if}} \approx m c_{\mathrm{c}\mathbf{k}} \mathbf{v}_{\mathrm{c}\mathbf{k}} + c_{\mathrm{v}\mathbf{k}} \mathbf{p}_{\mathrm{cv}\mathbf{k}} \tag{5.97}$$

where the subscripts c and v denote conduction and valence bands respectively. A deep-level impurity state can then be classified as having either $|\mathrm{s}\rangle$-like or $|\mathrm{p}\rangle$-like symmetry, and the appropriate term taken. Though rough, this approach is a useful conceptual step in the interpretation of spectra.

The coefficients $c_{\mathrm{n}\mathbf{k}}$ depend on the potential and energy of the state. In the previous cases we have considered the wavefunction, either hydrogenic or quantum defect, was known and the $c_{\mathrm{c}\mathbf{k}}$ could be obtained as Fourier components. If r_{T} denotes the effective radius of the wavefunction, the form of $c_{\mathrm{c}\mathbf{k}}$ is roughly $(V_{\mathrm{T}}/V)^{1/2}$, where V_{T} is the effective volume of the state, and is independent of k for $k^2 r_{\mathrm{T}}^2 \ll 1$ but falls off rapidly with k for $k^2 r_{\mathrm{T}}^2 \gg 1$. For deep-level states r_{T} will be of the order of 10 Å, and therefore $k^2 r_{\mathrm{T}}^2 \ll 1$ for excitations not too far above threshold. When $k^2 r_{\mathrm{T}}^2 \geqslant 1$ the final state is far above the conduction-band minimum and in a region where the conduction band may be quite complex. Interpretation of the spectrum in this region is likely to involve details of band structure which are as problematic as the details of the deep-level state. If the elucidation of the structure of the impurity state is the prime motivation, the region which is most useful is that in which the band structure is well known, i.e. when $k^2 r_{\mathrm{T}}^2 \ll 1$. Near the threshold, $E_{\mathrm{k}}/E_{\mathrm{T}} \ll 1$ and we can therefore take for an $|\mathrm{s}\rangle$-like impurity (Ridley 1980)

$$\mathbf{p}_{\mathrm{ik}} \approx (V_{\mathrm{T}}/V)^{1/2} m \mathbf{v}_{\mathbf{k}}. \tag{5.98}$$

Taking a parabolic approximation for the band we obtain for the photo-ionization cross-section for unpolarized light

$$\sigma_{\nu} = \tfrac{4}{3}\alpha a_{\mathrm{H}}^2 \left(\frac{R_{\mathrm{H}}}{\hbar\omega_{\nu}}\right) \frac{1}{\eta_{\mathrm{r}}} \frac{m}{m^*} V_{\mathrm{T}} \left(\frac{2m^*}{\hbar^2}\right)^{3/2} E_{\mathbf{k}}^{3/2} \qquad (|\mathrm{s}\rangle \text{ like}). \tag{5.99}$$

For a $|\mathrm{p}\rangle$-like impurity

$$\mathbf{p}_{\mathrm{if}} \approx (V_{\mathrm{T}}/V)^{1/2} \mathbf{p}_{\mathrm{cv}} \tag{5.100}$$

and

$$\sigma_\nu = \tfrac{4}{3}\alpha a_{\rm H}^2 \left(\frac{R_{\rm H}}{\hbar\omega_\nu}\right)\frac{1}{\eta_{\rm r}} V_{\rm T} \frac{p_{\rm cv}^2}{2m}\left(\frac{2m^*}{\hbar^2}\right)^{3/2} E_{\mathbf{k}}^{1/2} \qquad (|{\rm p}\rangle \text{ like}). \quad (5.101)$$

When $\alpha a_{\rm H}^2 \approx 2\times10^{-19}\,\text{cm}^2$, $R_{\rm H}/\hbar\omega_\nu \approx 10$, $\eta_{\rm r} \approx 4$, $p_{\rm cv}^2/2m \approx 0{\cdot}5\,\text{eV}$, and $(2m^*/\hbar^2)^{3/2}E_{\mathbf{k}}^{1/2} \approx 10^{21}\,\text{eV}^{-1}\,\text{cm}^{-3}$ at $E_{\mathbf{k}} \approx 0{\cdot}1\,\text{eV}$, the photo-ionization cross-section is about $10^3 V_{\rm T}\,\text{cm}^2$. Since $V_{\rm T}$ is about $10^{-21}\,\text{cm}^3$, $\sigma_\nu \approx 10^{-18}\,\text{cm}^2$.

Near threshold σ_ν is about $10^{-18}\,\text{cm}^2$ multiplied by the coulomb factor if the centre is charged. The radiative capture cross-section for a level about $0{\cdot}6\,\text{eV}$ below the conduction band is therefore only about $10^{-22}\,\text{cm}^2$.

Note that it is only near to threshold that a distinction between allowed and forbidden transitions can be made. Far above threshold the energy dependence of the matrix element blurs that distinction since the bands become non-parabolic and overlap bands with different parities. Thus the expressions in eqns (5.99) and (5.101) are applicable in just the region which provides most information about the centre. Neither expression predicts a maximum—a consequence of neglecting the tail of the localized wavefunction. However, since a maximum must occur well above threshold (if not as far as predicted by effective-mass theory, still far enough), its position may be significantly affected by the shape of the density-of-states function at high energies and consequently the interpretation of an observed peak several tenths of an electronvolt above threshold may be quite difficult to perform.

5.9. Summary of photo-ionization cross-sections

The expressions for the cross-sections derived in the previous sections have been collected together in Table 5.1.

5.10. Indirect transitions

Because of the small momentum associated with photons whose frequency lies in the visible or infrared regions of the spectrum, transitions in which both initial and final states are band states are allowed only if crystal momentum is conserved. Such processes are depicted by vertical lines in the E–$\mathbf{k}$ diagram and are termed direct transitions. For non-vertical transitions to occur momentum has to be supplied from another source such as an impurity centre or the pervasive phonon gas. Transitions involving a photon and a phonon or impurity are termed indirect. Two important examples of indirect radiative transitions are the inter-band transition from the top of the valence band to a conduction band

TABLE 5.1
Summary of photo-ionization cross-sections[a]

(1) Direct interband[b]	$2\sigma_0 V$
(2) Hydrogenic acceptor (to conduction band)	$\sigma_0 \dfrac{64\pi a^{*3}_{\mathrm{HV}}}{\{1+(E_{\mathbf{k}}/R^{*}_{\mathrm{HV}})\}^4}$
(3) Hydrogenic donor	$\sigma_0 \dfrac{E_{\mathbf{k}}(m/m^*)}{(p^2_{\mathrm{cv}}/2m)} \dfrac{64\pi a^{*3}_{\mathrm{HC}}}{\{1+(E_{\mathbf{k}}/R^{*}_{\mathrm{HC}})\}^4}$
(4) Quantum defect	$\sigma_0 \dfrac{E_{\mathbf{k}}(m/m^*)}{p^2_{\mathrm{cv}}/2m} \dfrac{8\pi 2^{2\mu}(\nu_{\mathrm{T}} a^{*}_{\mathrm{H}})^3 \sin^2\{(\mu+1)\tan^{-1}(E_{\mathbf{k}}/E_{\mathrm{T}})^{1/2}\}}{(E_{\mathbf{k}}/E_{\mathrm{T}})\{1+(E_{\mathbf{k}}/E_{\mathrm{T}})\}^{\mu+1}}$
(5) Deep level[c] ($\|s\rangle$ like)	$\sigma_0 \dfrac{E_{\mathbf{k}}(m/m^*)}{(p^2_{\mathrm{cv}}/2m)} V_{\mathrm{T}}$
(6) Deep-level[c]($\|p\rangle$ like)	$\sigma_0 V_{\mathrm{T}}$

$$\sigma_0 = \tfrac{4}{3}\alpha a^2_{\mathrm{H}} \left(\frac{R_{\mathrm{H}}}{\hbar\omega_\nu}\right) \frac{1}{\eta_{\mathrm{r}}} \frac{p^2_{\mathrm{cv}}}{2m} \left(\frac{2m^*}{\hbar^2}\right)^{3/2} E^{1/2}_{\mathbf{k}}$$

$$\alpha = \frac{e^2/4\pi\epsilon_0}{\hbar c}$$

$$R_{\mathrm{H}} = \frac{e^2/4\pi\epsilon_0}{2a_{\mathrm{H}}}$$

$$a_{\mathrm{H}} = \frac{\hbar^2/m}{e^2/4\pi\epsilon_0}$$

[a] Not including scattering or local-field correction. Where a coulomb field exists the above expressions are multiplied by the coulomb factor C_0 for transitions near threshold.
[b] Includes spin degeneracy of the initial state.
[c] Includes shallow negatively charged centres.

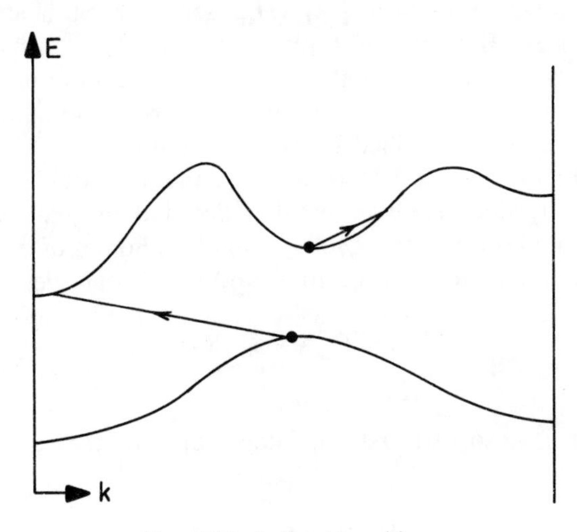

FIG. 5.11. Indirect transitions.

valley at or near the zone boundary, and the intra-valley transition responsible for free-carrier absorption (Fig. 5.11). Indirect transitions involving localized states are also possible, but since the defect centre can accommodate momentum differences without the help of a phonon such processes are usually in competition with direct processes which have a higher probability of occurring. In general, indirect transitions can occur alongside direct inter-band transitions and may contribute observable features in the sub-threshold region associated with the absorption, for example, of a long-wavelength optical phonon. We shall not discuss in detail all these possibilities but limit outselves to describing the transition rate in processes which without the phonon or impurity would be forbidden. For the sake of brevity we shall treat explicitly only the phonon case.

The transition rate is given by second-order perturbation theory (Schiff 1955):

$$W(\mathbf{k}) = \int_{\mathrm{f}} \frac{2\pi}{\hbar} \left| \sum_{\mathrm{n}} \frac{\langle \mathrm{f}| H_{\mathrm{per}} |\mathrm{n}\rangle\langle \mathrm{n}| H_{\mathrm{per}} |\mathrm{i}\rangle}{E_{\mathrm{i}} - E_{\mathrm{n}}} \right|^2 \delta(E_{\mathrm{f}} - E_{\mathrm{i}})\, \mathrm{d}S_{\mathrm{f}} \qquad (5.102)$$

$$H_{\mathrm{per}} = H_{\nu} + H_{\mathrm{ep}}. \qquad (5.103)$$

The optical perturbation H_{ν} (eqns (5.1), (5.11), and (5.12)) induces a transition from the initial state $|\mathrm{i}\rangle$ to a virtual intermediate state $|\mathrm{n}\rangle$, often conserving crystal momentum but not energy. The phonon perturbation H_{ep} (Chapter 3) completes the transition by taking the system from $|\mathrm{n}\rangle$ to the final state $|\mathrm{f}\rangle$, conserving momentum and overall energy. Alternatively, the first step can be accomplished by the phonon perturbation and the second step by the optical perturbation. Since we are not concerned with either two-phonon processes (H_{ep} active in both steps) or two-photon processes (H_{ν} active in both steps), the two alternatives are the only possible ways of making the transition. The intermediate state in each case is limited by the conservation of crystal momentum when all states considered are described by Bloch functions.

Let the initial state be a Bloch state of band a denoted by $|\mathrm{a}\mathbf{k}\rangle$ and the final state be a Bloch state of band b denoted by $|\mathrm{b}\mathbf{k}'\rangle$. Consider the absorption of a photon of energy $\hbar\omega_{\nu}$ and of a phonon of energy $\hbar\omega_{\mathbf{q}}$ and crystal momentum $\hbar\mathbf{q}$. The initial and final energies are (normal processes)

$$\begin{aligned} E_{\mathrm{i}} &= E_{\mathrm{a}\mathbf{k}} + \hbar\omega_{\nu} + \hbar\omega_{\mathbf{q}} \\ E_{\mathrm{f}} &= E_{\mathrm{b}\mathbf{k}+\mathbf{q}}. \end{aligned} \qquad (5.104)$$

If the phonon is absorbed first the intermediate energy is given by

$$\begin{aligned} &(\mathrm{a}) \quad E_{\mathrm{n}} = E_{\mathrm{a}\mathbf{k}+\mathbf{q}} + \hbar\omega_{\nu} \\ &(\mathrm{b}) \quad E_{\mathrm{n}} = E_{\mathrm{b}\mathbf{k}+\mathbf{q}} + \hbar\omega_{\nu}, \end{aligned} \qquad (5.105)$$

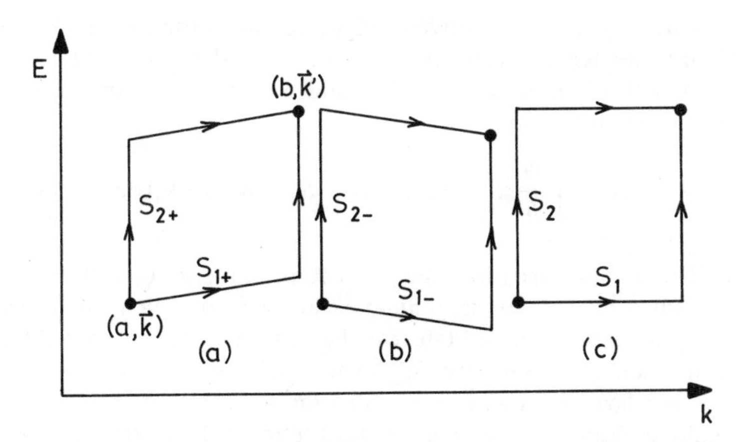

FIG. 5.12. Virtual processes in indirect transitions with the absorption of a photon: (a) accompanied by the absorption of a phonon; (b) accompanied by the emission of a phonon; (c) accompanied by impurity scattering. The $S_{i\pm}$ refer to matrix elements.

and if the photon is absorbed first

$$\begin{aligned} &\text{(a)} \quad E_{\text{n}} = E_{\text{b}\mathbf{k}} + \hbar\omega_{\mathbf{q}} \\ &\text{(b)} \quad E_{\text{n}} = E_{\text{a}\mathbf{k}} + \hbar\omega_{\mathbf{q}} \end{aligned} \tag{5.106}$$

(neglecting the photon momentum). Processes (a) and (b) are mutually distinct only when the initial and final states are in different bands. When this is the case the optical transitions for processes (b) are forbidden, since they depend upon a matrix element of the form $\langle \text{a}\mathbf{k}|\, H_\nu \,|\text{a}\mathbf{k}\rangle$ or $\langle \text{b}\mathbf{k}+\mathbf{q}|\, H_\nu \,|\text{b}\mathbf{k}+\mathbf{q}\rangle$, whereas processes (a) may entail allowed transitions, and so we need not consider processes (b) further (Fig. 5.12).

The sum S_{n} over intermediate states is therefore

$$S_{\text{n}} = \frac{\langle \text{b}\mathbf{k}+\mathbf{q}|\, H_\nu \,|\text{a}\mathbf{k}+\mathbf{q}\rangle\langle \text{a}\mathbf{k}+\mathbf{q}|\, H_{\text{ep}} \,|\text{a}\mathbf{k}\rangle}{E_{\text{a}\mathbf{k}} + \hbar\omega_{\mathbf{q}} - E_{\text{a}\mathbf{k}+\mathbf{q}}} + \frac{\langle \text{b}\mathbf{k}+\mathbf{q}|\, H_{\text{ep}} \,|\text{b}\mathbf{k}\rangle\langle \text{b}\mathbf{k}|\, H_\nu \,|\text{a}\mathbf{k}\rangle}{E_{\text{a}\mathbf{k}} + \hbar\omega_\nu - E_{\text{b}\mathbf{k}}}. \tag{5.107}$$

If $E_{\text{g}\mathbf{k}}$ is the direct energy gap between bands a and b at $\mathbf{k}$ we can use eqn (5.104) in the first denominator to write the sum thus:

$$S_{\text{n}} = \frac{\langle \text{b}\mathbf{k}+\mathbf{q}|\, H_\nu \,|\text{a}\mathbf{k}+\mathbf{q}\rangle\langle \text{a}\mathbf{k}+\mathbf{q}|\, H_{\text{ep}} \,|\text{a}\mathbf{k}\rangle}{E_{\text{g}\mathbf{k}+\mathbf{q}} - \hbar\omega_\nu} - \frac{\langle \text{b}\mathbf{k}+\mathbf{q}|\, H_{\text{ep}} \,|\text{b}\mathbf{k}\rangle\langle \text{b}\mathbf{k}|\, H_\nu \,|\text{a}\mathbf{k}\rangle}{E_{\text{g}\mathbf{k}} - \hbar\omega_\nu}. \tag{5.108}$$

To obtain the total rate associated with the absorption of a photon we must add two further terms to the sum, similar in form to those in eqn

(5.108) which describe the emission of a phonon. If the two terms in eqn (5.108) are labelled respectively S_{1+} and S_{2+} and the equivalent terms associated with phonon emission S_{1-} and S_{2-}, the rate for an indirect transition is given by

$$W(\mathbf{k}) = \int_{\mathrm{f}} \frac{2\pi}{\hbar} |S_{1+} + S_{2+} + S_{1-} + S_{2-}|^2 \, \delta(E_{\mathrm{f}} - E_{\mathrm{i}}) \, \mathrm{d}S_{\mathrm{f}}. \qquad (5.109)$$

Second-order perturbation theory leading to eqn (5.109) must be modified when any of the denominators are zero or when, more generally, a virtual intermediate state can become a real final state. Direct transitions then dominate anyway, so we shall assume that the photon energy is not large enough for this situation to occur.

The above expressions are also valid for impurity scattering if H_{ep} is interpreted as the appropriate interaction and $\hbar\mathbf{q}$ is the crystal momentum change in the elastic process.

5.11. Indirect interband transitions

In silicon the optical absorption edge is associated with a transition between the top of the valence band and one of the six Δ valleys, and in germanium between the top of the valence band and one of the four L valleys. Both of these are indirect transitions (Fig. 5.13). The full calculation of the transition rate is rather long since it involves eight matrix

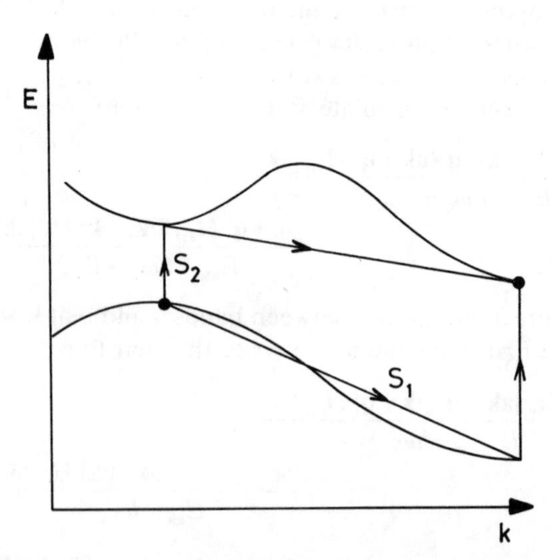

FIG. 5.13. Virtual transitions to real states in indirect inter-band absorption. Because of the larger band gap at the zone boundary S_1 processes are somewhat weaker than S_2 processes.

elements (two for each S). We shall limit ourselves to a calculation of the partial rate arising from the more important contributions which we assume to be embodied in the terms S_{2+} and S_{2-}. We choose S_{2+} and S_{2-} because they have smaller denominators. In silicon $E_{g0}=3{\cdot}4\,\mathrm{eV}$ and in germanium $E_{g0}=0{\cdot}81\,\mathrm{eV}$, whereas in both materials $E_{gX}\approx E_{gL}\approx 4\,\mathrm{eV}$. Thus $E_{g\mathbf{k}}-\hbar\omega_\nu$ is typically smaller than $E_{g\mathbf{k}+\mathbf{q}}-\hbar\omega_\nu$ in semiconductors, and in the case of germanium it is appreciably so. Since cross terms between S_{2-} and S_{2+} do not contribute to the rate, we can write

$$W_2(\mathbf{k})=\int_{\mathrm{f}}\frac{2\pi}{\hbar}(|S_{2+}|^2+|S_{2-}|^2)\delta(E_{\mathrm{f}}-E_{\mathrm{i}})\,\mathrm{d}S_{\mathrm{f}} \tag{5.110}$$

$$|S_{2\pm}|^2=\frac{|\langle\mathbf{ck}\pm\mathbf{q}|\,H^{\mathrm{abs}}_{\mathrm{ep\,em}}\,|\mathbf{ck}\rangle|^2\,|\langle\mathbf{ck}|\,H^{\mathrm{abs}}_{\nu}\,|\mathbf{vk}\rangle|^2}{(E_{g\mathbf{k}}-\hbar\omega_\nu)^2} \tag{5.111}$$

where c stands for the conduction band and v for the valence band.

The optical matrix element is identical to that for a direct transition. Thus in the long-wavelength approximation (eqn (5.11))

$$H^{\mathrm{abs}}_{\nu}=-\frac{e}{m}\left(\frac{\hbar n_\nu}{2V\epsilon_\nu\omega_\nu}\right)^{1/2}\mathbf{a}.\mathbf{p}. \tag{5.112}$$

Hence for non-polarized light, ignoring electron–hole scattering corrections close to threshold and neglecting any $\mathbf{k}$ dependence of the matrix element,

$$|\langle\mathbf{ck}|\,H^{\mathrm{abs}}_{\nu}\,|\mathbf{vk}\rangle|^2=M_\nu^2=\frac{8\pi\hbar}{3}\alpha a_{\mathrm{H}}^2\left(\frac{R_{\mathrm{H}}}{\hbar\omega_\nu}\right)\left(\frac{c\epsilon_0}{\epsilon_\nu}\right)\frac{n_\nu}{V}\frac{p_{\mathrm{cv}}^2}{2m} \tag{5.113}$$

where α is the fine-structure constant, a_{H} is the Bohr radius, R_{H} is the Rydberg energy, n_ν is the number of photons in the cavity of volume V, and p_{cv} is the interband momentum matrix element.

Compared with the rate for a direct optical transition at the zone centre, everything else being equal, the indirect transition rate is lower by a factor equal to the square of the phonon-scattering matrix element divided by $(E_{g0}-\hbar\omega_\nu)^2$. The phonon matrix element is that for an inter-valley scattering event in the conduction band, the valleys being non-equivalent. As such the phonons involved obey group theory selection rules (Table 3.3, p. 110). For example, in the case of germanium only the zone edge LO and LA phonons may participate.

According to eqn (3.14) the square of the phonon matrix element for a given $\mathbf{q}$ has the following general form

$$|\langle\mathbf{ck}\pm\mathbf{q}|\,H^{\mathrm{abs}}_{\mathrm{ep\,em}}\,|\mathbf{ck}\rangle|^2=M_{\mathbf{q}}^2=\frac{\hbar}{2NM'}\frac{C_{\mathbf{q}}^2I(\mathbf{k},\mathbf{k}+\mathbf{q})}{\omega_q}\begin{Bmatrix}n(\omega_q)\\ n(\omega_q)+1\end{Bmatrix}. \tag{5.114}$$

In the case of inter-valley scattering (generalizing from eqn (3.120))

$$\frac{C_{\mathbf{q}}^2 I(\mathbf{k}, \mathbf{k}+\mathbf{q})}{M'} = \frac{D_{ij}^2}{M_1 + M_2} \tag{5.115}$$

where $M_1 + M_2$ is the total mass in the unit cell and D_{ij} is a deformation potential constant. Thus

$$M_{\mathbf{q}}^2 = \frac{\hbar D_{ij}^2}{2\rho V \omega_{ij}} \left\{ \begin{matrix} n(\omega_{ij}) \\ n(\omega_{ij}) + 1 \end{matrix} \right\} \tag{5.116}$$

where ρ is the mass density and ω_{ij} is now the angular frequency of the appropriate phonon. (If more than one type of mode participates we must add the $M_{\mathbf{q}}^2$). We assume, as usual, that inter-valley scattering is isotropic and independent of q.

In the case of direct inter-band transitions only one final state was coupled to a given initial state by $\mathbf{k}$ conservation, and a finite transition rate existed only for the band of states lying between E_r and $E_r + \mathrm{d}E_r$ with density $N_{cv}(E_r)$ given by eqn (5.55) where $E_r = \hbar^2 k^2 / 2m_r^*$ and m_r^* is the reduced electron–hole mass. For indirect transitions the situation is different. Corresponding to a given initial state $|\mathrm{v}\mathbf{k}\rangle$ there is a spread of final states in the conduction band brought about by phonon scattering, and hence a transition rate exists given by

$$W_2(\mathbf{k}) = \frac{2\pi}{\hbar} \frac{M_\nu^2}{(E_{g0} - \hbar\omega_\nu)^2} \frac{\hbar D_{ij}^2 N_{val}}{2\rho\omega_{ij}} [n(\omega_{ij}) N_c(E_1 + \hbar\omega_{ij}) + \{n(\omega_{ij}) + 1\} N_c(E_1 - \hbar\omega_{ij})] \tag{5.117}$$

where N_{val} is the number of equivalent conduction band valleys containing final states, $N_c(E)$ is the density of states for a given spin (the transition does not flip spin) in a given valley, and

$$E_1 = \hbar\omega_\nu - E_g - E_{\mathbf{k}} \tag{5.118}$$

where E_g is the indirect gap and $E_{\mathbf{k}}$ is the hole energy of the initial state measured from the top of the valence band.

As far as the electron in state $|\mathrm{v}\mathbf{k}\rangle$ is concerned eqn (5.117) is the probability of making an indirect transition in unit time, and that is the end of the calculation. (Actually the end of the calculation should be a photo-ionization cross-section using eqn (5.24) because the square of the optical matrix element contains the arbitrary volume of the cavity.) However, we are usually interested more in the total rate induced by a given photon energy since this determines the absorption coefficient. We have therefore to sum all the $W_2(\mathbf{k})$ which correspond to allowed processes, keeping $\hbar\omega_\nu$ constant. This means summing over all possible initial states from $E_{\mathbf{k}}=0$ to $E_{\mathbf{k}\,\max}$, where

$$
\begin{aligned}
E_{\mathbf{k}\,\max} &= \hbar\omega_\nu - E_g + \hbar\omega_{ij} && \text{phonon absorption}\\
E_{\mathbf{k}\,\max} &= \hbar\omega_\nu - E_g - \hbar\omega_{ij} && \text{phonon emission.}
\end{aligned}
\tag{5.119}
$$

Thus we multiply $W_2(\mathbf{k})$ by $2VN_V(E_{\mathbf{k}})\,\mathrm{d}E_{\mathbf{k}}$, where $N_V(E_{\mathbf{k}})$ is the density of states of a given spin per unit energy interval in the valence band and the factor 2 accommodates spin degeneracy, and integrate between the appropriate limits. In the case of parabolic bands we note that

$$\int_0^{E_{\mathbf{k}\,\max}} E_{\mathbf{k}}^{1/2}(E_{\mathbf{k}\,\max}-E_{\mathbf{k}})^{1/2}\,\mathrm{d}E_{\mathbf{k}} = \frac{\pi E_{\mathbf{k}\,\max}^2}{8} \tag{5.120}$$

and hence we obtain

$$
\begin{aligned}
W_2(\hbar\omega_\nu) = {} & \frac{VM_\nu^2 D_{ij}^2 N_{\mathrm{val}}(m_c^* m_v^*)^{3/2}}{8\pi^2(E_{g0}-\hbar\omega_\nu)^2\hbar^6\rho\omega_{ij}}\\
& \times[n(\omega_{ij})(\hbar\omega_\nu - E_g + \hbar\omega_{ij})^2 + \{n(\omega_{ij})+1\}(\hbar\omega_\nu - E_g - \hbar\omega_{ij})^2].
\end{aligned}
\tag{5.121}
$$

Note that the cavity volume disappears from the expression (see eqn (5.113)). The absorption coefficient can then be obtained using eqn (5.58). Each type of allowed phonon contributes a rate as in eqn (5.121) with D_{ij} and ω_{ij} characteristic of the mode. Further contributions to the total rate come from the alternative route depicted by S_{1+} and S_{1-} in eqn (5.109).

Indirect inter-band absorption differs significantly from direct inter-band absorption in its dependence on photon energy. Near the threshold

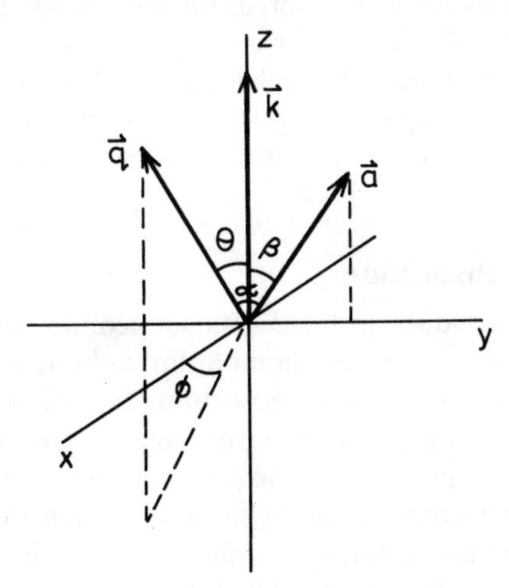

FIG. 5.14. Vector diagram for free-carrier absorption.

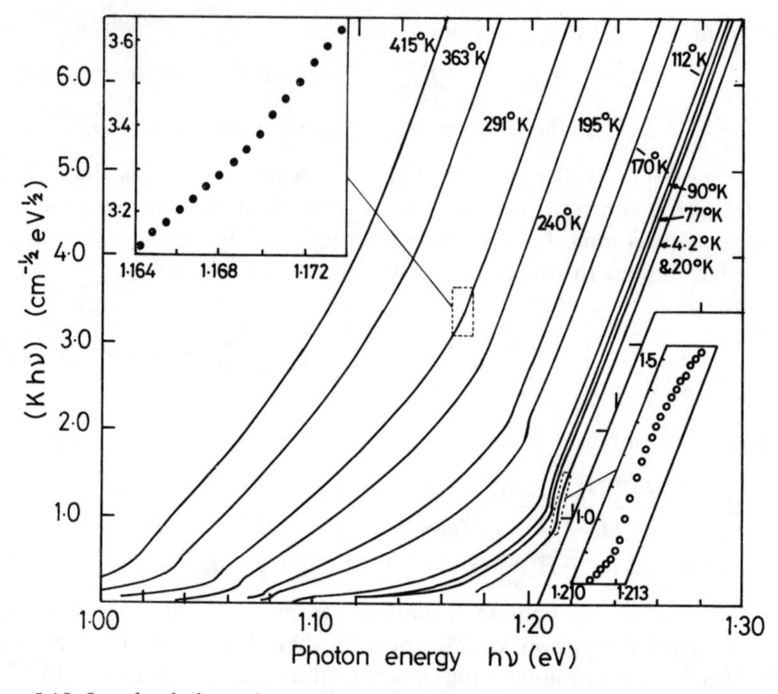

FIG. 5.15. Low-level absorption spectrum of high-purity silicon at various temperatures. The inserts indicate the accuracy with which the experimental points define the curves. (After MacFarlane, McLean, Roberts, and Quarrington 1958: Crown copyright.)

energy E_{th}, the former varies as $(\hbar\omega_\nu - E_{th})^2$ while the latter varies as (eqn (5.59)) $(\hbar\omega_\nu - E_{th})^{1/2}$ (as does excitonic absorption). A plot of the square root of the absorption coefficient versus photon energy near the threshold should give a straight line. Experimental results for silicon and germanium are shown in Figs. 5.14 and 5.15. Sharp rises which disrupt the general trend of the curves are associated with excitonic absorption. Indirect transitions to excitonic states have been discussed by McLean and Loudon (1960).

5.12. Free-carrier absorption

An electron in the conduction band of a semiconductor cannot absorb a photon and conserve energy and momentum without making a transition to another band, simultaneously absorbing or emitting a phonon, or simultaneously scattering off an impurity centre. The first process is the direct radiative transition already discussed in Section 5.4. The second and third are indirect processes involving a transition from a state $|\mathbf{k}\rangle$ to a state $|\mathbf{k}'\rangle$ in the same band, and it is such processes which are at the heart of the phenomenon of free-carrier absorption.

The transition rate is given by eqn (5.109) which we shall write in the following form:

$$W(\mathbf{k}) = \int_{\mathrm{f}} \frac{2\pi}{\hbar} |S_+ + S_-|^2 \, \delta(E_\mathrm{f} - E_\mathrm{i}) \, \mathrm{d}S_\mathrm{f}. \tag{5.122}$$

Since the energy gaps which appear in expressions like eqn (5.108) for S_+ are now zero we have

$$S_\pm = \frac{1}{\hbar\omega_\nu} (\langle \mathbf{k} \pm \mathbf{q} | \, H_\mathrm{s} \, | \mathbf{k} \rangle \langle \mathbf{k} | \, H_\nu^{\mathrm{abs}} \, | \mathbf{k} \rangle - \langle \mathbf{k} \pm \mathbf{q} | \, H_\nu^{\mathrm{abs}} \, | \mathbf{k} \pm \mathbf{q} \rangle \times \langle \mathbf{k} \pm \mathbf{q} | \, H_\mathrm{s} \, | \mathbf{k} \rangle) \tag{5.123}$$

where H_s is the scattering interaction.

The optical matrix elements, as we saw in Section 5.6, involve the total momentum of the Bloch states:

$$\langle \mathbf{k} | \, \mathbf{p} \, | \mathbf{k} \rangle = m \mathbf{v}_\mathbf{k} \tag{5.124}$$

where $\mathbf{v}_\mathbf{k}$ is the group velocity. Thus, from eqn (5.11)

$$S_\pm = \frac{\langle \mathbf{k} \pm \mathbf{q} | \, H_\mathrm{s} \, | \mathbf{k} \rangle}{\hbar\omega_\nu} \left(\frac{e^2 \hbar n_\nu}{2V\epsilon_\nu \omega_\nu} \right)^{1/2} \mathbf{a} \cdot (\mathbf{v}_{\mathbf{k}\pm\mathbf{q}} - \mathbf{v}_\mathbf{k}). \tag{5.125}$$

Summing over the final states is equivalent to a sum over $\mathbf{q}$, and so

$$W_\pm(\mathbf{k}) = \frac{2\pi}{\hbar} \frac{e^2 \hbar^2 n_\nu}{2V\epsilon_\nu (\hbar\omega_\nu)^3} \int |\langle \mathbf{k} \pm \mathbf{q} | \, H_\mathrm{s} \, | \mathbf{k} \rangle|^2 \{ \mathbf{a} \cdot (\mathbf{v}_{\mathbf{k}\pm\mathbf{q}} - \mathbf{v}_\mathbf{k}) \}^2 \times \delta(E_\mathrm{f} - E_\mathrm{i}) \, \mathrm{d}\mathbf{q} \frac{V}{8\pi^3} \tag{5.126}$$

The integrand requires the band structure and the scattering process to be defined. For simplicity we shall assume that the band is spherical and parabolic. In this case

$$\mathbf{v}_{\mathbf{k}\pm\mathbf{q}} - \mathbf{v}_\mathbf{k} = \frac{\hbar}{m^*} (\mathbf{k} \pm \mathbf{q} - \mathbf{k}) = \pm \frac{\hbar \mathbf{q}}{m^*} \tag{5.127}$$

whence

$$W_\pm(\mathbf{k}) = \frac{e^2 \hbar^3 n_\nu}{8\pi^2 \epsilon_\nu (\hbar\omega_\nu)^3 m^{*2}} \int |\langle \mathbf{k} \pm \mathbf{q} | \, H_\mathrm{s} \, | \mathbf{k} \rangle|^2 (\mathbf{a} \cdot \mathbf{q})^2 \delta(E_\mathrm{f} - E_\mathrm{i}) \, \mathrm{d}\mathbf{q}. \tag{5.128}$$

Let us choose spherical polar coordinates with the polar axis along $\mathbf{k}$ (Fig. 5.14). Then if the angle between $\mathbf{a}$ and $\mathbf{q}$ is α, the angle between $\mathbf{a}$ and $\mathbf{k}$ is β, and the angle between $\mathbf{q}$ and $\mathbf{k}$ is θ, we have

$$\mathbf{a} \cdot \mathbf{q} = q \cos \alpha = q(\cos \theta \cos \beta + \sin \theta \sin \beta \sin \phi) \tag{5.129}$$

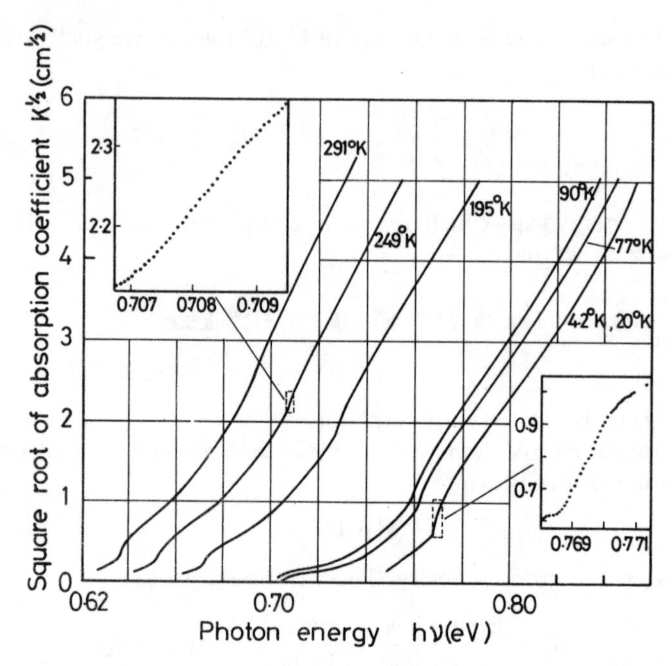

FIG. 5.16. Low-level absorption spectrum of high-purity germanium at various temperatures. The inserts illustrate the accuracy with which the experimental points define the curves. (After MacFarlane, McLean, Roberts, and Quarrington 1957: Crown copyright.)

where ϕ is the azimuthal angle. Nothing in the integrand other than $(\mathbf{a}.\mathbf{q})^2$ depends upon ϕ, and so the integration over ϕ can be carried out immediately. We obtain

$$W_{\pm}(\mathbf{k}) = \frac{e^2\hbar^3 n_\nu}{4\pi\epsilon_\nu(\hbar\omega_\nu)^3 m^{*2}}$$

$$\times \int_0^{q_{\text{ZB}}} \int_{-1}^{1} |\langle \mathbf{k} \pm \mathbf{q}| \, H_{\text{s}} \, |\mathbf{k}\rangle|^2 \, q^4(\cos^2\theta \cos^2\beta + \tfrac{1}{2}\sin^2\theta \sin^2\beta)$$

$$\times \delta(E_{\text{f}} - E_{\text{i}}) \, \mathrm{d}q \, \mathrm{d}(\cos\theta). \quad (5.130)$$

Limits on q are determined by conservation of energy and momentum.

5.12.1. *Energy and momentum*

Let the change in momentum $\hbar\mathbf{q}$ be accompanied by a change in energy $\hbar\omega_q$. The equations for momentum and energy conservation are as follows:

$$|\mathbf{k} \pm \mathbf{q}|^2 = k'^2 = k^2 + q^2 \pm 2kq \cos\theta$$

$$\frac{\hbar^2 k'^2}{2m^*} = \frac{\hbar^2 k^2}{2m^*} + \hbar\omega_\nu \pm \hbar\omega_{\mathbf{q}} \quad (5.131)$$

where the plus sign denotes absorption, and the minus sign emission. Following the steps of Chapter 3, Section 3.2.1, we obtain

$$\pm\cos\theta = -\frac{q}{2k} + \frac{m^*(\omega_\nu \pm \omega_{\mathbf{q}})}{\hbar kq}. \tag{5.132}$$

Let us assume that $\omega_{\mathbf{q}}$ is either very much less than ω_ν or, if not negligible, independent of $\mathbf{q}$. The former case will be a very good approximation for impurity scattering and generally useful for acoustic phonon and piezoelectric scattering; the latter case will be applicable to optical phonon scattering. Therefore both cases

$$\pm\cos\theta \approx -\frac{q}{2k} + \frac{m^*\omega_\pm}{\hbar kq} \tag{5.133}$$

where

$$\omega_\pm = \omega_\nu \pm \omega \tag{5.134}$$

is independent of q. Solutions exist provided that

$$q_{\min} \leqslant q \leqslant q_{\max}$$

$$q_{\pm\min} = k\left\{\left(1 + \frac{\hbar\omega_\pm}{E_{\mathbf{k}}}\right)^{1/2} - 1\right\} \qquad \omega_\pm \geqslant 0 \tag{5.135}$$

$$q_{\pm\max} = k\left\{\left(1 + \frac{\hbar\omega_\pm}{E_{\mathbf{k}}}\right)^{1/2} + 1\right\}.$$

The argument of the delta function can be changed as follows:

$$E_{\mathrm{f}} - E_{\mathrm{i}} = \frac{\hbar^2 q^2}{2m^*} \pm \frac{\hbar^2 kq}{m^*}\cos\theta + \hbar\omega_\pm. \tag{5.136}$$

5.12.2. Scattering matrix elements

In general the scattering matrix element depends upon the magnitude of $\mathbf{q}$ and its direction relative to $\mathbf{k}$. However, it is usually possible to take angular averages of parameters such as deformation potentials, elastic constants, and so on, and thereby to confine the angular dependence of the integral in eqn (5.130) to the effect of optical polarization and to the delta function. We shall therefore assume that the scattering matrix element contains no angular dependence.

Integration over $\cos\theta$ gives

$$W_\pm(\mathbf{k}) = \frac{e^2\hbar n_\nu}{4\pi\epsilon_\nu(\hbar\omega_\nu)^3 m^* k}\int_{q_{\pm\min}}^{q_{\pm\max}} |\langle\mathbf{k}\pm\mathbf{q}|\, H_{\mathrm{s}}\, |\mathbf{k}\rangle|^2\, q^3$$

$$\times\left\{\left(-\frac{q}{2k} + \frac{m^*\omega_\pm}{\hbar kq}\right)^2(\cos^2\beta - \tfrac{1}{2}\sin^2\beta) + \tfrac{1}{2}\sin^2\beta\right\}\mathrm{d}q \tag{5.137}$$

and we have used eqns (5.133) and (5.136) together with the identity

$$\int f(x)\delta(ax-b)\,\mathrm{d}x = \frac{1}{a} f\left(\frac{b}{a}\right). \tag{5.138}$$

In all the cases of phonon scattering (acoustic, optical, polar, and piezoelectric) and in some approximations of impurity scattering, the scattering element depends upon q in the form of a simple power law. We shall limit our discussion to situations where this simplification can be assumed. Thus we take

$$|\langle \mathbf{k} \pm \mathbf{q}| \, H_\mathrm{s} \, |\mathbf{k}\rangle|^2 = A_{\mathrm{S}\pm} q^r \tag{5.139}$$

where $A_{\mathrm{S}\pm}$ is a factor characteristic of the scattering process (Table 5.2). For acoustic phonon (equipartition) and optical phonon (including intervalley phonon) scattering $r=0$, for polar optical and piezoelectric (equipartition) scattering $r=-2$, and for charged-impurity scattering $r=-4$. (Scattering by neutral impurities does not appear to have received much attention in this context.)

TABLE 5.2
Scattering matrix elements

Scattering process	$|\langle \mathbf{k} \pm \mathbf{q}| \, H_\mathrm{s} \, |\mathbf{k}\rangle|^2 = A_{\mathrm{S}\pm} q^r$
Acoustic phonon (equipartition)	$\dfrac{\Xi^2 k_\mathrm{B} T}{2Vc_\mathrm{L}}$
Optical phonon	$\dfrac{D_0^2 \hbar}{2V\rho\omega_\mathrm{o}} \begin{Bmatrix} n(\omega_\mathrm{o}) \\ n(\omega_\mathrm{o})+1 \end{Bmatrix}$
Intervalley phonon	$\dfrac{D_\mathrm{i}^2 \hbar}{2V\rho\omega_\mathrm{i}} \begin{Bmatrix} n(\omega_\mathrm{o}) \\ n(\omega_\mathrm{o})+1 \end{Bmatrix}$
Polar optical phonon (unscreened)	$\dfrac{e^2 \hbar\omega_\mathrm{o}}{2V\epsilon_\mathrm{p} q^2} \begin{Bmatrix} n(\omega_\mathrm{o}) \\ n(\omega_\mathrm{o})+1 \end{Bmatrix}$
Piezoelectric (unscreened, equipartition)	$\dfrac{e^2 K_\mathrm{av}^2 k_\mathrm{B} T}{2V\epsilon q^2}$
Charged impurity (unscreened)	$\dfrac{Z^2 e^4 N_\mathrm{I}}{V\epsilon^2 q^4}$

5.12.3. *Electron scattering by photons*

Integration over q gives

$$W_\pm(\mathbf{k}) = \frac{e^2 \hbar n_\nu A_{\mathrm{S}\pm} k^{r+3}}{32\pi\epsilon_\nu (\hbar\omega_\nu)^3 m^*(r+6)} \{G_{\pm\mathrm{max}}(\omega_\pm, E_\mathbf{k}, \beta) - G_{\pm\mathrm{min}}(\omega_\pm, E_\mathbf{k}, \beta)\} \tag{5.140}$$

where

$$G_{\pm\mathrm{m}}(\omega_\pm, E_\mathbf{k}, \beta) = (3\cos^2\beta - 1)\left(\frac{q_\mathrm{m}}{k}\right)^{r+6} +$$

$$+4(r+6)\left\{1-\cos^2\beta - \frac{\hbar\omega_\pm}{2E_\mathbf{k}}(3\cos^2\beta - 1)\right\}\left\{\frac{1-\delta_{r,-4}}{r+4}\left(\frac{q_\mathrm{m}}{k}\right)^{r+4} + \delta_{r,-4}\log q_\mathrm{m}\right\}$$

$$+(r+6)\left(\frac{\hbar\omega_\pm}{E_\mathbf{k}}\right)^2(3\cos^2\beta - 1)\left\{\frac{1-\delta_{r,-2}}{r+2}\left(\frac{q_\mathrm{m}}{k}\right)^{r+2} + \delta_{r,-2}\log q_\mathrm{m}\right\}. \quad (5.141)$$

In this expression the subscript m stands for max or min; also $\delta_{r,x} = 1$ if $r = x$ and 0 otherwise, and $(1-\delta_{r,x})/(r-x) = 0$ if $r = x$ and $1/(r-x)$ otherwise.

If the light is unpolarized or if we average over the direction of the electron's motion we can replace $\cos^2\beta$ by $\frac{1}{3}$. This simplifies matters considerably, and we obtain

$$G_{\pm\mathrm{m}}(\omega_\pm, E_\mathbf{k}) = \frac{8(r+6)}{3}\left\{\frac{(1-\delta_{r,-4})}{r+4}\left(\frac{q_\mathrm{m}}{k}\right)^{r+4} + \delta_{r,-4}\log q_\mathrm{m}\right\} \quad (5.142)$$

and hence for the three cases

$$\frac{k^{r+3}}{r+6}\{G_{\pm\max}(\omega_\pm, E_\mathbf{k}) - G_{\pm\min}(\omega_\pm, E_\mathbf{k})\} = \frac{16}{3}k^3\left(1+\frac{\hbar\omega_\pm}{E_\mathbf{k}}\right)^{1/2}\left(2+\frac{\hbar\omega_\pm}{E_\mathbf{k}}\right) \qquad r = 0 \quad (5.143)$$

$$= \frac{16}{3}k\left(1+\frac{\hbar\omega_\pm}{E_\mathbf{k}}\right)^{1/2} \qquad r = -2 \quad (5.144)$$

$$= \frac{16}{3}\frac{1}{k}\coth^{-1}\left(1+\frac{\hbar\omega_\pm}{E_\mathbf{k}}\right)^{1/2}. \qquad r = -4. \quad (5.145)$$

The rates are then given by

$$W_\pm(\mathbf{k}) = \frac{e^2 n_\nu A_{\mathrm{S}\pm}(2m^*)^{1/2}}{3\pi\epsilon_\nu(\hbar\omega_\nu)^3\hbar^2}(E_\mathbf{k}+\hbar\omega_\pm)^{1/2}(2E_\mathbf{k}+\hbar\omega_\pm) \qquad r = 0 \quad (5.146)$$

$$= \frac{e^2 n_\nu A_{\mathrm{S}\pm}}{3\pi\epsilon_\nu(\hbar\omega_\nu)^3(2m^*)^{1/2}}(E_\mathbf{k}+\hbar\omega_\pm)^{1/2} \qquad r = -2 \quad (5.147)$$

$$= \frac{e^2 n_\nu A_{\mathrm{S}\pm}\hbar^4}{3\pi\epsilon_\nu(\hbar\omega_\nu)^3(2m^*)^{3/2}}\frac{1}{E_\mathbf{k}^{1/2}}\coth^{-1}\left(1+\frac{\hbar\omega_\pm}{E_\mathbf{k}}\right)^{1/2} \qquad r = -4. \quad (5.148)$$

These are the transition rates for an electron in a state $|\mathbf{k}\rangle$ absorbing a photon and either absorbing a phonon (+), emitting a phonon (−), or scattering off an impurity (in which case the ± sign can be dispensed with). The photon absorbed is one of the n_ν which inhabit the cavity V. Since the $A_{s\pm}$ are proportional to V^{-1} the rates are conveniently normalized for a radiation energy density $n_\nu \hbar\omega_\nu / V$, in which case rates per unit energy density are inversely proportional to the fourth power of the frequency. In this property free-carrier absorption is similar to Rayleigh scattering of light by bound electrons. The two processes are fairly closely related so it is not surprising that the characteristic dependence on frequency appears in both cases.

Unless the energy density of radiation is unusually high the above rates are very small in comparison with ordinary scattering rates associated with phonons and impurities. For example, in the case of acoustic phonon scattering (eqn (3.73), Chapter 3) the rate for absorption or emission is given by

$$W_{ac}(\mathbf{k}) = \frac{\Xi^2 k_B T (2m^*)^{3/2} E_\mathbf{k}^{1/2}}{4\pi\hbar^3 c_L} = A_{S\pm}\frac{(2m^*)^{3/2}E_\mathbf{k}^{1/2}V}{2\pi\hbar^4} \tag{5.149}$$

and hence

$$\frac{W_\pm(\mathbf{k})}{W_{ac}(\mathbf{k})} = \frac{e^2 n_\nu \hbar^2}{3\epsilon_\nu(\hbar\omega_\nu)^3 V m^*}\left(1+\frac{\hbar\omega_\pm}{E_\mathbf{k}}\right)^{1/2}(2E_\mathbf{k}+\hbar\omega_\pm)$$

$$= \frac{4\pi\alpha_0 E_\nu c}{3(\epsilon_\nu/\epsilon_0)\omega_\nu^3 m^*}\left(1+\frac{\hbar\omega_\nu}{E_\mathbf{k}}\right)^{1/2}\left(\frac{2E_\mathbf{k}}{\hbar\omega_\nu}+1\right) \tag{5.150}$$

where α_0 is the fine-structure constant, E_ν is the energy density, and ω_ν is the angular frequency of radiation, and we have neglected the acoustic phonon energy in relation to the photon energy. Taking typical values for ϵ_ν and m^* and regarding the brackets as contributing a factor of order unity, we calculate that for $\omega_\nu \approx 10^{13}\,\mathrm{s}^{-1}$ the ratio is of order unity only for an energy density about $100\,\mathrm{J\,m^{-3}}$ which is equivalent to an intensity of about $10\,\mathrm{GW\,m^{-2}}$. Therefore in most situations the presence of light in the semiconductor does not affect the mobility of electrons. The interesting physical quantity is usually not the rate at which electrons are scattered by photons but the rate at which photons are absorbed.

5.12.4. *Absorption coefficients*

The rate for all transitions involving the absorption of a photon of energy $\hbar\omega_\nu$ is obtained by summing over all initial electron states. Thus, for non-degenerate statistics

$$W_{\nu\pm} = \int_0^\infty W_\pm(\mathbf{k}) f(E_\mathbf{k}) 2VN(E_\mathbf{k})\,\mathrm{d}E_\mathbf{k} \tag{5.151}$$

where $f(E_{\mathbf{k}})$ is the occupation probability, $N(E_{\mathbf{k}})$ is the density of states in $\mathrm{d}E_{\mathbf{k}}$ for a given spin, and the factor of 2 reflects the spin degeneracy. The limits of the integration constitute a reasonable approximation provided that $f(E_{\mathbf{k}})$ falls off exponentially with energy. At thermodynamic equilibrium we can take the expression

$$f(E_{\mathbf{k}}) = \frac{n}{N_{\mathrm{c}}} \exp\left(-\frac{E_{\mathbf{k}}}{k_{\mathrm{B}}T}\right) \tag{5.152}$$

where N_{c} is the effective density of states for the non-degenerate gas of electrons and n is the electron density. Assuming a parabolic band we can take

$$\begin{aligned} N(E_{\mathbf{k}}) &= \frac{(2m^*)^{3/2} E_k^{1/2}}{4\pi^2\hbar^3} \\ N_{\mathrm{c}} &= \frac{2(2\pi m^* k_{\mathrm{B}} T)^{3/2}}{(2\pi)^3\hbar^3} \end{aligned} \tag{5.153}$$

and hence

$$W_{\nu\pm} = \frac{2nV}{\pi^{1/2}(k_{\mathrm{B}}T)^{3/2}} \int_0^\infty W_\pm(\mathbf{k}) E_k^{1/2} \exp\left(-\frac{E_k}{k_{\mathrm{B}}T}\right) \mathrm{d}E_k. \tag{5.154}$$

For $r = 0$ the integral over energy is (Gradshteyn and Ryzhik 1965)

$$\begin{aligned} &\int_0^\infty E_k (E_k + \hbar\omega_\pm)^{1/2} (2E_k + \hbar\omega_\pm) \exp\left(-\frac{E_k}{k_{\mathrm{B}}T}\right) \mathrm{d}E_k \\ &\quad = \tfrac{1}{2} k_{\mathrm{B}} T (\hbar\omega_\pm)^2 \exp\left(\frac{\hbar\omega_\pm}{2k_{\mathrm{B}}T}\right) \mathscr{K}_2\left(\frac{|\hbar\omega_\pm|}{2k_{\mathrm{B}}T}\right) \end{aligned} \tag{5.155}$$

where $\mathscr{K}_2(z)$ is a modified Bessel function of the second kind. For $r = -2$ the integral is (Gradshteyn and Ryzhik 1965)

$$\int_0^\infty E_k^{1/2} (E_k + \hbar\omega_\pm)^{1/2} \exp\left(-\frac{E_k}{k_{\mathrm{B}}T}\right) \mathrm{d}E_k = \tfrac{1}{2} k_{\mathrm{B}} T |\hbar\omega_\pm| \exp\left(\frac{\hbar\omega_\pm}{2k_{\mathrm{B}}T}\right) \mathscr{K}_1\left(\frac{|\hbar\omega_\pm|}{2k_{\mathrm{B}}T}\right) \tag{5.156}$$

For $r = -4$ (Gradshteyn and Ryzhik 1965)

$$\int_0^\infty \left\{\coth^{-1}\left(1 + \frac{\hbar\omega_\pm}{E_k}\right)^{1/2}\right\} \exp\left(-\frac{E_k}{k_{\mathrm{B}}T}\right) \mathrm{d}E_k = \tfrac{1}{2} k_{\mathrm{B}} T \exp\left(\frac{\hbar\omega_\pm}{2k_{\mathrm{B}}T}\right) \mathscr{K}_0\left(\frac{|\hbar\omega_\pm|}{2k_{\mathrm{B}}T}\right) \tag{5.157}$$

These expressions are valid even if $\hbar\omega_{\pm}$ is negative provided that the modulus of $\hbar\omega_{\pm}$ is taken as shown.

Finally we take into account the stimulated emission of photons by changing the sign of $\hbar\omega_{\nu}$ in $\hbar\omega_{\pm}$ to obtain the rate $W^{\mathrm{em}}_{\nu\pm}$. The net absorption rate is the difference between $W^{\mathrm{abs}}_{\nu\pm}$ and $W^{\mathrm{em}}_{\nu\pm}$. By using the relation between the absorption constant K_{ν} and W_{ν} (eqn (5.58)) we obtain

$$K_{\nu\pm}=\frac{1}{v_{\nu}n_{\nu}}(W^{\mathrm{abs}}_{\nu\pm}-W^{\mathrm{em}}_{\nu\pm}). \tag{5.158}$$

The total absorption coefficient is obtained by summing the effects of phonon absorption and emission, where phonons are involved.

The results from Table 5.2, eqns (5.146)–(5.148), and eqns (5.154)–(5.158) are as follows. Although we have not followed the usual path in deriving these absorption coefficients, we can be gently gratified by the fact that we obtain the same expressions (*cf.* Seegar 1973).

Acoustic phonons

$$K_{\nu}=\frac{8\alpha(2m^{*}k_{\mathrm{B}}T)^{1/2}\Xi^{2}n}{3\pi^{1/2}\eta_{\mathrm{r}}\hbar^{2}\omega_{\nu}c_{\mathrm{L}}}\sinh\left(\frac{\hbar\omega_{\nu}}{2k_{\mathrm{B}}T}\right)\mathcal{K}_{2}\left(\frac{\hbar\omega_{\nu}}{2k_{\mathrm{B}}T}\right) \tag{5.159}$$

where α is the fine-structure constant and η_{r} is the refractive index $(\epsilon_{\nu}/\epsilon_{0})^{1/2}$. The classical frequency dependence is obtained by letting $\hbar\omega_{\nu}$ tend to zero. We use the fact that

$$\mathcal{K}_{\nu}(z)\approx\tfrac{1}{2}\Gamma(\nu)(\tfrac{1}{2}z)^{-\nu} \qquad \nu>0, \quad z\to 0. \tag{5.160}$$

and obtain

$$K_{\nu}=\frac{32\alpha(2m^{*})^{1/2}(k_{\mathrm{B}}T)^{3/2}\Xi^{2}n}{3\pi^{1/2}\eta_{\mathrm{r}}\hbar^{3}\omega_{\nu}^{2}c_{\mathrm{L}}}=\frac{128\alpha e\hbar n}{9\eta_{\mathrm{r}}m^{*2}\omega_{\nu}^{2}\mu_{\mathrm{ac}}}. \tag{5.161}$$

The absorption coefficient is inversely proportional to the square of the frequency. It also has the classical dependence on mobility, μ_{ac} and indeed is identical to the classical expression provided that a factor $32/9\pi$, which is about 1.13, is equated to unity. The expression is invalid for very small frequencies since we have assumed that the photon energy greatly exceeds the phonon energy.

Optical phonons

$$K_{\nu}=\frac{4\alpha(2m^{*})^{1/2}D_{\mathrm{o}}^{2}n[n(\omega_{\mathrm{o}})\{n(\omega_{\mathrm{o}})+1\}]^{1/2}}{3\pi^{1/2}\eta_{\mathrm{r}}\hbar^{3}\omega_{\nu}^{3}\omega_{\mathrm{o}}\rho(k_{\mathrm{B}}T)^{1/2}}\sinh\left(\frac{\hbar\omega_{\nu}}{2k_{\mathrm{B}}T}\right)$$
$$\times\left\{(\hbar\omega_{+})^{2}\mathcal{K}_{2}\left(\frac{\hbar\omega_{+}}{2k_{\mathrm{B}}T}\right)+(\hbar\omega_{-})^{2}\mathcal{K}_{2}\left(\frac{|\hbar\omega_{-}|}{2k_{\mathrm{B}}T}\right)\right\} \tag{5.162}$$

where we have used the relation

$$n(\omega_o)+1=n(\omega_o)\exp(\hbar\omega_o/k_B T) \tag{5.163}$$

and $\omega_+=\omega_\nu+\omega_o$, $\omega_-=\omega_\nu-\omega_o$. A similar expression is obtained for inter-valley phonons.

Polar optical phonon

$$K_\nu=\frac{16\pi^{1/2}\alpha\alpha_{ep}\hbar^{1/2}\omega_o^{3/2}[n(\omega_o)\{n(\omega_o)+1\}]^{1/2}n}{3\eta_r m^*(k_B T)^{1/2}\omega_\nu^3}\sinh\left(\frac{\hbar\omega_\nu}{2k_B T}\right)$$
$$\times\left\{\hbar\omega_+\mathscr{K}_1\left(\frac{\hbar\omega_+}{2k_B T}\right)+|\hbar\omega_-|\,\mathscr{K}_1\left(\frac{|\hbar\omega_-|}{2k_B T}\right)\right\} \tag{5.164}$$

where α_{ep} is the polar coupling coefficient (see Table 3.7, p. 127) and $\epsilon_p^{-1}=\epsilon_\infty^{-1}-\epsilon^{-1}$.

Piezoelectric

$$K_\nu=\frac{2^{5/2}\alpha K_{av}^2(k_B T)^{1/2}n}{3\pi^{1/2}\eta_r\hbar\omega_\nu^2 m^{*1/2}\epsilon}\sinh\left(\frac{\hbar\omega_\nu}{2k_B T}\right)\mathscr{K}_1\left(\frac{\hbar\omega_\nu}{2k_B T}\right). \tag{5.165}$$

Charged impurity

$$K_\nu=\frac{2^{3/2}\alpha Z^2e^4N_I n}{3\pi^{1/2}\eta_r\epsilon^2\omega_\nu^3 m^{*3/2}(k_B T)^{1/2}}\sinh\left(\frac{\hbar\omega_\nu}{2k_B T}\right)\mathscr{K}_0\left(\frac{\hbar\omega_\nu}{2k_B T}\right). \tag{5.166}$$

5.13. Free-carrier scattering of light

The Hamiltonian describing interaction of a (non-relativistic) electron and the electromagnetic field is (see Chapter 2, Section 2.4)

$$H_\nu=-\frac{e}{m}\mathbf{A}.\mathbf{p}+\frac{e^2}{2m}\mathbf{A}^2. \tag{5.167}$$

The first part of the interaction has been used throughout this chapter to describe absorption processes of one sort or another, while the second part has been tacitly assumed to be negligible. Without a third entity being present to conserve momentum, neither term can induce the absorption of a photon by a free electron. However, both terms are capable of inducing the scattering of a photon by a free electron, the first term through a second-order process and the second term through a first-order process. It turns out that for free electrons the scattering rate produced by the $\mathbf{A}.\mathbf{p}$ interaction is smaller than the scattering rate produced by the $\mathbf{A}^2$ interaction by a factor of order $(v/c)^2$ where v is the group velocity of the electron and c is the velocity of light. Thus, the $\mathbf{A}^2$ term is dominant for free electrons. We might expect the effective-mass approximation to be applicable for electrons in a conduction band, and that the $\mathbf{A}^2$ term, with m replaced by m^*, would be a good approximation to the interaction. This indeed turns out to be the case, but only for photon energies well below the energy gap between bands. The $\mathbf{A}.\mathbf{p}$ term, far from being negligible, is of vital importance in converting eqn (5.167) to the effective-mass form (see Appendix, Section 5.14):

$$H_\nu \approx \frac{e^2}{2m^*}\mathbf{A}^2. \tag{5.168}$$

In discussing the scattering of light by an electron in a conduction band we shall assume that the photon energy is far removed from any inter-band separations, and therefore that scattering can be satisfactorily described by the $\mathbf{A}^2$ term in the effective-mass approximation.

The scattering rate is thus

$$W_\nu = \int_{\mathrm{f}} \frac{2\pi}{\hbar}\left|\left\langle \mathbf{k}' \left| \frac{e^2}{2m^*}\mathbf{A}^2 \right| \mathbf{k} \right\rangle\right|^2 \delta(E_{\mathrm{f}} - E_{\mathrm{i}})\,\mathrm{d}S_{\mathrm{f}} \tag{5.169}$$

where $\mathbf{k}$ and $\mathbf{k}'$ are respectively the wavevectors of the electron before and after scattering a photon. If the incident photon has angular frequency ω_ν and wavevector q_ν and the scattered photon has angular frequency ω'_ν and wavevector q'_ν, the vector potential can be written

$$\mathbf{A} = A\mathbf{a}\cos(\mathbf{q}_\nu.\mathbf{r} - \omega_\nu t) + A'\mathbf{a}'\cos(\mathbf{q}'_\nu.\mathbf{r} - \omega'_\nu t). \tag{5.170}$$

Consequently

$$\begin{aligned}\mathbf{A}^2 = \frac{A^2}{2}\{1 + \cos 2(\mathbf{q}_\nu.\mathbf{r} - \omega_\nu t)\} + \frac{A'^2}{2}\{1 + \cos 2(\mathbf{q}'_\nu.\mathbf{r} - \omega'_\nu t)\} \\ + AA'\mathbf{a}.\mathbf{a}'[\cos\{(\mathbf{q}_\nu + \mathbf{q}'_\nu).\mathbf{r} - (\omega_\nu + \omega'_\nu)t\} \\ + \cos\{(\mathbf{q}_\nu - \mathbf{q}'_\nu).\mathbf{r} - (\omega_\nu - \omega'_\nu)t\}]\end{aligned} \tag{5.171}$$

and hence the interaction has d.c., second-harmonic, sum, and difference components. The d.c. components leave the state unchanged and consequently produce zero rate. The second-harmonic components attempt to induce the electron to absorb or emit photons of energy $2\hbar\omega_\nu$ or $2\hbar\omega'_\nu$ but this is forbidden by simultaneous conservation of momentum and energy. The latter also forbids the absorption or emission of sum-frequency photons, and so only the difference-frequency photons are left.

For the incident wave (see eqn (5.9))

$$A = \left(\frac{2\hbar n_\nu}{V\epsilon_\nu\omega_\nu}\right)^{1/2} \tag{5.172}$$

and for the scattered wave, which is assumed to be spontaneously emitted,

$$A' = \left(\frac{2\hbar}{V\epsilon_\nu\omega'_\nu}\right)^{1/2}. \tag{5.173}$$

The scattering rate is therefore

$$W_\nu = \frac{2\pi}{\hbar}\left(\frac{e^2}{2m^*}\right)^2 \frac{\hbar^2 n_\nu}{V^2\epsilon_\nu^2\omega_\nu}\int\frac{(\mathbf{a}.\mathbf{a}')^2}{\omega'_\nu}\,\delta(E_\mathrm{f}-E_\mathrm{i})\,\mathrm{d}\mathbf{q}'_\nu\frac{V}{8\pi^3} \tag{5.174}$$

with momentum conservation

$$\mathbf{k}' = \mathbf{k} + \Delta\mathbf{q}_\nu \qquad \Delta\mathbf{q}_\nu = \mathbf{q}_\nu - \mathbf{q}'_\nu. \tag{5.175}$$

If α is the angle between $\mathbf{k}$ and $\Delta\mathbf{q}$, then for a parabolic band (Fig. 5.17)

$$k'^2 = k^2 + \Delta q_\nu^2 + 2k\,\Delta q_\nu \cos\alpha \tag{5.176}$$

$$\frac{\hbar^2 k'^2}{2m^*} = \frac{\hbar^2 k^2}{2m^*} + \hbar\,\Delta\omega_\nu$$
$$\Delta\omega_\nu = \omega_\nu - \omega'_\nu \tag{5.177}$$

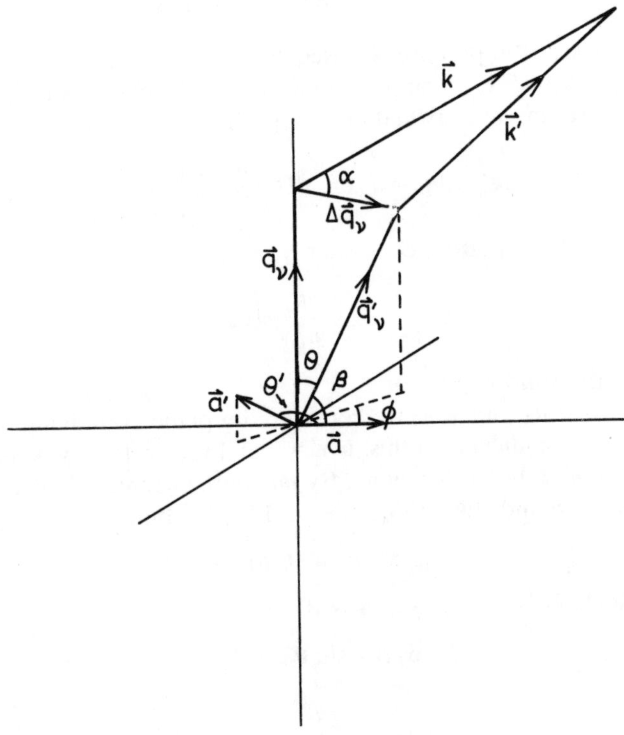

FIG. 5.17. Scattering diagram.

whence, for $\Delta q_\nu \ll k$,

$$\Delta\omega_\nu = v\,\Delta q_\nu \cos\alpha \tag{5.178}$$

where v is the group velocity of the electron. However,

$$\Delta\omega_\nu = v_\nu(q_\nu - q'_\nu) \tag{5.179}$$

where v_ν is the velocity of light in the semiconductor, and thus

$$q_\nu - q'_\nu = \frac{v}{v_\nu}\Delta q_\nu \cos\alpha \tag{5.180}$$

or since v/v_ν is very small, $q_\nu \approx q'_\nu$. Now

$$\Delta q_\nu^2 = q_\nu^2 + q_\nu'^2 - 2q_\nu q'_\nu \cos\theta \tag{5.181}$$

where θ is the angle through which the light is scattered, and so

$$\Delta q_\nu \approx 2q_\nu \sin(\theta/2). \tag{5.182}$$

From eqn (5.178) the fractional change of frequency on scattering is

$$\frac{\Delta\omega_\nu}{\omega_\nu} = 2\left(\frac{v}{v_\nu}\right)\cos\alpha\,\sin\left(\frac{\theta}{2}\right) \tag{5.183}$$

which is very small in practice. Consequently, the delta function conserving energy in eqn (5.174) can be taken to be independent of direction in an elastic scattering approximation. By putting

$$\mathrm{d}\mathbf{q}'_\nu \approx \mathrm{d}\mathbf{q}_\nu = q_\nu^2\,\mathrm{d}q_\nu\,\mathrm{d}\Omega = \frac{\omega_\nu^2\,\mathrm{d}\omega_\nu\,\mathrm{d}\Omega}{v_\nu^3} \tag{5.184}$$

we obtain, after integrating over energy,

$$W_\nu = \left(\frac{e^2}{4\pi\epsilon_\nu m^*}\right)^2 \frac{n_\nu}{v_\nu^3 V}\int(\mathbf{a}.\mathbf{a}')^2\,\mathrm{d}\Omega \tag{5.185}$$

where Ω is the solid angle.

In order to work out $\mathbf{a}.\mathbf{a}'$ we consider the plane containing $\mathbf{q}'_\nu$ and $\mathbf{a}$. Now if $\mathbf{a}'$ is perpendicular to this, $\mathbf{a}.\mathbf{a}' = 0$ and the scattering is forbidden. Thus we take $\mathbf{a}'$ to be in this plane. By taking the azimuthal angle ϕ to be in the plane perpendicular to $\mathbf{q}_\nu$ (Fig. 5.17) we obtain

$$\mathbf{q}'_\nu.\mathbf{a} = q_\nu \sin\theta\cos\phi \tag{5.186}$$

or, if the angle between $\mathbf{q}'_\nu$ and $\mathbf{a}$ is β,

$$\cos\beta = \sin\theta\cos\phi. \tag{5.187}$$

Thus

$$(\mathbf{a}.\mathbf{a}')^2 = \cos^2\theta' = \sin^2\beta = 1 - \sin^2\theta\cos^2\phi \tag{5.188}$$

and

$$\int_{-1}^{1}\int_{0}^{2\pi}(1-\sin^2\theta\cos^2\phi)\,\mathrm{d}(\cos\theta)\,\mathrm{d}\phi = 8\pi/3. \tag{5.189}$$

The scattering rate is therefore

$$W_\nu = \frac{8\pi}{3} r_e^{*2}\frac{v_\nu n_\nu}{V}$$
$$r_e^* = \frac{e^2}{4\pi\epsilon_0 m^* c^2} \tag{5.190}$$

where r_e^* is the effective classical electron radius. The scattering cross-section corresponding to this rate is

$$\sigma_\nu = \frac{8\pi}{3} r_e^{*2} = 6{\cdot}648\times 10^{-25}\left(\frac{m}{m^*}\right)^2 \text{ cm}^2 \tag{5.191}$$

which is the solid state analogue of the classical Thomson scattering solution and is rather small, although generally in semiconductors it is much larger than for free electrons.

According to eqn (5.183) the frequency shift of the scattered light is proportional to the group velocity, and a measurement of this can yield information about the electron distribution in hot-electron experiments. Perhaps more promising for the future is that the scattering of high-power coherent radiation may in principle yield information about the band structure.

5.13.1. Scattering of laser light

A high-power coherent laser beam in the infrared can be depicted as a classical electromagnetic wave with an electric field at a given position equal to

$$\mathscr{E} = \mathscr{E}_0 \cos \omega_\nu t. \tag{5.192}$$

For simplicity we regard the dimensions of the semiconductor to be small compared with the wavelength. This electric field accelerates the electron through the band, and if it is large enough and if the frequency is high enough so that collisions are unlikely in any one cycle the electron will traverse up and down the band and radiate light the spectrum of which will be indicative of the band structure.

If all collisions and all inter-band transitions are ignored, the electron motion will be determined by

$$\frac{\mathrm{d}\mathbf{k}}{\mathrm{d}t} = \frac{e\mathscr{E}}{\hbar} \tag{5.193}$$

and the classical scattered intensity at an angle θ and distance r is

$$S = \frac{e^2 \dot{v}^2 \sin^2\theta}{16\pi^2 \epsilon_\nu v_\nu^3 r^2}. \tag{5.194}$$

The acceleration $\dot{v}$ is obtained from

$$\mathbf{v} = \frac{1}{\hbar} \operatorname{grad}_{\mathbf{k}} E_{\mathbf{k}}. \tag{5.195}$$

In an isotropic band

$$\dot{\mathbf{v}} = \frac{\mathrm{d}\mathbf{k}}{\mathrm{d}t} \frac{1}{\hbar} \frac{\mathrm{d}^2 E_k}{\mathrm{d}k^2} = e\mathscr{E} \frac{1}{\hbar^2} \frac{\mathrm{d}^2 E_k}{\mathrm{d}k^2} \tag{5.196}$$

with a constant effective mass

$$\dot{\mathbf{v}} = e\mathscr{E}/m^* \tag{5.197}$$

and we return to the Thomson scattering solution. However, the mass is not a constant throughout the band and the scattering is no longer classical.

For simplicity let us assume an isotropic band with a minimum at $\mathbf{k} = 0$ and of shape given by

$$E_k = \sum_{n=0}^{\infty} E_n \cos nka \tag{5.198}$$

where a is the unit cell dimension. If the electron starts at $k = 0$, then from eqns (5.192) and (5.193)

$$k = \frac{e\mathscr{E}_0}{\hbar\omega_\nu} \sin \omega_\nu t \tag{5.199}$$

in the direction of the field.

From eqns (5.196) and (5.198)

$$\dot{v} = -e\mathscr{E} \sum_{n=0}^{\infty} \frac{E_n n^2 a^2}{\hbar^2} \cos\left(\frac{ne\mathscr{E}_0 a}{\hbar\omega_\nu} \sin \omega_\nu t\right) \tag{5.200}$$

$$= e\mathscr{E} \sum_{n=0}^{\infty} F_n \cos(z_n \sin \omega_\nu t) \tag{5.201}$$

$$F_n = \frac{E_n n^2 a^2}{\hbar^2} \qquad z_n = \frac{n\omega_z}{\omega_\nu} \qquad \omega_z = \frac{e\mathscr{E}_0 a}{\hbar} \tag{5.202}$$

where ω_z is the Bloch–Zener frequency. In terms of the frequency of scattered light, Fourier analysis gives

$$\dot{v} = -e\mathscr{E}_0 \sum_{n=0}^{\infty} \sum_{r=0}^{\infty} F_n \frac{2(2r+1)}{z_n} J_{2r+1}(z_n) \cos\{(2r+1)\omega_\nu t\} \tag{5.203}$$

where $J_{2r+1}(z_n)$ is a Bessel function. The square of this acceleration determines the spectrum of the scattered light, which consists of harmonics of the incident beam. The intensity of a harmonic is determined by the intensity of the incident beam through ω_z and hence z_n, and also by the band structure through F_n. In principle, therefore, we could 'look' at a band structure by examining the frequency distribution of scattered radiation from a high-powered laser the frequency and intensity of which satisfy the conditions

$$\omega_z \tau \gg \omega_\nu \tau \gg 1 \tag{5.204}$$

where τ is a scattering time constant and ω_ν is well below the absorption edge. As far as the author is aware experiments of this sort have not been proposed hitherto, but with the increasing availability of high-power lasers and associated measuring devices and techniques they may soon be attempted.

5.14. Appendix: Justification of effective-mass approximation in light scattering

The total rate is (ignoring the momentum of light)

$$W_\nu = \int_{\mathrm{f}} \frac{2\pi}{\hbar} \left| \left(\sum_n \frac{\langle 0\mathbf{k}| H_{1\nu} |n\mathbf{k}\rangle\langle n\mathbf{k}| H_{1\nu} |0\mathbf{k}\rangle}{E_{\mathrm{i}} - E_{\mathrm{int}}} \right) + \langle 0\mathbf{k}| H_{2\nu} |0\mathbf{k}\rangle \right|^2 \delta(E_{\mathrm{f}} - E_{\mathrm{i}})\, \mathrm{d}S_{\mathrm{f}} \tag{5.205}$$

where

$$\begin{aligned} H_{1\nu} &= -\frac{e}{m}(A\mathbf{a}\cos \mathbf{q}_\nu.\mathbf{r} + A'\mathbf{a}'\cos \mathbf{q}'_\nu.\mathbf{r}) \\ H_{2\nu} &= \frac{e}{2m}\mathbf{A}^2. \end{aligned} \tag{5.206}$$

E_{i} and E_{int} are the initial and intermediate energies respectively, and the bands are labelled by n with 0 denoting the initial band. The time-dependent factors have been subsumed in the delta function, and we have anticipated momentum conservation in labelling the states. Since no light corresponding to the scattered beam is incident, there can be no corresponding stimulated processes but only spontaneous emission. Consequently there are only two possible virtual transitions. The first involves the virtual absorption of an incident photon of energy $\hbar\omega_\nu$ (the electron makes a virtual 'vertical' transition to any state denoted by $|n\mathbf{k}\rangle$) followed by the spontaneous emission of a photon in the scattered beam with the electron returning to its initial state. The second process places the spontaneous emission first and the absorption second.

The matrix element is thus

$$M_\nu = \frac{e^2}{4m} AA'\langle 0\mathbf{k}|\,\mathbf{a}.\mathbf{a}'\,|0\mathbf{k}\rangle + \frac{e^2}{4m^2} AA' \sum_n \left(\frac{\mathbf{a}'.\langle 0.|\,\mathbf{p}\,|n\mathbf{k}\rangle\langle n\mathbf{k}|\,\mathbf{p}\,|0\mathbf{k}\rangle.\mathbf{a}}{E_{0\mathbf{k}} + \hbar\omega_\nu - E_{n\mathbf{k}}} + \frac{\mathbf{a}.\langle 0\mathbf{k}|\,\mathbf{p}\,|n\mathbf{k}\rangle\langle n\mathbf{k}|\,\mathbf{p}\,|0\mathbf{k}\rangle.\mathbf{a}'}{E_{0\mathbf{k}} - E_{n\mathbf{k}} - \hbar\omega'_\nu} \right). \quad (5.207)$$

If the photon energies are negligible

$$M_\nu \approx \frac{e^2}{4m} AA'\mathbf{a}.\mathbf{a}' \left(1 + \frac{2}{m} \sum_n \frac{\langle 0\mathbf{k}|\,\mathbf{p}\,|n\mathbf{k}\rangle\langle n\mathbf{k}|\,\mathbf{p}\,|0\mathbf{k}\rangle}{E_{0\mathbf{k}} - E_{n\mathbf{k}}}\right). \quad (5.208)$$

However, from **k.p** theory (Chapter 1, Section 1.11) we know that the term in parentheses can be replaced by m/m^* (assuming for simplicity an isotropic band). Consequently

$$M_\nu \approx \frac{e^2}{4m^*} AA'\mathbf{a}.\mathbf{a}' \quad (5.209)$$

which shows that the effective-mass approximation is justified provided that $\hbar\omega_\nu$ is small.

References

Altarelli, M. and Lipari, N. O. (1976). *Proc. 13th Int. Conf. on the Physics of Semiconductors,* p. 811. Tipographia Marves. North Holland, Amsterdam.
Amato, M. A. and Ridley, B. K. (1980). *J. Phys. C.* **13,** 2027.
Bebb, H. B. and Chapman, R. A. (1971). *Proc. 3rd Photoconductivity Conf.* (ed. E. M. Pell), p. 245. Pergamon, Oxford.
Dexter, D. L. (1958). *Solid State Phys.* **6,** 355.
Dumke, W. P. (1963). *Phys. Rev.* **132,** 1988.
Elliot, R. J. (1957). *Phys. Rev.* **108,** 1384.
Gibson, A. F. and Kimmitt, M. F. (1980). *Infrared Millimeter Waves* **3,** 181.
Gradshteyn, I. S. and Ryzhik, I. M. (1965). *Table of integrals, series, and products,* Nos. 3.391, 4.367, 8.432. Academic Press, New York.
Grimmeiss, H. G. and Ledebo, L.-A. (1975). *J. appl. Phys.* **46,** 2155.
Lucovsky, G. (1965). *Solid State Commun.* **3,** 299.
MacFarlane, G. G., McLean, T. P., Roberts, V., and Quarrington, J. E. (1957). *Phys. Rev.* **108,** 1377.
———, ———, ———, ——— (1958). *Phys. Rev.* **111,** 1245.
McLean, T. P. and Loudon, R. (1960). *J. Phys. Chem. Solids* **13,** 1.
Minkowski, H. (1910). *Math. Annln.* **68,** 472.
Ridley, B. K. (1980). *J. Phys. C.* **13,** 2015.
Schiff, L. (1955). *Quantum mechanics,* McGraw-Hill, New York.
Seegar, K. (1973). *Semiconductor physics,* Springer-Verlag, Vienna.
Sturge, M. D. (1962). *Phys. Rev.* **127,** 768.
Weisbuch, C. (1977) Ph.D. Thesis.

6. Non-radiative processes

6.1. Electron–lattice coupling

The one-electron Schrödinger equation including the electron–lattice interaction is

$$(H_e + H_{ep})\psi_{\mathbf{k}}(\mathbf{r}) = E_{\mathbf{k}}\psi_{\mathbf{k}}(\mathbf{r}). \tag{6.1}$$

H_{ep} is the interaction energy responsible for lattice scattering of electrons in band states and is a function of optical and acoustic strain in the lattice. The lattice displacements can be expanded in terms of dimensionless normal coordinates Q_i where

$$Q_i = \left(\frac{M\omega_i}{\hbar}\right)^{1/2} X_i. \tag{6.2}$$

X_i is a normal co-ordinate for the ith vibration mode, ω_i is its angular frequency, and M is the mass of the unit cell for acoustic vibrations or the reduced mass for optical vibrations. We assume, as usual, that the electron–lattice interaction is linear in Q_i:

$$H_{ep} = \sum_i V_i(\mathbf{r})Q_i = \mathbf{V}.\mathbf{Q} \tag{6.3}$$

where $\mathbf{V}$ is the interaction energy 'vector'.

Usually, we are interested in H_{ep} as an interaction inducing transitions between conduction band states at a rate determined by first-order time-dependent perturbation theory. This time we are interested in the time-independent perturbation of the electronic state since this reveals the dependence of the state on lattice distortion. Perturbations add to any one state components from all other states to a degree which, other things being equal, falls off with energy difference. For simplicity we shall consider a system in which only two states are mixed together. To obtain a first-order correction to energy we need wavefunctions to zero order only:

$$\begin{aligned} \psi_1(\mathbf{r}) &\approx \psi_{10}(\mathbf{r}) \qquad & E_1 &= E_{10} + H_{11} \\ \psi_2(\mathbf{r}) &\approx \psi_{20}(\mathbf{r}) \qquad & E_2 &= E_{20} + H_{22} \end{aligned} \tag{6.4}$$

where subscripts 1 and 2 label the states, subscript zero labels the solution in the absence of perturbation, and

$$\begin{aligned} H_{11} &= \langle 1|\,\mathbf{V}\,|1\rangle.\mathbf{Q} = \mathbf{V}_{11}.\mathbf{Q} \\ H_{22} &= \langle 2|\,\mathbf{V}\,|2\rangle.\mathbf{Q} = \mathbf{V}_{22}.\mathbf{Q} \end{aligned} \tag{6.5}$$

where $|1\rangle$ and $|2\rangle$ refer to the unperturbed states. In order to obtain the effect of the perturbation on the wavefunctions we must go to second-order perturbation theory for energy, whence

$$\psi_1(\mathbf{r}) \approx \psi_{10}(\mathbf{r}) + \frac{H_{12}}{E_{10}-E_{20}}\,\psi_{20}(\mathbf{r})$$
$$E_1 = E_{11} + \frac{H_{12}^2}{E_{10}-E_{20}} \tag{6.6}$$

and similarly for level 2, where $E_{11} = E_{10} + H_{11}$ and

$$H_{12} = \langle 2|\,\mathbf{V}\,|1\rangle.\mathbf{Q} = \mathbf{V}_{12}.\mathbf{Q}. \tag{6.7}$$

Carrying this procedure to infinite order one can obtain for the wavefunction (Kovarskii 1962)

$$\psi_1(\mathbf{r}) = \frac{\psi_{10}(\mathbf{r}) + \{H_{12}/(E_1-E_{21})\}\psi_{20}(\mathbf{r})}{\{1 + H_{12}^2/(E_1-E_{21})^2\}^{1/2}} \tag{6.8}$$

where $E_{21} = E_{20} + H_{22}$, and for energy

$$E_1 = E_{11} + H_{12}^2/(E_1 - E_{21}). \tag{6.9}$$

Therefore

$$E_1 = \tfrac{1}{2}[E_{21} + E_{11} \pm \{(E_{21}-E_{11})^2 + 4H_{12}^2\}^{1/2}]. \tag{6.10}$$

The double solution in eqn (6.10) indicates a crossing of levels when $E_{11} = E_{21}$ (Fig. 6.1). Except near this crossing E_1 and E_2 vary linearly with $\mathbf{Q}$ provided that the transition energy H_{12} is small. These energy shifts are determined by the interaction energies H_{11} and H_{22} respectively.

These shifts feed back on the lattice dynamics. In terms of dimensionless normal co-ordinates the lattice Hamiltonian is

$$H_L = \frac{1}{2}\sum_i \left(\frac{p_i^2}{M} + \hbar\omega_i Q_i^2\right). \tag{6.11}$$

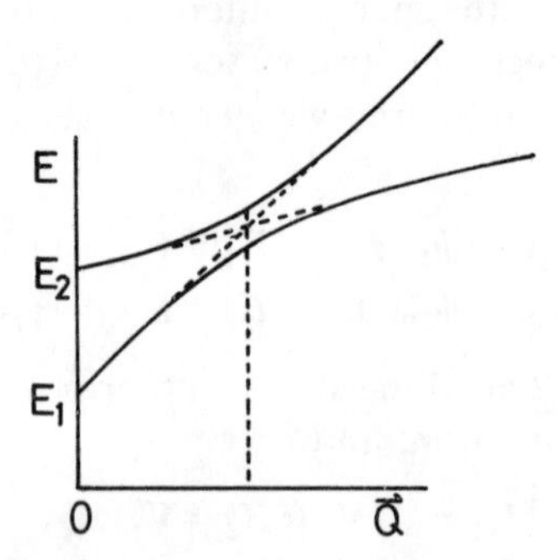

FIG. 6.1. Crossing of two levels by electron–lattice interaction.

The full adiabatic one-electron Schrödinger equation for the electron in level 1 is

$$\left\{\sum_i \left(\frac{p_i^2}{2M}+\tfrac{1}{2}\hbar\omega_i Q_i^2\right)+E_1\right\}\Phi = E_{\text{tot}}\Phi \tag{6.12}$$

where E_1 is the electron energy in level 1, $\mathbf{\Phi}$ are the harmonic oscillator wavefunctions, and E_{tot} is the total energy, vibrational plus electronic. E_1 is now a function of $\mathbf{Q}$ through the electron–lattice interaction. Substituting for E_1 from eqn (6.9) gives

$$\left\{\sum_i \left(\frac{p_i^2}{2M}+\tfrac{1}{2}\hbar\omega_i Q_i^2\right)+E_{10}+H_{11}+\frac{H_{12}^2}{E_1-E_{21}}\right\}\Phi = E_{\text{tot}}\Phi. \tag{6.13}$$

Now $H_{11}=\mathbf{V}_{11}\cdot\mathbf{Q}=\sum_i V_{11i}Q_i$, and $H_{12}=\sum_i V_{12i}Q_i$. The term in H_{12}^2 introduces a frequency shift through its introduction of Q_i^2. If $H_{12i}/H_{12}\ll\hbar\omega_i$ this shift will be negligible, even near cross-over, when $E_1-E_{21}\approx H_{12}$. Since H_{12i} is associated with one mode and H_{12} is the sum for all modes it is usually permissible to neglect frequency shifts entirely. Only if the interaction is strong for a few modes, e.g. local modes, will this neglect be questionable. The linear term H_{11i} introduces a displacement of the oscillator. In terms of displaced oscillators the equation becomes

$$\sum_i \left\{\frac{p_i^2}{2M}+\tfrac{1}{2}\hbar\omega_i (Q_i-Q_{i1})^2\right\}\Phi = \left(E_{\text{tot}}-E_{10}+\frac{1}{2}\sum_i \frac{|V_{11i}|^2}{\hbar\omega_i}\right)\Phi \tag{6.14}$$

where

$$Q_{i1}=-V_{11i}/\hbar\omega_i. \tag{6.15}$$

Q_{i1} is the displacement for mode i when the electron is in level 1. When it is in level 2 the displacement changes to $Q_{i2}=-V_{22i}/\hbar\omega_i$ (Fig. 6.2).

An important quantity is the vibrational energy associated with the shift from displacement Q_{i1} to displacement Q_{i2}, since this will be excited

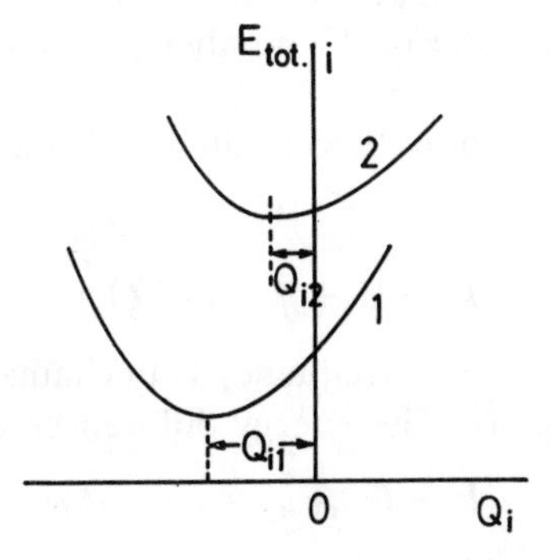

FIG. 6.2. Displacement of oscillator: vibrational plus electronic energy.

whenever the electron makes a transition from level 1 to level 2 or vice versa. This quantity is often called the Franck–Condon energy, and it is given by

$$E_{FC} = \sum_i \tfrac{1}{2}\hbar\omega_i (Q_2 - Q_1)_i^2. \tag{6.16}$$

When all modes have the same frequency it is useful to express this energy in terms of the number of phonons involved:

$$E_{FC} = S\hbar\omega \qquad S = \tfrac{1}{2}\sum_i (Q_2 - Q_1)_i^2. \tag{6.17}$$

The factor S was first introduced by Huang and Rhys (1950).

The main effect of electron–lattice coupling is to displace all the lattice oscillators and to introduce energy shifts of the electronic levels. With an interaction which is linear in the lattice co-ordinate, these shifts are linear also (except near cross-over). Although we have discussed a system with only two electronic levels, it is clear that our results for displacement and shift apply to any number of levels since they depend only on first-order perturbation and this does not mix states.

The total energy thus consists of three components: (a) a purely electronic energy E_{10}, (b) a strain energy $-\tfrac{1}{2}\hbar\omega \sum_i Q_{i1}^2$, and (c) the vibrational energy.

6.2. The configuration co-ordinate diagram

A useful fiction is to regard the energy of the system as depending only on a single normal co-ordinate Q known as the configuration co-ordinate. The correspondence with reality of this picture depends on how many vibrational modes contribute to the total energy. In the case of a diatomic molecule or an impurity centre with a strongly coupled local mode, a single co-ordinate may be a reasonable assumption. Generally, however, Q has to stand as a one-mode model co-ordinate for a real N-mode system where N is the order of the number of atoms in the material (Figs. 6.3 and 6.4).

The energies of the ground and excited states are

$$E_g = E_1 + \tfrac{1}{2}\hbar\omega(Q - Q_1)^2 \tag{6.18}$$

$$E_e = E_2 + \tfrac{1}{2}\hbar\omega(Q - Q_2)^2 \tag{6.19}$$

assuming that the vibrational frequency is unchanged and the interaction at cross-over is negligible. The energy differences at Q_1 and Q_2 are

$$(E_e - E_g)_{Q=Q_1} = E_0 + S\hbar\omega \tag{6.20}$$

$$(E_e - E_g)_{Q=Q_2} = E_0 - S\hbar\omega \tag{6.21}$$

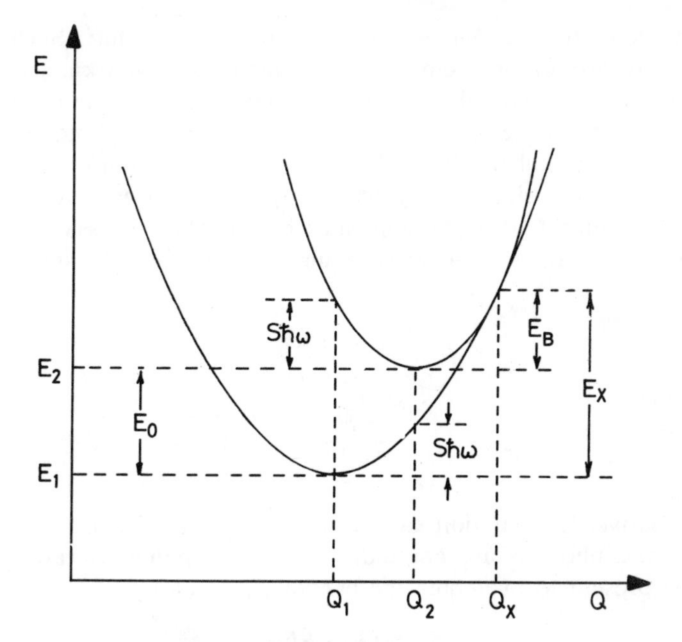

FIG. 6.3. Configuration-co-ordinate diagram.

where E_0 is the difference in energy between the ground and excited states and involves the strain energy of the distortion as well as a purely electronic part.

In an optically induced transition in which an electron moves from the ground to an excited state the movement of ions can be neglected to a good approximation because they are so much heavier. This approximation is usually referred to as the Franck–Condon principle. It implies that the favoured optical transitions on the configuration co-ordinate diagram occur vertically. Thus for absorption and emission

$$\hbar\omega_\nu^{\text{abs}} = E_0 + S\hbar\omega \tag{6.22}$$

$$\hbar\omega_\nu^{\text{em}} = E_0 - S\hbar\omega \tag{6.23}$$

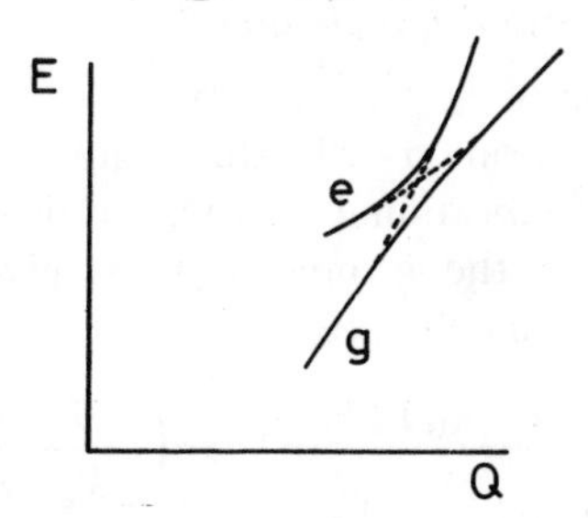

FIG. 6.4. Non-degenerate cross-over.

and the difference $2S\hbar\omega$ is known as the Stokes shift. Such a shift between absorption and luminescent frequencies is well known in luminescent materials. The absorption of a photon excites the electron and causes the appearance of S phonons. The emission of a photon de-excites the electron but still results in the emission of S phonons.

Non-radiative excitation may occur only if sufficient vibratory energy is present. At least $E_0/\hbar\omega$ phonons must be absorbed. Classically, enough energy to reach the cross-over is required. It can easily be shown that at cross-over

$$E_X = \frac{(E_0 + S\hbar\omega)^2}{4S\hbar\omega} \tag{6.24}$$

$$Q_X - Q_1 = \frac{E_0}{(2S)^{1/2}\hbar\omega} + \left(\frac{S}{2}\right)^{1/2}. \tag{6.25}$$

Non-radiative de-excitation can occur even in the absence of vibratory energy, and phonons are emitted. Classically, enough vibratory energy must be present to overcome the barrier E_B given by

$$E_B = \frac{(E_0 - S\hbar\omega)^2}{4S\hbar\omega}. \tag{6.26}$$

The classical expectations are more likely to be fulfilled at high temperatures when many phonons are excited.

The configuration co-ordinate diagram can therefore help to explain how optical transitions occur with the involvement of many phonons and how multi-phonon processes can explain thermal generation and capture. It is possible to obtain very simply semi-classical expresssions for the thermal generation and capture rates, as we shall show in the next section. We then go on to give the quantum theory of thermally broadened radiative transitions and to outline the quantum theory of thermal generation and capture.

6.2.1. *Semi-classical thermal broadening*

The thermal broadening of optical absorption bands can be described on the basis of the configuration co-ordinate diagram in a semi-classical way. If $\hbar\omega$ is the quantum of vibrational energy, the probability of there being vibrational energy E_v in the ground state is given at thermodynamic equilibrium by statistical mechanics as

$$P(E_v) = \frac{\exp(-E_v/k_B T)}{\sum_n \exp(-n\hbar\omega/k_B T)} = 2\exp\left(-\frac{E_v}{k_B T}\right)\sinh\left(\frac{\hbar\omega}{2k_B T}\right). \tag{6.27}$$

A vibrational energy E_v corresponds to an amplitude Q_v given by

$$Q_v = \pm\left(\frac{2E_v}{\hbar\omega}\right)^{1/2} + Q_1. \tag{6.28}$$

Given that optical transitions occur without change of Q, the energy in the excited state at $Q = Q_v$ is

$$E_e = E_2 + \tfrac{1}{2}\hbar\omega\left\{Q_1 - Q_2 \pm \left(\frac{2E_v}{\hbar\omega}\right)^{1/2}\right\}^2. \tag{6.29}$$

A transition from the vibrating ground state to the excited state requires a photon of energy given by

$$\hbar\omega_\nu = E_e - E_g = E_e - (E_1 + E_v). \tag{6.30}$$

Solving for E_v we obtain

$$E_v = \frac{\{\hbar\omega_\nu - (E_0 + S\hbar\omega)\}^2}{4S\hbar\omega} \tag{6.31}$$

and thus the probability of absorbing a photon $\hbar\omega_\nu$ is proportional to

$$P(\hbar\omega_\nu) = 2\sinh\left(\frac{\hbar\omega}{2k_BT}\right)\exp\left[-\frac{\{\hbar\omega_\nu - (E_0 + S\hbar\omega)\}^2}{4S\hbar\omega k_BT}\right] \tag{6.32}$$

which describes a Gaussian absorption band. Emission can be similarly treated, and $E_0 - S\hbar\omega$ replaces $E_0 + S\hbar\omega$ in the exponent.

6.3. Semi-classical thermal generation rate

Before embarking on the semi-classical argument let us look at how classical physics might be used to calculate the thermal rate of generation of an electron from an impurity centre. The only possibility of transition in the classical viewpoint occurs at cross-over (see Fig. 6.3). Now the system, if it vibrates with just the right amplitude, reaches cross-over once every cycle, i.e. at the rate of the frequency f. Let P_X be the probability of making the transition when the system is at threshold, and $P(E_X)$ be the probability of the system having energy E_X. Then the rate is

$$W = fP_XP(E_X). \tag{6.33}$$

In the phonon picture $P(E_X)$ is given at thermodynamic equilibrium by statistical mechanics as follows (eqn (6.27)):

$$P(E_X) = 2\exp\left(-\frac{E_X}{k_BT}\right)\sinh\left(\frac{\hbar\omega}{2k_BT}\right) \tag{6.34}$$

To obtain the classical result we can put $\hbar\omega = \Delta E$ and let ΔE become very small:

$$P(E_X)_{cl} = \frac{\Delta E}{k_B T} \exp\left(-\frac{E_X}{k_B T}\right). \tag{6.35}$$

If we take the probability for the transition to be just $\frac{1}{2}$ we obtain

$$W_{cl} = \frac{\omega}{4\pi} \frac{\Delta E}{k_B T} \exp\left(-\frac{E_X}{k_B T}\right). \tag{6.36}$$

In the limit $\Delta E \to 0$ the rate is zero. Thus, if the energies of the two states are defined with classical precision there is no possibility of a transition.

In the semi-classical model we relax this precision and consider the cross-over to be the close approach of two levels over a range ΔE. We can picture the vibration as one in which the system swings through the 'cross-over' in a time Δt. Time-dependent perturbation theory gives the transition rate at cross-over as

$$w = \frac{2\pi}{\hbar} V_X^2 N(E) \tag{6.37}$$

where V_X^2 is a squared-matrix element and $N(E)$ is the number of final states per unit energy. Let us take one final state; then

$$w = \frac{2\pi}{\hbar} V_X^2 \frac{1}{\Delta E}. \tag{6.38}$$

State 1 decays exponentially in time as the system transits through cross-over, and thus the probability of a transition is $1 - \exp(-w\,\Delta t)$. However, there is always the possibility of a back transition. The probability that this does *not* occur is just $1 - \exp(-w\,\Delta t)$ multiplied by $\exp(-w\,\Delta t)$. Consequently the net probability is

$$\exp(-w\,\Delta t)\{1 - \exp(-w\,\Delta t)\} \tag{6.39}$$

and since this applies as the system oscillates back through cross-over we finally obtain the net probability of a transition, given that the system at least reaches cross-over, of

$$P_X = 2\exp(-w\,\Delta t)\{1 - \exp(-w\,\Delta t)\}. \tag{6.40}$$

This result was obtained by Zener (1932) (see also Landau and Lifshitz 1965).

We now express Δt and ΔE in terms of the shape of the energy curves in the configuration co-ordinate diagram. In fact, since $w\,\Delta t \propto \Delta t/\Delta E$ we require their ratio. We can express this ratio as follows:

$$\frac{\Delta t}{\Delta E} = \Delta t \Big/ \Delta Q \left|\frac{dE_e}{dQ} - \frac{dE_g}{dQ}\right|_X. \tag{6.41}$$

Now

$$\left(\frac{\Delta Q}{\Delta t}\right)_{\mathrm{X}} = \left(\frac{2\omega}{\hbar}\right)^{1/2}(E-E_{\mathrm{X}})^{1/2}$$

$$\left(\frac{\mathrm{d}E_{\mathrm{g}}}{\mathrm{d}Q}\right)_{\mathrm{X}} = \hbar\omega(Q_{\mathrm{X}}-Q_1) \tag{6.42}$$

$$\left(\frac{\mathrm{d}E_{\mathrm{e}}}{\mathrm{d}Q}\right)_{\mathrm{X}} = \hbar\omega(Q_{\mathrm{X}}-Q_2).$$

Therefore

$$\frac{\Delta t}{\Delta E} = \left\{\left(\frac{2\omega}{\hbar}\right)^{1/2}(E-E_{\mathrm{X}})^{1/2}\hbar\omega(2S)^{1/2}\right\}^{-1} \tag{6.43}$$

and

$$w\,\Delta t = \frac{\pi V_{\mathrm{X}}^2}{S^{1/2}(\hbar\omega)^{3/2}(E-E_{\mathrm{X}})^{1/2}}. \tag{6.44}$$

The net probability of a transition maximizes at $\frac{1}{2}$ when $\exp(-w\,\Delta t)=\frac{1}{2}$. This occurs when $w\,\Delta t$ is of order unity. (To be precise $w\,\Delta t=\log 2$). Since V_{X} is a perturbation V_{X}^2 is a small quantity, and hence the maximum corresponds to the case when $E\approx E_{\mathrm{X}}$. This is the semi-classical analogue of the classical situation. The transition rate at high temperatures is therefore simply

$$W_{\mathrm{r}} = \frac{\omega}{4\pi}\frac{\hbar\omega}{k_{\mathrm{B}}T}\exp\left(-\frac{E_{\mathrm{X}}}{k_{\mathrm{B}}T}\right). \tag{6.45}$$

The subscript r denotes that this rate occurs only for the resonant case $E=E_{\mathrm{X}}$.

Thus a semi-classical approach can yield a transition at cross-over provided that $w\,\Delta t\approx\log 2$. However, since the vibrational energy E must change by at least one quantum $\hbar\omega$, a resonance is possible only if $\pi V_{\mathrm{X}}^2/S^{1/2}(\hbar\omega)^{3/2}$ is at least of the order of $(\hbar\omega)^{1/2}$, i.e. if $(V_{\mathrm{X}}/\hbar\omega)^2 \gtrsim S^{1/2}/\pi$. In practical cases of interest it turns out that $1\lesssim S\lesssim 20$, and so for resonance to be plausibly attainable $(V_{\mathrm{X}}/\hbar\omega)^2\gtrsim 1$. However, in the great majority of applications in solid state physics V_{X}, being associated with a perturbative electron–lattice interaction, is small compared with $\hbar\omega$ and the chance of a resonance is correspondingly small. Therefore in most cases we can rule out the resonant transition at cross-over as the principal mechanism. In this respect the semi-classical result agrees with the classical conclusion.

There are many other possibilities of transitions corresponding to $E>E_{\mathrm{X}}$. In these cases $w\,\Delta t\ll 1$ and

$$P_{\mathrm{X}}\approx 2w\,\Delta t. \tag{6.46}$$

By weighting each vibratory state with $P(E)$ and summing over all $E \geqslant E_X$ we obtain

$$W_{nr} = \omega \frac{V_X^2}{\hbar\omega}\left(\frac{\pi}{S\hbar\omega k_B T}\right)^{1/2} \exp\left(-\frac{E_X}{k_B T}\right) \tag{6.47}$$

where the subscript nr denotes the non-resonant case. This result was first obtained by Henry and Lang (1977).

The thermal capture rate is obtained from detailed balance at thermodynamic equilibrium:

$$W_c = W_{nr}\left(\frac{g_1}{g_2}\right)\exp\left(\frac{E_0}{k_B T}\right) \tag{6.48}$$

where g_1 and g_2 are the degeneracy factors for the ground and excited states respectively. (The occupation probability for a state is $\{1 + g^{-1}\exp(E-F)/k_B T\}^{-1}$ where F is the Fermi level.) Thus

$$W_c = \left(\frac{g_1}{g_2}\right)\omega \frac{V_X^2}{\hbar\omega}\left(\frac{\pi}{S\hbar\omega k_B T}\right)^{1/2} \exp\left(-\frac{E_B}{k_B T}\right) \tag{6.49}$$

where E_B is the vibrational barrier height. The capture cross-section is defined by

$$\sigma_{\mathbf{k}} = \frac{V}{v_{\mathbf{k}}} W_c \tag{6.50}$$

where V is the volume of the cavity and $v_{\mathbf{k}}$ is the group velocity of the electron.

6.4. Thermal broadening of radiative transitions

In the absence of an electron–lattice interaction radiative processes could proceed without the oscillator states being disturbed. With a finite electron–lattice interaction as described in Section 6.1 the oscillator states cannot avoid being disturbed because the oscillators become displaced when the electron changes its state. Even when the perturbation which causes the electronic transition does not act directly on the lattice modes, phonons are absorbed or emitted because of the displacement. The electromagnetic perturbation acts on the electron only (provided that the radiation frequency is far from the lattice absorption bands) but the transition matrix element must contain vibrational as well as electronic wavefunctions.

In a first-order radiative transition where the perturbation is H_v the rate is given by the expression (*cf.* Chapter 5, eqn (5.1))

$$W = \int_f \frac{2\pi}{\hbar} |\langle f| H_v |i\rangle|^2 \,\delta(E_f - E_i)\, dS_f. \tag{6.51}$$

In the adiabatic approximation we can express the state ket-vector as

$$|\mathrm{i}\rangle = |\mathrm{i_e}\rangle\,|\mathrm{i_L}\rangle \tag{6.52}$$

where $|\mathrm{i_e}\rangle$ is the electronic vector and $|\mathrm{i_L}\rangle$ the lattice vector. As we saw in Section 6.1 $|\mathrm{i_e}\rangle$ is a function of the oscillator co-ordinate Q, although to first order it is not. The electromagnetic interaction is not a function of Q, and so a simplification can be made by neglecting all Q dependence of the matrix element. This has been commonly termed the Condon approximation. Its validity rests on the Q dependence of the electron wavefunction being weak and on the initial and final states being close to the minima of the vibrational curves, so the spread of Q is small. What the Condon approximation allows us to do is to separate the matrix element into a purely electronic component and a purely vibratory component involving nothing more than the overlap of oscillator wavefunctions:

$$W = \int_{\mathrm{f}} \frac{2\pi}{\hbar}\,|\langle \mathrm{f_e}|\,H_\nu\,|\mathrm{i_e}\rangle|^2\,|\langle \mathrm{f_L}\,|\,\mathrm{i_L}\rangle|^2\,\delta(E_\mathrm{f}-E_\mathrm{i})\,\mathrm{d}S_\mathrm{f}. \tag{6.53}$$

If the oscillator were not displaced $\langle \mathrm{f_L}\,|\,\mathrm{i_L}\rangle$ would be unity provided that no phonons were absorbed or emitted and zero otherwise. Because they are displaced any one mode contributes a factor less than unity even if no phonon is absorbed or emitted, and it contributes a non-zero factor if phonons are absorbed or emitted.

With the possible change of phonon number taken into account, conservation of energy entails that, for the absorption of a photon $\hbar\omega_\nu$,

$$E_\mathrm{f} = E_\mathrm{i} + \hbar\omega_\nu + \sum_i (n_i - n_i')\hbar\omega_i \tag{6.54}$$

where n_i is the number of phonons in mode i initially, and n_i' is the number finally. Let us specify $\hbar\omega_\nu$ and E_f and work out the rate involving a given amount of phonon energy E_p emitted:

$$W(E_\mathrm{p}) = \frac{2\pi}{\hbar}\,|\langle \mathrm{f_e}|\,H_\nu\,|\mathrm{i_e}\rangle|^2\,J(E_\mathrm{p})\,VN(E_\mathrm{f}) \tag{6.55}$$

where

$$J(E_\mathrm{p}) = \int_{-\infty}^{\infty} |\langle \mathrm{f_L}\,|\,\mathrm{i_L}\rangle^2\,\delta(E-E_\mathrm{p})\,\mathrm{d}E. \tag{6.56}$$

It is convenient to replace the delta function by its integral representation

$$\delta(E-E_\mathrm{p}) = \frac{1}{2\pi\hbar}\int_{-\infty}^{\infty} \exp\left\{\frac{\mathrm{i}(E-E_\mathrm{p})t}{\hbar}\right\}\mathrm{d}t \tag{6.57}$$

and to express the overlap element in terms of oscillator wavefunctions $\phi_{ni}(Q_i)$ where n is the number of quanta excited:

$$\langle \mathrm{f_L} \,|\, \mathrm{i_L} \rangle = \prod_i \int_{-\infty}^{\infty} \phi_{n'i}(Q_i')\phi_{ni}(Q_i)\,\mathrm{d}Q_i. \tag{6.58}$$

There are several ways of computing the broadening factor $J(E_p)$. A convenient approach is to use the Slater sum:

$$\sum_n \phi_n(Q)\phi_n(\bar{Q})\exp\{-\gamma(n+\tfrac{1}{2})\}$$
$$= (2\pi \sinh \gamma)^{-1/2} \exp[-\tfrac{1}{4}\{(Q+\bar{Q})^2 \tanh \tfrac{1}{2}\gamma + (Q-\bar{Q})^2 \coth \tfrac{1}{2}\gamma\}]. \tag{6.59}$$

Thus we can put

$$|\langle \mathrm{f_L} \,|\, \mathrm{i_L} \rangle|^2 = \prod_i \int_{-\infty}^{\infty} \mathrm{d}Q_i \int_{-\infty}^{\infty} \mathrm{d}\bar{Q}_i \phi_{n'i}(Q_i')\phi_{n'i}(\bar{Q}_i')\phi_{ni}(Q_i)\phi_{ni}(\bar{Q}_i) \tag{6.60}$$

where the prime denotes final state quantities and Q_i' is the displaced co-ordinate, and sum over the initial states, weighting each state with its probability $P(E_i)$ where $E_i = (n+\frac{1}{2})\hbar\omega_i$ (see eqn (6.34)):

$$P(E_i) = 2 \sinh\left(\frac{\hbar\omega_i}{2k_B T}\right) \exp\left(-\frac{E_i}{k_B T}\right). \tag{6.61}$$

We can also convert the integral over energy in eqn (6.56) to a sum over n' by replacing $\mathrm{d}E$ by $\hbar\omega_i$. Thus

$$J(E_p) = \frac{1}{2\pi\hbar} \prod_i \sum_{n'} \hbar\omega_i \sum_n P(E_i) \int_{-\infty}^{\infty} |\langle \mathrm{f_L} \,|\, \mathrm{i_L} \rangle|^2 \exp\left\{\frac{\mathrm{i}(E_i - E_p)t}{\hbar}\right\} \mathrm{d}t. \tag{6.62}$$

Substituting from eqn (6.60) and using the Slater sums leads to straightforward integrations over Q_i and $\bar{Q}_i$, and we obtain

$$J(E_p) = \frac{1}{2\pi\hbar} \int_{-\infty}^{\infty} \prod_i \exp[(Q_0' - Q_0)_i^2\{\theta_i(t) - \theta_i(0)\} + \mathrm{i}\omega_p t]\hbar\omega_i \,\mathrm{d}t \tag{6.63}$$

where

$$\hbar\omega_p = E_p = \hbar\omega_\nu - (E_f - E_i) \tag{6.64}$$

$$\theta_i(t) = \frac{\cosh\{(\hbar\omega_i/2k_B T) + \mathrm{i}\omega_i t\}}{2 \sinh(\hbar\omega_i/2k_B T)} \tag{6.65}$$

and Q_o and Q'_0 are the co-ordinates of the centre of the oscillations in the initial and final states respectively (see eqn (6.15)). If the centres of oscillation coincide, $\omega_p = 0$ and $J(E_p) = 1$. Otherwise, a simple analytic expression for $J(E_p)$ can be obtained only if a single-frequency approximation is adopted. For this case

$$J(E_p) = \exp\{-2S(n+\tfrac{1}{2})\}\exp(p\hbar\omega/2k_BT)I_p(2S\{n(n+1)\}^{1/2}) \quad (6.66)$$

where S is the Huang–Rhys factor (eqn (6.17)), $p\hbar\omega = \hbar\,\omega_p$ (eqn (6.64)), $I_p(z)$ is a modified Bessel function of order p, and n is the Bose–Einstein factor (Fig. 6.5).

Some properties of the broadening function are worth noting. It is a maximum for $p = S$ at all temperatures. For absorption of phonons p changes sign, which affects the exponential but not the Bessel function

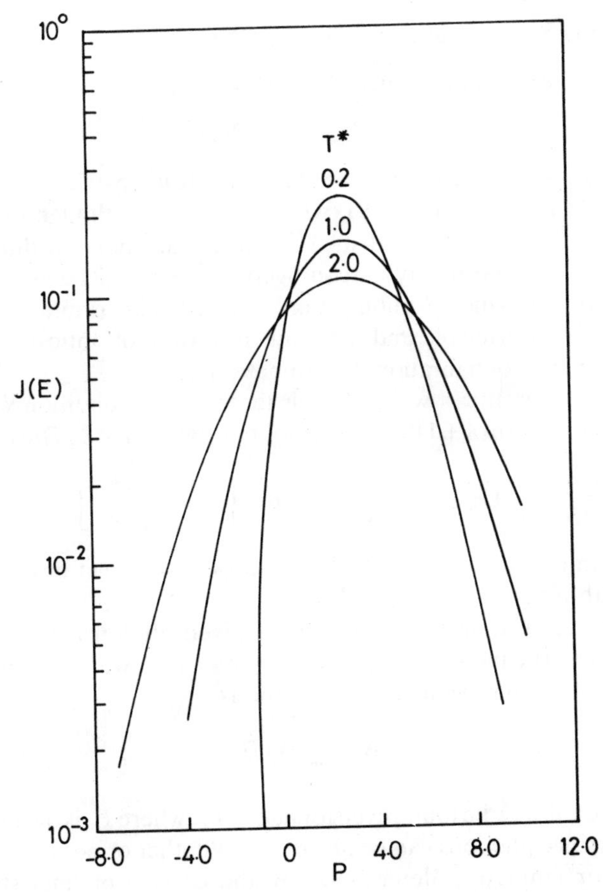

FIG. 6.5. *The thermal broadening function* $J(E_p): T^* = k_BT/\hbar\omega$. $S = 3$.

since $I_p(z) = I_{-p}(z)$. It is sometimes useful to note that

$$\exp\left(\frac{p\hbar\omega}{2k_B T}\right) = \left(\frac{n+1}{n}\right)^{p/2} \tag{6.67}$$

and that

$$\sum_{p=-\infty}^{\infty} J(E_p) = 1. \tag{6.68}$$

Simple expressions for the modified Bessel function can be obtained in the limiting cases of (a) low temperature–weak coupling and (b) high temperature–strong coupling:

$$\text{(a)}\ I_p(z) = \frac{(z/2)^p}{p!} \qquad z \ll p \tag{6.69}$$

$$\text{(b)}\ I_p(z) = (2\pi z)^{-1/2} \exp(z - p^2/2z) \qquad z \gg p. \tag{6.70}$$

Thus at low temperatures such that $2S\{n(n+1)\}^{1/2} \ll p$ and $n \ll 1$

$$J(E_p) \approx S^p e^{-S}/p! \tag{6.71}$$

which is a Poisson distribution about a mean at $p = S$. The probability of a zero-phonon process is $\exp(-S)$ (which is true even though the condition $2S\{n(n+1)\}^{1/2} \ll p$ is not satisfied). Thus, transitions without phonon participation are positively discouraged for $S \geqslant 1$. The most probable process is one in which S phonons are emitted. This is in accord with the Franck–Condon principle and with our discussion of optical processes on the basis of the configuration co-ordinate diagram.

At high temperatures we approach the classical condition when $n \gg 1$. Provided that $2S\{n(n+1)\}^{1/2} \gg p$ we obtain, with $n \approx k_B T/\hbar\omega$,

$$J(E_p) = \left(\frac{\hbar\omega}{4\pi S k_B T}\right)^{1/2} \exp\left\{-\frac{(p-S)^2\hbar\omega}{4k_B TS}\right\} \tag{6.72}$$

which again maximizes at $p = S$. This is now a Gaussian distribution over energy with a half-width $(k_B TS\hbar\omega)^{1/2}$.

Any electronic transition caused by a given photon is thus dependent on phonon participation. To obtain the total rate we must sum over all possible phonon emissions and absorptions:

$$W = \sum_p W(E_p). \tag{6.73}$$

A finite rate will now be found even for $\hbar\omega_\nu < E_0$, where E_0 is the phonon-free threshold, since phonons can be absorbed. Whether or not phonons can be absorbed or emitted is determined by the density of final states being non-zero. The most probable case is for S phonons to be emitted. For transitions such that $\hbar\omega_\nu > E_0 + S\hbar\omega$ we can approximate the sum in eqn (6.73) by using the normalizing property depicted in eqn (6.68) and taking the density of states to be that corresponding to S phonons having been emitted. In this way we return to the rates computed for the phonon-free case but with an effective threshold at

$$E_{\text{th}} \approx E_0 + S\hbar\omega \tag{6.74}$$

i.e.

$$W \approx \frac{2\pi}{\hbar} |\langle f_e| H_\nu |i_e\rangle|^2 \, VN(\hbar\omega_\nu - E_0 - S\hbar\omega). \tag{6.75}$$

Figures 6.6 and 6.7 illustrate the validity of this viewpoint for photo-ionization cross-sections. At and below this effective threshold we must

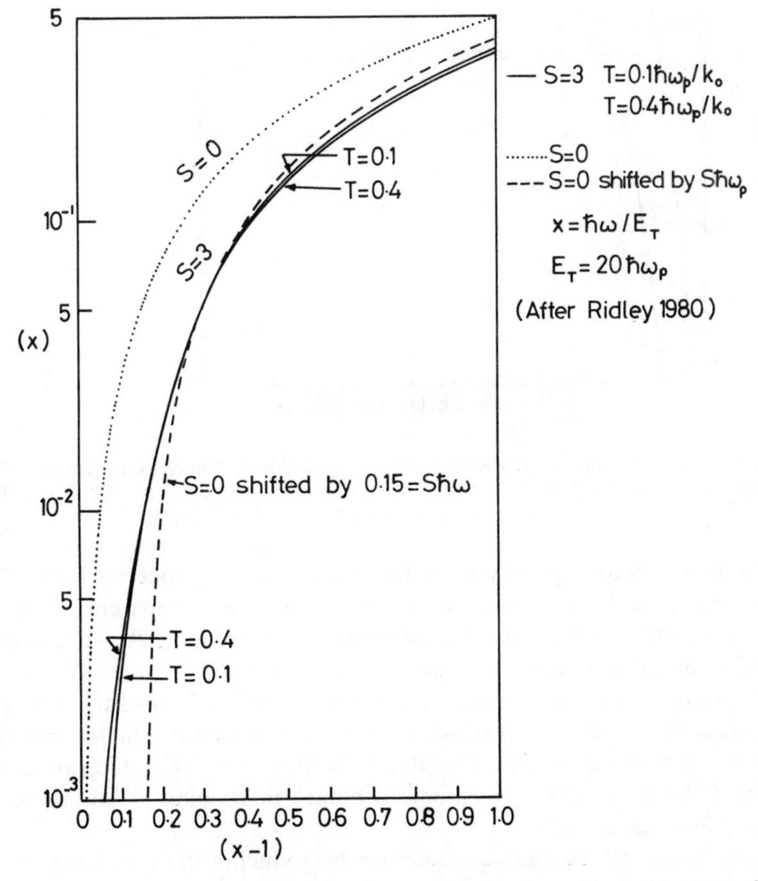

FIG. 6.6. Photo-ionization cross-section for neutral donor centre: solid curve, $S=3$, $T=0{\cdot}1\,\hbar\omega_p/k_D$ and $T=0{\cdot}4\,\hbar\omega_p/k_D$; dotted curve, $S=0$; broken curve, $S=0$ shifted by $S\hbar\omega_p$; $x=\hbar\omega/E_T$; $E_T=20\,\hbar\omega_p$. (After Ridley 1980.)

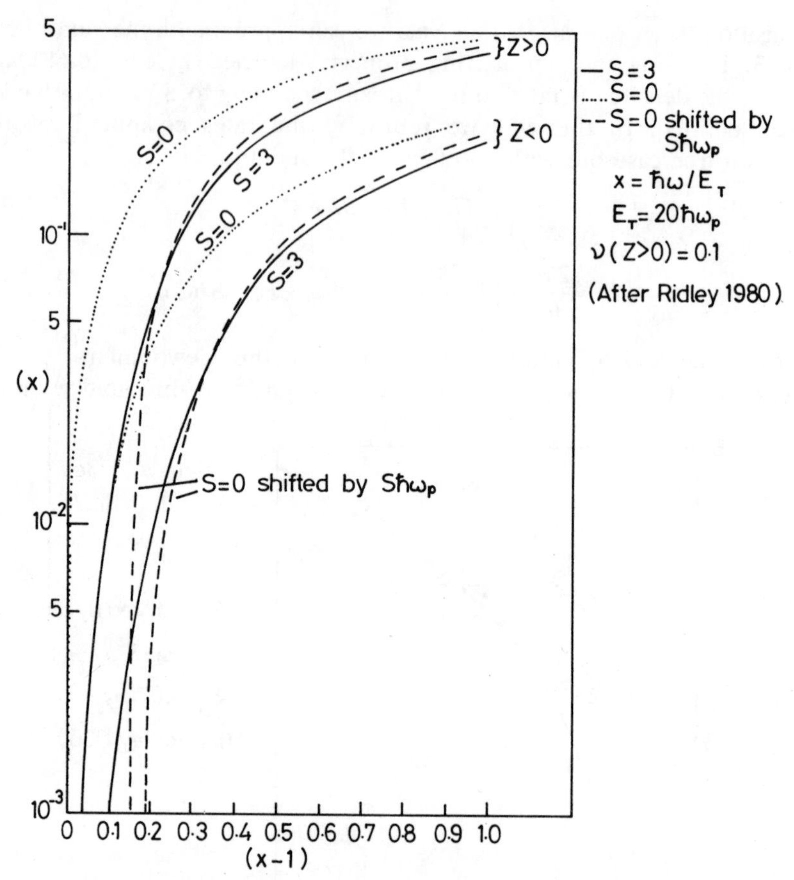

FIG. 6.7. Photo-ionization cross-section for charged donor centres: solid curves, $S = 3$; dotted curves, $S = 0$; broken curves, $S = 0$ shifted by $S\hbar\omega_p$; $x = \hbar\omega/E_T$; $E_t = 20\,\hbar\omega_p$; $\nu(Z<0) = 0{\cdot}1$. (After Ridley 1980.)

evaluate the sum explicitly in order to describe the spectral form. This approximation is also inadequate for describing apparent shifts of threshold with temperature. Usually, one has to compute the sum numerically to describe temperature dependence accurately (Fig. 6.8).

For optical transitions involving two discrete levels no sum over p is necessary since only one p satisfies energy conservation. The basic optical rate is given by eqn (5.20), Chapter 5. In this case $J(E_p)$ determines the form of the absorption (or emission) band—Poissonian at low temperatures, Gaussian at high.

Broadening affects radiative capture by reducing the frequency of the most probable light emission to

$$\hbar\omega_\nu^{\mathrm{em}} \approx E_0 - S\hbar\omega \tag{6.76}$$

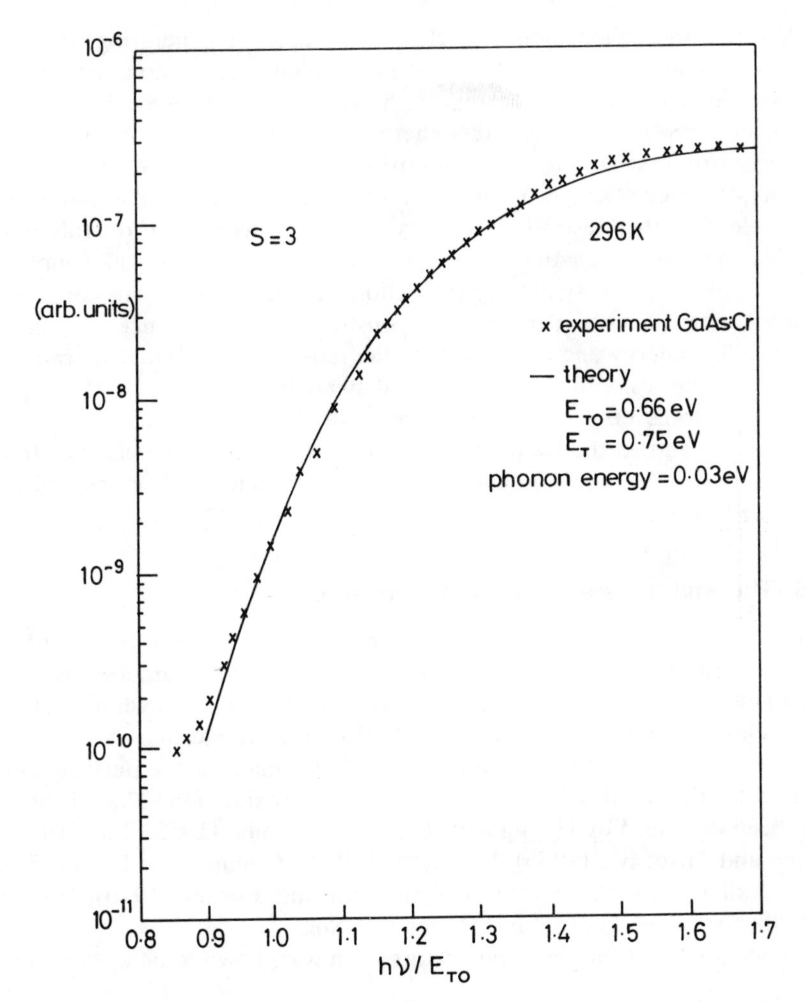

FIG. 6.8. Fit of theory to experiment in GaAs: Cr: points, experiment: curve, theory; $E_{\mathrm{TO}} = 0{\cdot}66$ eV; $E_{\mathrm{T}} = 0{\cdot}75$ eV; phonon energy, 0·03 eV. (After Amato *et al.* 1980)

which reduces the transition rate by a factor $E_0 - S\hbar\omega/E_0$ from the phonon-free rate. Relative to the photo-ionization cross-section, the capture cross-section becomes

$$\sigma_{\mathbf{k}}/\sigma_\nu = \frac{q_\nu^{\mathrm{abs}} q_\nu^{\mathrm{em}}}{k^2} \tag{6.77}$$

which replaces eqn (5.30), Chapter 5. Here q_ν^{abs} and q_ν^{em} are the wavevectors in the medium of the absorbed and emitted light for the most probable transitions, i.e. those in which S phonons are emitted.

A single-frequency model is clearly valuable in generating analytic expressions, and a surprisingly smooth dependence on photon energy can be generated in all cases. In reality, however, many frequencies will be involved. Even so, because either there is strongly coupled local mode or there is strongest coupling to the lattice modes which have the highest density of states, a single-frequency approximation can be expected to be plausible. In either case the frequency of importance will be roughly that at the zone edge for the given type of mode. Given equal coupling strengths $\mathbf{V}$ and zero symmetry restrictions, the most important contribution will come from the transverse acoustic (TA) modes since they have the smallest energy and are therefore the most readily excited thermally. However, centres with largely spherical symmetry are unlikely to couple strongly to transverse modes but are more likely to interact with longitudinal modes, and so the frequencies involved are likely to be close to that for LA and LO modes. Selection rules for coupling to phonons will be discussed in Section 6.7.

6.5. Thermal generation and capture rates

When photons are not involved in the transition a new form of perturbation is required. The electron–phonon interaction which induces scattering of electrons between band states and which has been shown in Section 6.1 to introduce displacements of the lattice oscillators cannot be invoked since its effect is already accounted for. What remains is the perturbation caused by the breakdown of the adiabatic approximation. This problem has been discussed by Huang and Rhys (1950), Kubo (1952), Lax (1952), Kubo and Toyozawa (1955), Rickayzen (1957), Gummel and Lax (1957), Kovarskii (1962), Pryce (1966), Engelman and Jortner (1970), Pässler (1974), Stoneham (1975), and Ridley (1978*a*).

In Section 1,2, Chapter 1 the perturbation was shown to be of the form

$$H_{\mathrm{NA}}\Psi(\mathbf{r},\mathbf{R})\Phi(\mathbf{R}) = H_{\mathrm{L}}\Psi(\mathbf{r},\mathbf{R})\Phi(\mathbf{R}) - \Psi(\mathbf{r},\mathbf{R})H_{\mathrm{L}}\Phi(\mathbf{R}). \tag{6.78}$$

In terms of a one-electron wavefunction and the displaced normal modes of the lattice this can be cast in the form

$$H_{\mathrm{NA}}\psi(\mathbf{r},\mathbf{Q})\Phi(\mathbf{Q}) = -\sum_i \hbar\omega_i\left\{\frac{\partial\psi(\mathbf{r},\mathbf{Q})}{\partial Q_i}\frac{\partial\Phi(\mathbf{Q})}{\partial Q_i} + \tfrac{1}{2}\Phi(\mathbf{Q})\frac{\partial^2\psi(\mathbf{r},\mathbf{Q})}{\partial Q_i^2}\right\} \tag{6.79}$$

where $\psi(\mathbf{r},\mathbf{Q})$ is the one-electron wavefunction—a function of electron position and of the normal co-ordinates through the linear adiabatic electron–lattice interaction—and $\Phi(\mathbf{Q})$ represents the product of harmonic wavefunctions, each a function of its displaced normal co-ordinate

Q_i. The transition rate can be expressed as follows:

$$W = \frac{1}{\hbar^2} \sum_f \int_{-\infty}^{\infty} |\langle f| H_{NA} |i\rangle|^2 \exp\left\{\frac{i(E_f - E_i)t}{\hbar}\right\} dt \tag{6.80}$$

where we have expressed the delta function in its integral representation.

The non-adiabatic perturbation is non-zero by virtue of the **Q** dependence of the electronic wavefunction. This dependence is linear in the lowest order of approximation which still contains **Q** as eqn (6.6) shows. Taking this linear dependence makes the last term in eqn (6.79) vanish and causes the **Q** dependence of the electronic part of the matrix element to disappear through the differentiation. Thus we obtain the Condon approximation, in which the electronic component of the matrix element is taken to be independent of the lattice co-ordinate. Such an approximation was adopted by Huang and Rhys in their pioneering calculation. Although the Condon approach may be valid when vertical photon transitions are involved, it cannot be a good assumption for purely thermal transitions in which relatively large amplitudes of vibration are required. Instead of a linear dependence of $\psi(\mathbf{r}, \mathbf{Q})$ on **Q** we shall therefore take the wavefunction to be of the form

$$\psi_1(\mathbf{r}, \mathbf{Q}) = \psi_{10}(\mathbf{r}) + \frac{H_{12}}{E_{11} - E_{21}} \psi_{20}(\mathbf{r}) \tag{6.81}$$

with the subscript notation of Section 6.1. The **Q** dependence in the numerator is

$$H_{12} = \langle 2| \mathbf{V} |1\rangle . \mathbf{Q} \tag{6.82}$$

and that in the denominator is given by

$$E_{21} - E_{11} = E_{20} - E_{11} + H_{22} - H_{11} = E_{20} - E_{10} - \sum_i \hbar\omega_i (Q_2 - Q_1)_i Q_i \tag{6.83}$$

(see eqns (6.4), (6.5), and (6.15)). Taking the denominator to be linear in **Q** is the natural next step in order of approximation, and it is valid to infinite order except close to cross-over. This approach was first adopted by Kovarskii (1962) and recently illuminated by Huang (1981) (see section 6.10).

To perform the calculations it is convenient to express the denominator as

$$\frac{1}{E_{21} - E_{11}} = \int_0^{\infty} \sin\{(E_{21} - E_{11})y\} \, dy \tag{6.84}$$

and follow the method of the Slater sum which was used to describe thermal broadening. The calculation is extremely laborious and we shall not follow it in detail. The adoption of a single-frequency model leads to the result for the thermal generation rate:

$$W = \omega \frac{\pi}{(\hbar\omega)^2} \exp\{-2S(n+\tfrac{1}{2})\} \exp\left(-\frac{p\hbar\omega}{2k_B T}\right) R \tag{6.85}$$

where S is the Huang Rhys factor, n the Bose–Einstein factor, and R is a complex sum of terms containing modified Bessel functions.

Essentially, the terms in eqn (6.85) represent the contributions from (a) the overlap of displaced oscillator functions which decreases with increasing S, (b) thermal weighting of initial vibrational energy, and (c) the probability of emitting and absorbing phonons such that overall p phonons are absorbed. The non-adiabatic interaction induces any given mode to absorb or emit up to four phonons and the overlap of the unaffected modes provides the balance. Stoneham (1975) has termed the mode being directly operated on by the perturbation as the promoting mode—the others react passively as they did in the thermal broadening problem.

In the limit of low temperatures and weak coupling such that

$$(p+4)^2 \gg 4S^2 n(n+1) \tag{6.86}$$

we obtain the simplified result (Ridley 1978)

$$W = W_0 n^p e^{-2nS} \tag{6.87}$$

where

$$W_0 = \omega \frac{\pi}{(\hbar\omega)^2} \frac{S^{p-1} e^{-S}}{(p-1)!} \left(\mathbf{V}_{12}^2 R_0 + |\mathbf{V}_{12} \cdot \mathbf{\Delta}|^2 \frac{p-1}{S} R_1\right). \tag{6.88}$$

In this equation

$$\mathbf{V}_{12} = \langle 2|\, \mathbf{V}\, |1\rangle \tag{6.89}$$

and we have introduced the N-dimensional vector $\mathbf{\Delta}$ whose components are given by

$$\mathbf{\Delta}_i = \frac{1}{\sqrt{2}} (Q_2 - Q_1)_i$$

so that

$$S = \sum_i \Delta_i^2 \tag{6.90}$$

and for $S \ll p$, $R_0 \approx 0{\cdot}26$ and $R_1 \approx 0{\cdot}18$.

From detailed balance the rate for capture is simply

$$W_c = \left(\frac{g_1}{g_2}\right) W_0 (n+1)^p e^{-2nS}. \tag{6.91}$$

These rates exceed those calculated on the basis of the Condon approximation by a factor $(p/2)^2$. Such an enhancement is a direct consequence of the increased mixing of the two electronic states induced by taking into account the **Q** dependence of the denominator.

In the limit of high temperatures and strong coupling such that

$$(p+4)^2 \ll 4S^2 n(n+1) \tag{6.92}$$

we obtain the simplified result (Ridley 1978)

$$W = \omega \frac{\mathbf{V}_{12}^2 p^2}{2S(\hbar\omega)^2} \left(\frac{\pi\hbar\omega}{Sk_B T}\right)^{1/2} \exp\left(-\frac{E_X}{k_B T}\right) \tag{6.93}$$

where E_X is the cross-over energy defined in eqn (6.24).

How does this result, which is derived rigorously within quantum mechanics, compare with the semi-classical result of eqn (6.47)? It is identical, provided that the semi-classical V_X is identified by

$$V_X^2 \rightarrow \frac{\mathbf{V}_{12}^2 p^2}{2S} = \mathbf{V}_{12}^2 \frac{E_0^2}{2S(\hbar\omega)^2} \tag{6.94}$$

Such an identification can indeed be made, since at cross-over the matrix element connecting the two states is in the quantum model

$$H_{12X} = \langle 2| \mathbf{V}_1 |1\rangle . \mathbf{Q}_X \tag{6.95}$$

and from eqn (6.25), for weak coupling,

$$Q_X \approx \frac{E_0}{(2S)^{1/2}\hbar\omega}. \tag{6.96}$$

Thus

$$H_{12X} = V_{12} \frac{E_0}{(2S)^{1/2}\hbar\omega} \tag{6.97}$$

which can be equated with the semi-classical interaction energy at cross-over. Thus the identification between the quantum-theoretical and semi-classical results is complete. It is not surprising that the agreement is with the non-resonant solution. Our model specifically deals with arbitrary vibrational energies not to near cross-over and is therefore a non-resonant model from the outset.

The capture rate at high temperatures is given by

$$W_c = \left(\frac{g_1}{g_2}\right) \omega \frac{\mathbf{V}_{12}^2 p^2}{2S(\hbar\omega)^2} \left(\frac{\pi\hbar\omega}{Sk_B T}\right)^{1/2} \exp\left(-\frac{E_B}{k_B T}\right) \tag{6.98}$$

where E_B is the height of the vibrational barrier.

The resonant case has been considered by Kovarskii and Sinyavskii (1963), but their result does not appear to agree with the semi-classical formula (eqn (6.45)). If we take the semi-classical formula to be valid, the ratio of non-resonant to resonant rates is

$$\frac{W_{\mathrm{nr}}}{W_{\mathrm{r}}}=4\pi\left(\frac{\pi k_{\mathrm{B}}T}{S\hbar\omega}\right)^{1/2}\frac{p^2\mathbf{V}_{12}^2}{2S(\hbar\omega)^2}=4\pi\left(\frac{\pi k_{\mathrm{B}}T}{S\hbar\omega}\right)^{1/2}\frac{V_{\mathrm{X}}^2}{(\hbar\omega)^2} \tag{6.99}$$

where V_{X} is the semi-classical interaction energy at cross-over (eqn (6.94)). We saw in Section 6.3 that for the resonant solution to be possible $(V_{\mathrm{X}}/\hbar\omega)^2$ had to be greater than or of order unity. Even if this were so, the non-resonant solution dominates in the high-temperature regime since $k_{\mathrm{B}}T \gg \hbar\omega$.

6.6. Electron–lattice coupling strength

The quantities which govern the strength of multi-phonon processes are the displacement factor $\Delta_i = 2^{-1/2}(Q_2 - Q_1)_i$ and the matrix element V_{12i} between ground and excited states. In order to calculate these we require the explicit form of the energy factor $V_i(\mathbf{r})$ in the electron–phonon perturbation. Continuation with the single-frequency model implies that the electron–phonon interaction should be taken to be of the form associated with long-wave optical phonons or zone-edge phonons, which means assuming that the principal electronic energy change is proportional to displacement rather than strain, as it is in the scattering problem:

$$H_{\mathrm{ep}} = \mathbf{U}\cdot\mathbf{u} \tag{6.100}$$

where $\mathbf{u}$ is the displacement and $\mathbf{U}$ is the energy change per unit displacement. By expanding $\mathbf{u}$ in plane waves we can transform eqn (6.100) to

$$H_{\mathrm{ep}} = \frac{1}{\sqrt{N}}\sum_i \{\mathcal{D}_i(q_i)Q_i \exp(\mathrm{i}\mathbf{q}_i\cdot\mathbf{r}) + \mathrm{c.c.}\} \tag{6.101}$$

where $\mathbf{q}_i$ is the wavevector of mode i and N is the number of modes. In the case of deformation potential coupling to optical modes and zone-edge phonons we can assume that

$$|\mathcal{D}_i(\mathbf{q}_i)|^2 = \frac{D^2}{M\omega/\hbar} \tag{6.102}$$

where D is the deformation potential constant and M is the mass in the unit cell. This form of coupling is the one used in the theory of lattice scattering of electrons (Section 3.4 and eqn (3.116)). Higher-order coupling proportional to acoustic strain is also possible, but we shall assume that for some modes D is non-zero and that these modes make the greatest contribution. Coupling to longitudinal optical modes is also possible through the polar interaction. In this case

$$|\mathscr{D}_i(\mathbf{q}_i)|^2 = \frac{e^2(\bar{M}/V_0)\omega^2}{(\bar{M}\omega/\hbar)}\left(\frac{1}{\epsilon_\infty}-\frac{1}{\epsilon}\right)\left(\frac{q_i}{q_i^2+q_0^2}\right)^2 \tag{6.103}$$

where $\bar{M}$ is the reduced mass, V_0 is the volume of the unit cell and is equal to a_0^3 where a_0 is $4^{-1/3}$ times the lattice constant in zinc blende lattices, ϵ_∞ and ϵ are the high frequency and static permittivities, and q_0 is the reciprocal Debye screening length. Piezoelectric coupling, which depends upon acoustic strain, is also possible but it is likely to be much weaker.

Let the final electronic state be one of the Bloch states $|\mathbf{k}\rangle$ in the conduction band and let the initial state be a localized state $|\mathrm{T}\rangle$ at the defect. The three matrix elements we must calculate are

$$Q_{2i} = \langle\mathbf{k}|\frac{\mathscr{D}_i(\mathbf{q}_i)}{\hbar\omega\sqrt{N}}\exp(\mathrm{i}\mathbf{q}_i\,.\mathbf{r})\,|\mathbf{k}\rangle \tag{6.104}$$

$$Q_{1i} = \langle\mathrm{T}|\frac{\mathscr{D}_i(\mathbf{q}_i)}{\hbar\omega\sqrt{N}}\exp(\mathrm{i}\mathbf{q}_i\,.\mathbf{r})\,|\mathrm{T}\rangle \tag{6.105}$$

$$V_{21i} = \langle\mathbf{k}|\frac{\mathscr{D}_i(\mathbf{q}_i)}{\hbar\omega\sqrt{N}}\exp(\mathrm{i}\mathbf{q}_i\,.\mathbf{r})\,|\mathrm{T}\rangle. \tag{6.106}$$

The matrix element Q_{2i} in eqn (6.104) is zero unless $\mathbf{q}_i$ is a reciprocal lattice vector. Since only a very small fraction of all modes can satisfy this criterion Q_{2i} can be regarded as negligible, and indeed infinitesimal as N tends to infinity. Thus the electron–lattice coupling does not displace oscillators when the electron is in a Bloch state. Even when a scattering state is taken instead of a simple Bloch state, the displacement will be small since most of the contribution to the matrix element comes from distant regions of space where the Bloch function is a good approximation.

The other matrix elements depend upon the localized wavefunction. In order to illuminate the principal features of the problem without becoming involved in the uncertainties which attend the assignation of a defect wavefunction, we shall follow the approximation used for the photoionization of deep-level impurities and assume a billiard-ball-like form for the wavefunction characterized by an effective radius r_T and a volume V_T. This means that integrations over space involving localized state are limited to radii between zero and r_T:

$$Q_{1i} \approx \frac{\langle \mathscr{D}_i(q_i)\rangle}{\hbar\omega\sqrt{N}} \frac{1}{V_T} \int_0^{r_T} 4\pi r^2 \left(\frac{\sin q_i r}{q_i r}\right) \mathrm{d}r \tag{6.107}$$

$$\approx \frac{\langle \mathscr{D}_i(q_i)\rangle}{\hbar\omega\sqrt{N}} \bigg|_{q_i r_T \leqslant \pi}. \tag{6.108}$$

Rapidly varying parts of the localized wavefunction, which are equivalent to the cell-periodic part of a Bloch function, are subsumed in the strength of the coupling constant.

The coupling between states is given by

$$V_{21i} \approx \frac{\langle \mathscr{D}_i(q_i)\rangle}{\sqrt{N}} \frac{1}{(V_T V)^{1/2}} \int_0^{r_T} 4\pi r^2 \left(\frac{\sin |\mathbf{q}_i - \mathbf{k}|\, r}{|\mathbf{q}_i - \mathbf{k}|\, r}\right) \mathrm{d}r \tag{6.109}$$

$$\approx \frac{\langle \mathscr{D}_i(q_i)\rangle}{\sqrt{N}} \left(\frac{V_T}{V}\right)^{1/2} \bigg|_{|\mathbf{q}_i - \mathbf{k}| r_T \leqslant \pi}. \tag{6.110}$$

Strictly $\langle \mathscr{D}_i(q_i)\rangle$ in this case is different from the equivalent term in eqns (6.107) and (6.108) since different rapidly varying components are involved, but for simplicity we shall assume them to be identical. Usually we are interested in states near a band edge and for most cases we can neglect $\mathbf{k}$ in the definition of the limits to q_i.

To perform the sums over the phonon modes we approximate the Brillouin zone by a sphere of radius $q_D = (6\pi^2 N/V)^{1/3}$ where V is the volume of the cavity. The Huang–Rhys factor is then given by eqn (6.17):

$$S = \int_0^{q_m} \frac{|\mathscr{D}_i(q_i)|^2}{2N(\hbar\omega)^2} \frac{4\pi q^2\, \mathrm{d}q V}{8\pi^3} \qquad q_m r_T = \pi. \tag{6.111}$$

For deformation coupling

$$S_D = \frac{\pi^3}{2(\hbar\omega)^2 (q_D r_T)^3} \frac{D^2}{M\omega/\hbar}. \tag{6.112}$$

For polar coupling, neglecting screening,

$$S_P = \frac{3\pi}{2(\hbar\omega)^2 (q_D r_T)} \left\{ \frac{e^2(\bar{M}/V_0)\omega^2}{\bar{M}\omega/\hbar} \left(\frac{1}{\epsilon_\infty} - \frac{1}{\epsilon}\right) \frac{1}{q_D^2} \right\}. \tag{6.113}$$

Maximum values of S_D and S_P occur when $q_D r_T = \pi$. These are shown in Table 6.1 for LO modes. (The radius q_D of the zone can be obtained by putting $V = N a_0^3$ where a_0 is $4^{-1/3}$ times the lattice constant, whence $q_D a_0 = (6\pi^2)^{1/3}$.)

TABLE 6.1
Limiting Huang–Rhys factors associated with the LO mode for some semiconductors

	a_0 (Å)	$\hbar\omega_{LO}$ (meV)	$\epsilon_\infty/\epsilon_0$	ϵ/ϵ_0	$(\bar{M}\omega_{LO}/\hbar)^{-1/2}$ (Å)	S_d [a]	S_P
GaAs	3·55	30	10·9	13·18	0·06212	53	2·66
InP	3·70	39	9·52	12·35	0·06629	35	2·98
InAs	3·82	20	11·8	14·55	0·06790	142	3·75
InSb	4·08	16	15·7	17·72	0·06649	212	1·99
GaSb	3·84	23	14·4	15·69	0·06402	95	1·15
GaP	3·43	50	9·04	11·1	0·06243	19	2·14
Ge	3·56	30	16·0	16·0	0·06196	26	0
Si	3.42	55	11·9	11·9	0·07355	11	0

[a] It is arbitrarily assumed that $D = 5 \times 10^8$ eV cm^{-1}.

Because the polar coupling is stronger at longer wavelengths and because the sum over modes weights short wavelengths heavily, the Huang–Rhys factor for polar coupling is relatively small even when a modest deformation coupling strength is assumed. In the case of intervalley scattering in silicon D is effectively of order 1×10^9 eV cm^{-1}†, and if this value were assumed the values for S_D would rise by a factor of 4. However, S_P is less sensitive to the radius of the localized state and so will be larger than S_D for extended states, though in absolute value it will be small. Practical strengths of coupling therefore rely mainly on deformation potential coupling.

Extended localized states such as those which can be described by effective mass theory—whether hydrogenic or quantum defect is unimportant—are only weakly coupled to the lattice. For such states $q_D r_T$ may be 10π or more and S_D will be reduced from its maximum value by the order of 10^3 to a small value. Figures 6.9(a) and 6.9(b) illustrate this effect for quantum-defect centres. In contrast, deep-level impurities will have $q_D r_T \approx 3\pi$ and significant magnitudes for S_D and S_P will be obtained. In that r_T can be expected to vary only weakly among deep-level impurities in a given semiconductor, the strength of coupling ought not to differ very significantly from one deep-level centre to another provided that the same modes are involved.

The magnitude of S_D is sensitive to the phonon energy of the mode. The values in Table 6.1 are for LO modes. They would be higher for LA modes and especially so for TA modes, given equal coupling strengths.

† This figure is not the deformation potential constant for a single mode but rather a quantity describing the result of three modes scattering to five equivalent silicon valleys. When this is taken into account it turns out that the average D associated with a single mode of high frequency is 5×10^8 eV cm^{-1}.

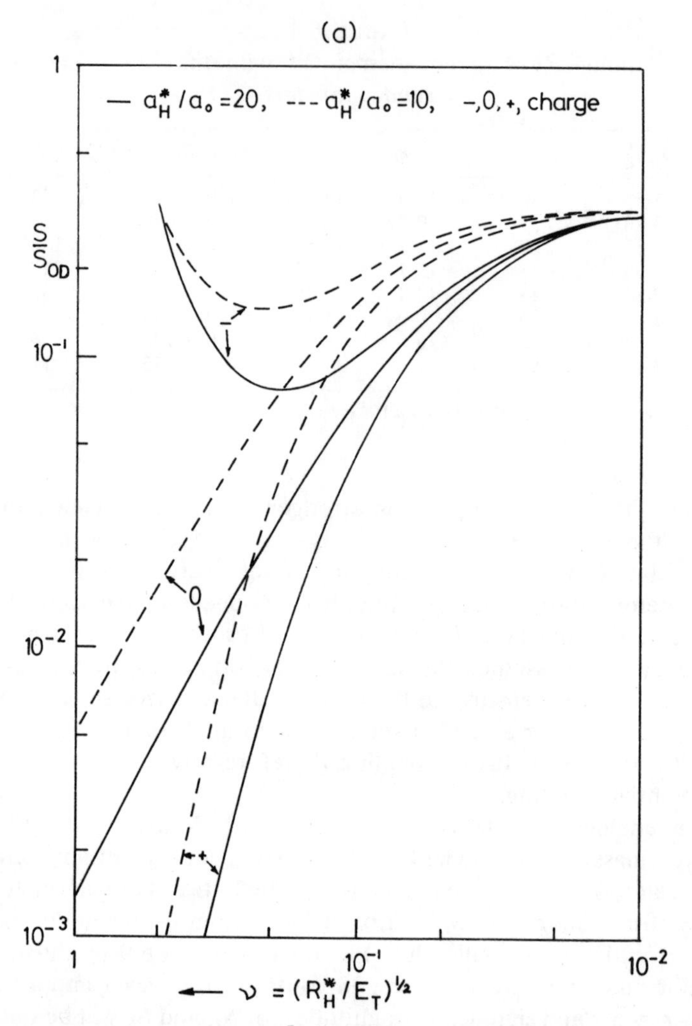

FIG. 6.9. Huang–Rhys factor (as a ratio of maximum value) as a function of level depth for quantum-defect states for $a_H^*/a_0 = 20$ (solid curves) and $a_H^*/a_0 = 10$ (broken curves) and for centres with charge +1, 0, and −1. a_H^* is the effective Bohr radius, a_0 is the unit cell dimension and is $1/4^{1/3}$ times the lattice constant. (a) Deformation potential interaction; (b) polar interaction. (After Ridley 1978*b*.)

From eqn (6.110) and summing over the modes, we obtain the transition factor for deformation coupling:

$$\mathbf{V}_{12}^2 = \left(\frac{\pi}{q_D r_T}\right)^3 \frac{D^2}{m\omega/\hbar}\frac{V_T}{V} = 2S_D(\hbar\omega)^2\frac{V_T}{V}. \tag{6.114}$$

The latter equality assumes that the same modes contribute to both V_{12}^2

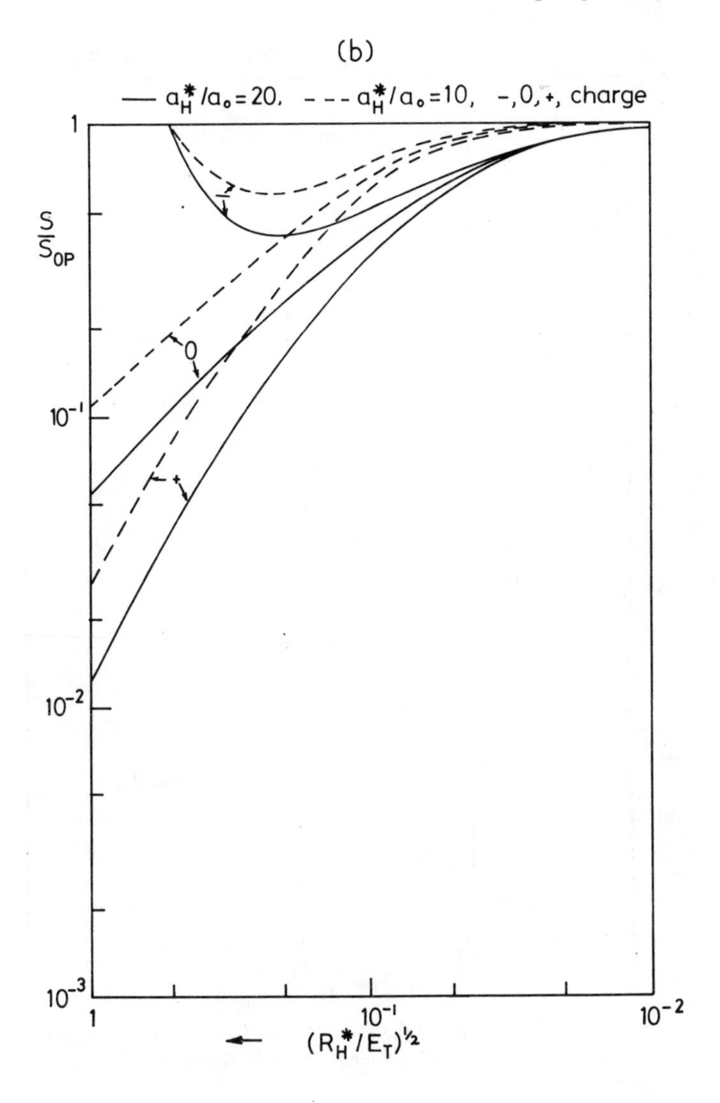

and to S_D. For polar coupling

$$\mathbf{V}_{12}^2 = 2S_P(\hbar\omega)^2 \frac{V_T}{V}. \tag{6.115}$$

Finally, the transition factor $|\mathbf{V}_{12} \cdot \mathbf{\Delta}|^2$ is, for deformation coupling,

$$|\mathbf{V}_{12} \cdot \mathbf{\Delta}|^2 = \left\{ \left(\frac{\pi}{q_D r_T} \right)^3 \frac{1}{2^{1/2}\hbar\omega} \frac{D^2}{M\omega/\hbar} \left(\frac{V_T}{V} \right)^{1/2} \right\}^2 = \mathbf{V}_{12}^2 S_D \tag{6.116}$$

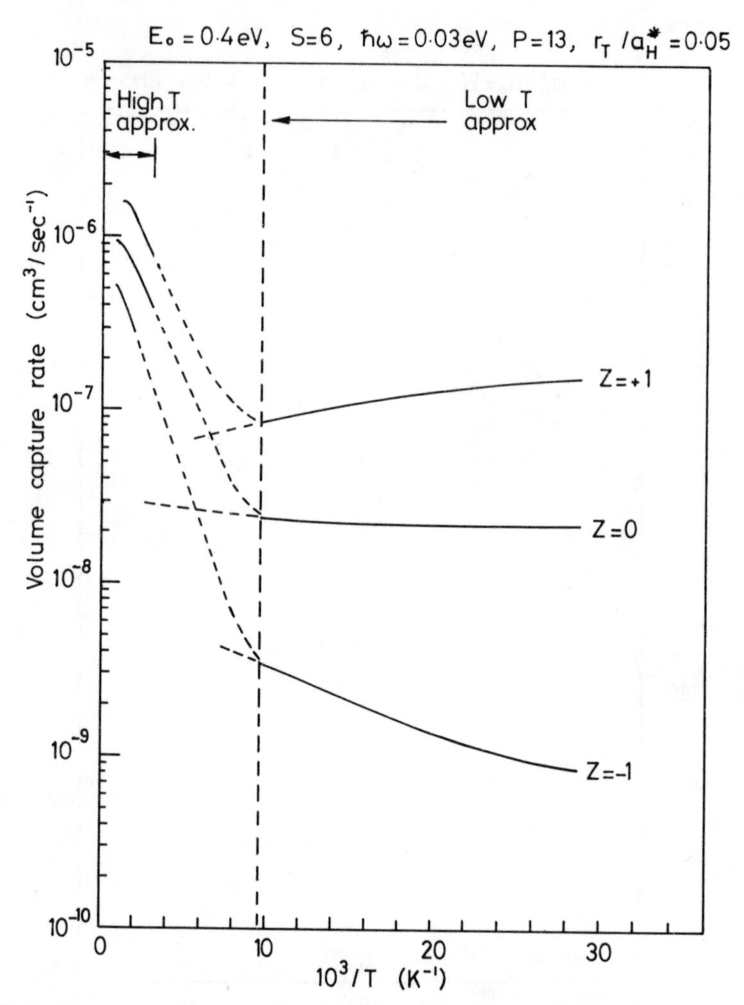

FIG. 6.10. Temperature dependence of the capture rate: $E_0 = 0{\cdot}4$ eV; $S = 6$; $\hbar\omega = 0{\cdot}03$ eV; $p = 13$; $r_T/a_H^* = 0{\cdot}05$. (After Ridley and Amato 1981)

where again the latter equality assumes that the same modes contribute to both $\mathbf{V}_{12}$ and $\mathbf{\Delta}$. For polar coupling

$$|\mathbf{V}_{12} \cdot \mathbf{\Delta}|^2 = \mathbf{V}_{12} S_P. \tag{6.117}$$

We have expressed these elements in terms of the Huang–Rhys factor, since it is this factor which is most easily obtained from measurement. However, it should be noted that in our simple model $\mathbf{V}_{12}^2$ does not depend upon the extent of the localized wavefunction in the case of deformation potential coupling, which can be seen by substituting $V_T = 4\pi r_T^3/3$ in eqn (6.114). Moreover, both elements are infinitesimal because of the factor V_T/V, which rules out any sort of resonant transition.

Substituting these model values for $\mathbf{V}_{12}^2$ and $|\mathbf{V}_{12}.\mathbf{\Delta}|^2$ in eqn (6.88) gives

$$W_0 \approx \omega 2\pi \frac{S^p e^{-S}}{p!} p(p-1) \frac{V_T}{V} R_1 \tag{6.118}$$

with $\omega = 5 \times 10^{13}\,\text{s}^{-1}$ ($\hbar\omega \approx 0{\cdot}033$ eV), $S = 3$, $p = 20$ ($E_0 = 0{\cdot}66$ eV), and $V_T = 10^{-21}\,\text{cm}^3$, the volume rate $W_0 V$ is $1{\cdot}5 \times 10^{-15}\,\text{cm}^3\,\text{s}^{-1}$ which is insignificant, but if $S = 6$ the rate is about $10^{-10}\,\text{cm}^3\,\text{s}^{-1}$ which *is* significant. This demonstrates how sensitive the rate is to S. If $p = 15$ ($E_0 = 0{\cdot}50$ eV) the volume rate with $S = 3$ becomes $6{\cdot}4 \times 10^{-12}\,\text{cm}^3\,\text{s}^{-1}$ and the volume rate with $S = 6$ becomes $1{\cdot}0 \times 10^{-8}\,\text{cm}^3\,\text{s}^{-1}$, both of which are significant rates. This demonstrates how sensitive that rate is to the depth of the level. Note that the corresponding cross-sections (obtained by dividing by the thermal velocity of the electron, i.e. about $10^7\,\text{cm}\,\text{s}^{-1}$) lie between about 10^{-22} and $10^{-15}\,\text{cm}^2$. Such magnitudes are commonly observed in experiments. Figure 6.10 shows the temperature dependence of the volume capture rate based on the above model.

If the final state is an extended excited state of the same centre the treatment follows through as before, but with V replaced by V_{ex} in the matrix elements where V_{ex} is the volume of the excited state and is assumed to be much larger than V_T. Equation (6.118) then becomes

$$W_0 \approx \omega 2\pi \frac{S_p e^{-S}}{p!} p(p-1) \frac{V_T}{V_{ex}} R_1. \tag{6.119}$$

Thus, with $V_{ex} \approx 10^{-18}\,\text{cm}^3$, $p = 20$, and the other parameters as before, W_0 is $10^3\,\text{s}^{-1}$ for $S = 3$ and $10^8\,\text{s}^{-1}$ for $S = 6$. With $p = 15$ the corresponding rates are about $10^6\,\text{s}^{-1}$ and $10^{10}\,\text{s}^{-1}$ respectively.

6.7. Selection rules for phonon–impurity coupling

In our discussion of the squared matrix elements S, $\mathbf{V}_{12}^2$, and $|\mathbf{V}_{12}.\mathbf{\Delta}|^2$ we have implicitly assumed that any selection rule for phonon participation applies equally to all three parameters. Moreover, we have seen that the magnitudes of these quantities, and of the transition rate, are very sensitive to phonon energy, and therefore it is very important to determine which phonon interactions are allowed and which are forbidden in the case of deformation potential coupling. The density of states throws emphasis on phonon modes near the zone edge, and that has been used to justify a single-frequency approach. A guide to which phonons can participate can be obtained from group-theoretic arguments involving zone-edge phonons at the high symmetry points L and X (Loudon 1964; Birman, Lax, and Loudon 1966; Amato and Ridley 1980).

Table 6.2 gives the space-group representations of phonons, and Tables 6.3 and 6.4 show how these can be reduced to a sum of point-group representations at a cubically symmetric site in diamond and zinc blende lattices. Matrix elements involving a substitutional impurity in a tetrahedral site can then be examined using elementary group theory. For a transition to be allowed the matrix element ⟨f| H |i⟩ must contain the

TABLE 6.2
Phonon symmetries in diamond and zinc blende lattices.

Lattice	Γ	X	L
Diamond	$\Gamma_{25}(O)+\Gamma_{15}(A)$	$X_1(LO, LA)+X_3(TA)+X_4(TO)$	$L_{1'}(LO)+L_{3'}(TA)$ $+L_2(LA)+L_3(TO)$
Zinc blende	$2P_4(A, O)$	$X_1(LA \text{ or } LO)+X_3(LO \text{ or } LA)$ $+2X_5(TA, TO)$	$2L_1(LA, LO)$ $+2L_3(TA, TO)$

With the origin at the A site in compound AB, LA has X_1 symmetry and LO has X_3 symmetry if $M_A<M_B$ and vice versa if $M_A>M_B$ (M is the mass).

TABLE 6.3
Space-group to point-group reduction coefficients for phonons in diamond

	$P_1(A_1)$	$P_2(A_2)$	$P_3(E)$	$P_4(T_2)$	$P_5(T_1)$
Γ_{25}'				1	
Γ_{15}				1	
X_1	1		1	1	
X_3				1	1
X_4				1	1
L_1	1			1	
L_2	1			1	
L_3			1	1	1
$L_{3'}$			1	1	1

TABLE 6.4
Space-group to point-group reduction coefficients for phonons in zinc blende

	P_1	P_2	P_3	P_4	P_5
X_1	1		1		
X_3				1	
X_5				1	1
L_1	1			1	
L_3			1	1	1

TABLE 6.5
Character table for T_d

	E	$8C_3$	$3C_2$	$6\sigma_d$	$6S_4$
$P_1(A_1)$	1	1	1	1	1
$P_2(A_2)$	1	1	1	−1	−1
$P_3(E)$	2	−1	2	0	0
$P_4(T_2)$	3	0	−1	1	−1
$P_5(T_1)$	3	0	−1	−1	1

From Koster 1957.

totally symmetrical representation P_1. If the symmetries of $|f\rangle$, H, and $|i\rangle$ are P_f, P_H, and P_i respectively, then that requirement implies that the direct product $P_H \times P_i$ must contain P_f. The direct products can be worked out from Table 6.5 which is the character table for the site group T_d of the diamond and zinc blende lattices. The results for the principal matrix elements $\mathbf{\Delta}$ and $\mathbf{V}_{12}$ are given in Table 6.6 for some simple cases. If a given mode is allowed at both X and L points it is a good indication that it will participate at other points in the zone, but if a mode is forbidden at both X and L its participation is almost certainly weak or negligible.

Table 6.6 shows that the assumption made in eqns (6.114) and (6.116) that the same phonon modes contribute to both matrix elements is true only if the initial and final states have the same symmetry. Where these symmetries are different, different phonons contribute. Thus, in a $P_1 \rightarrow P_4$ ionization, only the LO phonon or the LA phonon contributes to both $\mathbf{\Delta}$ and $\mathbf{V}_{12}$, and hence it is the only mode to contribute to $\mathbf{V}_{12}.\mathbf{\Delta}$. Another general point which emerges is that a localized state with P_4 symmetry ($|p\rangle$ state) couples to more modes than a P_1 ($|s\rangle$) state and is therefore likely to be more strongly coupled. However, it may be expected that in many cases the ground state of the impurity centre has the spherically symmetric P_1 symmetry, in which case the important phonons are the LA and LO phonons and possibly only those with X_1 symmetry. If the latter is the case a determination of the energy of the phonon most strongly coupled would indicate whether the impurity resided preferentially on the A site or the B site in a compound semiconductor.

6.8. Phonon-cascade capture

Capture cross-sections associated with radiative transitions are only of order 10^{-21} cm^2 (Chapter 5, Section 5.2). Those associated with multiphonon transitions span many orders of magnitude depending on coupling strength, depth of level, and temperature, but they are not expected

TABLE 6.6
Selection Rules for X and L phonons in multiphonon coupling

Symmetry		Phonons allowed in matrix elements							
		Impurity on site V				Impurity on site III			
		$M_V > M_{III}$		$M_{III} > M_V$		$M_V > M_{III}$		$M_{III} > M_V$	
Localized $\|l\rangle$ state	Band state $\|b\rangle$	Δ	V	Δ	V	Δ	V	Δ	V
P_1	P_1	LO+LA (L only)	LO+LA (L only)	LA+LO (L only)	LA+LO (L only	LA+LO (L only)	LA+LO (L only)	LO+LA (L only)	LO+LA (L only)
P_1	P_4	LO+LA (L only)	LA+LO (L only)+ TA+TO	LA+LO (L only)	LA (L only)+ LO+TA+ TO	LA+LO (L only)	LA (L only)+ LO+TA+ TO	LO+LA (L only)	LA+LO (L only)+ TA+TO
P_4	P_1	LA+LO+ TA+TO	LA+LO (L only)+ TA+TO	LA+LO+ TA+TO	LA (L only)+ LO+TA+ TO	LA+LO+ TA+TO	LA (L only)+ LO+TA+ TD	LA+LO+ TA+TO	LA+LO (L only)+ TA+TD
P_4	P_4	LA+LO+ TA+TO	LA+LO+ TA+TO	LA+LO+ TA+TO	LA+LO+ TA+TO	LA+LO+ TA+TO	LA+LO+ TA+TO	LA+LO+ TA+TO	LA+LO+ TA+TO

Amato and Ridley 1980.

to exceed, say, $10^{-14}\,\mathrm{cm}^2$. Neither mechanism can explain the observations made in germanium by Koenig (1958) and by other workers in silicon of cross-sections at low temperatures associated with the capture of electrons by shallow hydrogenic donors which range in magnitude from $10^{-13}\,\mathrm{cm}^2$ to $10^{-11}\,\mathrm{cm}^2$ as the temperature is reduced from about 10 K with a temperature variation of roughly T^{-n} ($1 \leqslant n \leqslant 4$). Such giant cross-sections are explained in terms of an electron sequentially emitting phonons and cascading down through the excited states of the hydrogenic centre. This theory was first proposed by Lax (1960), whose essentially classical treatment was subsequently extended by Hamann and McWhorter (1964), and quantum-mechanical formulations have been made by Ascarelli and Rodriguez (1961), Brown (1966), and Smith and Landsberg (1966). The phonon-cascade mechanism is fraught with statistical and quantum-mechanical problems (Abakumov *et al.* 1991). Here we shall merely outline the essential elements.

Lax's approach was an adaptation of the theory of recombination in gases developed by Thomson (1924). The essential idea is that an electron whose energy is greater than $\frac{3}{2}k_{\mathrm{B}}T$ will on average lose energy in a phonon collision. An electron will gain energy by being accelerated by the coulomb field of the ionized donor, so all collisions within a critical radius r_0 defined by

$$\frac{Ze^2}{4\pi\epsilon r_0} = \tfrac{3}{2}k_{\mathrm{B}}T \tag{6.120}$$

will be energy losing. The cross-section σ_{c} is then just πr_0^2 weighted by the probability that an energy-losing collision occurs every r_0/λ, where λ is an appropriate length. Energy and momentum conservation limits the amount of energy an electron can lose in a single collision, which can be accounted for by using the concept of the energy-relaxation mean free path l_{E}, i.e. $\lambda = l_{\mathrm{E}}$. If l is the collision mean free path (eqn 3.79) then the energy-relaxation mean-free path is $l_{\mathrm{E}} = l(k_{\mathrm{B}}T/2m^* v_{\mathrm{s}}^2)$, which is energy and temperature independent. With $4r_0/3$ as the average distance across the sphere, the cross-section becomes

$$\sigma_{\mathrm{c}} = \frac{4\pi r_0^3}{3l_{\mathrm{E}}} = \frac{4}{3l_{\mathrm{E}}}\left(\frac{Ze^2}{6\pi\epsilon k_{\mathrm{B}}T}\right)^3. \tag{6.121}$$

Thus, σ_{c} is proportional to T^{-3}. (Note that Lax's result is incorrect in predicting a T^{-4} dependence (Abakumov *et al.* 1991). With an energy-relaxation mean free path for acoustic phonons estimated to be about 10^{-3} cm at 10 K, $Z = 1$, and $\epsilon = 16\epsilon_0$, this cross-section turns out to be about $10^{-12}\,\mathrm{cm}^2$, which is indeed the right order of magnitude. The emission of an optical phonon of energy $\hbar\omega_0$ is possible within a radius r_{op} defined by

$$\frac{Ze^2}{4\pi\epsilon r_{\mathrm{op}}} = \hbar\omega_0 \tag{6.122}$$

whence the cross-section becomes

$$\sigma_{\mathrm{cp}} = \frac{4\pi}{3l_{\mathrm{op}}}\left(\frac{Ze^2}{4\pi\epsilon\hbar\omega_0}\right)^3. \tag{6.123}$$

Since l_{op} is usually larger than the acoustic phonon mean free path at low temperatures and since $\hbar\omega_0 \gg k_B T$, the contribution from optical modes is not expected to be the dominant one. In piezoelectric materials collisions with acoustic modes via the piezoelectric interaction will be important at low temperatures.

This classical theory is defective in a number of ways. It assumes the availability of a continuum of bound states. It takes no account of the energy of the incident particle, it assumes that scattering into a bound state occurs with the same probability as scattering into a free state, and it assumes that once a collision occurs the electron becomes bound. A more complete theory must take into account quantum-mechanical transition rates into and between hydrogenic excited states, and it must calculate a sticking probability for the electron in a given state. From the experimental point of view it is enough that the electron be removed from the conduction process: how it gets into the ground state is another matter. Yet that means that the theory must take account of impurity conduction mechanisms. It is not surprising, therefore, that phonon-cascade theories have hitherto resorted to numerical outputs.

Though very large cross-sections are predicted by the phonon-cascade process, it is interesting to note that they could in principle be even larger. The energy-loss mechanism associated with what is the sequential emission of single phonons is as fast, or almost as fast, as scattering events allow, but in practice it is still the factor which limits the capture rate. However, suppose this scattering rate to increase such that many collisions occur as the electron moves towards the centre. In such a case the rate of capture is limited not by the energy-loss process but by the rate at which the electrons move. This situation is described by a theory due to Langevin (1903). The electrons drift with velocity v_d given by the product of the mobility and field:

$$v_d = \mu \frac{Ze}{4\pi\epsilon r^2}. \tag{6.124}$$

The rate of influx through a spherical surface of radius r per unit electron density is just the capture rate per unit density, or in alternative nomenclature the volume capture rate c_c:

$$c_c = 4\pi r^2 v_d = Ze\mu/\epsilon. \tag{6.125}$$

Dividing by the thermal velocity v_{th} gives the capture cross-section: thus

$$\sigma_c = Ze\mu/\epsilon v_{th}. \tag{6.126}$$

With $Z = 1$, $\epsilon = 16\epsilon_0$, $v_{th} = 10^6$ cm s^{-1}, and $\mu = 10^4$ cm^2 V^{-1} s^{-1} we obtain a cross-section of 1×10^{-9} cm^2. Observed capture cross-sections are at least two orders of magnitude smaller than this, which indicates that the capture process is not drift-limited.

6.9. The Auger effect

The idea that a two-electron collision in which a substantial energy change occurs could be a recombination or capture mechanism in non-metals goes back to the suggestion of Fröhlich and O'Dwyer (1950). Five basic processes are depicted in Fig. 6.11. In process (1) two electrons collide in the vicinity of a hole, one recombines with the hole, and the energy released is absorbed by the other electron. We speak of a direct or an indirect Auger effect depending on whether the energy gap is direct or indirect. Process (2) incorporates a phonon whose momentum helps to relax conditions on the energy and momentum of the colliding particles. In processes (3) and (4) a localized state is involved, and process (5) describes an internal transition in a donor–acceptor pair. The inverse of each process corresponds to an impact-ionization mechanism, which in semiconductors is usually the mechanism which brings about electrical breakdown at high electric fields, no matter how low the carrier concentration may be. However, under normal conditions a high concentration is necessary for the Auger effect to compete significantly with other recombination processes.

The rate for the direct Auger effect is closely related to the electron–electron scattering rate discussed in Chapter 4, Section 4.7. The difference is that one of the electrons ends the collision in another band. The rate is given by

$$W_{\text{recomb}} = \int_{\text{f}} \frac{2\pi}{\hbar} |M|^2 \, \delta(E_{\text{f}} - E_{\text{i}}) \, \mathrm{d}S_{\text{f}}. \tag{6.127}$$

In the case of electron–electron scattering we were interested in obtaining a net collision cross-section which incorporated the effects of spin and exchange, and $|M|^2$ was obtained accordingly (see Chapter 4, eqn (4.134)). In computing the rate for the Auger effect we shall employ the matrix elements for each distinct situation without taking an average.

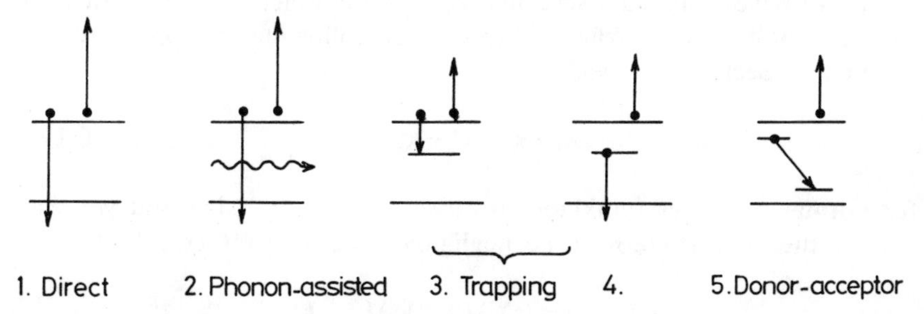

FIG. 6.11. Five basic Auger processes. Five more exist in which a hole carries away the energy. The reverse of these processes leads to ten basic processes of impact ionization.

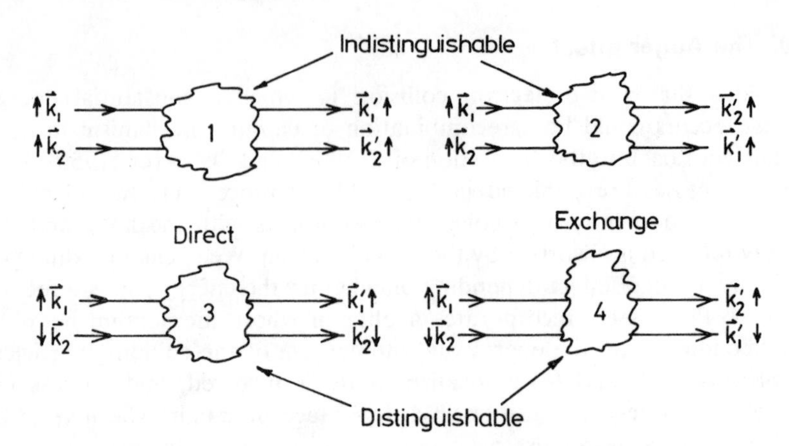

FIG. 6.12. Four types of two-electron collision with conservation of spin for an incident electron of a given spin.

We shall suppose that an incident electron in the conduction band, which has wavevector $\mathbf{k}_1$ and a defined spin, collides with a second electron in the conduction band, which has a wavevector $\mathbf{k}_2$ and either an identical spin or its opposite. The four possibilities for the collision in which spin angular momentum is conserved are shown in Fig. 6.12. Let the wavevector $\mathbf{k}_1'$ refer to the valence band state and the wavevector $\mathbf{k}_2'$ refer to the high-energy state in the conduction band.

The matrix element for process (1), the direct collision of two electrons of identical spin, is

$$M_{12}=\frac{1}{V^2}\iint u^*_{v\mathbf{k}_1'}(\mathbf{r}_1)\exp(-i\mathbf{k}_1'.\mathbf{r}_1)u^*_{c\mathbf{k}_2'}(\mathbf{r}_2)\exp(-i\mathbf{k}_2'\,\mathbf{r}_2)$$
$$\times\frac{e^2\exp\{-q_0\,|\mathbf{r}_1-\mathbf{r}_2|\}}{4\pi\epsilon\,|\mathbf{r}_1-\mathbf{r}_2|}u_{c\mathbf{k}_1}(\mathbf{r}_1)\exp(i\mathbf{k}_1.\mathbf{r}_1)u_{c\mathbf{k}_2}(\mathbf{r}_2)\exp(i\mathbf{k}_2.\mathbf{r})\,d\mathbf{r}_1\,d\mathbf{r}_2 \quad (6.128)$$

where we have assumed a screened coulomb interaction. Transformation to the centre-of-mass frame of reference following the procedure of Chapter 4, Section 4.6, leads to

$$\mathbf{k}_1+\mathbf{k}_2-\mathbf{k}_1'-\mathbf{k}_2'=0 \quad (6.129)$$

for normal processes (umklapp processes are also possible but we shall assume their contribution to be negligible here) and (*cf.* eqn 4.129)

$$M_{12}=\frac{(e^2/\epsilon V)I(\mathbf{k}_1,\mathbf{k}_1')I(\mathbf{k}_2,\mathbf{k}_2')}{|\mathbf{k}_1'-\mathbf{k}_1|^2+q_0^2} \quad (6.130)$$

where

$$I(\mathbf{k}_1, \mathbf{k}_1') = \int_{\text{cell}} u^*_{v\mathbf{k}_1'}(\mathbf{r}_1) u_{c\mathbf{k}_1}(\mathbf{r}_1)\, d\mathbf{r}_1$$

$$I(\mathbf{k}_2, \mathbf{k}_2') = \int_{\text{cell}} u_{c\mathbf{k}_2'}(\mathbf{r}_2) u_{c\mathbf{k}_2}(\mathbf{r}_2)\, d\mathbf{r}_2 \tag{6.131}$$

are overlap integrals, and we have used the fact that in the notation of Section 4.6, Chapter 4, in which $\mathbf{K}_{12} = \frac{1}{2}(\mathbf{k}_1 - \mathbf{k}_2)$ (see eqn (4.129)) the following is true as a consequence of momentum conservation:

$$|\mathbf{K}_{12}' - \mathbf{K}_{12}|^2 = |\mathbf{k}_1' - \mathbf{k}_1|^2. \tag{6.132}$$

For process (2) (the exchange collision) the matrix element is given by

$$M_{21} = \frac{(e^2/\epsilon V) I(\mathbf{k}_1, \mathbf{k}_2') I(\mathbf{k}_2, \mathbf{k}_1')}{|\mathbf{k}_1' - \mathbf{k}_2|^2 + q_0^2} \tag{6.133}$$

where the overlap integrals are

$$I(\mathbf{k}_1, \mathbf{k}_2') = \int_{\text{cell}} u^*_{c\mathbf{k}_2'}(\mathbf{r}_1) u_{c\mathbf{k}_1}(\mathbf{r}_1)\, d\mathbf{r}_1$$

$$I(\mathbf{k}_2, \mathbf{k}_1') = \int_{\text{cell}} u^*_{v\mathbf{k}_1'}(\mathbf{r}_2) u_{c\mathbf{k}_2}(\mathbf{r}_2)\, d\mathbf{r}_2. \tag{6.134}$$

Since the particles have identical spins they are indistinguishable and processes (1) and (2) interfere. The rate for the two processes therefore involves the difference $M_{12} - M_{21}$ since the exchange of an electron changes the sign of the wavefunction.

Process (3) involves M_{12} and its exchange partners; process (4) involves M_{21}. Since the spins are opposed, processes (3) and (4) are distinguishable and no interference occurs. Thus

$$|M|^2 = |M_{12} - M_{21}|^2 + |M_{12}|^2 + |M_{21}|^2. \tag{6.135}$$

(Note that Landsberg (1966) multiplies the right-hand side by 2 to take into account the fact that there are two ways of choosing the spin of the incident particle. We prefer not to incorporate that aspect at this point since it more rightly belongs to a different calculation, that of the total rate for the electron population. Here our primary concern is to calculate the rate for a single electron which has one or other spin, but cannot have both.)

We cannot avoid bringing in statistics at this point for it is necessary to weigh the rate with the probability that state $|\mathbf{k}_2\rangle$ is full, $|\mathbf{k}_1'\rangle$ is empty (occupied by a hole since it is in the valence band), and $|\mathbf{k}_2'\rangle$ is empty. (The incident electron state is full by hypothesis!) Assuming non-degenerate statistics we can take the weighting factor $P(\mathbf{k}_2, \mathbf{k}_1')$ to be

$$P(\mathbf{k}_2, \mathbf{k}_1') = f(\mathbf{k}_2)(1 - f(\mathbf{k}_1')) \tag{6.136}$$

where for near thermodynamic equilibrium

$$f(\mathbf{k}_2) = \frac{n}{N_c} \exp\left(-\frac{E_{c\mathbf{k}2}}{k_B T}\right) \tag{6.137}$$

$$1 - f(\mathbf{k}_1') = \frac{p}{N_v} \exp\left(-\frac{E_{v\mathbf{k}_1'}}{k_B T}\right) \tag{6.138}$$

where n and p are the electron and hole densities, N_c and N_v are the effective densities of states in the conduction and valence bands, $E_{c\mathbf{k}_2}$ is the energy of state $|\mathbf{k}_2\rangle$ measured from the conduction band edge, and $E_{v\mathbf{k}_1'}$ is the hole energy of the state $|\mathbf{k}_1'\rangle$ measured downwards from the valence band edge. Thus

$$P(\mathbf{k}_2, \mathbf{k}_1') = \frac{np}{N_c N_v} \exp\left(-\frac{E_{c\mathbf{k}_2} + E_{v\mathbf{k}_1'}}{k_B T}\right) = \exp\left(-\frac{E_g + E_{c\mathbf{k}_2} + E_{v\mathbf{k}_1'}}{k_B T}\right) \tag{6.139}$$

where E_g is the band gap. $P(\mathbf{k}_2, \mathbf{k}_1')$ provides the major part of the temperature dependence of the transition rate.

If an additional weighting factor of $f(\mathbf{k}_1)$ is added to take into account the probability of there being an incident electron in state $|\mathbf{k}_1\rangle$, it can be shown in the case of parabolic bands that for the most probable transition as determined solely by these weighting factors, which is denoted by $P(\mathbf{k}_1, \mathbf{k}_2, \mathbf{k}_1')$, the following are true:

$$E_{c\mathbf{k}_1} = E_{c\mathbf{k}_2} = \mu E_{v\mathbf{k}_1'} = \left(\frac{\mu^2}{1 + 3\mu + 2\mu^2}\right) E_g \qquad \mu = \frac{m_c^*}{m_v^*} \tag{6.140}$$

whence (Beattie and Landsberg 1959)

$$P(\mathbf{k}_1, \mathbf{k}_2, \mathbf{k}_1') = \frac{n}{N_c} \exp\left(-\frac{1+2\mu}{1+\mu} \frac{E_g}{k_B T}\right). \tag{6.141}$$

Also

$$E_{c\mathbf{k}_2'} = \frac{1+2\mu}{1+\mu} E_g. \tag{6.142}$$

The most probable transition therefore involves incident electrons and a hole, each of energy given by eqn (6.140), which for $\mu = 1$ is $E_g/6$, and the surviving electron is lifted to an energy given by eqn (6.142), which for $\mu = 1$ is $\frac{3}{2}E_g$. Equation (6.142) is the threshold energy for the reverse process of impact ionization. Usually $\mu < 1$. In GaAs $\mu \approx 0{\cdot}1$ and so the incident energy is about $0{\cdot}01\, E_g$ ($\approx 0{\cdot}015$ eV) and the threshold for impact ionization is $1{\cdot}1\, E_g$.

Several important simplifications can be made by exploiting the conditions for the most probable transition and the fact that the probability factor varies rapidly. To begin with, the overlap integrals vary comparitively slowly and so we can put

$$I(\mathbf{k}_1, \mathbf{k}_1') = I(\mathbf{k}_2, \mathbf{k}_1') \qquad I(\mathbf{k}_2, \mathbf{k}_2') = I(\mathbf{k}_1, \mathbf{k}_2'). \tag{6.143}$$

As a consequence of these equalities and of $k_1 \approx k_2$ we can further take

$$M_{12} \approx M_{21}. \tag{6.144}$$

This means that we can neglect collisions between like spins, which is a considerable simplification. Thus we assume

$$|M_{12} - M_{21}|^2 \ll |M_{12}|^2 + |M_{21}|^2 \tag{6.145}$$

and take

$$|M|^2 \approx 2\, |M_{12}|^2. \tag{6.146}$$

In this approximation the transition rate for a given electron (although now our incident electron must have properties not too far removed from average thermal equilibrium ones) is

$$W_{\text{recomb}} = \frac{4\pi}{\hbar} \left\{ \frac{e^2}{(2\pi)^3 \epsilon} \right\}^2 \frac{np}{N_c N_v} \times \iint \frac{|I(\mathbf{k}_1, \mathbf{k}_1') I(\mathbf{k}_2, \mathbf{k}_2')|^2 \exp\{-(E_{c\mathbf{k}_2} + E_{v\mathbf{k}_1'})/k_B T\}}{(|\mathbf{k}_1' - \mathbf{k}_1|^2 + q_0^2)^2} \times \delta(E_{c\mathbf{k}_2'} - E_g - E_{v\mathbf{k}_1'} - E_{c\mathbf{k}_1} - E_{c\mathbf{k}_2})\, d\mathbf{k}_2\, d\mathbf{k}_1' \tag{6.147}$$

and each integration is over the states of a given spin.

Where all states in a scattering problem lie in a single band we can approximate the overlap integrals by equating each to unity. In the Auger effect this can be done in the case of $I(\mathbf{k}_2, \mathbf{k}_2')$ (eqn (6.131)), but it certainly cannot be done in the case of $I(\mathbf{k}_1, \mathbf{k}_1')$. Because of the orthogonality of the $u_{\mathbf{k}}(\mathbf{r})$ for different bands but the same $\mathbf{k}$, $I(\mathbf{k}_1, \mathbf{k}_1')$ is, in the crudest approximation, zero. By using $\mathbf{k}.\mathbf{p}$ theory we can obtain $I(\mathbf{k}_1, \mathbf{k}_1')$ to the first order in the form

$$I(\mathbf{k}_1, \mathbf{k}_1') \approx \frac{\hbar}{mE_g} (\mathbf{k}_1 . \mathbf{p}_{cv} - \mathbf{k}_1' . \mathbf{p}_{cv}^*) \tag{6.148}$$

where $\mathbf{p}_{cv}$ is the momentum matrix element for the conduction and valence bands and m is the mass of the free electron. Unfortunately, simple $\mathbf{k}.\mathbf{p}$ theory is not adequate to describe the heavy-hole band (see

Section 1.11, Chapter 1), but by using the f-sum rule we can express the squared overlap integral in terms of the heavy-hole mass m_v^* as follows (see eqn (1.90), Chapter 1):

$$|I(\mathbf{k}_1, \mathbf{k}_1')|^2 = \frac{\hbar^2 p_{cv}^2}{m^2 E_g^2} |\mathbf{k}_1 - \mathbf{k}_1'|^2 = \frac{\hbar^2}{2E_g}\left(\frac{1}{m} + \frac{1}{m_v^*}\right) |\mathbf{k}_1 - \mathbf{k}_1'|^2. \quad (6.149)$$

With the further simplification that screening is negligible we obtain

$$W_{\text{recomb}} = \frac{2\pi}{\hbar}\left(\frac{e^2}{8\pi^3\epsilon}\right)^2 \frac{\hbar^2}{E_g}\left(\frac{1}{m} + \frac{1}{m_v^*}\right)\frac{np}{N_c N_v}$$
$$\times \iint \frac{\exp\{-(E_{c\mathbf{k}_2} + E_{v\mathbf{k}_1'})/k_B T\}}{|\mathbf{k}_1' - \mathbf{k}_1|^2} \delta(E_{c\mathbf{k}_2'} - E_g - E_{v\mathbf{k}_1'} - E_{c\mathbf{k}_1}$$
$$- E_{c\mathbf{k}_2})\, d\mathbf{k}_2\, d\mathbf{k}_1'. \quad (6.150)$$

The integrations are rather awkward and some simplifying assumptions are usually made in order to obtain analytic results. We shall assume that the effective mass in the conduction band is much smaller than that in the valence band, and that $\mathbf{k}_1' \gg \mathbf{k}_1$. The latter assumption is consistent with the former for the most probable transition, and in many direct-gap semiconductors the conduction band mass is much smaller than the valence band mass so the approximation has a practical application.

If the angle between $\mathbf{k}_2$ and $\mathbf{k}_1' - \mathbf{k}_1$ is θ_2 (Fig. 6.13) the argument of the delta function can be written

$$\frac{\hbar^2 |\mathbf{k}_1' - \mathbf{k}_1|^2}{2m_c^*} - E_g - \frac{\hbar^2 k_1'^2}{2m_v^*} - \frac{\hbar^2 k_1^2}{2m_c^*} + \frac{\hbar^2 k_2 |\mathbf{k}_1' - \mathbf{k}_1|}{m_c^*} \cos\theta_2. \quad (6.151)$$

Integration over the azimuthal angle of $\mathbf{k}_2$ yields a factor of 2π, and integration over $\cos\theta_2$ puts limits on the magnitude of $\mathbf{k}_2$ through the delta function. Integrating over k_2 finally converts the double integral in eqn (6.150) to the following:

$$\frac{2\pi m_c^{*2} k_B T}{\hbar^4} \int \frac{1}{|\mathbf{k}_1' - \mathbf{k}_1|^3} \exp\left[-\left\{E_{v\mathbf{k}_1'} + \frac{m_c^*}{2\hbar^2 |\mathbf{k}_1' - \mathbf{k}_1|^2}\left(E_g + \frac{\hbar^2 k_1'^2}{2m_v^*}\right.\right.\right.$$
$$\left.\left.\left. + \frac{\hbar^2 k_1^2}{2m_c^*} - \frac{\hbar^2}{2m_c^*}|\mathbf{k}_1' - \mathbf{k}_1|^2\right)^2\right\} \Big/ k_B T\right] d\mathbf{k}_1' \quad (6.152)$$

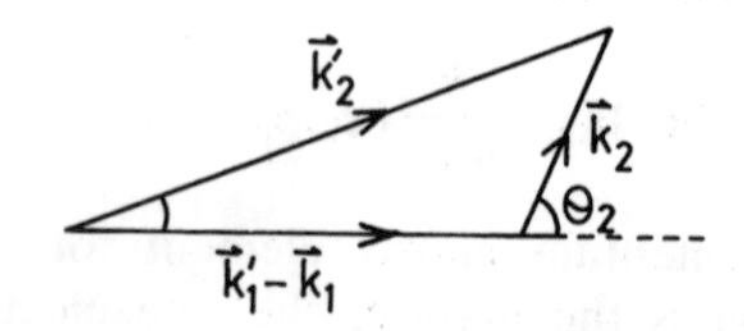

FIG. 6.13. Wavevectors in an Auger collision.

We now use our approximation $\mathbf{k}_1' \gg \mathbf{k}_1$ which makes the integration over $\mathbf{k}_1'$ straightforward. We obtain

$$\frac{8\pi^2 m_c^{*2} k_B T}{\hbar^4} \exp\left\{\frac{\frac{1}{2}(1-\mu)E_g}{k_B T}\right\} \mathcal{K}_0\left(\frac{1}{2}\frac{(1+2\mu+4\mu^2)^{1/2}E_g}{k_B T}\right) \tag{6.153}$$

where $\mathcal{K}_0(z)$ is a zero-order modified Bessel function of the second kind. Since $E_g/k_B T$ is usually large, we can approximate $\mathcal{K}_0(z)$ by its asymptotic form. If in addition $\mu \ll 1$, we obtain finally for the integration

$$\frac{8\pi^{5/2} m_c^* (k_B T)^{3/2}}{\hbar^4 (1+\mu)^{1/2} E_g^{1/2}} \exp\left(-\frac{\mu E_g}{k_B T}\right) \tag{6.154}$$

and for the rate (with $m_c^*/m \ll 1$ and substituting for np)

$$W_{\text{recomb}} = \frac{e^4 m_c^* (k_B T)^{3/2} (m_c^*/m + \mu)}{4\pi^{5/2} \epsilon^2 \hbar^3 (1+\mu)^{1/2} E_g^{3/2}} \exp\left\{-\frac{(1+\mu)E_g}{k_B T}\right\}. \tag{6.155}$$

As far as the author is aware this simple expression has not been given before. Although some fairly crude approximations have been used in its derivation, it provides a useful expression for estimating the direct Auger rate and for exhibiting the dependence upon band gap and temperature. Thus for GaAs at 300 K ($E_g = 1{\cdot}43$ eV) the rate is about $5\times10^{-17}\,\text{s}^{-1}$ which is entirely negligible, but for InSb ($E_g = 0{\cdot}18$ eV) the calculated rate at 300 K is $8\times10^6\,\text{s}^{-1}$ which is very significant. Equation (6.155) exhibits the same functional dependence on material parameters as the expression for recombination rate given by Beattie and Landsberg (1959).

Auger recombination is an important loss process in semiconductor lasers in which carrier densities are of order $10^{18}\,\text{cm}^{-3}$, particularly at room temperature or above. In order to recombine, excess carriers have to surmount the energy barrier imposed by momemtum and energy conservation, and this becomes increasingly difficult to do towards low temperatures. Ultimately the Auger recombination rate becomes determined by phonon-assisted processes which allow the carriers to bypass the energy barrier.

The theory of Auger recombination and impact ionization has been extended to indirect-gap transitions by Hill and Landsberg (1976), and there are several treatments of the Auger effect involving traps (e.g. Landsberg 1970; Robbins and Landsberg 1980). A comprehensive treatment of impact ionization including phonon-assisted processes has been given by Robbins (1980). It is clear that Auger processes can be important non-radiative processes in narrow-gap semiconductors at room temperature and at higher temperatures. This is mainly due to the high carrier concentrations which are present in such materials at elevated temperatures. When the concentration of electron and end state—whether the latter be hole or trap—reaches or exceeds about $10^{16}\,\text{cm}^{-3}$ Auger processes may become significant. However the trapping rate via the Auger effect at a hydrogenic centre is, according to Landsberg, Rhys-Roberts, and Lal (1964) (recasting their expression)

$$W_{\mathrm{trap}} = \frac{(e^2/4\pi\epsilon_0)^2 m}{\hbar^3} \frac{m_c^*}{m} \left(\frac{\epsilon_0}{\epsilon}\right)^2 (4\pi a_H^{*2})^3 n p_T \tag{6.156}$$

where a_H^* is the effective Bohr radius, n is the electron density, and p_T is the density of empty hydrogenic centres. In the case of GaAs for instance $a_H^* \approx 100$ Å and thus $W_{\mathrm{trap}} \approx 3{\cdot}2 \times 10^{-20} n p_T\ \mathrm{s}^{-1}$. Significant rates are thus predicted for densities of order only $10^{14}\ \mathrm{cm}^{-3}$ or even less, but they would have practical significance only at exceedingly low temperatures. Equation (6.160) can be generalized heuristically to apply to capture at an impurity with a bound-state radius r_T:

$$W_{\mathrm{trap}} = \frac{(e^2/4\pi\epsilon_0)^2 m}{\hbar^3} \frac{m_c^*}{m} \left(\frac{\epsilon_0}{\epsilon}\right)^2 (4\pi r_T^2)^3 n p_T. \tag{6.157}$$

For levels much deeper than hydrogenic, r_T in GaAs is likely to be of order 10 Å and thus $W_{\mathrm{trap}} \approx 3{\cdot}2 \times 10^{-26} n p_T\ \mathrm{s}^{-1}$. Densities of order $10^{16}\ \mathrm{cm}^{-3}$ are required to obtain significant rates in such a case.

6.10. Impact ionization

The inverse process of Auger recombination is impact ionization. Statistical factors are no longer necessary since the electron wavevector $\mathbf{k}_2$, to be excited across the gap, is deemed with certainty to be in the valence band, and the final states $\mathbf{k}_1'$ and $\mathbf{k}_2'$ are deemed to be entirely empty, in a non-degenerate semiconductor. Thus

$$W_{\mathrm{imp}} = \int \frac{2\pi}{\hbar} |M|^2 \delta(E_f - E_i)\, \mathrm{d}S_f \tag{6.158}$$

and incorporating the approximations of eqns (6.144) to (6.146) we obtain

$$W_{\mathrm{imp}} = \frac{4\pi}{\hbar} \left(\frac{e^2}{(2\pi)^3 \epsilon}\right)^2 \int\int \frac{I^2(\mathbf{k}_1, \mathbf{k}_1') I^2(\mathbf{k}_2, \mathbf{k}_1')}{|\mathbf{k}_1 - \mathbf{k}_1'|^4} \delta(E_f - E_i)\, \mathrm{d}\mathbf{k}_1'\, \mathrm{d}\mathbf{k}_2. \tag{6.159}$$

At threshold, all wavevectors lie along the same direction and

$$\begin{aligned} E_1 &= E_T = E_2 + E_1' + E_2' + E_g \\ k_1 &= -k_2 + k_1' + k_2' \end{aligned} \tag{6.160}$$

and for spherical parabolic bands

$$E_T = \left(\frac{1+2\mu}{1+\mu}\right) E_g, \qquad k_2 = -\frac{k_1}{1+2\mu}, \qquad k_1' = k_2' = \frac{\mu k_1}{1+2\mu} \tag{6.161}$$

where as before $\mu = m_c^*/m_v^*$. Regarding the matrix element as virtually constant near threshold, we remove it from the integrand, and using the above relationships, we obtain

$$W_{\text{imp}} = \frac{2}{\hbar}\left(\frac{e^2}{\epsilon}\right)^2 \frac{1}{(2\pi)^5} \frac{\hbar^4}{(2m_c^*)^2} \frac{I_c^2 I_v^2}{E_T^2} \left(\frac{1+2\mu}{1+\mu}\right)^4 \iint \delta(E_f - E_i)\, d\mathbf{k}_1' d\mathbf{k}_2. \quad (6.162)$$

Where we have put $I_c^2 = I^2(\mathbf{k}, \mathbf{k}_1')$ and $I_v^2 = I^2(\mathbf{k}_2, \mathbf{k}_1')$. The integrals can be evaluated exactly (Robbins 1980),† viz:

$$\iint \delta(E_f - E_i)\, d\mathbf{k}_1' d\mathbf{k}_2 = \left(\frac{2m_c^*}{\hbar^2}\right)^3 \frac{\pi^3}{2} \frac{(1+\mu)^2}{(1+2\mu)^{7/2}} (E_1 - E_T)^2, \quad (6.163)$$

whence

$$W_{\text{imp}} = W_0 \left(\frac{\epsilon_0}{\epsilon}\right)^2 \frac{m_c^*}{m} \frac{I_c^2 I_v^2}{(1+2\mu)^{3/2}} \left(\frac{E_1 - E_T}{E_g}\right)^2, \quad (6.164)$$

where

$$W_0 = \left(\frac{e^2}{4\pi\epsilon_0}\right) \frac{m}{\hbar^3} = 4{\cdot}14 \times 10^{16}\text{s}^{-1}. \quad (6.165)$$

The quadratic relation on energy was first obtained by Keldysh (1960).

In the simplest approximation of the overlap integrals $I_c = 1$ and, for heavy holes $I_v = 0$. Using **k.p** theory and *f*-sum rules to estimate *I*, as we did in the previous section, (eqn (6.149)) yields at threshold,

$$I_v^2 = 2\left(\frac{1+\mu}{1+2\mu}\right)\left(\frac{m_c^*}{m} + \frac{m_c^*}{m_v^*}\right). \quad (6.166)$$

For GaAs, the magnitude of I_v^2 is 0·38, on this basis. This is high compared with the result of calculations using a 15-band **k.p** model and a wavevector difference of 0·15 $(2\pi/a)$, which yields $I_v \approx 10^{-2}$ (Burt and Smith 1984; Burt *et al.* 1984) so, unfortunately we cannot take eqn (6.166) as reliable. For light holes, straightforward **k.p** theory gives $I_v = 0{\cdot}91$ for GaAs. Weighting this with the density of states ratio for the light holes gives $I_v^2 \approx 0{\cdot}06$. With $I_c^2 \approx 1$ then for GaAs,

$$W_{\text{imp}} \approx 5 \times 10^{11} \left(\frac{E_1 - E_T}{E_g}\right)^2 \text{s}^{-1}. \quad (6.167)$$

The smallness of the overlap integrals make the impact ionization rate rather small compared with the lattice scattering rates. It is usual to write the ionization rate, following Keldysh, as follows

$$W_{\text{imp}} = W_{\text{ph}}(E_T) P \left(\frac{E - E_T}{E_T}\right)^2. \quad (6.168)$$

(The difference between having E_T instead of E_g is not large.) The factor P measures whether the threshold is hard or soft, $P \gg 1$ being very hard.

† Equation (6.163) is derived assuming a parabolic band. For a simple generalization that does not make this assumption for the *impacting* electron, see Ridley (1998).

Since phonon scattering rates at high energies are of order $10^{14}\,\mathrm{s}^{-1}$, eqn (6.165) suggests that the threshold is very soft, with $P=5\times10^{-3}$. However it has been pointed out that what matters is a comparison with the energy relaxation rate rather than with the scattering rate (Ridley 1987), viz.

$$W_{\mathrm{imp}}=\frac{P_{\mathrm{E}}}{\tau_{\mathrm{E}}(E_{\mathrm{T}})}\left(\frac{E-E_{\mathrm{T}}}{E_{\mathrm{T}}}\right)^2. \tag{6.169}$$

The relation between P_{E} and P is obtained assuming optical-phonon scattering is dominant, viz.

$$\frac{E}{\tau_{\mathrm{E}}}=\frac{\hbar\omega}{2n(\omega)+1}W_{\mathrm{ph}} \tag{6.170}$$

whence

$$P_{\mathrm{E}}=\frac{E_{\mathrm{T}}(2n(\omega)+1)}{\hbar\omega}P. \tag{6.171}$$

Thus for GaAs $P_{\mathrm{E}}\approx 50P$, i.e. $P_{\mathrm{E}}\approx 0{\cdot}3$. The observed value for electrons is 3 (Ridley 1987). In view of the large uncertainties in our estimation of I_{v}^2 and of the phonon scattering rate, agreement within an order of magnitude is to be regarded as satisfactory. We may conclude that the ionization threshold is certainly not hard. Using the Keldysh criterion it is soft, but using the energy-relaxation criterion it is neutral.

Beattie (1988) has recently analysed the energy dependence near threshold for ellipsoidal bands and obtains a more rapid variation with energy. The same author has also pointed out a method for tackling awkward multidimensional integrals without making simple approximations (Beattie 1985).

6.11. Appendix: The multiphonon matrix element

Recently Huang (1981) has shown how the calculation may be simplified without sacrifice of accuracy. Let

$$\psi_1(\mathbf{r},\mathbf{Q})=\psi_{10}(\mathbf{r})+f(\mathbf{Q})\psi_{20}(\mathbf{r}). \tag{6.172}$$

If the diagonal elements of the electron–phonon interaction, i.e. H_{11} and H_{22}, are incorporated into the zero-order electronic hamiltonian, and the off-diagonal element H_{12} is regarded as the perturbation, then

$$f(\mathbf{Q})=\frac{H_{12}}{E_{10}+H_{11}-E_{20}-H_{22}}, \tag{6.173}$$

exactly as in eqn (6.81). The diagonal element in each case merely shifts the energy in the zero-order electronic equation, viz.

$$(H_e + H_{11})\psi_{10}(\mathbf{r}) = (E_{10} + H_{11})\psi_{10}(\mathbf{r}) \tag{6.174}$$

and similarly for the excited state. So far nothing is different. The simplification occurs in the calculation of the matrix element, for it turns out that

$$\langle f_e f_v | H_{NA} | i_e i_v \rangle = \langle f_v | H_{12} | i_v \rangle, \tag{6.175}$$

where H_{NA} is the non-adiabatic operator, i and f refer to initial and final states, electronic (e) and vibrational (v).

The proof of eqn (6.165) is simple. We first observe that (eqn (6.79))

$$H_{NA}\psi_1(\mathbf{r},\mathbf{Q})\phi_1(\mathbf{Q}) = -\sum_i \hbar\omega_i \left\{ \frac{\partial f(\mathbf{Q})}{\partial Q_i}\frac{\partial \phi_1(\mathbf{Q})}{\partial Q_i} + \tfrac{1}{2}\phi_1(\mathbf{Q})\frac{\partial^2 f(\mathbf{Q})}{\partial Q_i^2} \right\} \psi_{20}(\mathbf{r}). \tag{6.176}$$

Using the fact that

$$\frac{\partial^2}{\partial Q_i^2}[f(\mathbf{Q})\phi_1(\mathbf{Q})] = \phi_1(\mathbf{Q})\frac{\partial^2 f(\mathbf{Q})}{\partial Q_i^2} + 2\frac{\partial f(\mathbf{Q})}{\partial Q_i}\frac{\partial \phi_1(\mathbf{Q})}{\partial Q_i} + f(\mathbf{Q})\frac{\partial^2 \phi_1(\mathbf{Q})}{\partial Q_i^2}, \tag{6.177}$$

we may recast eqn (6.166) in the form

$$H_{NA}\psi_1(\mathbf{r},\mathbf{Q})\phi_1(\mathbf{Q}) = -\frac{1}{2}\sum_i \hbar\omega_i \left\{ \frac{\partial^2[f(\mathbf{Q})\phi_1(\mathbf{Q})]}{\partial Q_i^2} - f(\mathbf{Q})\frac{\partial^2 \phi_1(\mathbf{Q})}{\partial Q_i^2} \right\} \psi_{20}(\mathbf{r}), \tag{6.178}$$

whence

$$\begin{aligned} \langle f_e f_v | H_{NA} | i_e i_v \rangle &= \int \psi_2(\mathbf{r})\phi_2(\mathbf{Q}) H_{NA}\psi_1(\mathbf{r},\mathbf{Q})\phi_1(\mathbf{Q})\,\mathrm{d}\mathbf{Q}\,\mathrm{d}\mathbf{r} \\ &= -\frac{1}{2}\sum_i \hbar\omega_i \int \phi_2(\mathbf{Q}) \left\{ \frac{\partial^2[f(\mathbf{Q})\phi_1(\mathbf{Q})]}{\partial Q_i^2} - f(\mathbf{Q})\frac{\partial^2 \phi_1(\mathbf{Q})}{\partial Q_i^2} \right\} \mathrm{d}Q_i. \end{aligned} \tag{6.179}$$

The trick is to observe that the operator $\sum_i (\partial^2/\partial Q_i^2)$ is Hermitian and so

$$\langle f_e f_v | H_{NA} | i_e i_v \rangle = -\frac{1}{2}\sum_i \hbar\omega_i \int f(\mathbf{Q}) \left\{ \phi_1(\mathbf{Q})\frac{\partial^2 \phi_2(\mathbf{Q})}{\partial Q_2^2} - \phi_2(\mathbf{Q})\frac{\partial^2 \phi_1(\mathbf{Q})}{\partial Q_i^2} \right\} \mathrm{d}Q_i \tag{6.180}$$

But this operator is (apart from a constant factor) the squared-momentum operator. From the Schrödinger equation for the displaced oscillators (eqn (6.13)) we may therefore infer that, neglecting frequency shifts,

$$-\frac{1}{2}\sum_i \hbar\omega_i \frac{\partial^2 \phi_1(\mathbf{Q})}{\partial Q_i^2} = \left[E_{\mathrm{tot}\,1} - \left(\sum_i \tfrac{1}{2}\hbar\omega_i Q_i^2 \right) - E_{10} - H_{11} \right] \phi_1(\mathbf{Q}) \tag{6.181}$$

and hence

$$\langle f_e f_v | H_{NA} | i_e i_v \rangle = \int f(\mathbf{Q})\phi_2(\mathbf{Q})\phi_1(\mathbf{Q}) \times [E_{tot\,2} - E_{20} - H_{22} - E_{tot\,1} + E_{10} + H_{11}]d\mathbf{Q}. \tag{6.182}$$

Conservation of energy entails that $E_{tot\,2} = E_{tot\,1}$, and therefore the energy term in eqn (6.172) exactly cancels the energy denominator in the expression for $f(\mathbf{Q})$ in eqn (6.163). Thus

$$\langle f_e f_v | H_{NA} | i_e i_v \rangle = \int \phi_2(\mathbf{Q}) H_{12} \phi_1(\mathbf{Q}) d\mathbf{Q} = \langle f_v | H_{12} | i_v \rangle. \tag{6.183}$$

In this form, the matrix element is much simpler to calculate.

As pointed out by Huang, it is identical to the transition matrix element assumed in the so-called static coupling theory of Helmis (1956), and Passler (1974). Huang's proof shows that Kovarskii's non-Condon approximation and the static coupling scheme are entirely equivalent.

References

ABAKUMOV, V. N., PEREL, V. I., and YASSIEVICH, I. N. (1991). *Non-radiative recombination in semiconductors*. North-Holland, Amsterdam.
AMATO, M. A., and RIDLEY, B. K. (1980). *Phys. Lett.* **78A**, 170.
——, ARIKAN, M. C., and RIDLEY, B. K. (1980). *Semi-Insulating III–V Materials* p. 249 (ed. G. J. Rees, Shiva, Orpington, U.K.).
ASCARELLI, G. and RODRIGUEZ, S. (1961). *Phys. Rev.* **124**, 1321.
BEATTIE, A. R. (1985). *J. Phys. C. Solid State Phys.* **18**, 6501.
—— (1988). *Semicond. Sci. Technol.* (in press).
—— and LANDSBERG, P. T. (1958). *Proc. R. Soc. A* **249**, 16.
BIRMAN, J., LAX, M., and LOUDON, R. (1966). *Phys. Rev.* **145**, 620.
BROWN, R. A. (1966). *Phys. Rev.* **148**, 974.
BURT, M. G. and SMITH, C. (1984). *J. Phys. C: Solid State Phys.* **17**, L47.
——, BRAND, S., SMITH, C., and ABRAM, R. A. (1984). *J. Phys. C: Solid State Phys.* **17**, 6385.
ENGELMAN, R. and JORTNER, J. (1970). *Mol. Phys.* **18**, 145.
FRÖHLICH, H. and O'DWYER, J. (1950). *Proc. Phys. Soc. A* **63**, 81.
GUMMEL, H. and LAX, M. (1957). *Ann. Phys. N.Y.* **2**, 28.
HAMANN, D. R. and MCWHORTER, A. L. (1964). *Phys. Rev.* **134A**, 250.
HELMIS (1956) *Ann. Physik* **19**, 41.
HENRY, C. H. and LANG, D. V. (1977). *Phys. Rev. B* **15**, 989.
HILL, D. and LANDSBERG, P. T. (1976). *Proc. R. Soc. A*, **347**, 547.
HUANG, K. (1981). *Scientia Sinica* **24**, 27.
—— and RHYS, A. (1950). *Proc. R. Soc. A* **204**, 406.
KELDYSH, L. V. (1960). *Sov. Phys. —J.E.T.P* **10**, 509.
KOENIG, S. H. (1958). *Phys. Rev.* **110**, 988.
KOSTER, G. F. (1957). *Solid State Phys.* **5**, 194.
KOVARSKII, V. A. (1962). *Sov. Phys.—Solid State* **4**, 1200.
KOVARSKII, V. A. and SINYAVSKII, E. P. (1963). *Sov. Phys—Solid State* **4**, 2345.
KUBO, R. (1952). *Phys. Rev.* **86**, 929.
—— and TOYOZAWA, Y. (1955). *Prog. theor. Phys.* **13**, 160.

LANDAU, L. D. and LIFSHITZ, E. M. (1965). *Quantum mechanics*, Pergamon, Oxford.
LANDSBERG, P. T. (1966). *Lectures in theoretical physics*, Vol. **8.A,** p. 313, Univ. Colorado Press, Boulder.
—— (1970) *Phys. Status solidi* **41,** 457.
——, RHYS-ROBERTS, C., and LAL, P. (1964). *Proc. Phys. Soc.* **84,** 915; *Int. Conf. on Physics of Semiconductors, Paris*, p. 803. Dunod, Paris.
LANGEVIN, P. (1903). *Ann. Chem. Phys.* **28,** 289, 433.
LAX, M. (1952). *J. Chem. Phys.* **20,** 1752.
—— (1960). *Phys. Rev.* **119,** 1502.
LOUDON, R. (1964). *Proc. Phys. Soc.* **84,** 379.
PÄSSLER, R. (1974). *Czech. J. Phys.* **B24,** 322.
PRYCE, M. H. L. (1966). In *Phonons* (ed. R. W. H. Stevenson), p. 403. Oliver and Boyd, Edinburgh.
RICKAYZEN, G. (1957). *Proc. R. Soc. A* **241,** 480.
RIDLEY, B. K. (1978*a*). *J. Phys. C.* **11,** 2323.
—— (1978*b*). *Solid State Electron.* **21,** 1319.
—— (1980). *J. Phys. C.* **13,** 2015.
—— (1998). *J. Phys. Condens. Matter* **10,** L607.
ROBBINS, D. J. (1980). *Phys. Status Solidi B* **97,** 9, 387; **98,** 11.
—— and LANDSBERG, P. T. (1980). *J. Phys. C.* **13,** 2425.
SMITH, E. F. and LANDSBERG, P. T. (1966). *J. Phys. Chem. Solids* **27,** 1727.
STONEHAM, A. M. (1975). *Theory of defects in solids.* Clarendon Press, Oxford.
THOMSON, J. J. (1924). *Philos. Mag.* **47,** 337.
ZENER, C. (1932). *Proc. R. Soc. A* **137,** 696.

7. Quantum processes in a magnetic field

7.1. Introduction

IN A strong magnetic field an electron is constrained to move in orbits in the plane perpendicular to the field, though its motion parallel to the field is unaffected. If the orbital field is comparable or much shorter than the scattering time, the motion is coherent and, as described in Section 2.4, is quantized into Landau levels. Scattering is then no longer a matter of transitions between plane-wave states, but rather a transition in which the electron can jump from one orbit to another and it can also change its kinetic energy associated with the motion parallel to the field. The effect of the field is to localize the electron in the transverse plane and scattering takes on many of the attributes of a hopping mechanism. Indeed, in the transverse configuration in which an electric field is applied perpendicular to the magnetic field, transport of charge can occur only through the scattering mechanisms. This state of affairs is diametrically opposed to the situation in the absence of a quantizing magnetic field, in which charge transport occurs *in spite of* the scattering mechanisms. In the longitudinal configuration, in which the fields are parallel, transport is normal except that because of the transverse confinement, the motion is quasi-one-dimensional. In both configurations, however, magnetic quantization changes the wavefunction of the electron and hence changes the scattering matrix elements.

We will begin by looking at the collision-free situation and go over some of the ground of Section 2.4 in order to obtain the eigenfunctions that will be used to determine scattering rates, and to familiarize ourselves with the parameters that enter the physics, and their order of magnitude. The transverse configuration will be treated first, since it contains all the new physics that appears when a quantizing magnetic field is present. In order to understand more fully the peculiar new properties that appear we will not follow the usual treatments of the problem, which tend to be rather inaccessible mathematically and very often opaque. Instead we will reduce the description to its simplest component—that of a single-electron scattering between levels with the same quantum number—before introducing (a) statistics and (b) many quantum levels. In doing so we will discover that under certain circumstances the electron can behave highly anomalously in that it can drift the wrong way, an effect completely obscured if equilibrium statistics are introduced too early in the calculation, which is what is conventionally done.

The many-level situation is best illustrated theoretically and experimentally by the Shubnikov–de Haas effect, which is the appearance of oscillations in the magnetoresistance as a function of magnetic field. The theory of the Shubnikov–Haas effect is unavoidably complex mathematically, but an attempt has been made to simplify the calculation as far as possible and to clarify obscurities found in the literature.

Most of the discussion will assume that the scattering mechanisms are elastic, a not unrealistic assumption for low temperatures where most of the effects of magnetic quantization have been observed. The case of scattering by optical phonons is deferred to our discussion of the magnetophonon effect.

7.2. Collision-free situation

7.2.1. Quantum states in a magnetic field

We begin by solving Schrodinger's equation in order to describe one-electron states in the presence of a magnetic field **B** and electric field, viz:

$$H_0\Psi(\mathbf{r}) = E\Psi(\mathbf{r}) \tag{7.1}$$

$$H_0 = \frac{1}{2m^*}(\mathbf{p} - \bar{e}\mathbf{A})^2 + \bar{e}\phi \tag{7.2}$$

where we assume that the condition discussed in Section 2.4 applies, and that the effective-mass equation is valid. In eqn (7.2) $\mathbf{A}$ is the vector potential and ϕ is the scalar potential of the electromagnetic field, and $\bar{e}$ is the elementary charge containing the sign. (As usual the symbol e without the bar will be used to denote magnitude only.) Expanding the bracket in eqn (7.2) we obtain

$$\begin{aligned}(\mathbf{p} - \bar{e}\mathbf{A})^2 &= \mathbf{p}^2 - \bar{e}\mathbf{p}\,.\,\mathbf{A} - \bar{e}\mathbf{A}\,.\,\mathbf{p} + e^2A^2 \\ &= \mathbf{p}^2 + \bar{e}(i\hbar\nabla\,.\,\mathbf{A}) - 2\bar{e}\mathbf{A}\,.\,\mathbf{p} + e^2A^2\end{aligned} \tag{7.3}$$

In order to illustrate that the gauge chosen in Section 2.4 is not unique, we choose the gauge

$$\mathbf{A} = (0, Bx, 0) \tag{7.4}$$

which still makes $\nabla\,.\,\mathbf{A} = 0$, and H_0 may be written as follows:

$$H_0 = \frac{1}{2m^*}(p_x^2 + p_z^2) + \frac{1}{2m^*}(p_y - \bar{e}Bx)^2 - \bar{e}\xi x \tag{7.5}$$

where ξ is the electric field, taken to be along the x direction (Fig. 7.1) in the transverse configuration. (We will look at the longitudinal configuration later.)

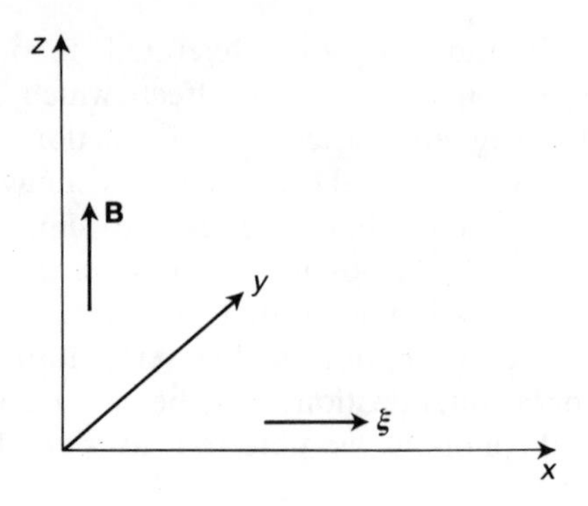

FIG. 7.1. Field directions.

We can show that, in this gauge, p_y and p_z are constants of the motion. The equation of motion is

$$m^* \frac{d\mathbf{v}}{dt} = \bar{e}\xi + \bar{e}\mathbf{v} \times \mathbf{B}. \tag{7.6}$$

Integration gives the time-dependence of the three components of the momentum. (Here, as elsewhere, we assume that the effective mass is scalar.) Thus

$$m^* v_x = \bar{e}\xi t + \bar{e} y B + C_x \tag{7.7}$$

$$m^* v_y = -\bar{e} x B + C_y \tag{7.8}$$

$$m^* v_z = C_z \tag{7.9}$$

Since $p_x = m^* v_x$, $p_y = m^* v_y + \bar{e} B x$, $p_z = m^* v_z$, it is clear that p_x is not a constant of the motion, but p_y and p_z *are* constants. Motion along the z direction is unaffected by the magnetic field, and so p_z is determined, as usual, by the relation

$$p_z = \hbar k_z \tag{7.10}$$

which determines C_z. To determine C_y we return to eqn (7.6) and note that in the x direction

$$m^* \frac{dv_x}{dt} = \bar{e}\xi + \bar{e} v_y B \tag{7.11}$$

We may define a position at $x = X$ where dv_x/dt is zero, and hence

$$v_y(X) = -\frac{\xi}{B}. \tag{7.12}$$

This is the Hall velocity, which we denote v_B. Thus we can identify the constant of motion C_y with position X and velocity v_B as follows

$$p_y = \bar{e}BX - m^* \frac{\xi}{B} \tag{7.13}$$

Inserting this relationship into the expression for H_0, eqn (7.5), we obtain

$$H_0 = \frac{1}{2m^*}(p_x^2 + p_z^2) + \frac{e^2B^2}{2m^*}(x - X)^2 - \bar{e}\xi X + \tfrac{1}{2}m^* v_B^2. \tag{7.14}$$

With the introduction of the cyclotron frequency

$$\omega_c = \frac{eB}{m^*} \tag{7.15}$$

we obtain finally

$$H_0 = \frac{1}{2m^*}(p_x^2 + p_z^2) + \tfrac{1}{2}m^*\omega_c^2(x - X)^2 - \bar{e}\xi X + \tfrac{1}{2}m^* v_B^2. \tag{7.16}$$

The solution of Schrodinger's equation is now straightforward viz:

$$\Psi_n(\mathbf{r}) = C\exp(lk_y y)\exp(lk_z z)\phi_n(x - X), \tag{7.17}$$

where $\phi_n(x - X)$ is a harmonic oscillator wavefunction. The energy is

$$E = (n + \tfrac{1}{2})\hbar\omega_c + \frac{\hbar^2 k_z^2}{2m^*} - \bar{e}\xi X + \tfrac{1}{2}m^* v_B^2, \tag{7.18}$$

and k_y and X are related by

$$\hbar k_y = \bar{e}BX + m^* v_B. \tag{7.19}$$

In the extreme quantum limit (E.Q.L.) only the ground state ($n = 0$) is of importance. The oscillator wavefunction is given by

$$\phi_0 = \frac{1}{\pi^{1/4}R^{1/2}}\exp\left[\frac{1}{2}\left(\frac{x - X}{R}\right)^2\right] \tag{7.20}$$

where

$$R = \left(\frac{\hbar}{eB}\right)^{1/2}. \tag{7.21}$$

The general expression for ϕ is given in eqn (7.133).

7.2.2. *Magnitudes*

The characteristic energy is

$$\hbar\omega_c = \frac{\hbar eB}{m^*} = 0{\cdot}1152\left(\frac{m}{m^*}\right)B \text{ meV}. \tag{7.22}$$

The characteristic length is

$$R = \left(\frac{\hbar}{eB}\right)^{1/2} = \frac{256{\cdot}7}{B^{1/2}}\,\text{Å}. \tag{7.23}$$

Note that R is independent of the properties of the material. The energy has usually to be compared with $k_B T$, which is given by

$$k_B T = 0{\cdot}08617T \text{ meV} \tag{7.24}$$

and with the voltage drop over the characteristic length,

$$e\xi R = 2{\cdot}567 \times 10^{-3} \frac{\xi}{B^{1/2}} \text{meV} \qquad (\xi \text{ in V cm}^{-1}). \tag{7.25}$$

For example, if $m^*/m = 0{\cdot}067$ (GaAs), $\hbar\omega_c = 1{\cdot}72B$ meV. In a field of 1 tesla this energy corresponds to a temperature of 20 K, or to an electric field of 669 V cm^{-1}.

7.2.3. Density of states

Magnetic quantization changes the distribution of states over energy from the three-dimensional variation proportional to $E^{1/2}$ to the one-dimensional variation proportional to $E^{-1/2}$. In a single magnetic state of a given spin there are two variables, k_z and k_y. The number of states associated with a given magnitude of k_z in the interval dk_z is

$$N(k_z)\,\mathrm{d}k_z = 2\frac{L_z}{2\pi}\,\mathrm{d}k_z \tag{7.26}$$

where L_z is the cavity length and k_z is quantized as usual by periodic boundary conditions. The factor 2 arises because we have to count states with $+k_z$ and $-k_z$. The number of states associated with k_y in the interval dk_y is

$$N(k_y)\,\mathrm{d}k_y = \frac{L_y}{2\pi}\,\mathrm{d}k_y \tag{7.27}$$

where again periodic boundary conditions are assumed, but now the range of k_y is determined by eqn (7.19), and hence

$$N(k_y)\,\mathrm{d}k_y = \frac{L_y}{2\pi}\frac{eB}{\hbar}\,\mathrm{d}X \tag{7.28}$$

Since the energy is independent of k_y when there is no electric field we can add up all allowed states and obtain

$$N_y = \frac{eB}{\hbar}\frac{L_x L_y}{2\pi} \tag{7.29}$$

Another way of calculating this degeneracy factor is to follow the method used in Section 2.4 and sum all free-electron states with wave vector $\mathbf{k}_\perp$

in the plane perpendicular to B whose energies lie over the range $\hbar\omega_c$. This number is

$$N_y = 2\pi k_\perp \, \mathrm{d}k_\perp \frac{L_x L_y}{(2\pi)^2} \tag{7.30}$$

with

$$\frac{\mathrm{d}}{\mathrm{d}k_\perp}\left(\frac{\hbar^2 k_\perp^2}{2m^*}\right) \mathrm{d}k_\perp = \hbar\omega_c \tag{7.31}$$

whence

$$\frac{\hbar^2 k_\perp}{m^*} \mathrm{d}k_\perp = \frac{\hbar e B}{m^*}$$

and thus, as before,

$$N_y = \frac{eB}{\hbar} \frac{L_x L_y}{2\pi} \tag{7.32}$$

The total number of states with a given kinetic energy of motion along the z direction, E_z, is therefore

$$N_y N(k_z) \, \mathrm{d}k_z = 2 \frac{L_x L_y L_z}{(2\pi)^2} \cdot \frac{eB}{\hbar} \cdot \mathrm{d}k_z. \tag{7.33}$$

Converting to energy using eqn (7.18) we obtain finally the density of states of a given spin in unit energy range

$$N(E_z) = \frac{eB(2m^*)^{1/2}}{(2\pi)^2 \hbar^2 E_z^{1/2}} \tag{7.34}$$

The form of the energy dependence of the density-of-states function gives rise to problems connected with the divergence at zero energies in subsequent calculations. In reality collisions broaden the states and thus produce a smooth transition at $E_z = 0$ (Fig. 7.2).

7.2.4. Spin

In addition to the energy components in eqn (7.18) there exists a magnetic component associated with spin, i.e.

$$E_B = \pm\tfrac{1}{2} g \mu_\mathrm{B} B \tag{7.35}$$

where μ_B is the free-electron Bohr magneton ($e\hbar/2mc$), and the g factor is 2 for free electrons, but its magnitude can be much larger in semiconductors because of strong spin–orbit coupling, viz. (Lax *et al.* 1958):

$$g = 2\left[1 + \left(\frac{m^* - m}{m^*}\right) \frac{\Delta_0}{3E_g + 2\Delta_0}\right] \tag{7.36}$$

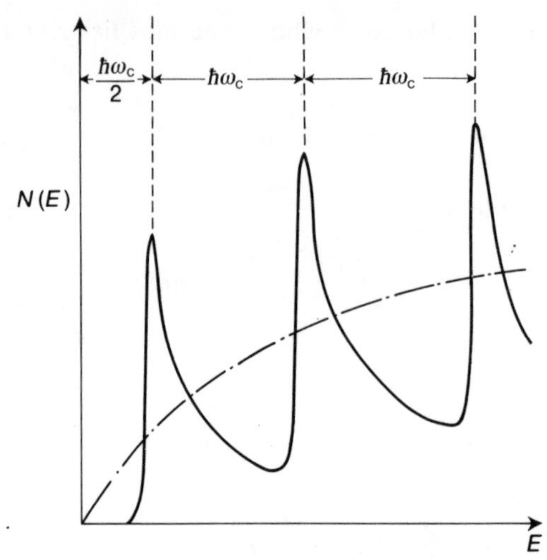

FIG. 7.2. Density of states. Without broadening $N(E)$ would be infinite at $E = (n + \frac{1}{2})\hbar\omega_c$. The chained line indicates the density of states when $B = 0$.

where Δ_0 is the spin–orbit splitting and E_g is the energy gap. In most cases E_B is small compared with other components, and we shall forget about splitting due to spin for the moment.

7.2.5. *Phenomenological quantities*

What we measure is the conductivity tensor, depending upon the experimental arrangement. Looking first at the conductivity tensor, viz.

$$j_\mu = \sum_\nu \sigma_{\mu\nu}\xi_\nu \tag{7.37}$$

we can conclude from the foregoing theory that the only current which flows is the Hall current, proportional to v_B. This means that, with $\boldsymbol{\xi}$ in the x direction

$$\sigma_{xx} = 0, \qquad \sigma_{xy} = \frac{\bar{e}nv_B}{\xi} = -\frac{\bar{e}n}{B}, \qquad \sigma_{xz} = 0, \tag{7.38}$$

where n is the carrier density. Quite often, but rather confusingly, σ_{xx} is referred to as transverse conductivity. It deviates from zero only when collisions are introduced into the description. Thus, collisions, which in the conventional situation limit the current, in this case actually initiate it.

A conductivity tensor component is measured whenever the direction of field is well-defined, but in the usual experimental arrangement for measuring the Hall effect and magnetoresistance it is the current direction which is well-defined. Consequently it is the resistivity tensor that is measured. In the situation envisaged the resistivity components are related to the conductivity components as follows:

$$\xi_\mu = \sum_\nu \rho_{\mu\nu} j_\nu \tag{7.39}$$

from which it is straightforward to show that

$$\rho_{xx} = \frac{\sigma_{xx}}{\sigma_{xx}^2 + \sigma_{xy}^2}, \qquad \rho_{xy} = \frac{\sigma_{xy}}{\sigma_{xx}^2 + \sigma_{xy}^2}. \tag{7.40}$$

In the absence of collisions $\rho_{xx} = 0$, $\rho_{xy} = \sigma_{xy}^{-1}$. Where collisions produce a finite conductivity, but so small that $\sigma_{xx}^2 \ll \sigma_{xy}^2$,

$$\rho_{xx} \approx \frac{\sigma_{xx}}{\sigma_{xy}^2} = \frac{\sigma_{xx} B^2}{e^2 n^2}, \qquad \rho_{xy} = -\frac{B}{\bar{e}n}. \tag{7.41}$$

7.3. Collision-induced current

7.3.1. Expression for the scattering rate in the extreme quantum limit

If the scattering mechanism is weak we can regard the transition rate as being given by first-order perturbation theory. The scattering rate is thus

$$W(E_i) = \int \frac{2\pi}{\hbar} |\langle f|\, H_s \,|i\rangle|^2\, \delta(E_f - E_i)\, \mathrm{d}S_f \tag{7.42}$$

where the integral is over final states. The interaction hamiltonian can be taken to be of the form

$$H_s = \sum_{\mathbf{q}} (V_{\mathbf{q}} \exp(i\mathbf{q}\,.\,\mathbf{r}) + V_{\mathbf{q}} \exp(-i\mathbf{q}\,.\,\mathbf{r})) \tag{7.43}$$

whence

$$W(E_i) = \sum_{\mathbf{q}} \int \frac{2\pi}{\hbar} |V_{\pm\mathbf{q}}|^2\, |M_{\pm\mathbf{q}}|^2\, \delta(E_f - E_i)\, \mathrm{d}S_f \tag{7.44}$$

$$M_{\pm\mathbf{q}} = \int \Psi_f^*(\mathbf{r}) \exp(\pm i\mathbf{q}\,.\,\mathbf{r}) \Psi_i(\mathbf{r})\, \mathrm{d}\mathbf{r}. \tag{7.45}$$

To evaluate $M_{\pm\mathbf{q}}$ we substitute the magnetic state wavefunction (eqn (7.17)) and perform the integration. We will consider the E.Q.L. and deal only with ground states (eqn (7.20)). In this case

$$M_{\pm\mathbf{q}} = \frac{1}{L_y L_z \pi^{1/2} R} \int\limits_{-L_x/2}^{+L_x/2} \int\limits_{-L_y/2}^{+L_y/2} \int\limits_{-L_z/2}^{+L_z/2} \mathrm{d}x\, \mathrm{d}y\, \mathrm{d}z\, \exp[i(k_y \pm q_y - k_y')y]$$
$$\times \exp[i(k_z \pm q_z - k_z')z] \exp[\pm i q_x x]$$
$$\times \exp\left[-\frac{(x - X')^2}{2R^2}\right] \exp\left[-\frac{(x - X)^2}{2R^2}\right], \tag{7.46}$$

where the integration extends over the cavity $L_x \,.\, L_y \,.\, L_z$. Primes denote final-state quantities. The quantity R is the characteristic dimension of the quantum state given by eqn (7.21). The integrations over y and z give zero unless crystal momentum is conserved (umklapp process can be neglected here), i.e.

$$k_y \pm q_y - k'_y = 0, \tag{7.47}$$

$$k_z \pm q_z - k'_z = 0. \tag{7.48}$$

Thus

$$|M_{\pm\mathbf{q}}|^2 = \delta_{k_y \pm q_y, k'_y}\, \delta_{k_z \pm q_z, k'_z}\, |I(X'-X)|^2 \tag{7.49}$$

where the overlap integral between the two magnetic states is readily shown to be

$$I(X'-X) = \exp\left[-\frac{(X'-X)^2}{4R^2} - \frac{(q_x R)^2}{4}\right] \exp\left[\mp i q_x \frac{(X+X')}{2}\right]. \tag{7.50}$$

Using the relation between X and k_y, eqn (7.19), to eliminate X we see that

$$X' - X = \frac{\hbar}{\bar{e}B}(k'_y - k_y) = \pm \frac{e}{\bar{e}} q_y R^2 \tag{7.51}$$

where the last step follows from eqn (7.47).

The sum over the final states is effectively a sum over k'_y and k'_z. But because of eqns (7.47) and (7.48) there is only one choice for each value of q_y and q_z. Thus we need only sum over $\mathbf{q}$ to encompass all possibilities. Converting this to an integration, we obtain

$$W(E_i) = \int \frac{2\pi}{\hbar} \frac{V}{8\pi^3} |V_{\pm\mathbf{q}}|^2$$
$$\times \exp[-\tfrac{1}{2}(q_x^2 + q_y^2)R^2]\, \delta_{k_y \pm q_y, k'_y}\, \delta_{k_z + q_z, k'_z}\, \delta(E_f - E_i)\, d\mathbf{q} \tag{7.52}$$

where V is the volume of the cavity

7.3.2. *Energy and momentum conservation*

Where phonons are responsible for scattering, the upper sign corresponds to the absorption of a phonon of energy $\hbar\omega_{\mathbf{q}}$, the lower sign to an emission. Elastic scattering can always be recovered by restricting attention to the upper sign only and putting $\hbar\omega_{\mathbf{q}} = 0$. Thus

$$E_f - E_i = (n' - n)\hbar\omega_c + \frac{\hbar^2}{2m^*}(k_z'^2 - k_z^2) - \bar{e}\xi(X' - X) \mp \hbar\omega_{\mathbf{q}}. \tag{7.53}$$

In the case of E.Q.L. under consideration, $n' = n = 0$. Using eqns (7.48) and (7.51) we obtain

$$E_f - E_i = \frac{\hbar^2}{2m^*}[(k_z \pm q_z)^2 - k_z^2] \mp e\xi q_y R^2 \mp \hbar\omega_{\mathbf{q}} \tag{7.54}$$

Conservation of energy implies that $E_f - E_i = 0$, and this imposes limits on the components of $\mathbf{q}$.

The general case is quite complex, and it is worthwhile looking for simplifying features at this stage. One problem is the $\mathbf{q}$-dependence of $\omega_{\mathbf{q}}$. In the case of impurity scattering collisions are elastic so we can put $\hbar\omega_{\mathbf{q}} = 0$ and the problem disappears. In the case of acoustic modes, $\omega_q = v_s q = v_s(q_x^2 + q_y^2 + q_z^2)$ (for an isotropic solid). If we can regard $\hbar\omega_q$ as negligible, the problem again disappears. The upper limit to the magnitude of phonon wavevector which can produce appreciable scattering is going to be determined by the overlap integral. Roughly we can expect that

$$q_\perp^2 R^2 \lesssim 1 \tag{7.55}$$

where

$$q_\perp^2 = q_x^2 + q_y^2. \tag{7.56}$$

On the other hand, provided $|e\xi q_y R^2|$ and $\hbar\omega_{\mathbf{q}}$ are small compared to $\hbar^2 k_z^2/2m^*$

$$q_z^2 \lesssim 4k_z^2. \tag{7.57}$$

Thus if

$$4k_z^2 R^2 \ll 1, \tag{7.58}$$

we can neglect the dependence of phonon energy on q_z. Equation (7.58) is equivalent to

$$\frac{8(\hbar^2 k_z^2/2m^*)}{\hbar\omega_c} \ll 1, \tag{7.59}$$

which is certainly consistent with the condition for E.Q.L. We may therefore assume that for acoustic modes $\omega_{\mathbf{q}}$ is $\omega_{\mathbf{q}_\perp}$, independent of q_z.

However we must satisfy ourselves that $|e\xi q_y R^2|$ and $\hbar\omega_{q_\perp}$ for acoustic modes are indeed small compared with, effectively, $k_B T$. Since $q_y R \lesssim 1$, eqn (7.25) shows already that the field term is small, and in most cases very small compared with $k_B T$. A typical sound velocity is 3×10^5 cm s^{-1}, and consequently

$$\hbar\omega_{q_\perp} \lesssim \frac{\hbar v_s}{R} \lesssim 10^{-1} B^{1/2}\,\text{meV}. \tag{7.60}$$

For $B = 9T$ the maximum phonon energy would be as high as $k_B T$ **at** about 3.5 K. In many situations, therefore, the assumption that $\omega_{\mathbf{q}}$ is largely independent of q_z is not going to be seriously invalidated.

In the case of optical phonons $\hbar\omega_{\mathbf{q}}$ can be taken to be independent of $\mathbf{q}$, so the problem once again disappears.

We may conclude therefore that the principal dependence of the energy on q_z in eqn (7.54) is the term in the square brackets. Thus

$$k_z \pm q_z = (\pm)k_z\left[1 \pm \frac{(e\xi q_y R^2 + \hbar\omega_{q_\perp})}{E_z}\right]^{1/2} \tag{7.61}$$

where $E_z = \hbar^2 k_z^2/2m^*$. The signs in brackets i.e. $(\pm)$ apply to both absorption and emission possibilities and they correspond to maximum and minimum values for q_z in each case. In situations where the change in potential energy in the electric field is negligible, and where the collisions are elastic or nearly so, the limits of q_z are 0 and $\mp 2k_z$.

7.3.3. *Integrations*

It is convenient to adopt cylindrical coordinates $q_\perp$, q_z, θ in the integration over $\mathbf{q}$ in eqn (7.52), viz.

$$W(E_i) = \frac{V}{4\pi^2\hbar}\int_0^{q_{\perp D}}\int_0^{2\pi}\int_{-q_{zD}}^{+q_{zD}} |V_{\pm\mathbf{q}}|^2 \exp(-\tfrac{1}{2}q_\perp^2 R^2)\,\delta\left[\frac{\hbar^2}{2m^*}([k_z \pm q_z]^2 - k_z^2)\right.$$
$$\left.\mp (e\xi R^2 q_\perp \cos\theta + \hbar\omega_q)\right] q_\perp\, dq_\perp\, d\theta\, dq_z, \tag{7.62}$$

where the subscript D denotes maximum wavevectors in the spherical-Brillouin-zone approximation. Integration over q_z gives

$$W(E_i) = \frac{V}{4\pi^2\hbar}\cdot\frac{m^{*1/2}}{2^{1/2}\hbar}\int_{q_{\perp\min}}^{q_{\perp\max}}\int_{\theta_{\min}}^{\theta_{\max}} \frac{(|V_{\pm q_\perp, q_{z\max}}|^2 + |V_{\pm q_\perp, q_{z\min}}|^2)}{[E_z \pm (e\xi R^2 q_\perp \cos\theta + \hbar\omega_q)]^{1/2}}$$
$$\times \exp(-\tfrac{1}{2}q_\perp^2 R^2) q_\perp\, dq_\perp\, d\theta \tag{7.63}$$

where the limits ensure that the square-root remains real, corresponding to conservation of energy.

The integration over direction depends upon the angular dependence of the interaction strength $V_{\mathbf{q}}$ as well as the explicit dependence associated with the applied electric field. Frequently $V_{\mathbf{q}}$ can be taken to be isotropic, or if not strictly so, an angular average can be taken. We will assume that $|V_{\mathbf{q}}|^2$ is independent of angle. Unless the electric field is high, or the energy is very low, the angular dependence under the square-root sign will be very weak, and we will neglect it. Thus the integration over angle yields a factor of 2π, viz:

$$W(E_i) = \frac{Vm^{*1/2}}{2^{3/2}\pi\hbar^2}\int_{q_{\perp\min}}^{q_{\perp\max}} \frac{(|V_{\pm q_\perp, q_{z\max}}|^2 + |V_{\pm q_\perp, q_{z\min}}|^2)}{(E_z \pm \hbar\omega_q)^{1/2}}\cdot \exp(-\tfrac{1}{2}q_\perp^2 R^2) q_\perp\, dq_\perp, \tag{7.64}$$

where we have assumed that the electric field term is negligible, and the limits of $q_\perp$ are as shown.

Another simplification can be made in the spirit of the approximation in which q_z is considered small compared with $q_\perp$, namely

$$|V_{\pm q_\perp, q_{z\max}}|^2 \approx |V_{\pm q_\perp, q_{z\min}}|^2, \tag{7.65}$$

whence

$$W(E_i) = \frac{Vm^{*1/2}}{2^{1/2}\pi\hbar^2} \int\limits_{q_{\perp\min}}^{q_{\perp\max}} \frac{|V_{\pm q_\perp}|^2}{(E_z \pm \hbar\omega_q)^{1/2}} \cdot \exp(-\tfrac{1}{2}q_\perp^2 R^2) q_\perp \, \mathrm{d}q_\perp. \tag{7.66}$$

This as far as one can go without knowing the $\mathbf{q}$-dependence of $V_\mathbf{q}$. Before discussion of explicit scattering mechanisms we investigate the scattering-induced current.

7.3.4. *General expression for the drift velocity*

A particle in the initial state scatters, and changes its average coordinate in the electric field direction by an amount $(X' - X)$ as shown in Fig. 7.3. If the average value of $(X' - X)$ is non-zero over all scattering possibilities the particle possesses a velocity. The velocity can readily be calculated by inserting $(X' - X)$ into the integrand for the scattering rate.

The quantity $(X' - X)$ is not dependent on q_z as eqn (7.51) shows, but it does depend upon angle. Thus we insert $(X' - X)$ into eqn (7.63) and incorporate the approximation of eqn (7.65), and obtain

$$v(E_i) = \frac{Vm^{*1/2}}{2^{3/2}\pi^2\hbar^2} \cdot \int\limits_{q_{\pm\min}}^{q_{\perp\max}} \int\limits_{\theta_{\min}}^{\theta_{\max}} \frac{|V_{\pm q_\perp}|^2 \exp(-\tfrac{1}{2}q_\perp^2 R^2)(\pm)(e/\bar{e})}{[E_2 \pm (e\xi R^2 q_\perp \cos\theta + \hbar\omega_q)]^{1/2}}$$
$$\times q_\perp^2 R^2 \cos\theta \, \mathrm{d}q_\perp \, \mathrm{d}\theta \tag{7.67}$$

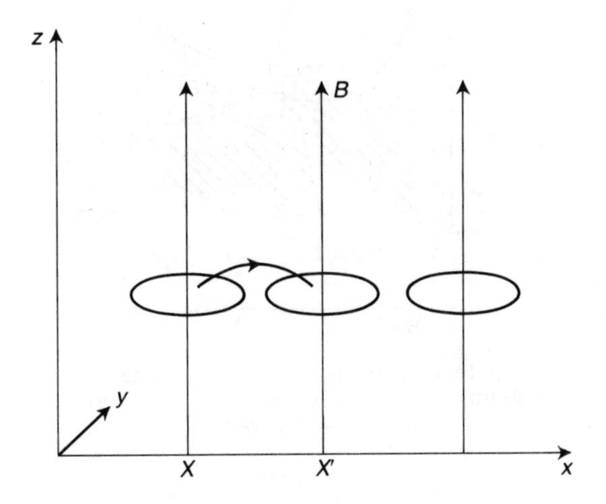

FIG. 7.3. Transport by scattering from one state to another.

Because of the density-of-states dependence on total energy (kinetic energy plus potential energy in the electric field), as quantified by the term in the electric field in the denominator, the scattering rate is not isotropic, and therefore a drift velocity exists, though it turns out to be highly anomalous. To show this we expand the denominator and retain terms up to those linear in field. Thus

$$v(E_i) = \frac{Vm^{*1/2}}{2^{3/2}\pi^2\hbar^2}\int_0^{\infty}\int_0^{2\pi}\frac{(\pm)(e/\bar{e})|V_{\pm q_\perp}|^2}{(E_z \pm \hbar\omega_q)^{1/2}} q_\perp^2 R^2 \exp(-\tfrac{1}{2}q_\perp^2 R^2)$$
$$\times \cos\theta\left[1 \mp \frac{1}{2}\frac{e\xi R^2 q_\perp \cos\theta}{E_z \pm \hbar\omega_q}\right] \mathrm{d}q_\perp \mathrm{d}\theta. \tag{7.68}$$

Integrating over θ we obtain a negative drift velocity viz.

$$v(E_i) = -\frac{Vm^{*1/2}(e^2/\bar{e})\xi}{2^{5/2}\pi\hbar^2}\int_0^{\infty}\frac{|V_{\pm q_\perp}|^2 q_\perp^3 R^4 \exp(-\tfrac{1}{2}q_\perp^2 R^2)}{(E_z \pm \hbar\omega_q)^{3/2}} \cdot \mathrm{d}q_\perp. \tag{7.69}$$

This is an extraordinary result. It predicts that carriers drift up the potential gradient. It clearly arises from the greater density of states—and therefore greater scattering probability—upstream than downstream at constant total energy (Ridley 1983), as shown in Fig. 7.4.

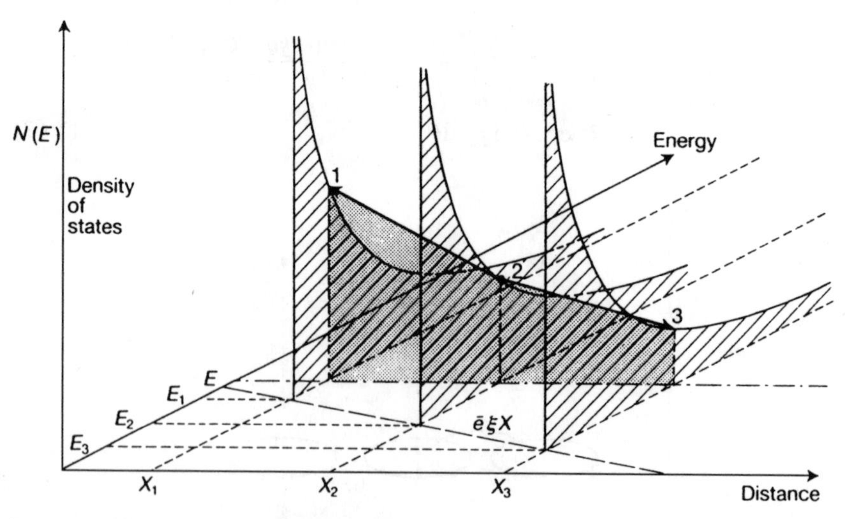

FIG. 7.4. Elastic scattering between ground-state Landau levels in the presence of an electric field. 1. Higher density of states upstream. 2. Scattering particle. 3. Lower density of states downstream.

Such anomalous behaviour disappears at low fields when an average is performed over a carrier population at thermodynamic equilibrium. In an elastic scattering event the probability has to be weighted by the factor $f(E_i)\,(1-f(E_f))$, where $f(E_i)$ and $f(E_f)$ are the occupation probabilities of the initial and final states. The net drift is then computed by weighting the integrand by half of the difference between the probabilities of the scattering event and its reverse, that is, by the factor

$$F(E_i)=\tfrac{1}{2}[f(E_i)(1-f(E_f))-f(E_f)(1-f(E_i))] \tag{7.70}$$

The factor $\frac{1}{2}$ is necessary to avoid counting each hopping twice (Davydov and Pomeranchuk 1940). Usually the occupation probabilities are a function only of the kinetic energy of motion in the z direction and

$$E_{fz}=E_{iz}\pm e\xi R^2 q_\perp \cos\theta \tag{7.71}$$

so if the electric field is small

$$f(E_{fz})\approx f(E_{iz})\pm\frac{\mathrm{d}f(E_{iz})}{\mathrm{d}E_z}\,e\xi R^2 q_\perp \cos\theta \tag{7.72}$$

and

$$F(E_i)=\mp\frac{1}{2}\frac{\mathrm{d}f(E_{iz})}{\mathrm{d}E_z}\,e\xi R^2 q_\perp \cos\theta. \tag{7.73}$$

For the case of inelastic phonon collisions, see Section 7.4.4. Returning to eqn (7.67) we can obtain the drift velocity for an energy E_i by weighting the integrand by $F(E_i)$, and retaining only terms linear in the electric field:

$$f(E_i)v(E_i)=-\frac{Vm^{*1/2}(e^2/\bar{e})\xi}{2^{5/2}\pi^2\hbar^2}\cdot\frac{\mathrm{d}f(E_{iz})}{\mathrm{d}(E_z)}$$

$$\times\int_{q_{\perp\min}}^{q_{\perp\max}}\int_0^{2\pi}\frac{|V_{\pm q_\perp}|^2\exp(-\frac{1}{2}q_\perp^2R^2)}{(E_z\pm\hbar\omega_q)^{1/2}}\,q_\perp^3R^4\cos^2\theta\,\mathrm{d}q_\perp\,\mathrm{d}\theta. \tag{7.74}$$

Since $\mathrm{d}f(E_{iz})/\mathrm{d}E_z$ is negative at thermodynamic equilibrium the drift velocity is no longer anomalous. Integration over θ gives

$$f(E_i)v(E_i)=-\frac{Vm^{*1/2}(e^2/\bar{e})\xi}{2^{5/2}\pi\hbar^2}\cdot\frac{\mathrm{d}f(E_{iz})}{\mathrm{d}E_z}$$

$$\times\int_{q_{\perp\min}}^{q_{\perp\max}}\frac{|V_{\pm q_\perp}|^2\exp(-\frac{1}{2}q_\perp^2R^2)}{(E_z\pm\hbar\omega_q)^{1/2}}\,.\,q_\perp^3R^4\,\mathrm{d}q_\perp. \tag{7.75}$$

This is as far as it is possible to go without consideration of the specific form of the scattering.

7.3.5. Diffusion

Weighting the scattering integrand by the distance $(X'-X)$ to obtain the drift velocity was first done by Davydov and Pomeranchuk (1940), but a different approach was made by Adams and Holstein (1959). Using a density-matrix approach they showed that the expression obtained for conductivity was equivalent to the Einstein relation between diffusion and drift. A diffusion constant $D(E_i)$ can be defined in terms of the mean square hopping distance $\langle (X'-X)^2 \rangle$. Thus

$$D(E_i) = \tfrac{1}{2}\langle (X'-X)^2 W(E_i) \rangle. \tag{7.76}$$

Weighting the integrand of eqn (7.63) with $\frac{1}{2}(X'-X)^2$, expressed in terms of q following eqn (7.51), we obtain the following expression for the diffusion constant

$$D(E_i) = \frac{Vm^{*1/2}}{2^{5/2}\pi^2\hbar^2} \int_{q_{\perp\min}}^{q_{\perp\max}} \int_{\theta_{\min}}^{\theta_{\max}} \frac{|V_{\pm q_\perp}|^2 \exp(-\frac{1}{2}q_\perp^2 R^2) q_\perp^3 R^4 \cos^2\theta}{[E_z \pm (e\xi R^2 q_\perp \cos\theta + \hbar\omega_q)]^{1/2}} \, dq_\perp \, d\theta. \tag{7.77}$$

In the limit of small electric fields

$$D(E_i) = \frac{Vm^{*1/2}}{2^{5/2}\pi\hbar^2} \int_{q_{\perp\min}}^{q_{\perp\max}} \frac{|V_{\pm q_\perp}|^2 \exp(-\frac{1}{2}q_1^2 R^2)}{(E_z \pm \hbar\omega_q)^{1/2}} q_\perp^3 R^4 \, dq_\perp. \tag{7.78}$$

Comparing eqns (7.75) and (7.78) we can obtain the Einstein relation between mobility, $\mu(E_i) = v(E_i)/\xi$, and diffusion coefficient

$$\mu(E_i) = -\left(\frac{e^2}{\bar{e}}\right) \frac{df(E_{iz})}{dE_z} \cdot \frac{D(E_{iz})}{f(E_{iz})} = \frac{e^2}{\bar{e}} \frac{(1-f(E_{iz}))}{k_B T} D(E_{iz}). \tag{7.79}$$

7.4. Scattering mechanisms

7.4.1. Acoustic phonon scattering

In the case of scattering by acoustic modes via the unscreened deformation potential, the interaction strength is given by

$$|V_{\pm q}|^2 = \frac{\hbar \Xi^2 q^2}{2MN\omega_q} (n(\omega_1) + \tfrac{1}{2} \mp \tfrac{1}{2}), \tag{7.80}$$

where Ξ is the deformation potential, M is the mass of a unit cell, N is the number of cells, and the upper and lower signs refer to absorption and emission of a phonon respectively. For simplicity, we assume isotropic scattering, which is true only for spherical bands.

The simplest case is to assume equipartition i.e. $\hbar\omega_q \ll k_B T$. Since the average value of E_z is $k_B T/2$ we can assume that $E_z \gg \hbar\omega_q$. Noting that

$\omega_q = v_s q$, and $MN/V = \rho$, the mass density we obtain for the scattering rate from eqn (7.66),

$$W(E_i) = \frac{\Xi^2 m^{*1/2} k_B T}{2^{3/2} \pi \hbar^2 \rho v_s^2 E_z^{1/2}} 2 \int_0^\infty \exp(-\tfrac{1}{2} q_\perp^2 R^2) q_\perp \, \mathrm{d}q_\perp \tag{7.81}$$

where the factor of 2 arises from the sum of absorption and emission processes, and the effective limits of $q_\perp$ are those shown. Thus

$$W(E_i) = \frac{\Xi^2 m^{*1/2} k_B T}{2^{1/2} \pi \hbar^2 \rho v_s^2 E_z^{1/2} R^2} = \frac{\Xi^2 m^{*1/2} k_B T e B}{2^{1/2} \pi \hbar^3 \rho v_s^2 E_z^{1/2}} = \frac{2\pi \Xi^2 k_B T}{\hbar \rho v_s^2} N(E_z). \tag{7.82}$$

The last form is exactly the form for three-dimensional and two-dimensional scattering. In all cases, the equipartition acoustic-phonon scattering rate is proportional to the density of states with the same constant of proportionality. In the particular case under consideration, the scattering rate is proportional to the magnetic field.

The anomalous drift velocity associated with a single particle, eqn (7.69), becomes

$$\begin{aligned} v(E_i) &= -\frac{(e^2/\bar{e}) \xi \Xi^2 m^{*1/2} k_B T}{2^{5/2} \pi \hbar^2 \rho v_s^2 E_z^{3/2}} \int_0^\infty q_\perp^3 R^4 \exp(-\tfrac{1}{2} q_\perp^2 R^2) \, \mathrm{d}q_\perp \\ &= -\frac{(e^2/\bar{e}) \xi \Xi^2 m^{*1/2} k_B T}{2^{3/2} \pi \hbar^2 \rho v_s^2 E_z^{3/2}}, \end{aligned} \tag{7.83}$$

which is independent of magnetic field. (This expression has not been reported previously). The drift velocity in the case of a distribution at thermal equilibrium, eqn (7.75), is

$$f(E_i) v(E_i) = -\frac{(e^2/\bar{e}) \xi \Xi^2 m^{*1/2} k_B T}{2^{3/2} \pi \hbar^2 \rho v_s^2 E_z^{1/2}} \cdot \frac{\mathrm{d}f(E_{iz})}{\mathrm{d}E_z} \tag{7.84}$$

$$= -(e^2/\bar{e}) \xi \left(\frac{R^2}{2}\right) W(E_i) \frac{\mathrm{d}f(E_{iz})}{\mathrm{d}E_z}, \tag{7.85}$$

whence the diffusion constant, via eqn (7.79), is simply

$$D(E_{iz}) = (R^2/2) W(E_i). \tag{7.86}$$

Equation (7.85) shows that the average hopping distance is $R/\sqrt{2}$. However eqn (7.84) shows that the equilibrium drift velocity is, like the anomalous drift velocity, independent of magnetic field. Their ratio is

$$\frac{v_{\text{anomalous}}}{v_{\text{normal}}} = \left[\frac{E_{iz}}{f(E_{iz})} \cdot \frac{\mathrm{d}f(E_{iz})}{\mathrm{d}E_z}\right]^{-1} = \frac{\mathrm{d}(\ln E_z)}{\mathrm{d}(\ln f(E_z))}, \tag{7.87}$$

which has a magnitude of order unity for a non-degenerate distribution.

The conductivity for a thermal distribution of carriers is obtained from an integration of eqn (7.84) over initial states. Thus

$$\sigma_{xx} = -\frac{e^2\Xi^2 m^{*1/2} k_B T}{2^{3/2}\pi\hbar^2\rho v_s^2} \int_{E_{z\min}}^{E_{z\max}} \frac{df(E_z)}{dE_z} \cdot 2\frac{N(E_z)}{E_z^{1/2}} dE_z. \tag{7.88}$$

In the case of a non-degenerate distribution of n carriers per unit volume,

$$f(E_z) = \frac{n}{a(\pi k_B T)^{1/2}} \exp(-E_z/k_B T), \qquad a = 2N(E_z)E_z^{1/2}. \tag{7.89}$$

Note that 'a' is a constant by virtue of the energy dependence of the density of states (eqn (7.34)). The factor 2 in eqn (7.88) accounts for spin degeneracy. The upper limit of energy can safely be taken to be ∞, but the lower limit cannot be taken to be zero on account of the logarithmic divergence of the integral. This is a general problem of a one-dimensional density of states. In the situation under consideration the divergence arises in the case of absorption only because we have neglected the finite phonon energy, through it still remains for emission, even if we re-introduce $\hbar\omega_{\mathbf{q}}$. What is usually done is to cut off the integration at

$$E_z = \hbar\omega_{\mathbf{q}} \approx \frac{\hbar v_s}{R} \tag{7.90}$$

whence

$$\sigma_{xx} = \frac{ne^2\Xi^2 m^{*1/2}}{2^{3/2}\pi^{3/2}\hbar^2\rho v_s^2 (k_B T)^{1/2}} \ln\left(\frac{Rk_B T}{\hbar v_s}\right). \tag{7.91}$$

This result agrees with that of Gurevitch and Firsov (1961), but is a factor 2 greater than that quoted by Kubo *et al.* (1965) who, perhaps, did not take into account the two allowed values of q_z. According to eqn (7.41) the corresponding resistivity is given by

$$\rho_{xx} = \frac{\Xi^2 m^{*1/2} B^2}{(2\pi)^{3/2}\hbar^2\rho v_s^2 (k_B T)^{1/2} n} \ln\left(\frac{Rk_B T}{\hbar v_s}\right). \tag{7.92}$$

The dependence on B and T agrees with that given by Adams and Holstein (1959).

At low temperatures and high magnetic fields the assumptions of equipartition and quasi-elasticity break down. The orbital radius R becomes small and the favoured phonon wavevector becomes large. Only emission processes are important, and then only for sufficiently energetic particles. As the temperature is lowered fewer phonons are around to initiate transitions and fewer particles have energy sufficient to spontaneously emit the short wavelength phonons which are most effective in causing transitions. Consequently the contribution made by acoustic phonons falls off roughly exponentially at temperatures below T_c, where T_c is given by (cf. eqn (7.60)):

$$T_c \approx \frac{\hbar v_s}{R k_B} \approx B^{1/2} K \qquad (B \text{ in tesla}). \tag{7.93}$$

The approximate temperature range in which the assumptions of equipartition, quasi-elasticity and E.Q.L. all apply is therefore given by

$$B^{1/2} \leqslant T \leqslant \frac{1}{2}\left(\frac{m}{m^*}\right) B \tag{7.94}$$

where the upper limit is determined by the condition for E.Q.L., which is roughly $\hbar\omega_c \gtrsim 3k_B T$ (see eqn (7.22)).

7.4.2. Piezoelectric scattering

In the case of scattering by acoustic modes via the piezoelectric interaction, which occurs whenever the crystal lattice lacks a centre of inversion symmetry, the interaction strength is given by

$$|V_{\pm\mathbf{q}}|^2 = \frac{\hbar e^2 \langle e_{14}\rangle^2}{2MN\omega_q \epsilon^2} \cdot \frac{q^4}{(q^2 + q_0^2)^2} \cdot (n(\omega_q) + \tfrac{1}{2} \mp \tfrac{1}{2}) \tag{7.95}$$

where $\langle e_{14}\rangle$ is an effective piezoelectric coefficient obtained by suitable averaging over direction (see Section 3.6), and q_0 is the reciprocal screening length. We will again assume equipartition and quasi-elasticity and E.Q.L.

The scattering rate, eqn (7.66), is

$$W(E_i) = \frac{e^2 \langle e_{14}\rangle^2 m^{*1/2} k_B T}{2^{1/2} \pi \hbar^2 \epsilon^2 \rho v_s^2 E_z^{1/2}} \int_0^\infty \exp(-\tfrac{1}{2} q_\perp^2 R^2) \frac{q_\perp^3}{(q_\perp^2 + q_0^2)^2} \, dq_\perp. \tag{7.96}$$

Once again, we have ignored the contribution of q_z to q. In the absence of screening the integral diverges (as it does in the classical regime). In the limit of small screening ($q_0^2 R^2 \ll 1$),

$$W(E_i) = \frac{e^2 K^2 m^{*1/2} k_B T}{2^{1/2} \pi \hbar^2 \epsilon E_z^{1/2}} \ln\left(\frac{2^{1/2}}{q_0 R}\right) \tag{7.97}$$

where we have introduced the averaged electromechanical coupling coefficient K, which is dimensionless. (This expression has not been reported previously.)

The anomalous drift velocity, eqn (7.69), is

$$v(E_i) = -\frac{(e^4/\bar{e}) \xi K^2 m^{*1/2} k_B T}{2^{5/2} \pi \hbar^2 \epsilon E_z^{3/2}} \int_0^\infty \exp(-\tfrac{1}{2} q_\perp^2 R^2) \frac{q_\perp^5 R^4}{(q_\perp^2 + q_0^2)^2} \, dq_\perp \tag{7.98}$$

which does not diverge in the absence of screening. Ignoring screening we obtain,

$$v(E_i) = -\frac{(e^3/\bar{e})\xi K^2 m^{*1/2} k_B T}{2^{5/2}\pi\hbar\epsilon E_z^{3/2} B} \tag{7.99}$$

an expression not reported previously.

The drift velocity at thermal equilibrium, eqn (7.75), is

$$f(E_i)v(E_i) = -\frac{(e^3/\bar{e})\xi K^2 m^{*1/2} k_B T}{2^{5/2}\pi\hbar\epsilon E_z^{1/2} B}\frac{\mathrm{d}f(E_z)}{\mathrm{d}E_z}. \tag{7.100}$$

This is no longer expressible in terms of the scattering rate, which depends critically on the degree of screening. From eqn (7.79) the diffusion coefficient is

$$D(E_i) = \frac{eK^2 m^{*1/2} k_B T}{2^{5/2}\pi\hbar\epsilon E_z^{1/2} B}. \tag{7.101}$$

The conductivity for non-degenerate distribution is

$$\sigma_{xx} = \frac{ne^3 K^2 m^{*1/2}}{2^{5/2}\pi^{3/2}\hbar\epsilon(k_B T)^{1/2} B}\ln\left(\frac{Rk_B T}{\hbar v_s}\right) \tag{7.102}$$

where the cut-off energy is $\hbar v_s/R$, as before. And, as before, this is just a factor 2 greater than the expression of Kubo *et al.* (1965), presumably for the same reason as before. The resistivity is, from eqn (7.41),

$$\rho_{xx} = \frac{eK^2 m^{*1/2} B}{2^{5/2}\pi^{3/2}\hbar\epsilon(k_B T)^{1/2} n}\cdot\ln\left(\frac{Rk_B T}{\hbar v_s}\right) \tag{7.103}$$

and the dependence on B and T agrees with that given by Adams and Holstein (1959). Whereas in the case of deformation-potential scattering the resistivity is proportional to B^2, for piezoelectric scattering it is proportional to B.

These results are valid in the regime roughly given by eqn (7.94).

7.4.3. *Charged-impurity scattering*

In the case of a point-charge at r the screened interaction energy is

$$H_s = \frac{e^2}{4\pi\epsilon\,|\mathbf{r}-\mathbf{r}_0|}e^{-q_0(r-r_0)} \tag{7.704}$$

or, alternatively,

$$H_s = \sum_{\mathbf{q}} V_{\mathbf{q}} \exp[i\mathbf{q}\cdot(\mathbf{r}-\mathbf{r}_0)] \tag{7.105}$$

where

$$V_{\mathbf{q}} = \frac{e^2}{V\epsilon(q^2+q_0^2)}. \tag{7.106}$$

Here V is the volume of the cavity, and q_0 is the reciprocal screening length. Equation (7.105) is an appropriate form to use eqns (7.66), (7.69), and (7.75) for, respectively, the scattering rate, the anomalous velocity, and the thermal drift velocity, taking into account that the scattering events are elastic, and only 'absorption' processes are to be taken into account. We are therefore working in the Born approximation, and we will ignore statistical screening.

Accordingly, the scattering rate is given by

$$W(E_i) = \frac{e^4 m^{*1/2} N_I}{2^{1/2}\epsilon^2 \pi \hbar^2 E_z^{1/2}} \int_0^\infty \frac{\exp(-\frac{1}{2}q_\perp^2 R^2)}{(q_\perp^2 + q_0^2)^2} q_\perp \, dq_\perp \tag{7.107}$$

$$= \frac{e^4 m^{*1/2} N_I}{2^{1/2}\pi\epsilon^2 \hbar^2 E_z^{1/2}} \cdot \frac{R^2}{4}\left[\frac{2}{q_0^2 R^2} + \exp(\tfrac{1}{2}q_0^2 R^3) Ei(-\tfrac{1}{2}q_0^2 R^2)\right] \tag{7.108}$$

where $Ei(x)$ is the exponential integral and N_I is the density of charged impurities. In the limit of weak screening, i.e.

$$\tfrac{1}{2}q_0^2 R^2 \ll 1 \tag{7.109}$$

$$W(E_i) \approx \frac{e^4 m^{*1/2} N_I}{2^{3/2}\pi\epsilon^2 \hbar^2 q_0^2 E_z^{1/2}}. \tag{7.110}$$

The scattering rate is very sensitive to screening as it is in the normal regime.

The anomalous drift velocity from eqn (7.69) is

$$v(E_i) = -\frac{(e^6/\bar{e}) m^{*1/2} N_I \xi}{2^{5/2}\pi\epsilon^2 \hbar^2 E_z^{3/2}} \int_0^\infty \frac{q_\perp^3 R^4 \exp(-\frac{1}{2}q_\perp^2 R^2)}{(q_\perp^2 + q_0^2)^2} dq_\perp \tag{7.111}$$

$$= \frac{(e^6/\bar{e}) m^{*1/2} N_I \xi}{2^{7/2}\pi\epsilon^2 \hbar^2 E_z^{3/2}} \cdot \frac{R^4}{2}[(1 + \tfrac{1}{2}q_0^2 R^2)\exp(\tfrac{1}{2}q_0^2 R^2) Ei(-\tfrac{1}{2}q_0^2 R^2) + 1]. \tag{7.112}$$

In the small screening limit

$$v(E_i) \approx \frac{(e^4/\bar{e}) m^{*1/2} N_I \xi}{2^{7/2}\pi\epsilon^2 E_z^{3/2} B^2} \cdot \ln(\tfrac{1}{2}q_0^2 R^2). \tag{7.113}$$

Equations (7.110) and (7.113) have not been reported before.

The equilibrium drift velocity, from eqn (7.75), is in the same limit.

$$f(E_i)v(E_i) = \frac{(e^4/\bar{e}) m^{*1/2} N_I \xi}{2^{7/2}\pi\epsilon^2 E_z^{1/2} B^2} \cdot \frac{df(E_z)}{dE_z} \cdot \ln(\tfrac{1}{2}q_0^2 R^2). \tag{7.114}$$

All of these expressions entail the assumption that $q_z \ll q_\perp$, but although that assumption simplifies matters in the case of scattering by phonons, it is not at all useful in the case of impurity scattering. If the assumption is not made the integrals split into two components, one in which $q_z = 0$, and the other in which $q_z = 2k_z$. The modified expressions for the scattering rate and drift velocities in the limit of weak screening are as follows:

$$W(E_i) = \frac{e^4 N_I}{2^{7/2}\pi\epsilon^2 m^{*1/2} E_z^2}\left(\frac{1}{E_s} + \frac{1}{E_s + 4E_z}\right) \tag{7.115}$$

$$v(E_i) = \frac{(e^4/\bar{e})m^{*1/2}N_I\xi}{2^{9/2}\pi\epsilon^2 E_z^{3/2} B^2}\ln\left[\frac{E_s(E_s + 4E_z)}{(\hbar\omega_c)^2}\right] \tag{7.116}$$

$$f(E_i)v(E_i) = \frac{(e^4/\bar{e})m^{*1/2}N_I\xi}{2^{9/2}\pi\epsilon^2 E_z^{1/2} B^2} \cdot \frac{\mathrm{d}f(E_z)}{\mathrm{d}E_z} \cdot \ln\left[\frac{E_s(E_s + 4E_z)}{(\hbar\omega_c)^2}\right]. \tag{7.117}$$

In these expressions we have introduced a screening energy, thus

$$E_s = \frac{\hbar^2 q_0^2}{2m^*}, \qquad \tfrac{1}{2}q_0^2 R^2 = \frac{E_s}{\hbar\omega_c}. \tag{7.118}$$

Thus, for a non-degenerate distribution,

$$\sigma_{xx} = \frac{ne^4 m^{*1/2} N_I}{2^{9/2}\pi^{3/2}\epsilon^2 B^2 (k_B T)^{3/2}}\ln\left[\frac{(\hbar\omega_c)^2}{E_s(E_s + 4k_B T)}\right] \cdot \ln\left[\frac{k_B T}{E_c}\right]. \tag{7.119}$$

In the derivation of this expression the kinetic energy E_z appearing in the logarithm has been replaced by $k_B T$, and the integration over energy has been curtailed at an energy E_c, as yet undefined. The corresponding resistivity, according to eqn (7.41), is

$$\rho_{xx} = \frac{e^2/m^{*1/2}N_I}{2^{9/2}\pi^{3/2}\epsilon^2 n^2 (k_B t)^{3/2}} \cdot \ln\left[\frac{(\hbar\omega_c)^2}{E_s(E_s + 4k_B T)}\right]\ln\left[\frac{k_B T}{E_c}\right] \tag{7.120}$$

which is independent of magnetic field. This expression has the same dependence of temperature, magnetic field and screening as that quoted by Adams and Holstein (1959).

In the case of phonon collisions a cut-off energy was prescribed by the magnitude of the phonon energy. Where no energy is exchanged, the cut-off energy may be taken to be of magnitude of the level-broadening induced by collisions. Sharply defined energies are replaced according to

$$\delta(E - E_0) \rightarrow \frac{1}{\pi}\frac{\hbar\Gamma}{(E - E_0)^2 + \hbar^2\Gamma^2} \tag{7.121}$$

which has the effect of defining the cut-off energy as

$$E_c \approx \Gamma. \tag{7.122}$$

Since according to the uncertainty principle $\Gamma \approx \hbar W(E_i)$, where $W(E_i)$ is the scattering rate, and when $W(E_i) = W_0 E_z^{-1/2} \approx W_0\Gamma^{-1/2}$

$$\Gamma \approx (\hbar W_0)^{2/3}. \tag{7.123}$$

This approach was made by Davydov and Pomeranchuk (1940) to estimate broadening. If collisions are dominated by impurity scattering $W(E_i) = W_0 E_z^{-2}$ as eqn (7.115) indicates, whence

$$\Gamma \approx (\hbar W_0)^{1/3}. \tag{7.124}$$

This is exactly the argument used in the non-quantum situation to avoid the notorious divergence associated with small-angle scattering (see Section 4.2.3.) In that case it is known that the approach leads to a cross-section too large for the two-body collision processes envisaged by the theory to be valid, so it is not clear how useful the collision-broadening theory is in obtaining finite expressions.

Another criticism which may be levelled is that the theory is based on the Born approximation, whereas a partial-wave treatment may be more appropriate, as Kubo *et al.* (1965) discuss.

7.4.4. *Statistical weighting for inelastic phonon collisions*

In the case of inelastic phonon scattering we have to distinguish absorption processes and emission processes when the statistical weighting is carried out. The interaction term $|V_{\mathbf{q}}|^2$ contains the factor $n(\omega_q)$ for absorption and $(n(\omega_q)+1)$ for emission and these must be incorporated into the factor $F(E_i)$ which, for absorption, becomes

$$F_a(E_i) = \tfrac{1}{2}[n(\omega_q)f(E_i)(1-f(E_f)) - (n(\omega_q)+1)f(E_f)(1-f(E_i))] \tag{7.125}$$

where now

$$f(E_f) = f(E_{iz} + \hbar\omega_q) + \frac{\mathrm{d}f(E_{iz})}{\mathrm{d}E_z}\, e\xi R^2 q_\perp \cos\theta \tag{7.126}$$

whence

$$F_a(E_i) = \tfrac{1}{2}[n(\omega_q)(f(E_{iz}) - f(E_{iz}+\hbar\omega_q)) - f(E_{iz}+\hbar\omega_q)(1-f(E_{iz})) \\ - \Delta f(n(\omega_q)+1-f(E_{iz}))] \tag{7.127}$$

with Δf representing the field term. The corresponding expression for emission is

$$F_e(E_i) = \tfrac{1}{2}[n(\omega_q)(f(E_{iz}) - f(E_{iz}-\hbar\omega_q)) + f(E_{iz})(1-f(E_{iz}-\hbar\omega_q)) \\ + \Delta f(n(w_q)+f(E_{iz}))]. \tag{7.128}$$

At thermodynamic equilibrium we can use the identity

$$f(E_{iz})(1-f(E_{iz}-\hbar\omega_q)) = (f(E_{iz}-\hbar\omega_q) - f(E_{iz}))n(\omega_q) \tag{7.129}$$

to eliminate the field-independent terms—basically, an example of detailed balance. We have finally

$$F_{a,e}(E_i) = \mp\frac{1}{2}\frac{\mathrm{d}f(E_{iz})}{\mathrm{d}E_z}\, e\xi R^2 q_\perp \cos\theta(n(\omega_q) + \tfrac{1}{2} \pm \tfrac{1}{2} \mp f(E_{iz})). \tag{7.130}$$

This expression will be useful in our discussion of magnetophonon oscillations later

7.5. Transverse Shubnikov–de Haas oscillations

7.5.1. Magnetoconductivity in the presence of many Landau levels

In degenerate material the electrical resistance oscillates with magnetic field—the Shubnikov–de Haas effect. This phenomenon is caused by the changing occupation of the Landau levels in the vicinity of the Fermi level. The description is quite complex since the effect is one involving many magnetic states. However, no new physical concepts additional to those already encountered in our discussion of conductivity in the extreme quantum limit are required.

The squared matrix element that describes the transition from an initial state $|k_y, k_z, n\rangle$ to a final state $|k_y', k_z', m\rangle$ is a generalization of eqn (7.49), namely,

$$|M_{\pm\mathbf{q}}|^2_{nm} = \delta_{k_y \pm q_y, k_y'}\, \delta_{k_z \pm q_z, k_z'}\, |I_{nm}(X' - X)|^2 \tag{7.131}$$

where

$$I_{nm}(X' - X) = \int_{-L_x/2}^{+L_x/2} \phi_m(x - X')\exp(\pm i q_x x)\phi_n(x - X)\,\mathrm{d}x \tag{7.132}$$

and

$$\phi_n(x - X) = (2^n n!\pi^{1/2}R)^{-1/2}\exp[-(x - X)^2/2R^2]H_n[(x - X)/R] \tag{7.133}$$

where $H_n[(x - X)/R]$ is a Hermite polynomial and R is, as before, $(\hbar/eB)^{1/2}$. Using eqn (7.51) to relate $X' - X$ to q_y, and putting $q_\perp^2 = q_x^2 + q_y^2$, we obtain [see Gradshtein and Ryzhik (1965) 7.377, p. 838]

$$|I_{nm}(X' - X)|^2 = \frac{m!}{n!}\left(\frac{q_\perp^2 R^2}{2}\right)^{n-m}\exp(-\tfrac{1}{2}q_\perp^2 R^2)[L_m^{n-m}(\tfrac{1}{2}q_\perp^2 R^2)]^2 \tag{7.134}$$

where $L_m^{n-m}(q_\perp^2 R^2/2)$ is a Laguerre polynomial. [This comparatively straightforward result is not found in the foundation literature of the subject, e.g. Titeica (1935), Adams and Holstein (1959)].

For simplicity we will restrict discussion to elastic scattering. The conductivity can then be expressed in terms familiar from eqn (7.75): thus

$$\begin{aligned}\sigma_{xx} &= \sum_n \frac{\bar{e}}{\xi}\int f(E_i)v_n(E_i)\,.\,2N_n(E_i)\,\mathrm{d}E_i \\ &= -\frac{e^2 m^{*1/2}V}{2^{5/2}\pi\hbar^2}\sum_n \int \frac{\mathrm{d}f(E_i)}{\mathrm{d}E_i}N_n(E_i)\,\mathrm{d}E_i \sum_m \int_0^\infty \frac{|V_{\pm q_\perp, q_{z\max}}|^2 + |V_{\pm q_\perp, q_{z\min}}|^2}{[E_z + (n - m)\hbar\omega_c]^{1/2}} \\ &\quad \times |I_{nm}(X' - X)|^2 q_\perp^3 R^4\,\mathrm{d}q_\perp\end{aligned} \tag{7.135}$$

where $N_n(E_i)$ is the density of states for a given spin. The connection with eqn (7.75) becomes clear when the E.Q.L. conditions are imposed, i.e. $n = m = 0$, $L_0^0(q_\perp^2 R^2/2) = 1$.

The integration in eqn (7.135) is straightforward only in the case for acoustic phonon scattering for which $|V_q|^2$ is independent of q. Limiting attention to this case we obtain [Gradshtein and Ryzhik (1965) 7.414 No. 12, p. 845]

$$\int_0^\infty [|V_{\pm q_\perp, q_{z\max}}|^2 + |V_{\pm q_\perp, q_{z\min}}|^2]\,|I_{nm}(X' - X)|^2\, q_\perp^3 R^4\, \mathrm{d}q_\perp = \frac{4\Xi^2 k_\mathrm{B} T}{\rho v_s^2 V}(1 + n + m). \tag{7.136}$$

This result, hitherto obtained explicitly only for the case of large quantum numbers, is exact, independent of the magnitudes of n and m (as asserted by Adams and Holstein without proof). The conductivity becomes

$$\sigma_{xx} = -\frac{e^2\Xi^2 k_\mathrm{B} T m^{*1/2}}{2^{1/2}\pi\hbar^2\rho v_s^2}\sum_n \int \frac{\mathrm{d}f(E_i)}{\mathrm{d}E_i} N_n(E_i)\,\mathrm{d}E_i \sum_m \frac{1 + n + m}{[E_z + (n - m)\hbar\omega_c]^{1/2}}. \tag{7.137}$$

In the degenerate case with $\hbar\omega_c \ll E_\mathrm{F}$ and $T \approx 0$ we can take

$$\frac{\mathrm{d}f(E_i)}{\mathrm{d}E_i} \approx -\delta(E_i - E_\mathrm{F}) \tag{7.138}$$

where E_F is the Fermi level, and substituting

$$E_z = E_i - (n + \tfrac{1}{2})\hbar\omega_\mathrm{c} \tag{7.139}$$

we obtain

$$\sigma_{xx} = \frac{e^2\Xi^2 k_\mathrm{B} T m^*}{4\pi^3\hbar^3\rho v_s^2 R^2}\sum_n \frac{1}{[E_\mathrm{F} - (n + \frac{1}{2})\hbar\omega_c]^{1/2}} \sum_m \frac{1 + n + m}{[E_\mathrm{F} - (m + \frac{1}{2})\hbar\omega_c]^{1/2}}. \tag{7.140}$$

The sums are taken over m and n such that the denominators remain real.

This expression predicts an oscillatory variation with magnetic field of over-dramatic proportions, being of infinite amplitude! The infinities arise as usual from the density of states function, in this case compounded by having the product of the density of final states and the density of initial states. The conductivity is infinite when

$$E_\mathrm{F} - (n + \tfrac{1}{2})\hbar\omega_c = 0. \tag{7.141}$$

Expressed in a form suitable for use in experiment, the condition,

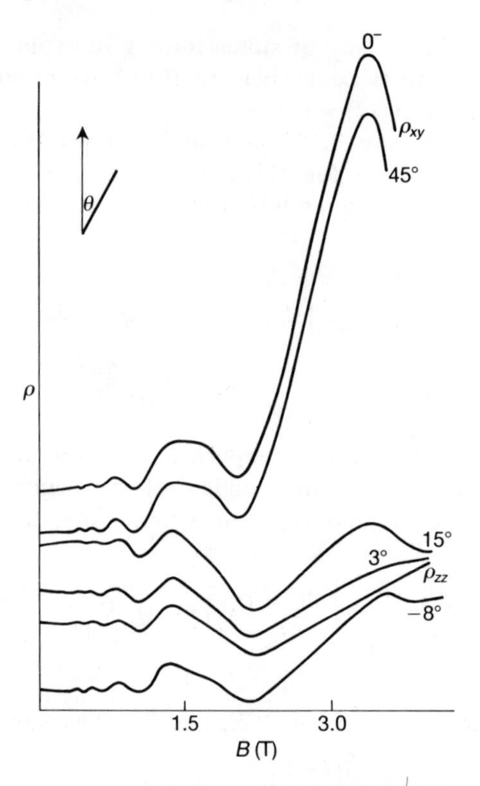

FIG. 7.5. Shubnikov–de Haas oscillations in transverse configuration. Experimental recordings of resistance against magnetic field showing the Shubnikov–de Haas effect for sample of InSb at various angles of the magnetic field with respect to the current direction. (Staromylnska *et al.* 1983.)

rewritten, is

$$\frac{1}{B}=\frac{e\hbar}{m^*E_{\mathrm{F}}}(n+\tfrac{1}{2}). \tag{7.142}$$

A plot of B^{-1} versus oscillation number will give a straight line of slope $e\hbar/m^*E_{\mathrm{F}}$, from which the carrier density n_e can be extracted since at $T=0$

$$m^*E_{\mathrm{F}}=\tfrac{1}{2}(3\pi^2)^{2/3}\hbar^2 n_e^{2/3}. \tag{7.143}$$

Thus the Shubnikov–de Haas effect provides a direct measure of carrier density.

Needless to say infinite amplitude oscillations are not observed (Fig. 7.5). The oscillations are not even spiky. Moreover, the amplitude of the oscillations decreases with increasing temperature, but eqn (7.140) does not describe this effect. Finally, it is possible to distinguish spin-splitting in some of the oscillations, and again eqn (7.140) is deficient. These deficiencies can be rectified by including collision broadening, thermal broadening, and spin energy into the description. But before discussing these aspects we observe that eqn (7.140) as it stands reduces to the classical conductivity:

$$\sigma_{xx} = \frac{e^2 n_e}{m^* \omega_c^2 \tau} \tag{7.144}$$

in the limit $\hbar\omega_c/E_F \to 0$, when the sums can be replaced by integrals. The proof is straightforward, and the scattering time τ is just the reciprocal of the rate given in eqn (3.78). We may therefore regard the conductivity as consisting of a non-oscillatory component σ_0, which is just the classical conductivity, and an oscillatory component $\tilde{\sigma}$, viz.

$$\sigma_{xx} = \sigma_0 + \tilde{\sigma} \tag{7.145}$$

where the oscillatory component arises from the movement of the uppermost occupied Landau level.

7.5.2. *The oscillatory component*

In order to describe this component we express the conductivity as follows:

$$\sigma_{xx} = \sigma_0 \frac{3}{16} \frac{1}{\eta^2} \sum_{n,m} \frac{1+n+m}{[\eta - (n+\frac{1}{2})]^{1/2}[\eta - (m+\frac{1}{2})]^{1/2}} \tag{7.146}$$

where $\eta = E_F/\hbar\omega_c$. If we assume that $\eta \gg 1$ we can exploit the Poisson sum formula in the form used by Dingle (1952):

$$\sum_{n=0}^{\infty} f(n+\tfrac{1}{2}) = \sum_{r=-\infty}^{+\infty} (-1)^r \int_0^{\infty} f(n) \exp(2\pi i n r)\, dn. \tag{7.147}$$

Thus

$$\sum_m [\eta - (m+\tfrac{1}{2})]^{-1/2} \approx 2\left[\eta^{1/2} + \sum_{r=1}^{\infty} \frac{(-1)^r}{(2r)^{1/2}} \cos(2\pi\eta r - \pi/4)\right] \tag{7.148}$$

$$\sum_m \frac{m+\frac{1}{2}}{[\eta - (m+\frac{1}{2})]^{1/2}} \approx \tfrac{4}{3}\eta^{3/2} + \sum_{r=1}^{\infty} \left[\frac{2\eta(-1)^r}{(2r)^{1/2}} \cos\left(2\pi\eta r - \frac{\pi}{4}\right) - \frac{1}{(2r)^{3/2}} \cos\left(2\pi\eta r - \frac{3\pi}{4}\right)\right]. \tag{7.149}$$

Neglecting products of the summations we obtain finally

$$\sigma_{xx} = \sigma_0 \left\{1 + \sum_{r=1}^{\infty} (-1)^r \left[\frac{5}{2} \cdot \frac{1}{(2\eta r)^{1/2}} \cos\left(2\pi\eta r - \frac{\pi}{4}\right) - \frac{3}{4} \frac{1}{(2\eta r)^{3/2}} \cos\left(2\pi\eta r - \frac{3\pi}{4}\right)\right]\right\}. \tag{7.150}$$

The leading term in the summation is that given by Adams and Holstein (1959). (Note that the factor 5/2 is often left out erroneously in descriptions of the transverse Shubnikov–de Haas effect.) The second term is much smaller and should be neglected since it is even smaller than some of the cross-product terms. Thus we end up with

$$\sigma_{xx} = \sigma_0\left(1 + \frac{5}{2}\sum_{r=1}^{\infty}\frac{(-1)^r}{(2\eta r)^{1/2}}\cos\left(2\pi\eta r - \frac{\pi}{4}\right)\right). \tag{7.151}$$

7.5.3. *Collision broadening*

Without collision broadening we would require all terms in the infinite sum to describe the oscillations, since the density of states goes to infinity at small energies of motion along the z direction. To incorporate collision broadening we note that the most rapidly varying functions of energy in eqn (7.151) are the cosines in the sum. We may regard them as having come about as follows

$$\exp(i2\pi r\eta) = \int_{-\infty}^{+\infty} \exp[i2\pi r(E/\hbar\omega_c)]\,\delta(E - E_F)\,dE. \tag{7.152}$$

We now replace the delta function as follows (cf. eqn (4.37))

$$\delta(E - E_F) = \frac{1}{\pi}\frac{\hbar\Gamma}{(E - E_F)^2 + \hbar^2\Gamma^2} \tag{7.153}$$

where Γ is the scattering rate. The integration is straightforward and we obtain

$$\frac{\hbar\Gamma}{\pi}\int_{-\infty}^{+\infty}\frac{\exp[i2\pi r(E/\hbar\omega_c)]}{(E - E_F)^2 + \hbar^2\Gamma^2}\,dE = \exp(i2\pi r\eta)\exp(-2\pi r\Gamma/\omega_c). \tag{7.154}$$

All terms in the sum are multiplied by the last exponential factor in eqn (7.154).

Broadening is often described by the Dingle temperature defined by

$$k_B T_D = \frac{\hbar\Gamma}{\pi} \tag{7.155}$$

$$\therefore \quad \frac{\Gamma}{\omega_c} = \frac{\pi k_B T_D}{\hbar\omega_c}. \tag{7.156}$$

Thus with collision broadening only the first few terms in the sum will be important. Indeed usually only the first is used.

7.5.4. *Thermal broadening*

In order to describe the temperature dependence of the oscillations it is necessary to relax our prescription of eqn (7.138). Once again only the oscillating functions need be considered. Thus instead of effectively taking

$$-\int_{-\infty}^{+\infty} \frac{\mathrm{d}f(E)}{\mathrm{d}E} \exp[-i2\pi r(E/\hbar\omega_c)]\,\mathrm{d}E = \exp[-i2\pi r(E_{\mathrm{F}}/\hbar\omega_c)] \quad (7.157)$$

we take

$$\frac{\mathrm{d}f(E)}{\mathrm{d}E} = -\frac{\exp[(E-E_{\mathrm{F}})/k_{\mathrm{B}}T_e]}{k_{\mathrm{B}}T_e[1+\exp[(E-E_{\mathrm{F}})/k_{\mathrm{B}}T_e]]^2} \quad (7.158)$$

where T_e is the electron temperature, perform the integration, and obtain

$$-\int_{-\infty}^{+\infty} \frac{\mathrm{d}f(E)}{\mathrm{d}E} \exp[-i2\pi r(E/\hbar\omega_c)]\,\mathrm{d}E = \exp(-i2\pi r\eta)\cdot\frac{2\pi^2 rk_{\mathrm{B}}T_e/\hbar\omega_c}{\sinh(2\pi^2 rk_{\mathrm{B}}T_e/\hbar\omega_c)}. \quad (7.159)$$

This temperature-dependent factor must also be incorporated into the summation.

7.5.5. *Spin-splitting*

So far we have neglected the effect on the energy of a state caused by the spin of the electron. All previous energies become

$$E = E_0 \pm \tfrac{1}{2}g\mu_B B. \quad (7.160)$$

In particular,

$$\begin{aligned}\exp(-i2\pi rE_F/\hbar\omega_c) &= \tfrac{1}{2}\exp[-i2\pi r(\eta \pm \tfrac{1}{2}g\mu_{\mathrm{B}}B/\hbar\omega_c)]\\ &= \exp(-i2\pi r\eta)\cos(\pi grm^*/2mc) \quad (7.161)\end{aligned}$$

where we have substituted $\mu_B = e\hbar/2mc$ for the Bohr magneton.

7.5.6. *Shubnikov–de Haas formula*

Incorporating collision broadening, thermal broadening, and spin-splitting, we obtain finally

$$\sigma_{xx} = \sigma_0\left(1 + \sum_{r=1}^{\infty} b_r \cos\left(2\pi\eta r - \frac{\pi}{4}\right)\right) \quad (7.162)$$

$$\begin{aligned}b_r = (-1)^r \frac{5}{2}\frac{1}{(2\eta r)^{1/2}}\cdot\frac{2\pi^2 rk_{\mathrm{B}}T_e/\hbar\omega_c}{\sinh(2\pi^2 rk_{\mathrm{B}}T_e/\hbar\omega_c)}\\ \times\exp(-2\pi^2 rk_{\mathrm{B}}T_D/\hbar\omega_c)\cos(\pi grm^*/2mc). \quad (7.163)\end{aligned}$$

Collision broadening weakens the effect of harmonics higher than the fundamental while thermal broadening reduces the amplitude of the oscillations with increasing temperature. This relationship of amplitude

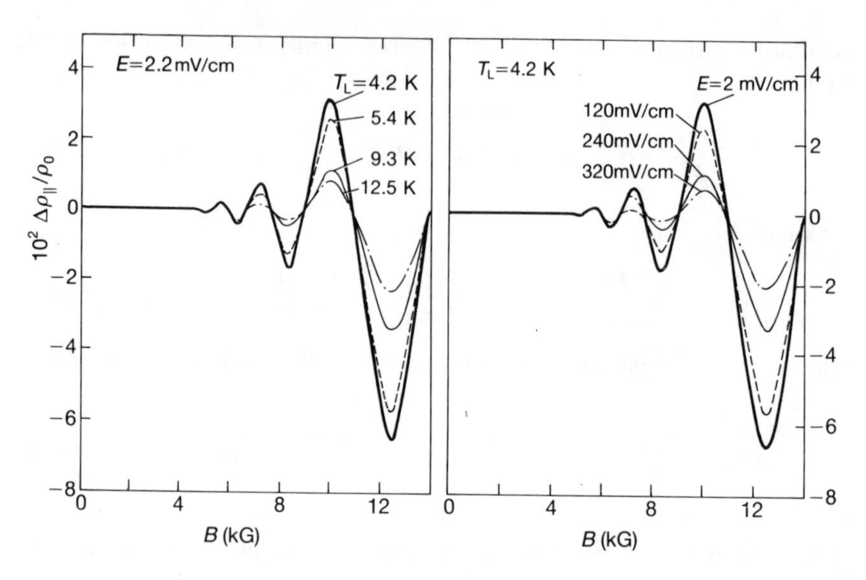

FIG. 7.6. Hot-electron Shubnikov–de Haas oscillation. Oscillatory component of the longitudinal magnetoresistance of InAs; left-hand side: measurement under Ohmic conditions and different lattice temperatures between 4.2 and 12.5 K; right-hand side: measurements at a constant lattice temperature of 4.2 K and different electric fields between 2 and 320 mV/cm. (Bauer and Kahlert 1972.)

with temperature is exploited in hot-electron experiments to obtain the electron temperature as a function of electric field by comparing the amplitude as a function of electric field with the amplitude as a function of lattice temperature (Fig. 7.6), an approach that relies on the distribution function remaining a Fermi–Dirac one.

The amount whereby each oscillation is split by spin depends on the g factor. In semiconductors, in which the effect of remote bands can be neglected, the g factor is given by

$$g = 2\left[1 - \frac{\left(\dfrac{m}{m^*} - 1\right)}{2 + 3E_g/\Delta_0}\right] \tag{7.164}$$

where E_g is the band gap and Δ_0 is the spin–orbit splitting at the zone centre. For GaAs $g = 0{\cdot}32$, compared with 2 for free electrons.

7.6. Longitudinal Shubnikov–de Haas oscillations

When the magnetic field and the electric field are parallel, both along the z-direction, the Hamiltonian becomes

$$H_0 = \frac{1}{2m^*}(p_x^2 + p_z^2) + \frac{1}{2m^*}(p_y - \bar{e}Bx)^2 - \bar{e}\xi z. \tag{7.165}$$

The magnetic part is as before since we have not changed gauge. Integration of the equation of motion, as before, now gives

$$m^*v_x = \bar{e}yB + c_x \quad (7.166)$$

$$m^*v_y = -\bar{e}xB + c_y \quad (7.167)$$

$$m^*v_2 = \bar{e}\xi t + c_z. \quad (7.168)$$

Since $p_x = m^*v_x$, $p_y = m^*v_y + \bar{e}Bx$, and $p_z = m^*v_z$, only p_y remains a constant of the motion. Putting $c_y = \bar{e}BX$ allows us to write the Hamiltonian as follows:

$$H_0 = \frac{1}{2m^*}(p_x^2 + p_z^2) + \frac{1}{2m^*}(x - X)^2e^2B^2 - \bar{e}\xi z. \quad (7.169)$$

Solutions of the Schrodinger equation are then

$$\Psi_n(\mathbf{r}) = C\exp(ik_y y)\exp(ik_z z)\phi_n(x - X) \quad (7.170)$$

$$E = (n + \tfrac{1}{2})\hbar\omega_c + \frac{\hbar^2k_z^2}{2m^*} \quad (7.171)$$

$$\hbar\frac{\mathrm{d}k_z}{\mathrm{d}t} = \bar{e}\xi \quad (7.172)$$

$$\hbar k_y = \bar{e}BX. \quad (7.173)$$

Classically, the electrons execute a spiralling motion along the field axis. Unlike the crossed-fields case, the electron is accelerated along the field axis in the normal way. Its motion is no longer dependent upon scattering from one Landau level to another down the potential gradient; instead its motion is impeded by scattering events, as is the usual case in the absence of a magnetic field.

The current is then described in the usual way in terms of the axial perturbation of the distribution function. Thus if the latter is taken to be of the form

$$f(\mathbf{k}) = f_0(E) + f_1(E)\cos\theta \quad (7.174)$$

where θ is the angle $\mathbf{k}$ makes with the electric field, then the current density is given by

$$j_z = \bar{e}\int f_1(E)\cos\theta v_z\,\mathrm{d}S \quad (7.175)$$

where the integral is over all allowed states. Solving the Boltzmann equation yields

$$f_1(E)\cos\theta = -\bar{e}\frac{\hbar k_z}{m^*}\xi\tau_m(E)\frac{\mathrm{d}f_0(E)}{\mathrm{d}E} \quad (7.176)$$

where $\tau_m(E)$ is the reciprocal of the mommentum relaxation rate $W_m(E)$, which for elastic processes can be written (see eqn (7.44))

$$W_m(E) = \sum_{\mathbf{q}} \frac{2\pi}{\hbar} \int |V_{\pm\mathbf{q}}|^2 \, |M_{\pm\mathbf{q}}|^2 \, \delta(E_f - E_i)(1 - \cos\theta_k) \, \mathrm{d}S \quad (7.177)$$

where θ_k is the angle between $\mathbf{k}$ and the scattered wavevector $\mathbf{k}'$, and $|M_{\pm\mathbf{q}}|^2$ is given by eqn (7.131), viz.

$$|M_{\pm\mathbf{q}}|^2_{nm} = \delta_{k_y \pm q_y, k_y'} \, \delta_{k_z \pm q_z, k_z'} \, |I_{nm}(X' - X)|^2 \quad (7.178)$$

with $|I_{nm}(X' - X)|^2$ given by eqn (7.134).

For acoustic-phonon scattering via an unscreened deformation potential the integration in eqn (7.177) is straightforward. In the crossed-field case we were presented with the integral

$$\int_0^\infty x I^2_{nm}(x) \, \mathrm{d}x = 1 + n + m \quad (7.179)$$

(see eqn (7.136)). This time the integral is easier viz:

$$\int_0^\infty I^2_{nm}(x) \, \mathrm{d}x = 1 \quad (7.180)$$

and we obtain, finally,

$$W_n = \sum_m \frac{\Xi^2 k_{\mathrm{B}} T m^{*1/2}}{2^{1/2} \pi \hbar^2 \rho v_s^2 R^2} \frac{1}{[E - (m + \frac{1}{2})\hbar\omega_2]^{1/2}}. \quad (7.181)$$

Note that this reduces properly to the extreme-quantum-limit form given in eqn (7.82).

Returning to the current density, we can obtain the conductivity in the form

$$\sigma_{xx} = -e^2 \sum_n \int v_z^2 W_n^{-1} \frac{\mathrm{d}f_0(E)}{\mathrm{d}E} 2N(E) \, \mathrm{d}E. \quad (7.182)$$

At low temperatures with Fermi–Dirac statistics prevailing this becomes

$$\sigma_{xx} = \frac{2e^2 \hbar \rho v_s^2}{\pi \Xi^2 k_{\mathrm{B}} T m^*} \sum_n [E_{\mathrm{F}} - (n + \tfrac{1}{2})\hbar\omega_c]^{1/2} \left[\sum_m [E_{\mathrm{F}} - (m + \tfrac{1}{2})\hbar\omega_c]^{-1/2} \right]^{-1} \quad (7.183)$$

which shows oscillating behaviour (Fig. 7.7). When $\hbar\omega_c/E_{\mathrm{F}} \ll 1$, Argyres (1958) has shown that

$$\sigma_{xx} = \frac{2e^2 \hbar \rho v_s^2}{\pi \Xi^2 k_{\mathrm{B}} T m^*} \cdot \frac{E_{\mathrm{F}}}{3} \left[1 - \sum_{r=1}^\infty \frac{(-1)^r}{(2\eta r)^{1/2}} \cos\left(2\pi r \eta - \frac{\pi}{4}\right) \times \frac{2\pi^2 r k_{\mathrm{B}} T_e / \hbar\omega_c}{\sinh(2\pi^2 r k_{\mathrm{B}} T_e / \hbar\omega_c)} \right]. \quad (7.184)$$

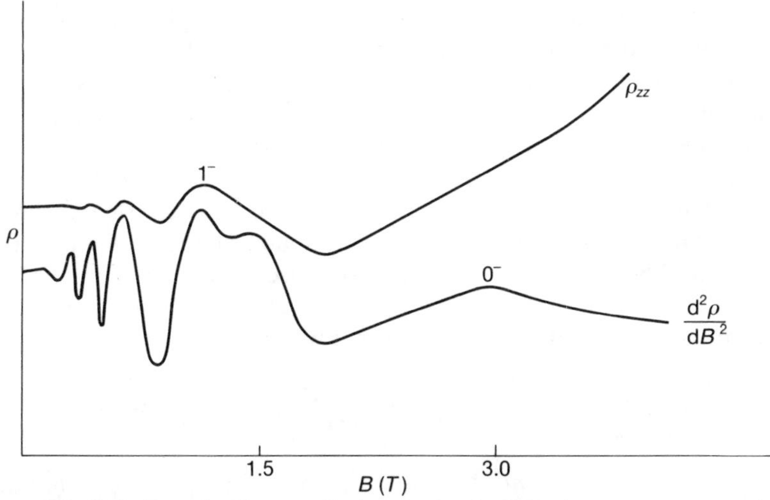

FIG. 7.7. Shubnikov–de Haas oscillation in longitudinal configuration. Experimental recordings of the resistance and its second derivative against magnetic field for sample of InSb. (Staromylnska *et al.* 1983.)

Incorporating collision broadening and spin splitting leads finally to

$$\sigma_{xx} = \sigma_{0L}\left(1 - \sum_{r=1}^{\infty} b_r \cos\left(2\pi\eta r - \frac{\pi}{4}\right)\right) \tag{7.185}$$

where

$$b_r = \frac{(-1)^r}{(2\eta r)^{1/2}} \cdot \frac{2\pi^2 r k_B T_e/\hbar\omega_c}{\sinh(2\pi^2 r k_B T_e/\hbar\omega_c)} \times \exp(-2\pi^2 r k_B T_D/\hbar\omega_c)\cos(\pi g r m^*/2mc) \tag{7.186}$$

with $\eta = E_F/\hbar\omega_c$. Note that the conductivity σ_{0L} is quite different from the corresponding conductivity in the transverse configuration.

7.7. Magnetophonon oscillations

In the previous two sections we have discussed oscillations in the magnetoresistance which arise as a result of the beating together of the cyclotron frequency and that associated with the Fermi level. The latter component entails that the Shubnikov–de Haas effect is fundamentally a property of a degenerate electron gas. Oscillations in the magnetoresistance can, however, occur in a non-degenerate electron gas if the dominant scattering mechanism is via optical phonons, for then the phonon energy $\hbar\omega$ replaces E_F. This phenomenon is known as the magnetophonon effect, and it is useful in providing an experiment to determine the optical phonon energy [see for example Stradling and Wood (1968) and Harper *et al.* (1973)]. It was first proposed by Gurevitch and Firsov (1961) for the transverse configuration. Since the magnetophonon effect turns out to be much weaker in the longitudinal configuration we will concentrate on the transverse case.

The conductivity expressed in eqn (7.135) was derived for the case of elastic scattering but it can readily be modified to describe inelastic scattering via absorption or emission of a phonon of energy $\hbar\omega$ with the help of eqn (7.54), which describes energy conservation, and eqn (7.130), which describes the statistical weighting factor. In the case of a non-degenerate gas and small electric field we may write

$$\sigma_{xx} = -\frac{e^2 m^{*1/2} V}{2^{5/2}\pi\hbar^2} \sum_{n,m} \int \frac{\mathrm{d}f(E_i)}{\mathrm{d}E_i} N_n(E_i)\,\mathrm{d}E_i \int_0^\infty \frac{|V_{\pm q_\perp, q_{z\max}}|^2 + |V_{\pm q_\perp, q_{z\min}}|^2}{[E_z + (n-m)\hbar\omega_c \pm \hbar\omega]^{1/2}} \times (n(\omega) + \tfrac{1}{2} \pm \tfrac{1}{2})\,|I_{nm}(X' - X)|^2 q_\perp^3 R^4\,\mathrm{d}q_\perp \quad (7.187)$$

where for absorption

$$\begin{aligned} q_{z\min} &= -k_z\left[\left\{1 + \frac{1}{E_z}((n-m)\hbar\omega_c + \hbar\omega)\right\}^{1/2} + 1\right] \\ q_{z\max} &= k_z\left[\left\{1 + \frac{1}{E_z}((n-m)\hbar\omega_c + \hbar\omega)\right\}^{1/2} - 1\right] \end{aligned} \quad (7.188)$$

and for emission

$$\begin{aligned} q_{z\min} &= k_z\left[1 - \left\{1 + \frac{1}{E_z}((n-m)\hbar\omega_c - \hbar\omega)\right\}^{1/2}\right] \\ q_{z\max} &= k_z\left[1 + \left\{1 + \frac{1}{E_z}((n-m)\hbar\omega_c - \hbar\omega)\right\}^{1/2}\right] \end{aligned} \quad (7.189)$$

and $|I_{nm}(X' - X)|^2$ is given by eqn (7.134). For deformation-potential scattering

$$|V_q|^2 = \frac{\hbar D_0^2}{2\rho\omega V} \quad (7.190)$$

and for polar scattering (unscreened)

$$|V_q|^2 = \frac{e^2\hbar\omega}{2Vq^2}\left(\frac{1}{\epsilon_\infty} - \frac{1}{\epsilon_s}\right). \quad (7.191)$$

The dependence of the latter on wavevector makes the calculation for this case difficult. To begin with we will therefore deal with deformation-potential scattering.

Considering only the case $n(\omega) \ll 1$ and absorption [in the statistical factor of eqn (7.130)] we obtain, using the integration involved in eqn (7.136) and

$$f(E_i) = \frac{n_e \sinh(\hbar\omega_c/2k_B T_e)}{(\pi k_B T_e)^{1/2} N_n(E_z) E_z^{1/2}} \cdot \exp[-(E_z + (n + \tfrac{1}{2})\hbar\omega_c)/k_B T_e] \quad (7.192)$$

$$\sigma_{xx} = \frac{e^2 m^* D_0^2 n_e \sinh(\hbar\omega_c/2k_B T_e)}{2^{5/2} \rho (\pi k_B T_e)^{3/2} \hbar\omega} \sum_{n=0}^{\infty} \sum_{m=0}^{m_{max}}$$

$$\times \int_0^{\infty} \frac{(1 + n + m)\exp[-(E_z + (n + \tfrac{1}{2})\hbar\omega_c)/k_B T_e]}{E_z^{1/2}(E_z + (n - m)\hbar\omega_c + \hbar\omega)^{1/2}} \mathrm{d}E_z \quad (7.193)$$

where n_e is the electron density, and $m_{max} \leqslant n + \hbar\omega/\hbar\omega_c$. This given on integration

$$\sigma_{xx} = \frac{e^2 m^* D_0^2 n_e \sinh(\hbar\omega_c/2k_B T_e)}{2^{5/2} \rho (\pi k_B T_e)^{3/2} \hbar\omega} \sum_{n,m} (1 + n + m)\exp\left[\frac{\hbar\omega + (n - m)\hbar\omega_c}{2k_B T_e}\right]$$

$$\times \exp\left[-\frac{(n + \tfrac{1}{2})\hbar\omega_c}{k_B T_e}\right] K_0\left(\frac{\hbar\omega + (n - m)\hbar\omega_c}{2k_B T_e}\right) \quad (7.194)$$

where $K_0(x)$ is the zero-order modified Bessel function. A divergence occurs for

$$(m - n)\hbar\omega_c = \hbar\omega. \quad (7.195)$$

This divergence is logarithmic since $K_0(x) \to -\ln(x/2)$ as $x \to 0$. In order to obtain a finite result broadening must be incorporated at the outset.

A similar result can be obtained for polar scattering if it is assumed that $q_z \ll q_\perp$. In this case the dependence on q is no problem and we obtain

$$\sigma_{xx} = \frac{e^4 m^* \omega n_e \sinh(\hbar\omega_c/2k_B T_e)}{2^{5/2} (\pi k_B T_e)^{3/2} \hbar} \left(\frac{1}{\epsilon_\infty} - \frac{1}{\epsilon_s}\right) \sum_{n,m} \exp\left[\frac{\hbar\omega + (n - m)\hbar\omega_c}{2k_B T_e}\right]$$

$$\times \exp\left[-\frac{(n + \tfrac{1}{2})\hbar\omega_c}{k_B T_e}\right] K_0\left(\frac{\hbar\omega + (n - m)\hbar\omega_c}{2k_B T_e}\right). \quad (7.196)$$

Use of the Poisson formula leads to the form (Barker 1972)

$$\sigma_{xx} = \sigma_0\left(1 + \sum_{r=1}^{\infty} r^{-1} \exp(-2\pi r\Gamma/\omega_c)\cos(2\pi r\omega/\omega_c)\right) \quad (7.197)$$

where Γ is the scattering rate associated with the broadening of the energy levels near $E_z = 0$.

In view of the uncertainties and difficulties that enter the theory it is usually assumed that eqn (7.197), if not totally accurate, provides a good empirical formula to describe the effect, and this indeed appears to be borne out in experiment (Fig. 7.8).

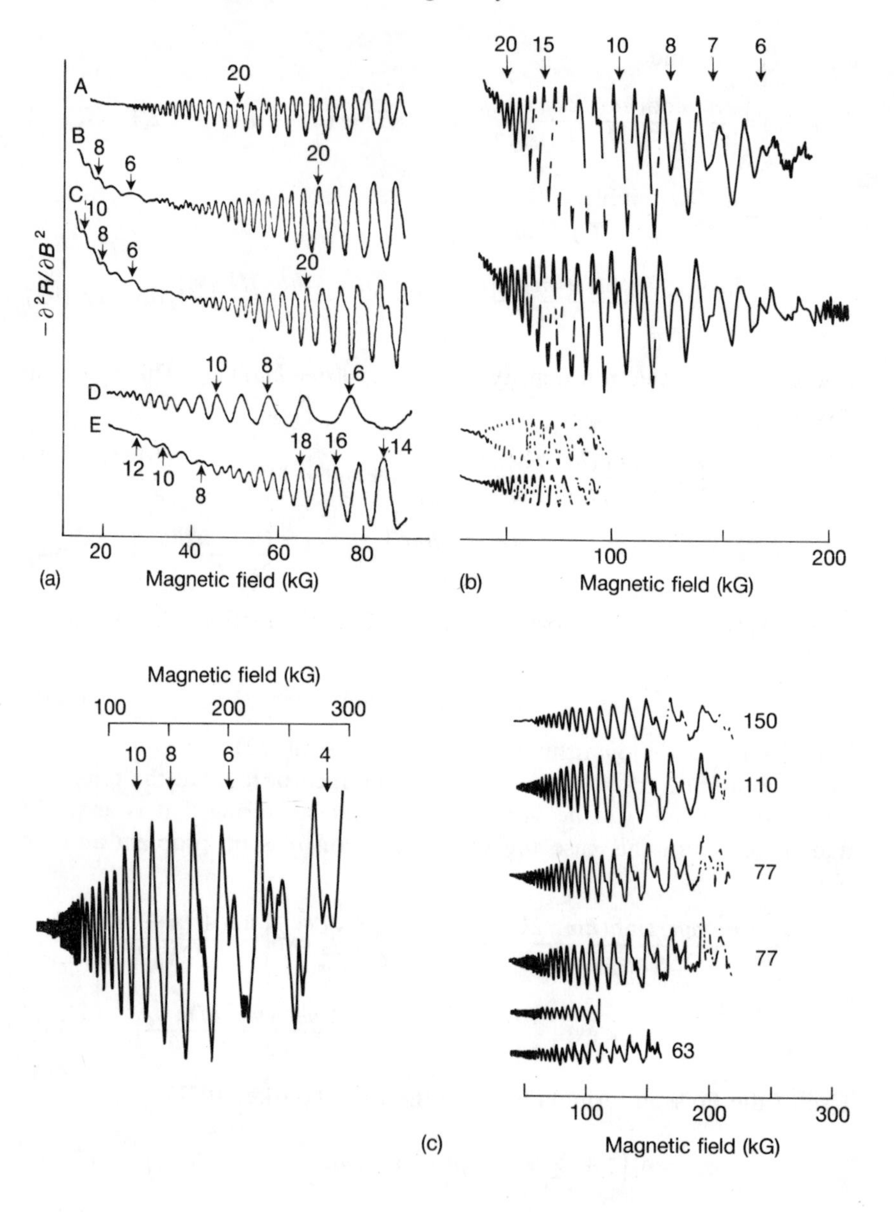

FIG. 7.8. Experimental recordings of magnetophonon peaks observed in the second derivative of the transverse magnetoresistance of germanium at about 120 K. (*a*) A p-Ge, $B \parallel \langle 100 \rangle$; B p-Ge, $B \parallel \langle 111 \rangle$; C p-Ge, $B \parallel \langle 110 \rangle$; D n-Ge, $B \parallel \langle 100 \rangle$; E n-Ge, $B \parallel \langle 110 \rangle$; (*b*) pulsed field measurements on p-Ge with $B \parallel \langle 100 \rangle$ showing that $N = 6$ peak splits into three components; (*c*) pulsed field measurements with p-Ge with $B \parallel \langle 111 \rangle$. (Harper *et al.* 1973.)

When the electric field is taken into account the magnetoresistance peaks split into two, one above and one below the small-field position. According to Mori *et al.* (1987) the expression becomes, for small electric fields

$$\sigma_{xx} = \sigma_0\Big(1 + \sum_{r=1}^{\infty} r^{-1}\exp(-2\pi r\Gamma/\omega_c)$$

$$\times \{\cos(2\pi r[(\omega/\omega_c) + \gamma(\xi)]) + \cos(2\pi r[(\omega/\omega_c) - \gamma(\xi)])\}\Big) \quad (7.198)$$

where

$$\gamma(\xi) = \sqrt{\left(\frac{3}{2}\right)} \cdot \frac{e\xi R}{\hbar\omega_c}\sqrt{[(\omega/\omega_c) + 1]}. \quad (7.199)$$

References

ADAMS, E. N. and HOLSTEIN, T. D. (1959). *J. Phys. Chem. Solids* **10,** 254.
ARGYRES, P. N. (1958). *J. Phys. Chem. Solids* **4,** 19.
BARKER, J. R. (1972). *J. Phys. C. Solid State Phys.* **5,** 1657.
BAUER, G. and KAHLERT, H. (1972). *Phys. Rev. B* **5,** 566
DAVYDOV, B. and POMERANCHUK, I. (1940). *J. Phys. Moscow* **2,** 147.
DINGLE, R. B. (1952). *Proc. R. Soc.* **A211,** 517.
GRADSHTEIN, I. S. and RYZHIK, I. M. (1965). *Tables of Integrals, Series and Products* Academic Press, New York.
GUREVITCH, V. L. and FIRSOV, Yu. A. (1961). *Sov. Phys. J.E.T.P.* **13,** 137.
HARPER, P. G., HODBY, J. W., and STRADLING, R. A. (1973). *Rep. Progr. Phys.* **37,** 1.
KUBO, R., MIYAKE, S. J., and HASHITSUME, N. (1965). *Solid State Phys.* (ed. Seitz and Turnbull) p. 270.
LAX, B., ROTH, L. M., and ZWERDLING, S. (1958). *J. Phys. Chem. Solids* **8,** 311.
MORI, N., NAKAMURA, N., TANIGUCHI, K. and HAMAGUCHI, C. (1987). *J. Phys. Soc. Jpn.* (in press).
RIDLEY, B. K. (1983). *J. Phys. C: Solid State Phys.* **16,** 2261.
STAROMYLNSKA, J., FINLAYSON, D. M., and STRADLING, R. A. (1983). *J. Phys. C: Solid State Phys.* **16,** 6373.
STRADLING, R. A. and WOOD, R. A. (1968). *Solid State Commun.* **6,** 701.
TITEICA, S. (1935). *Ann. Phys. Leipzig* **22,** 128.

8. Scattering in a degenerate gas

8.1. General equations

MANY of the foregoing sections on scattering have dealt with the situation in which the occupation probability of the final state in a scattering process is negligible. Above a concentration of carriers determined by the effective density of states in the band the situation is changed radically by the operation of Pauli exclusion. Scattering rates become dependent on the probability of occupancy of the final state, and statistics cannot be ignored in calculating them. But if high concentrations produce this complexity they also allow us to assume that in many cases carrier–carrier interactions are strong enough for the distribution function to maintain the Fermi–Dirac form, even for a temperature above that of the lattice in the case of hot electrons. This allows scattering rates to be formulated in terms of electron temperature rather than electron energy.

For degenerate systems eqn (3.16) becomes

$$\begin{aligned}\frac{\mathrm{d}f(\mathbf{k})}{\mathrm{d}t} = &\int [W(\mathbf{k}'', \mathbf{k})f(\mathbf{k}'')(1-f(\mathbf{k})) - W(\mathbf{k}, \mathbf{k}'')f(\mathbf{k})(1-f(\mathbf{k}''))] \\ &\times \delta(E_{\mathbf{k}''} - E_{\mathbf{k}} - \hbar\omega_{\mathbf{q},b})\, \mathrm{d}\mathbf{k}'' \\ &+ \int [W(\mathbf{k}', \mathbf{k})f(\mathbf{k}')(1-f(\mathbf{k})) - W(\mathbf{k}, \mathbf{k}')f(\mathbf{k})(1-f(\mathbf{k}'))] \\ &\times \delta(E_{\mathbf{k}'} - E_{\mathbf{k}} + \hbar\omega_{\mathbf{q},b})\, \mathrm{d}\mathbf{k}'. \end{aligned} \tag{8.1}$$

Expanding the distribution function in spherical harmonics and retaining the first two terms we obtain

$$f(\mathbf{k}) = f_0(E) + f_1(E)\cos\theta, \qquad f_1(E) \ll f_0(E) \tag{8.2}$$

where θ is the angle between $\mathbf{k}$ and the force applied to the electron by external fields. If θ' is the corresponding angle for $\mathbf{k}'$ we may effectively make the substitution

$$\cos\theta' = \cos\theta\cos\alpha' \tag{8.3}$$

where α' is the angle between $\mathbf{k}$ and $\mathbf{k}'$. In eqn (8.3) the term involving sine has been suppressed since it will give zero contribution in subsequent integrations.

Substitution of the $f(\mathbf{k})$ using eqns (8.2) and (8.3) splits eqn (8.1) into two equations (since spherical harmonics are mutually orthogonal functions), viz.

$$\frac{df_0(E)}{dt} = \int [W(\mathbf{k}'', \mathbf{k}) f_0(E'')(1 - f_0(E) - W(\mathbf{k}, \mathbf{k}'') f_0(E)(1 - f_0(E''))]$$
$$\times\, \delta(E_{\mathbf{k}''} - E_{\mathbf{k}} - \hbar\omega_{\mathbf{q}})\, d\mathbf{k}''$$
$$+ \int [W(\mathbf{k}', \mathbf{k}) f_0(E')(1 - f_0(E)) - W(\mathbf{k}, \mathbf{k}') f_0(E)(1 - f_0(E'))]$$
$$\times\, \delta(E_{\mathbf{k}'} - E_{\mathbf{k}} + \hbar\omega_{\mathbf{q}})\, d\mathbf{k}'. \tag{8.4}$$

$$\frac{df_1(E)}{dt} = \int [W(\mathbf{k}'', \mathbf{k})(f_1(E'')(1 - f_0(E))\cos\alpha'' - f_0(E'')f_1(E))$$
$$+ W(\mathbf{k}, \mathbf{k}'')(f_1(E'')f_0(E)\cos\alpha'' - f_1(e)(1 - f_0(E'')))]$$
$$\times\, \delta(E_{\mathbf{k}''} - E_{\mathbf{k}'} - \hbar\omega_{\mathbf{q}})\, d\mathbf{k}''$$
$$+ \int [W(\mathbf{k}', \mathbf{k})(f_1(E')(1 - f_0(E))\cos\alpha' - f_0(E')f_1(E))$$
$$+ W(\mathbf{k}, \mathbf{k}')(f_1(E')f_0(E)\cos\alpha' - f_1(E)(1 - f_0(E')))]$$
$$\times\, \delta(E_{\mathbf{k}'} - E_{\mathbf{k}} + \hbar\omega_{\mathbf{q}})\, d\mathbf{k}. \tag{8.5}$$

Equation (8.4) is related to energy relaxation and eqn (8.5) to momentum relaxation. Thus, the power input per unit volume, P, is given by

$$P = \frac{d}{dt}\left[\int E f_0(E) N(E)\, dE\right] \tag{8.6}$$

which, at steady state, equals the energy relaxation rate. From the Boltzmann equation for a uniform system with electric field ξ at steady state

$$-\frac{\bar{e}\xi}{\hbar} \cdot \nabla_{\mathbf{k}} f(\mathbf{k}) + \frac{df(\mathbf{k})}{dt} = 0 \tag{8.7}$$

we can obtain for the momentum balance

$$\frac{df_1(E)}{dt} = \bar{e}\xi v(E) \frac{df_0(e)}{dE} \tag{8.8}$$

where $\bar{e}$ is the elementary charge containing the sign of the charge, and $v(E)$ is the group velocity of the particle.

Finally, we may assume that electron–electron scattering is strong, and that the spherical part of the distribution function maintains the Fermi–Dirac form:

$$f_0(E) = \frac{1}{1 + \exp[(E - E_F)/k_B T_e]} \tag{8.9}$$

where E_F is the Fermi energy and T_e is the electron temperature. As a consequence of this form it is useful to note that

$$\frac{\mathrm{d}f_0(E)}{\mathrm{d}E} = -\frac{f_0(E)(1-f_0(E))}{k_B T_e} \tag{8.10}$$

and that this strongly peaked at the Fermi surface and small elsewhere, especially at low temperatures.

8.2. Elastic collisions

Strongly inelastic collisions such as those involving optical and zone-edge phonons are affected by degeneracy in a fairly straightforward way. To a good approximation, the better the lower the temperature. Absorption processes are possible only for electrons lying no deeper in the Fermi sea than $\hbar\omega_{\mathbf{q}}$ from the Fermi level, and emission processes are possible for electrons only with energies at least $\hbar\omega_{\mathbf{q}}$ above the Fermi level. The rates are then given by the expressions derived in Chapter 3. Elastic and approximately elastic processes, however, cannot be treated in this way since both initial and final states are close to the Fermi surface, and it is to these processes we turn.

Truly elastic processes such as impurity scattering are easy to deal with. The right-hand side of eqn (8.4) is then clearly zero and eqn (8.5) becomes

$$\frac{\mathrm{d}f_1(E)}{\mathrm{d}t} = \int W(\mathbf{k}', \mathbf{k})(\cos\alpha' - 1)f_1(E)\,\delta(E_{\mathbf{k}'} - E_{\mathbf{k}})\,\mathrm{d}\mathbf{k}'. \tag{8.11}$$

This defines a momentum-relaxation time thus:

$$\frac{1}{\tau_m(E)} = \int W(\mathbf{k}', \mathbf{k})(1 - \cos\alpha')\,\delta(E_{\mathbf{k}'} - E_{\mathbf{k}})\,\mathrm{d}\mathbf{k}' \tag{8.12}$$

which is the same as for a non-degenerate system. Equation (8.8) becomes

$$f_1(E) = -\bar{e}\xi v(E)\tau_m(E)\frac{\mathrm{d}f_0(E)}{\mathrm{d}E}. \tag{8.13}$$

When the current is worked out the peakiness of $\mathrm{d}f_0(E)/\mathrm{d}E$ constrains all other energy dependent, but less rapidly varying, quantities to take their values at the Fermi surface. Thus the formulae for elastic scattering worked out for a non-degenerate system can be used with $k = k_F$, the Fermi wavevector, and $E = E_F$.

8.3. Acoustic phonon scattering

Collisions with acoustic modes involve an energy exchange. If this energy exchange is small compared with $k_B T_e$, such collisions can be treated like impurity scattering and the rates are just those at the Fermi surface. But at low temperatures $k_B T_e$ becomes comparable with the energy exchanged and the situation becomes more complicated. Acoustic phonons which scatter electrons right across the Fermi surface have wavevectors around $2k_F$. These may be substantial enough for the corresponding phonon energy to be greater than $k_B T_e$, with the result that, for emission, Pauli exclusion tends to inhibit the transition, and for absorption few phonons are excited (Fig. 8.1). The effect introduces an added temperature dependence into the scattering rate—as the temperature lowers there is an increasing tendency for large-angle scattering to disappear, with the result that momentum relaxation becomes strikingly weaker.

As in the case of truly elastic scattering we resort to the calculation of the rates averaged over the Fermi–Dirac distribution and rely on the peaky quality of functions like $f(E)(1-f(E))$ at the Fermi surface to allow the integrals to be approximately evaluated. Since the collisions are nearly elastic we can define a momentum relaxation time as was done in eqn (8.13) and substitute for the $f_1(E)$ in eqn (8.5). We then, as a first step, integrate over E', exploiting the delta function, and then over E

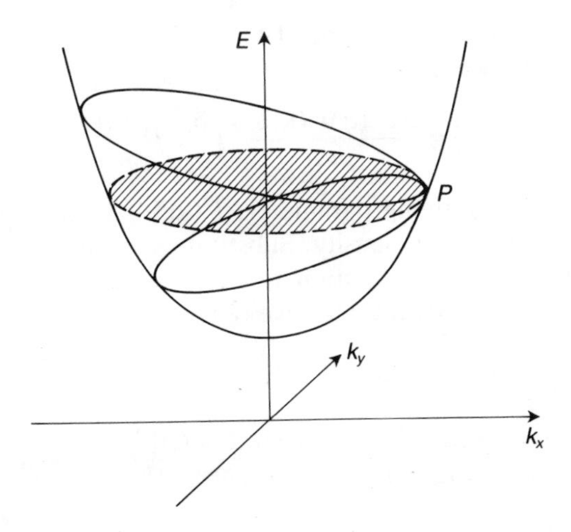

FIG. 8.1. Inelastic scattering by acoustic phonons. The shaded circle is the Fermi surface in the k_x k_y plane. States to which the particle at P scatters by absorption and emission are depicted by the upper and lower circles respectively. (The energies of the acoustic phonons have been exaggerated enormously for clarity of presentation.)

using the identities

$$\int_{-\infty}^{+\infty} \frac{\exp(-x)}{[\exp(\hbar\omega/k_B T_e)+\exp(-x)][1+\exp(-x)]}\,dx = \frac{\hbar\omega/k_B T_e}{\exp(\hbar\omega/k_B T_e)-1}$$

$$\int_{-\infty}^{+\infty} \frac{x\exp(-x)}{[\exp(\hbar\omega/k_B T_e)+\exp(-x)][1+\exp(-x)]}\,dx = \frac{1}{2}\frac{(\hbar\omega/k_B T_e)^2}{\exp(\hbar\omega/k_B T_e)-1}. \tag{8.14}$$

We rely on the form of the functions like $f(E\pm\hbar\omega)(1-f(E))$ to allow the limits of the integration to be $\pm\infty$, and take all other functions of energy in the integrand to have their values at the Fermi surface. Equations (8.5) and (8.6) with eqn (8.13) then provide expressions for the momentum and energy relaxation rates

$$\frac{1}{\tau_m(E_F)} = \frac{1}{2k_B T_e}\int\left[\frac{W(\mathbf{k}',\mathbf{k})\exp(-\hbar\omega/2k_B T_e)+W(\mathbf{k},\mathbf{k}')\exp(\hbar\omega/2k_B T_e)}{\sinh(\hbar\omega/2k_B T_e)}\right] \times \hbar\omega(1-\cos\alpha')\frac{V}{8\pi^3}\frac{d\mathbf{k}'}{dE'} \tag{8.15}$$

$$P = N(E_F)\int\left[\frac{W(\mathbf{k}',\mathbf{k})\exp(-\hbar\omega/2k_B T_e)-W(\mathbf{k},\mathbf{k}')\exp(\hbar\omega/2k_B T_e)}{\sinh(\hbar\omega/2k_B T_e)}\right] \times (\hbar\omega)^2\frac{V}{8\pi^3}\frac{d\mathbf{k}'}{dE'}. \tag{8.16}$$

Small differences in the contributions associated with scattering from above and scattering from below have been ignored. (For convenience the subscript on ω has been dropped.)

From section 3.1 we have

$$\left.\begin{matrix} W(\mathbf{k}',\mathbf{k}) \\ W(\mathbf{k},\mathbf{k}') \end{matrix}\right\} = \frac{\pi C^2(q) I^2(\mathbf{k},\mathbf{k}')}{\rho\omega V}\left\{\begin{matrix} \delta_{\mathbf{k}-\mathbf{q},\mathbf{k}'}(n(\omega)+1) \\ \delta_{\mathbf{k}+\mathbf{q},\mathbf{k}'}n(\omega) \end{matrix}\right. \tag{8.17}$$

where $C^2(q)$ is the coupling parameter, $I^2(\mathbf{k},\mathbf{k}')$ is the squared overlap integral, and ρ is the mass density. Substituting in eqns (8.15) and (8.16) and taking the phonon occupation to be determined by thermal equilibrium at a lattice temperature T_L, we obtain

$$\frac{1}{\tau_m(E_F)} = \frac{\hbar}{16\pi^2\rho k_B T_e}\int C^2(q) I^2(\mathbf{k},\mathbf{k}') \times\frac{\cosh[(\hbar\omega/2k_B T_L)-(\hbar\omega/2k_B T_e)]}{\sinh(\hbar\omega/2k_B T_L)\,.\,\sinh(\hbar\omega/2k_B T_e)}(1-\cos\alpha')\frac{d\mathbf{k}'}{dE'} \tag{8.18}$$

$$P = \frac{\hbar N(E_F)}{16\pi^2\rho}\int C^2(q) I^2(\mathbf{k},\mathbf{k}') \times\frac{\sinh[(\hbar\omega/2k_B T_L)-(\hbar\omega/2k_B T_e)]}{\sinh(\hbar\omega/2k_B T_L)\,.\,\sinh(\hbar\omega/2k_B T_e)}\,.\,\hbar\omega\frac{d\mathbf{k}'}{dE'}. \tag{8.19}$$

Note that eqn (8.19) can be derived directly (with change of sign) from the energy relaxation rate, defined as follows:

$$\frac{\mathrm{d}E}{\mathrm{d}t} = \int A\hbar\omega[n(\omega)(1-f(E+\hbar\omega)) - (n(\omega)+1)(1-f(E-\hbar\omega))]f(E)N(E)\,\mathrm{d}E \quad (8.20)$$

where A represents all the other parameters which enter. Conservation of crystal momentum entails that

$$1-\cos\alpha' = \frac{q^2}{2k_\mathrm{F}^2}. \quad (8.21)$$

In general these integrals have to be solved numerically. However it is useful to obtain explicit expressions by making certain approximations. For parabolic, spherical bands we can take the overlap integral to be unity and

$$\frac{\mathrm{d}E'}{\mathrm{d}k'} = \frac{\hbar^2 k_\mathrm{F}}{m^*}. \quad (8.22)$$

Thus, using eqn (8.21), we obtain

$$\frac{\mathrm{d}\mathbf{k}'}{\mathrm{d}E'} = \frac{m^*}{\hbar^2 k_\mathrm{F}} \cdot k_\mathrm{F}^2\,\mathrm{d}(-\cos\alpha')\,\mathrm{d}\phi = \frac{m^*}{\hbar^2 k_\mathrm{F}} \cdot q\,\mathrm{d}q\,\mathrm{d}\phi. \quad (8.23)$$

In the absence of screening the coupling parameters are (Sections 3.3 and 3.6)

$$C^2(q) = \begin{cases} \Xi^2 q^2 & \text{non-polar} \\ \dfrac{e^2 K_\mathrm{av}^2 c_\mathrm{L}}{\epsilon} & \text{piezoelectric.} \end{cases} \quad (8.24)$$

Evaluation of the integral can be carried out analytically in the low-temperature limit, i.e. $\hbar\omega/k_\mathrm{B}T_\mathrm{e} > 1$, and in the high-temperature limit $\hbar\omega/k_\mathrm{B}T_\mathrm{L} < 1$, and $T_\mathrm{e} \geqslant T_\mathrm{L}$ is implied.

8.3.1. Low-temperature limit

At low temperatures we can approximate the hyperbolic sine in the denominator by the exponential function and use the identities

$$\int_0^\infty x^{\mu-1}\exp(-\beta x)\cosh\gamma x\,\mathrm{d}x = \tfrac{1}{2}\Gamma(\mu)[(\beta-\gamma)^{-\mu} + (\beta+\gamma)^{-\mu}]$$

$$\int_0^\infty x^{\mu-1}\exp(-\beta x)\sinh\gamma x\,\mathrm{d}x = \tfrac{1}{2}\Gamma(\mu)[(\beta-\gamma)^{-\mu} - (\beta+\gamma)^{-\mu}]. \quad (8.25)$$

We obtain for deformation-potential scattering

$$\frac{1}{\tau_{\rm m}(E_F)}=\frac{15\Xi^2 m^*}{\pi\rho\hbar^7 v_s^6 k_F^3}\left[\frac{(k_{\rm B}T_{\rm e})^6+(k_{\rm B}T_{\rm L})^6}{k_{\rm B}T_{\rm e}}\right], \tag{8.26}$$

$$P=\frac{6\Xi^2 m^{*2}}{\pi^3\rho\hbar^7 v_s^4}[(k_{\rm B}T_{\rm e})^5-(k_{\rm B}T_{\rm L})^5], \tag{8.27}$$

and for piezoelectric scattering

$$\frac{1}{\tau_{\rm m}(E_{\rm F})}=\frac{3e^2 m^* e_{14}^2}{4\pi\hbar^5 k_{\rm F}^3\epsilon^2\rho v_{sT}^4}\left(\frac{16}{35}+\frac{12v_{sT}^4}{35v_{s\rm L}^4}\right)\left[\frac{(k_{\rm B}T_{\rm e})^4+(k_{\rm B}T_{\rm L})^4}{k_{\rm B}T_{\rm e}}\right], \tag{8.28}$$

$$P=\frac{e^2 m^{*2} e_{14}^2}{2\pi^3\hbar^5\epsilon^2 v_{sT}^2}\left(\frac{16}{35}+\frac{12v_{sT}^2}{35v_{s\rm L}^2}\right)[(k_{\rm B}T_{\rm e})^3-(k_{\rm B}T_{\rm L})^3]. \tag{8.29}$$

When $T_{\rm e}=T_{\rm L}$, eqn (8.26) exhibits the Gruneisen–Bloch T^5-dependence familiar in metal physics (see Ziman (1963); also Kogan (1963)).

The low-temperature regime is entered as soon as the maximum allowed phonon energy equals $k_{\rm B}T_{\rm e}$, viz:

$$\frac{2\hbar v_s k_{\rm F}}{k_{\rm B}T_{\rm e}}=1. \tag{8.30}$$

In the case of GaAs for an electron density of $5\times10^{17}\,{\rm cm}^{-3}$ the above equality holds when $T_{\rm e}\sim 18\,{\rm K}$.

8.3.2. High-temperature limit

For temperatures much higher than the demarcation implied by eqn (8.30), but not so high that degeneracy is significantly weakened, the integration becomes straightforward.

For deformation-potential scattering we obtain

$$\frac{1}{\tau_{\rm m}(E_{\rm F})}=\frac{\Xi^2 m^* k_{\rm F} k_{\rm B}T_{\rm L}}{\pi\rho v_s^2\hbar^3} \tag{8.31}$$

$$P=\frac{\Xi^2 m^{*2}k_{\rm F}^4}{\pi^3\rho\hbar^3}(k_{\rm B}T_{\rm e}-k_{\rm B}T_{\rm L}) \tag{8.32}$$

and for piezoelectric scattering:

$$\frac{1}{\tau_{\rm m}(E_{\rm F})}=\frac{e^2 K_{\rm av}^2 m^* k_{\rm B}T_{\rm L}}{2\pi\epsilon\hbar^3 k_{\rm F}} \tag{8.33}$$

$$P=\frac{2e_{14}^2 e^2 m^{*2}k_{\rm F}^2}{5\rho\epsilon^2\pi^3\hbar^3}(k_{\rm B}T_{\rm e}-k_{\rm B}T_{\rm L}). \tag{8.34}$$

The temperature-dependence of the Fermi wavevector is given by

$$k_{\rm F}(T_{\rm e})\approx k_{\rm F}(0)\left[1-\frac{\pi^2}{24}\left(\frac{k_{\rm B}T_{\rm e}}{E_{\rm F}(0)}\right)^2\right] \tag{8.35}$$

and $k_F(0)$ is related to carrier density via

$$n = \frac{k_F^3(0)}{3\pi^2}. \tag{8.36}$$

The momentum relaxation rates are identical in form to the non-degenerate expression of eqns (3.78) and (3.180), the latter in the limit of large energy. In other words, in the high-temperature limit the collisions can be regarded as elastic as far as momentum relaxation is concerned and treated as discussed in Section 8.2. The same is true of energy relaxation. For instance, if eqn (3.110) is averaged over a Maxwell–Boltzmann distribution characterized by a temperature T_e the equivalent of eqn (8.32) is recovered.

8.3.3. Strong screening

The effect of screening by the mobile carriers will increase towards low temperatures, as scattering of acoustic modes involves smaller and smaller wavevectors. In this case the coupling parameters become

$$C^2(q) = \begin{cases} \dfrac{\Xi^2 q^2}{\left[1 + (q_0^2/q^2)\right]^2} \xrightarrow{q \to 0} \Xi^2 \dfrac{q^6}{q_0^4} & \text{non-polar} \\[2ex] \dfrac{e^2 K_{\text{av}}^2 c_L}{\epsilon\left[1 + (q_0^2/q^2)\right]^2} \xrightarrow{q \to 0} \dfrac{e^2 K_{\text{av}}^2 c_L q^4}{\epsilon q_0^4} & \text{polar.} \end{cases} \tag{8.37}$$

where q_0 is the reciprocal screening length in the static screening approximation (see Sections 4.9 and 9.5). Repeating the calculation for the non-polar interaction gives

$$\frac{1}{\tau_m(E_F)} = \frac{\Gamma(10)\Xi^2 m^*}{8\pi\rho\hbar^{11} v_{sL}^{10} k_F^3 q_0^4} \left[\frac{(k_B T_e)^{10} + (k_B T_L)^{10}}{k_B T_e}\right], \tag{8.38}$$

$$P = \frac{\Gamma(9)\Xi^2 m^{*2}}{4\pi^3 \rho\hbar^{11} v_{sL}^8 q_0^4} \left[(k_B T_e)^9 - (k_B T_L)^9\right], \tag{8.39}$$

and the polar interaction

$$\frac{1}{\tau_m(E_F)} = \frac{\Gamma(8) e^2 m^*}{8\pi\rho\hbar^9 v_{sT}^8 k_F^3 q_0^4} \left(\frac{e_{14}}{\epsilon}\right)^2 \left(\frac{16}{35} + \frac{12 v_{sT}^8}{35 v_{sL}^8}\right) \left[\frac{(k_B T_e)^8 + (k_B T_L)^8}{k_B T_e}\right], \tag{8.40}$$

$$P = \frac{\Gamma(7) e^2 m^{*2}}{4\pi^3 \rho\hbar^9 v_{sT}^6 q_0^4} \left(\frac{e_{14}}{\epsilon}\right)^2 \left(\frac{16}{35} + \frac{12 v_{sT}^6}{35 v_{sL}^6}\right) \left[(k_B T_e)^7 - (k_B T_L)^7\right]. \tag{8.41}$$

In these equations

$$q_0^2 = -\frac{e^2}{\epsilon} \sum_i N_i \frac{df(E_i)}{dE_i}, \tag{8.42}$$

(see eqn (4.149)). Taking the sum over band states only we can exploit the delta-function quality of $df(E_i)/dE_i$ by replacing the sum with an

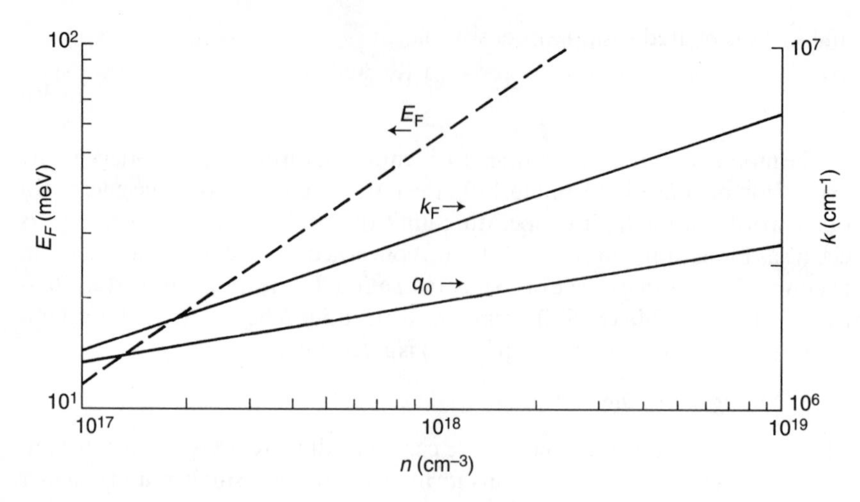

FIG. 8.2. Fermi energy, Fermi wavevector, and screening wavevector for electrons in GaAs at $T = 0$, assuming a parabolic band.

integral, viz.

$$q_0^2 = -\frac{e^2}{\epsilon}\int_0^\infty N(E)\frac{\mathrm{d}f(E)}{\mathrm{d}E}\cdot \mathrm{d}E$$
$$= \frac{e^2}{\epsilon}N(E_F) \tag{8.43}$$

and obtain the Thomas–Fermi screening parameter.

Figure 8.2 shows how k_F and q_0 compare in magnitude for the case of n-GaAs at $T = 0$ K. Clearly, screening cannot be neglected at very low temperatures. At elevated temperatures typical wavevectors will be of order 10^7 cm^{-1} and screening can then be ignored. At moderately low temperatures we may note that in the general case when the unscreened coupling parameter varies as q^n, the strong-screening result may be obtained by multiplying the unscreened result by the factor

$$S = \frac{\Gamma(n+8)}{\Gamma(n+4)}\frac{1}{(\hbar v_{sL,T} q_0)^4}\frac{(k_B T_e)^{n+8} \pm (k_B T_L)^{n+8}}{(k_B T_e)^{n+4} \pm (k_B T_L)^{n+4}} \tag{8.44}$$

where the upper sign is taken for the momentum-relaxation rate, the lower for power. The condition for strong screening implies the $S \ll 1$. Thus S is a measure of effective screening strength for the degenerate, low-temperature case. The phonon energy $\hbar v_s q_0$ is typically 0.5 meV, corresponding to a temperature of about 6 K, but the numerical factor effectively reduces this temperature to about 1 K. Strong screening is thus expected to be effective only at temperatures below 1 K. Measurements, for example of the Shubnikov–de Haas effect, are normally carried out over a temperature range above 1 K, and so we would not expect screening to have a major effect in that case.

8.4. Energy relaxation time

An energy-relaxation time τ_E can be defined within the energy balance equation as follows:

$$\frac{d\langle E\rangle}{dt} = \frac{P}{n} - \frac{[\langle E\rangle - \langle E_0\rangle]}{\tau_E} \tag{8.45}$$

where $\langle E\rangle$ is the average energy of an electron, P/n is the power input per electron, and $\langle E_0\rangle$ is the average energy at equilibrium. Standard theory for a degenerate electron gas gives

$$\langle E\rangle \approx \tfrac{3}{5}E_F(0)\left[1 + \tfrac{5}{12}\pi^2\left(\frac{k_B T_e}{E_F(0)}\right)^2\right] \tag{8.46}$$

where $E_F(0)$ is the Fermi energy at $T = 0$ K, and consequently

$$\langle E\rangle - \langle E_0\rangle = \frac{\pi^2}{4E_F(0)}[(k_B T_e)^2 - (k_B T_L)^2]. \tag{8.47}$$

At steady state we may obtain τ_E from

$$\frac{1}{\tau_E} = \frac{6P}{[(k_B T_e)^2 - (k_B T_L)^2]N(E_F(0))} \tag{8.48}$$

where we have used $n = (2/3)N(E_F(0))E_F(0)$, $N(E_F(0))$ is the density of states at the $T = 0$ K Fermi level. The power, P, may be substituted from eqns (8.27) or (8.29) in the low-temperture limit, from eqns (8.32) or (8.34) in the high-temperature limit, or from eqns (8.39) or (8.41) when screening is strong.

References

KOGAN, S. M. (1963). *Sov. Phys.–Solid St.* **4,** 1813 (F.T.T. **4,** 2474).
ZIMAN, J. M. (1963). *Electrons and Phonons.* Oxford University Press.

9 Dynamic screening

9.1. Introduction

THE scattering of electrons in a semiconductor by a polar interaction is one of the most important physical processes in the subject. The principle sources of polar scattering are charged impurities, piezoelectric modes, holes, other electrons, and optical phonons. All are susceptible to electrical screening by the mobile electron gas. Hitherto in our discussion of these processes we have assumed that screening is effectively instantaneous, any time-dependence of the scattering potential being assumed to be slow enough for the electron gas to respond and form a screening pattern. In other words, the screening has been assumed to be what it would be for a static potential. But no physical effect happens instantaneously. In order to screen, the electrons must move under the influence of the electric field, which is the raw source of the scattering, and because they have inertia, this takes time. Whenever the scattering potential varies in time, screening becomes a dynamic process.

The assumption of static screening is, of course, valid for charged-impurity scattering since the potential is truly static. It turns out that static screening is an adequate description for the screening of the piezoelectric interaction, since the frequencies of acoustic phonons which can interact with electrons are quite low. (This is of course also true for screening the deformation-potential interaction.) It is usually assumed that because carrier–carrier scattering is describable in the static centre-of-mass frame of reference, static screening is also applicable in this case. Where the assumption of static screening truly breaks down is in the description of the interaction between electrons and polar optical phonons, since the frequency of the latter is high. We therefore need to revise critically our previous discussion of scattering by polar optical phonons.

But the dynamic aspect of screening is not the only factor. Screening becomes important only when the density of the electron gas is high, e.g. $10^{17}\,\mathrm{cm}^{-3}$ and above in GaAs. At such densities the collective motion of the electron gas cannot be ignored. Thus dynamic screening and plasma effects become inextricably mixed and have to be treated together. The screening of a polar optical mode involves the forced oscillation of electrons at the phonon frequency ω, which in general is different in magnitude from the natural frequency of the electron gas, i.e. the plasma frequency, ω_P. If $\omega > \omega_P$ phase lags occur which produces an anti-screening effect, with the electrons piling up on the potential peaks instead of the troughs, and the interaction between the polar optical mode and an electron is actually enhanced; whereas if $\omega < \omega_P$ the electrons respond more rapidly, and an approach to the static screening limit can be made. In both cases the screening effect modifies the restorting forces involved in the lattice vibration and leads to a change in frequency. In effect, we have to deal with coupled plasmon/polar-optical-phonon modes.

The problem thus boils down to describing the scattering of an electron by either of two coupled modes, one of them phonon-like, the other plasmon-like. The situation is quite complex, and made more so by the strong electron–plasmon interaction, which blurs the definition of distinct modes in the regime where this interaction is allowed, i.e. the regime of Landau damping. In what follows we will attempt to describe the essential physics of the situation without getting too involved with the inevitably complex theory, some of it fairly impenetrable, which is to be found in the literature of the field. A good account of screening can be found in Harrison's book (1970), and a useful review of coupled modes has been given by Richter (1982).

9.2. Polar optical modes

Following the general approach of Born and Huang (1956) we can write down the equation of motion for ions in the primitive unit cell as follows:

$$\ddot{\mathbf{u}}_L = -\omega_0^2 \mathbf{u}_L + \frac{e_i}{M} \xi_i \tag{9.1}$$

where $\mathbf{u}_L$ is the relative (optical) displacement of the ions, ω_0 is the natural angular frequency (determined by non-polar force constant and mass), e_i is the charge on the ion, M is the reduced mass of the ions, and ξ_i is the electric field in the unit cell. The latter is a superposition of an average field ξ plus a local component associated with the average ionic polarization, $\mathbf{P}_L$, and the shape of the cavity containing the ions, viz.

$$\xi_i = \xi + a\frac{\mathbf{P}_L}{\epsilon_0} \tag{9.2}$$

where 'a' is a numerical factor determined by cavity shape. (For a spherical cavity $a = \frac{1}{3}$). The average ionic polarization is determined partly by electronic polarization and partly by ionic displacement, viz.

$$\mathbf{P}_L = \epsilon_0 \chi \xi_i + \frac{e_i}{V_0} \mathbf{u}_L \tag{9.3}$$

where χ is the susceptibility and V_0 is the volume of the unit cell. As usual ϵ_0 is the permittivity of free space.

These equations describe ionic vibrations with no dependence on wavevector. They are therefore appropriate for describing optical modes with wavevectors near the zone-centre, i.e. long-wavelength modes, which are the ones with which electrons can interact with via the polar interaction. To specify that the modes are longitudinally polarized we must impose the condition that the electrical displacement vanish, viz:

$$\mathbf{D} = \epsilon_0 \boldsymbol{\xi} + \mathbf{P} = 0. \tag{9.4}$$

We need a relationship between $\boldsymbol{\xi}$ and P to insert in eqn (9.4) and describe the allowed modes. This can readily be obtained by putting $\mathbf{u}_{\mathrm{L}} = \mathbf{u}_{\mathrm{L}0} \exp(i\omega t)$ in eqn (9.1) and eliminating ξ_1 and $\mathbf{u}_{\mathrm{L}}$ from eqns (9.1), (9.2) and (9.3) to obtain

$$[(\omega^2 - \omega_0^2)(1 - a\chi) + a\omega_i^2]\mathbf{P}_{\mathrm{L}} = [(\omega^2 - \omega_0^2)\chi - \omega_i^2]\epsilon_0 \xi \tag{9.5}$$

where $\omega_i^2 = e_i^2/MV_0\epsilon_0$. The quantities ω_0, a, χ, and ω_i can now be related to the high-frequency and the low-frequency permittivities, ϵ_∞ and ϵ_{S} respectively, using $\mathbf{D} = \epsilon_\infty \boldsymbol{\xi}$ for $\omega \to \infty$ and $\mathbf{D} = \epsilon_{\mathrm{S}} \boldsymbol{\xi}$ for $\omega \to 0$, with $\mathbf{P} = \mathbf{P}_{\mathrm{L}}$ in each case. We obtain

$$\chi = \frac{(\epsilon_\infty - \epsilon_0)}{a\epsilon_\infty + (1-a)\epsilon_0}, \qquad \omega_i^2 = \omega_0^2 \frac{(\epsilon_{\mathrm{S}} - \epsilon_\infty)\epsilon_0}{[a\epsilon_{\mathrm{S}} + (1-a)\epsilon_0][a\epsilon_\infty + (1-a)\epsilon_0]} \tag{9.6}$$

Defining

$$\omega_{\mathrm{T}}^2 = \omega_0^2 \frac{a\epsilon_\infty + (1-a)\epsilon_0}{a\epsilon_{\mathrm{S}} + (1-a)\epsilon_0} \tag{9.7}$$

we convert eqn (9.5) to the desired relationship

$$(\omega^2 - \omega_{\mathrm{T}}^2)\mathbf{P}_{\mathrm{L}} = [\omega^2(\epsilon_\infty - \epsilon_0) - \omega_{\mathrm{T}}^2(\epsilon_{\mathrm{S}} - \epsilon_0)]\boldsymbol{\xi}. \tag{9.8}$$

It is now useful to introduce the lattice permittivity ϵ_{L} defined by $D = \epsilon_{\mathrm{L}} \boldsymbol{\xi}$, which implies that $\mathbf{P}_{\mathrm{L}} = (\epsilon_{\mathrm{L}} - \epsilon_0)\boldsymbol{\xi}$. From eqn (9.8) we therefore obtain

$$\epsilon_{\mathrm{L}} = \epsilon_\infty \frac{\omega^2 - \omega_{\mathrm{L}}^2}{\omega^2 - \omega_{\mathrm{T}}^2}, \tag{9.9}$$

in which we have introduced the frequency ω_{L}, defined by

$$\omega_{\mathrm{L}}^2 = \frac{\epsilon_{\mathrm{S}}}{\epsilon_\infty} \omega_{\mathrm{T}}^2. \tag{9.10}$$

The condition $\mathbf{D} = 0$ applied to this purely lattice vibration is satisfied by $\epsilon_{\mathrm{L}} = 0$, whence $\omega = \omega_{\mathrm{L}}$. Thus ω_{L} is the angular frequency of the longitudinally polarized polar optical phonon. The condition $\boldsymbol{\xi} = 0$ describes the transversely polarized optical phonon, satisfied when $\omega = \omega_{\mathrm{T}}$ for then $\epsilon_{\mathrm{L}} = \infty$. The relation between longitudinal and transverse frequencies embodied in eqn (9.10) is the Lyddane–Sax–Teller formula. Note that the relationship between field and spatial displacement, which may be derived from eqns (9.1)–(9.7) is,

$$\boldsymbol{\xi} = -\frac{e_L^*}{V_0\epsilon_0}\left(\frac{\omega^2-\omega_T^2}{\omega_L^2-\omega_T^2}\right)\mathbf{u}_L \tag{9.11}$$

where

$$e_L^{*2} = MV_0\epsilon_0^2\omega_L^2\left(\frac{1}{\epsilon_\infty}-\frac{1}{\epsilon_S}\right) \tag{9.12}$$

is the effective charge introduced to describe scattering strength in Chapter 3, as may be seen by putting $\omega = \omega_L$ in eqn (9.11).

9.3. Plasma modes

Continuing the spirit of the previous section we can describe plasma modes by an equation of motion as follows

$$m^*\ddot{\mathbf{u}}_e = -e\boldsymbol{\xi} - \mathbf{F}_D + \mathbf{F}_P \tag{9.13}$$

where $\mathbf{u}_e$ is the electron displacement, m^* is the effective mass, $\mathbf{F}_D$ is a damping force, and $\mathbf{F}_P$ is the pressure force of the electron gas. In a relaxation-time approximation

$$\mathbf{F}_D = \Gamma\dot{\mathbf{u}}_e \tag{9.14}$$

where Γ is a reciprocal time-constant, whose significance we will discuss later. The pressure force is

$$\mathbf{F}_P = -\frac{1}{n_0}\,\mathrm{grad}\,p \tag{9.15}$$

where n_0 is the uniform density and p is the pressure. Kinetic theory gives for the latter

$$p = \tfrac{1}{3}nm^*v^2 \tag{9.16}$$

where v^2 is the mean square velocity of the electrons, to be regarded as a spatially uniform quantity. The gradient of density, n, can be related to the electron field via Gauss's theorem, and for a plane wave, wavevector $\mathbf{q}$ aligned along $\boldsymbol{\xi}$, we may express the pressure force as follows

$$\mathbf{F}_P = -\frac{m^*v^2\epsilon_0 q^2}{3en_0}\boldsymbol{\xi} = -\frac{ev^2q^2}{3\kappa_\infty\omega_P^2}\boldsymbol{\xi} \tag{9.17}$$

where we have introduced the plasma frequency

$$\omega_P^2 = \frac{e^2n_0}{\epsilon_\infty m^*} \tag{9.18}$$

and the high-frequency dielectric constant $\kappa_\infty = \epsilon_\infty/\epsilon_0$. Equation (9.13) then yields

$$\mathbf{u}_e = \frac{e}{m^*} \cdot \frac{(\omega_P^2 + \omega_q^2)}{\omega_P^2 \omega(\omega - i\Gamma)} \cdot \boldsymbol{\xi} \tag{9.19}$$

where we have introduced the frequency $\omega_q^2 = v^2 q^2 / 3\kappa_\infty$.

The electronic polarization is given by $\mathbf{P}_e = -en_0\mathbf{u}_e$ and so

$$\mathbf{P}_e = -\frac{(\omega_P^2 + \omega_q^2)}{\omega(\omega - i\Gamma)} \epsilon_\infty \boldsymbol{\xi} \tag{9.20}$$

whence an electronic contribution to the permittivity may be derived viz.

$$\epsilon_e = \epsilon_0 - \epsilon_\infty \frac{(\omega_P^2 + \omega_q^2)}{\omega(\omega - i\Gamma)}. \tag{9.21}$$

For pure undamped plasma modes ($\Gamma = 0$), $\epsilon_e = 0$, and so

$$\omega^2 = \kappa_\infty(\omega_P^2 + \omega_q^2). \tag{9.22}$$

For long wavelengths $(q \to 0)\omega^2 \to e^2 n_0/\epsilon_0 m^*$, which is the free-space plasma frequency.

The parameter Γ, which describes the damping, will in general be dependent on wavevector. For plasma modes whose frequency and wavevector are such that they cannot interact with individual electrons, interaction being forbidden by the conditions of momentum and energy conservation described in Section 3.2, Γ is the scattering rate associated with scattering via impurities and acoustic phonons, etc. Well-defined plasma modes are possible provided $\omega > \Gamma$, a condition normally satisfied for the moderate to high carrier densities envisaged here. On the other hand, when single-particle excitation is possible, the modes can be severely damped. In this case Γ is high and well-defined modes with a plasma character cease to exist.

9.4. Coupled modes

Let us now couple together the plasma and polar-optical modes. The total permittivity is given by

$$\epsilon_T = \epsilon_L + \epsilon_e - \epsilon_0 \tag{9.23}$$

which, for longitudinal modes, must vanish. Substituting from eqns (9.9) and (9.21) and setting $\Gamma = 0$, we obtain

$$\omega^4 - \omega^2(\omega_L^2 + \omega_P^2 + \omega_q^2) + \omega_T^2(\omega_P^2 + \omega_q^2) = 0. \tag{9.24}$$

Figure 9.1 illustrates the solutions for $q = 0$. At low electron densities $\omega_+ \approx \omega_L$ and $\omega_- \approx \omega_P(\epsilon_\infty/\epsilon_S)^{1/2}$ corresponding to a pure optical mode

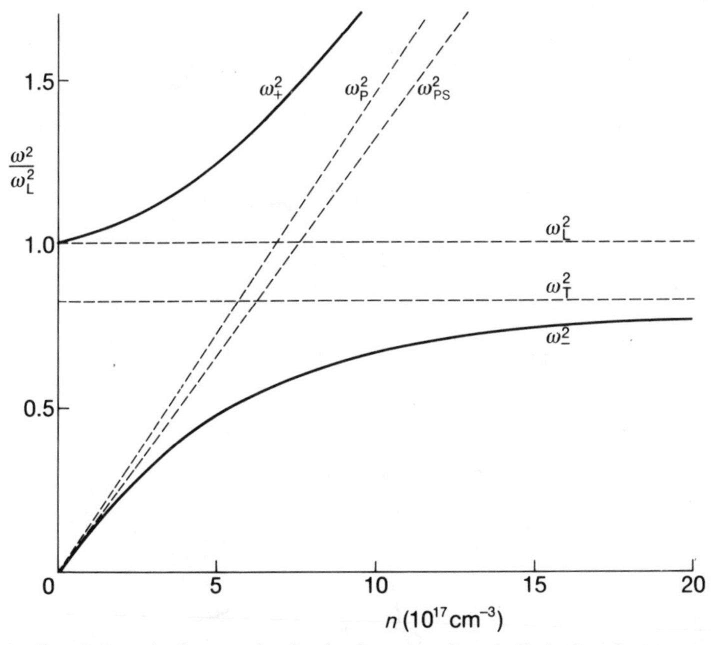

FIG. 9.1. Coupled-mode frequencies in the long-wavelength limit for electrons and LO phonons in GaAs.

plus a pure low-frequency plasma mode. At high densities, $\omega_+ \approx \omega_P$ and $\omega_- \approx \omega_T$, corresponding to a pure high-frequency plasma mode and a totally screened optical mode which, as consequence of screening, oscillates at the transverse-mode frequency. At intermediate densities the modes assume a mixed plasma/optical-mode character.

The scattering strength of these coupled modes can be obtained using the usual concept of effective charge plus a new concept, namely, the effective displacement. Thus we may put

$$\mathbf{P} = \frac{e^*}{V_0}\mathbf{u} \tag{9.25}$$

where e^* is the coupled-mode effective charge and $\mathbf{u}$ is the effective displacement. The latter may be chosen so that the mechanical energy is the sum of the ionic and electronic components. Thus

$$Mu^2 = Mu_L^2 + nV_0m^*u_e^2. \tag{9.26}$$

We have taken the mass of the effective oscillator to be that of the ion oscillator, ignoring the small contribution of the electrons. The displacements u_L and u_e are related through their dependencies on ξ as in eqns

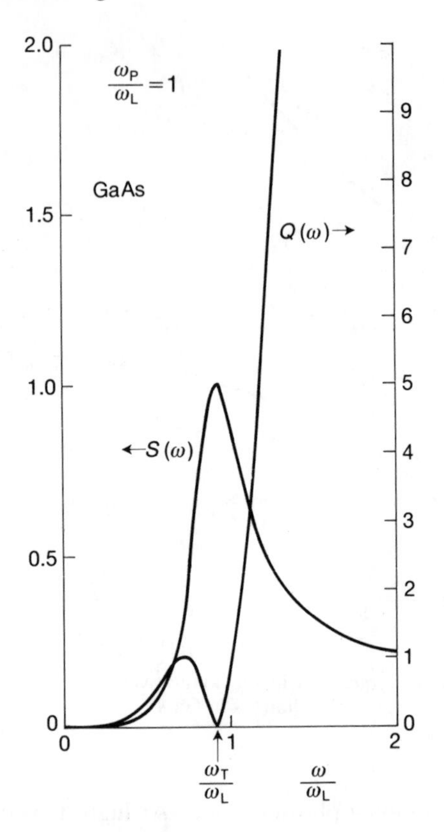

FIG. 9.2. Phonon ($S(\omega)$) and charge ($Q(\omega)$) factors for n-GaAs when $\omega_P = \omega_L$.

(9.11) and (9.19). Eliminating u_e in favour of u_L we can write

$$u_L^2 = Su^2 \tag{9.27}$$

where S is a factor measuring the phonon content of the coupled mode, given by

$$S = \left[1 + \frac{(\omega_P^2 + \omega_q^2)^2(\omega^2 - \omega_T^2)^2}{\omega_P^2\omega^4(\omega_L^2 - \omega_T^2)}\right]^{-1}. \tag{9.28}$$

Figure 9.2 depicts S for $q = 0$. S is unity (pure phonon) only for $\omega = \omega_T$ and falls away rapidly as ω departs from ω_T. This expression reduces to that derived by Kim *et al.* (1978) in the limit of $q \to 0$ when the condition which determines the frequency, i.e. eqn (9.24), is used.

The effective charge is obtained using $\mathbf{P} = \epsilon_0 \boldsymbol{\xi}$ in eqn (9.25), the relation between $\boldsymbol{\xi}$ and $\mathbf{u}_L$ i.e. eqn (9.11), and the relation between u_L

and u, i.e. eqn (9.27). The result is

$$e^{*2} = e_{\mathrm{L}}^{*2} Q \tag{9.29}$$

where Q is a scaling factor measuring the amount by which the effective charge is increased. It is given by

$$Q = \left(\frac{\omega^2 - \omega_{\mathrm{T}}^2}{\omega_{\mathrm{L}}^2 - \omega_{\mathrm{T}}^2}\right)^2 S \tag{9.30}$$

and depicted in Fig. 9.2.

The scattering rate for an electron interacting with a coupled mode is given by

$$W(\mathbf{k}) = \frac{e^2}{8\pi^2 M V_0 \epsilon_0^2} \int \frac{e^{*2}}{\omega q^2} (1 - f(\mathbf{k}'))\delta_{\mathbf{k}\pm\mathbf{q}+\mathbf{k}',0} \times (n(\omega) + \tfrac{1}{2} \mp \tfrac{1}{2})\delta(E_{\mathbf{k}'} - E_{\mathbf{k}} \mp \hbar\omega)\, \mathrm{d}\mathbf{q} \tag{9.31}$$

from which it may be seen that the coupling strength factor measuring the increased coupling is given by

$$R = \frac{\omega_{\mathrm{L}}}{\omega} Q. \tag{9.32}$$

This is shown in Fig. 9.3 for $q = 0$ as a function of electron density for GaAs. The coupling strength of the ω_- mode is comparable with that for the bare phonon up to densities at which strong screening occurs ($\sim 10^{18}\,\mathrm{cm}^{-3}$).

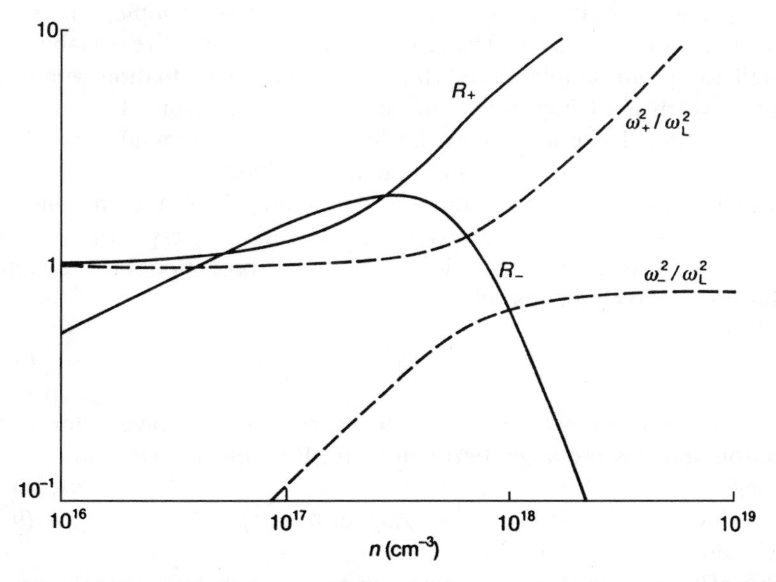

FIG. 9.3. Interaction strengths in the long-wavelength limits for n-GaAs. The subscript + refers to the higher frequency mode, the subscript − to the lower frequency mode.

That for the ω_+ mode rises monotonically with density—a marked anti-screening effect. The result of increasing the wavevector will be to increase the frequencies of the two modes and this will modify the coupling strength, causing an enhancement for the ω_+ mode and a reduction for the ω_- mode in the regime where $\omega_- \leqslant \omega_T$. Interaction with electrons occurs in general at wavevectors large enough for dispersion to be an important factor. Nevertheless, depictions of the situation at long wavelengths are useful in illustrating trends without involving complicated expressions.

We can draw a few conclusions from the above analysis. Note first of all that the coupling strength, R_-, remains significant even at low densities, where the frequency of the plasma-like mode is more than an order of magnitude less than the phonon frequency. This implies that in the single-particle excitation regime the plasma wave will be relatively heavily damped, i.e. $\omega_-\tau_- < 1$, where τ_- is the lifetime of the mode. In these circumstances the mode cannot be well-defined and its power to exchange momentum and energy thereby weakens, along with its importance as a scattering mechanism. Similar remarks apply to the upper branch. The enhanced coupling strength, R_+, implies that heavy damping of plasma-like modes at high densities will occur in the single-particle excitation regime, once again leading to loss of mode identity, and of scattering power.

It is clear, therefore, that any valid description of coupled modes in the single-particle excitation regime must incorporate damping in a self-consistent way. Outside this regime the situation is reasonably well described by our simple model, but it is exactly the situation within this regime which is of importance in determining scattering. Before seeing what our model can say about this situation let us remind ourselves of how the single-particle excitation regime is defined.

The interaction of a quantum $\hbar\omega$ with an electron in a spherical, parabolic band is determined by the conservation of crystal momentum and energy. As pointed out in Section 3.2 absorption of a quantum entails the following equality

$$\hbar\omega = \frac{\hbar^2}{2m^*}(q^2 + 2kq\cos\theta) \tag{9.33}$$

where q is the wavevector of the quantum, k is the wavevector of the electron and θ is the angle between them. For emission,

$$\hbar\omega = \frac{\hbar^2}{2m^*}(2kq\cos\theta - q^2). \tag{9.34}$$

The single-particle excitation regime in each case is defined as the area in (ω, q) space which satisfies the condition for energy and momentum

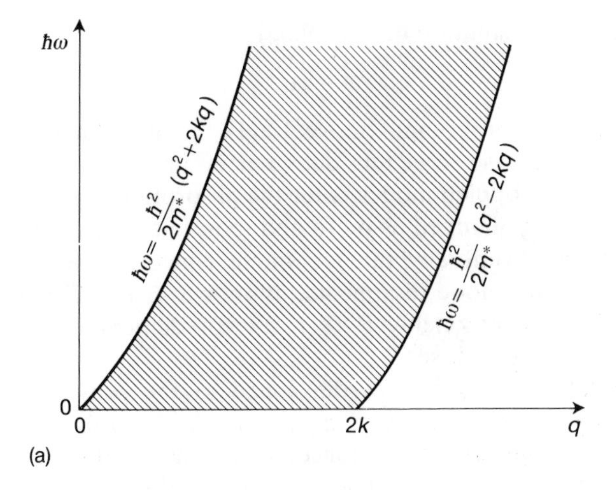

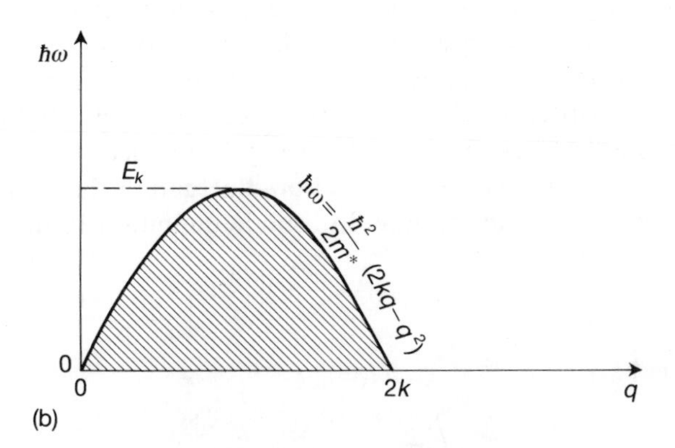

FIG. 9.4. Regimes for single-particle excitation: (a) absorption (b) emission. At $T = 0$ in the case of degenerate material k can be taken to be k_F, the Fermi wavevector.

conservation. These are shown in Fig. 9.4. In the case of a degenerate electron gas, at $T = 0$, the intercept along the q axis occurs at $q = 2k_F$.

If a mode in the absorption regime is heavily damped, that implies that the scattering rate of the electron is high as a consequence. We can describe this situation crudely by making Γ, the damping rate for electron motion, high. (In principle we could choose Γ in a self-consistent way.) Incorporating Γ in the equation $\epsilon_T = 0$ leads to a dispersion relation in place of eqn (9.24) given by

$$\omega^4 - i\omega^3\Gamma - \omega^2(\omega_L^2 + \omega_P^2 + \omega_q^2) + i\omega_L^2\omega\Gamma + \omega_T^2(\omega_P^2 + \omega_q^2) = 0 \quad (9.35)$$

In general the allowed frequencies are now complex. However if Γ is large an undamped solution of eqn (9.35) exists, viz.

$$\omega = \omega_L. \tag{9.36}$$

Obviously, if electrons cannot move freely the phonon mode emerges undressed. In this case screening does not occur at all. Thus we would expect that as the coupled modes enter the absorption region both will be increasingly heavily damped as more and more electrons find themselves able to interact. Eventually the ω_+ mode converts into pure phonon mode while the ω_- mode disappears, as shown in Fig. 9.5. It turns out that this picture is reasonably close to the prediction of more sophisticated theories, as we will see.

The classical model we have developed in the previous sections is adequate only for the lattice component. As regards the electronic contribution to the dielectric constant, it takes no account of the distribution of electrons over the available energy states, nor does it describe screening in the presence of single-particle excitation. Finally, its insistence on defining precise modes makes it incapable of dealing with the situation when Landau damping is strong and modes are not well-defined. In order to deal with the electronic contribution to the dielectric constant we use Lindhard's theory to replace the discussion of Section 9.3, and we invoke the dissipation–fluctuation theory to generalize our description of scattering and of mode definition in the excitation regime.

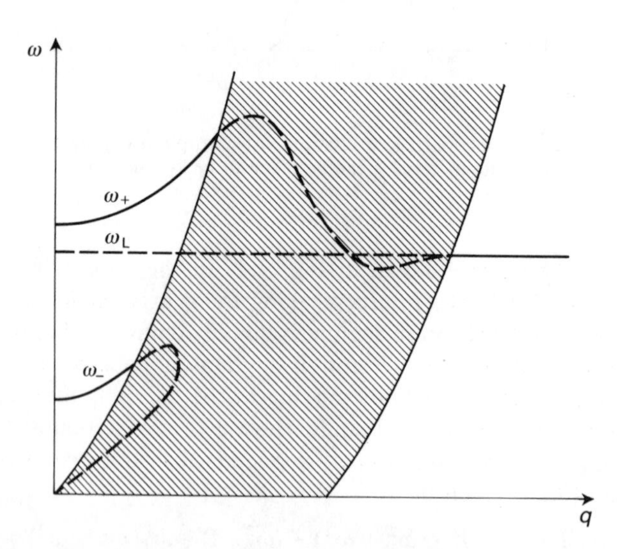

FIG. 9.5. Dispersion of coupled modes. In the single-particle excitation regime for absorption (shaded) the modes become ill-defined, and this is depicted by dashed lines.

9.5. The Lindhard dielectric function

We need to describe the response of the electron system to a weak potential of the form $V_0 \exp[i(\mathbf{q}\cdot\mathbf{r} - \omega t)]$. We assume that the response is linear and so the electron density varies spatially and temporally as does the potential, with an amplitude given by

$$n_s = F(q, \omega) V_0 \tag{9.37}$$

where $F(q, \omega)$ is the density-response function. This variation of density gives rise to a screening potential $V_s \exp[i(\mathbf{q}\cdot\mathbf{r} - \omega t)]$, according to Poisson's equation, and so

$$V_s = \frac{e^2}{\epsilon_0 q^2} n_s = \frac{e^2}{\epsilon_0 q^2} F(q, \omega) V_0. \tag{9.38}$$

The total potential seen by the electrons is therefore

$$V_T = V_0 + V_s = \left(1 + \frac{e^2}{\epsilon_0 q^2} F(q, \omega)\right) V_0. \tag{9.39}$$

The linear response to this potential is, say

$$n_s = G(q, \omega) V_T = \frac{F(q, \omega)}{1 + \dfrac{e^2}{\epsilon_0 q^2} F(q, \omega)} . V_T \tag{9.40}$$

and a dielectric (strictly permittivity) function $\epsilon_e(q, \omega)$ can be defined such that

$$V_T = \frac{\epsilon_0}{\epsilon_e(q, \omega)} V_0 \tag{9.41}$$

whence

$$\epsilon_e(q, \omega) = \epsilon_0 - \frac{e^2}{q^2} G(q, \omega). \tag{9.42}$$

The function $G(q, \omega)$ can readily be calculated using the Liouville equation for the time evolution of the quantum mechanical density matrix, and we arrive at the Lindhard formula (Lindhard 1954)[†]

$$\epsilon_e(q, \omega) = \epsilon_0 - \frac{e^2}{q^2 V} \sum_{\mathbf{k}} \frac{f(E_{k+q}) - f(E_k)}{E_{k+q} - E_k - \hbar\omega - i\hbar\alpha} \tag{9.43}$$

where V is the cavity volume and α is a collision-damping parameter which is usually allowed to approach zero. An often more convenient form is

$$\epsilon_e(q, \omega) = \epsilon_0 + \frac{e^2}{q^2 V} \sum_{\mathbf{k}} f(E_k) \times \left(\frac{1}{E_{k-q} - E_k + \hbar\omega + i\hbar\alpha} + \frac{1}{E_{k+q} - E_k - \hbar\omega - i\hbar\alpha}\right). \tag{9.44}$$

[†] See Section 11.2 for the derivation of this important formula.

In these equations $f(E_k)$ is the distribution function. This equation replaces eqn (9.21) as the electronic contribution to the total dielectric function.

The sum can be replaced by an integral and an analytical solution found for the case of a degenerate electron gas at $T = 0$ K. The result is

$$\epsilon_e(q, \omega) = \epsilon_0 + \frac{e^2 N(E_F)}{2q^2}\left[1 - \frac{1}{4\eta^3}\left\{[(\gamma+\eta^2)^2 - \eta^2]\ln\left|\frac{\eta+\eta^2+\gamma}{\eta-\eta^2-\gamma}\right| + [(\gamma-\eta^2)^2 - \eta^2]\ln\left|\frac{\eta+\eta^2-\gamma}{\eta-\eta^2+\gamma}\right|\right\}\right] \tag{9.45}$$

where $\eta = q/2k_F$, $\gamma = (\hbar\omega + i\alpha)/4E_F$, E_F = Fermi energy and $N(E_F)$ = density of states at the Fermi surface. $(= m^* k_F/\pi^2\hbar^2)$.

When $q \to 0$, and $\alpha = 0$, this reduces (after much algebra) to

$$\epsilon_e(q, \omega) \approx \epsilon_0 - \epsilon_\infty \frac{\omega_P^2}{\omega^2}\left(1 + \frac{3}{5}\frac{v_F^2 q^2}{\omega^2}\right). \tag{9.46}$$

The quantity $3v_F^2/5$ is the mean square velocity of the Fermi distribution. Putting $\omega \approx \omega_P$ in the bracket leads to the result obtained in our simple model, if allowance is made for the difference in statistics. The condition under which eqn (9.46) is valid is $qv_F \ll \omega$, or, if v_q is the phase velocity, $v_F \ll v_q$. Under this condition electrons cannot respond quickly enough to bunch in the troughs of electrical potential, which would produce screening; instead they lag 180° out of phase, bunch on the potential peaks and produce anti-screening, as indicated by the minus sign in eqn (9.46).

In the opposite limit when $q \to \infty$, $\epsilon \to \epsilon_0$, corresponding to no screening at all. More precisely, this condition occurs when $q \gg 2k_F$ and $q^2 \gg q_0^2$, where q_0 is the wavevector at which v_q equals the group velocity of the electron, a situation which occurs for $\hbar^2 q_0^2/2m^* = \hbar\omega$. In many cases $q_0 \approx k_F$, and so the two inequalities imply the same restriction.

We should note in passing that the static screening limit $(\omega \to 0)$ gives

$$\epsilon_e(q, 0) = \epsilon_0 + \frac{e^2 N(E_F)}{2q^2}\left[1 + \frac{1}{2\eta}(1-\eta^2)\ln\left|\frac{1+\eta}{1-\eta}\right|\right]. \tag{9.47}$$

The frequency must satisfy the conditions implied by $\gamma \ll \frac{1}{2}$ and $\gamma \ll \eta(1+\eta)$. In the limit $q \to 0$ the permittivity reduces to the usual Thomas–Fermi screening solution.

The Lindhard function has singularities of the form $x \ln x$ near where $\gamma = (\eta^2 \pm \eta)$ and $\gamma = (\eta - \eta^2)$. These correspond to the boundaries of the single-particle excitation regime, the former associated with absorption, the latter with emission, as eqns (9.33) and (9.34) show. Because

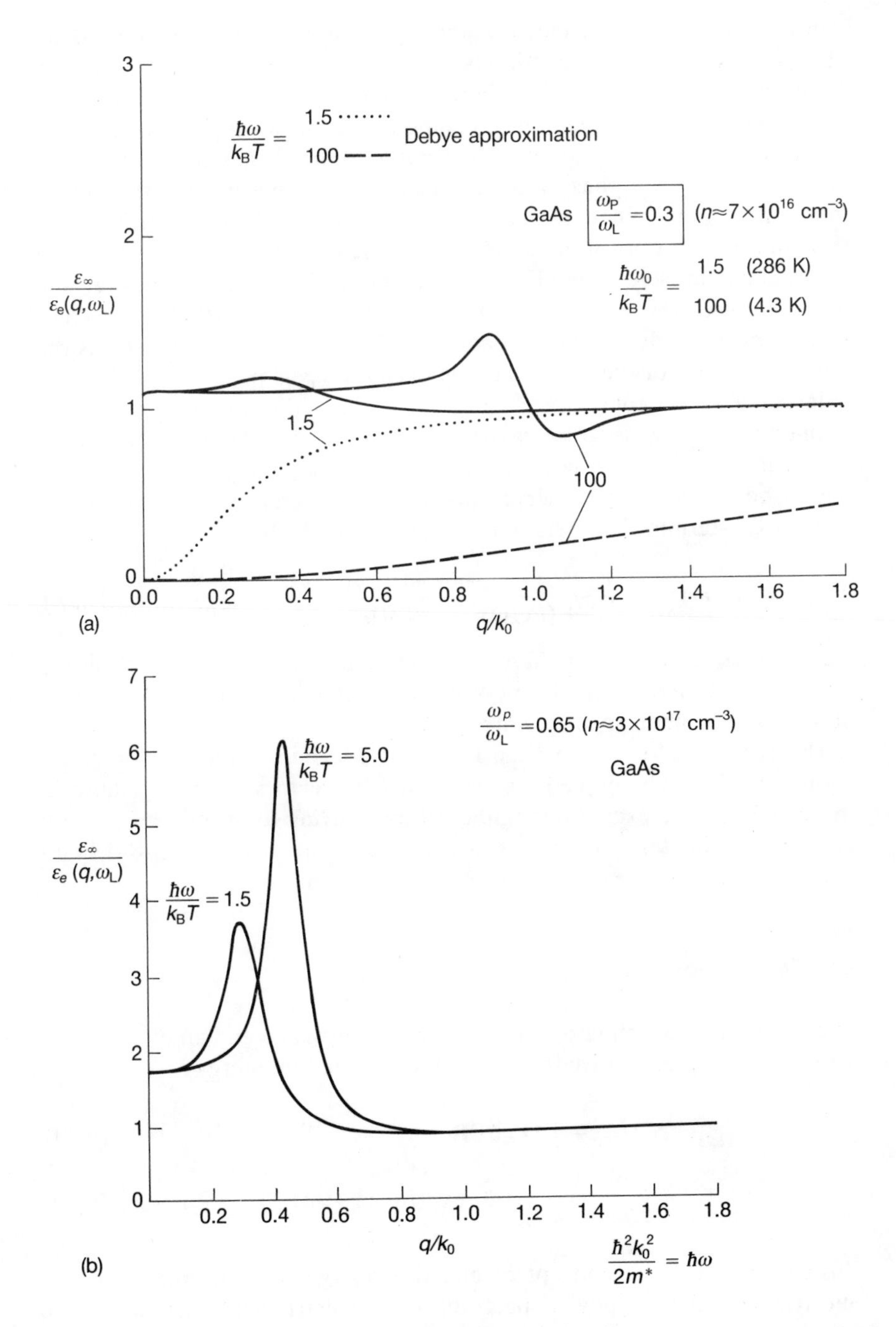

FIG. 9.6. The Lindhard dielectric function.

$x \ln x \to 0$ as $x \to 0$ the dielectric function remains continuous, but a sharp change occurs at these boundaries.

Figure 9.6 illustrates the variation of $\epsilon_\infty/\epsilon_e(q, \omega)$ with q for a frequency $\omega = \omega_L$. Large anti-screening can occur for $q < q_0$, where q_0 is the wavevector such that $\hbar^2 q_0^2/2m^* = \hbar\omega_L$, corresponding to the wavevector at which the phase velocities of the electron wave and the optical mode are equal. When $q < q_0$ the phase velocity of the electron is less than that of the optical mode and the phase of the response is therefore to produce anti-screening. The anti-screening effect becomes weaker and more diffuse towards higher temperatures. It nevertheless remains important in producing heavy damping of the ω_+ mode as it crosses the low q boundary of the single-particle excitation regime.

If collision damping (e.g. caused by charged impurity or acoustic phonon scattering) is to be incorporated we can interpret the parameter α in eqn (9.44) as the damping rate. In this case the Lindhard formula has to be adjusted to conserve particle number and we have to use the Lindhard–Mermin dielectric function (Mermin 1970):

$$\epsilon_e(q, \omega) = \epsilon_0 + \frac{(1 + i\alpha/\omega)(\epsilon_{\mathscr{L}}(q, \omega) - \epsilon_0)}{1 + (i\alpha/\omega)(\epsilon_{\mathscr{L}}(q, \omega) - \epsilon_0)/(\epsilon_{\mathscr{L}}(q, 0) - \epsilon_0)} \tag{9.48}$$

where $\epsilon_{\mathscr{L}}(q, \omega)$ is the Lindhard function. This turns out to be useful for describing dispersion and line-shapes of the coupled mode system observed by Raman spectroscopy.

The Lindhard formula (or the Lindhard–Mermin formula) is powerful enough to describe coupled modes at all wavevectors, including those in the single-particle excitation regime without having to introduce arbitrary damping parameters. It remains now to tackle the problem of ill-defined modes and the interaction of electrons with them.

9.6. Fluctuations

The Coulombic coupling between a polar disturbance and an electron, as we saw in Chapter 3, is described by the interaction energy

$$H_{ep} = \int \mathbf{D} \cdot \xi \, d\mathbf{R} \tag{9.49}$$

$$= \frac{e}{q} \sum_q (i\xi_q \exp(i\mathbf{q} \cdot \mathbf{r}) + cc). \tag{9.50}$$

Instead of using the concept of effective charge we concentrate on the electric field and the power spectrum of the electric field fluctuations. To do this we turn to the dissipation–fluctuation theorem, which states that in thermal equilibrium the power spectrum of fluctuations of a quantity X is connected with the linear response function T in the following way:

$$\langle XX^* \rangle = \frac{\hbar}{\pi}(n(\omega) + 1)\,\mathrm{Im}\,T \tag{9.51}$$

where $n(\omega)$ is the Bose–Einstein function and Im T stands for the imaginary part of T. The linear response function, T, is defined such that under the influence of an external force F,

$$X = TF. \tag{9.52}$$

From the point of view of the electric field the external force is the polarization and

$$\xi = -\frac{\mathbf{P}}{\epsilon(q, \omega)} \tag{9.53}$$

for longitudinally polarized waves. Consequently

$$\langle \xi\xi^* \rangle = \frac{\hbar}{\pi}(n(\omega) + 1)\,\mathrm{Im}\left(-\frac{1}{\epsilon(q, \omega)}\right). \tag{9.54}$$

We therefore obtain for the scattering rate in a parabolic band

$$W(\mathbf{k}) = \frac{2\pi}{\hbar}\int \frac{\mathrm{d}\mathbf{q}}{(2\pi)^3}\int_{-\infty}^{+\infty} \frac{\mathrm{d}(\hbar\omega)}{\pi}(1 - f(\mathbf{k} + \mathbf{q}))(n(\omega) + 1) \times \frac{e^2}{q^2}\,\mathrm{Im}\left(-\frac{1}{\epsilon(q, \omega)}\right)\delta(E_{\mathbf{k}+\mathbf{q}} - E_{\mathbf{k}} + \hbar\omega) \tag{9.55}$$

and it is understood that crystal momentum is conserved. This expression is entirely equivalent to eqn (9.31).

To show that eqn (9.55) yields the standard result when plasma coupling is absent we use the lattice permittivity given by eqn (9.9), expressed in the following form

$$\frac{1}{\epsilon_{\mathrm{L}}} = \frac{1}{\epsilon_\infty} - \frac{\omega_{\mathrm{L}}}{2}\left(\frac{1}{\epsilon_\infty} - \frac{1}{\epsilon_{\mathrm{s}}}\right)\left(\frac{1}{\omega_{\mathrm{L}} - \omega + is} + \frac{1}{\omega + \omega_{\mathrm{L}} + is}\right). \tag{9.56}$$

Using

$$\lim_{s\to 0} \frac{1}{x \pm is} = P\left(\frac{1}{x}\right) \mp i\pi\delta(x) \tag{9.57}$$

we obtain

$$\mathrm{Im}\left(-\frac{1}{\epsilon_{\mathrm{L}}}\right) = \frac{\omega_{\mathrm{L}}\pi}{2}\left(\frac{1}{\epsilon_\infty} - \frac{1}{\epsilon_{\mathrm{s}}}\right)[\delta(\omega - \omega_{\mathrm{L}}) - \delta(\omega + \omega_{\mathrm{L}})]. \tag{9.58}$$

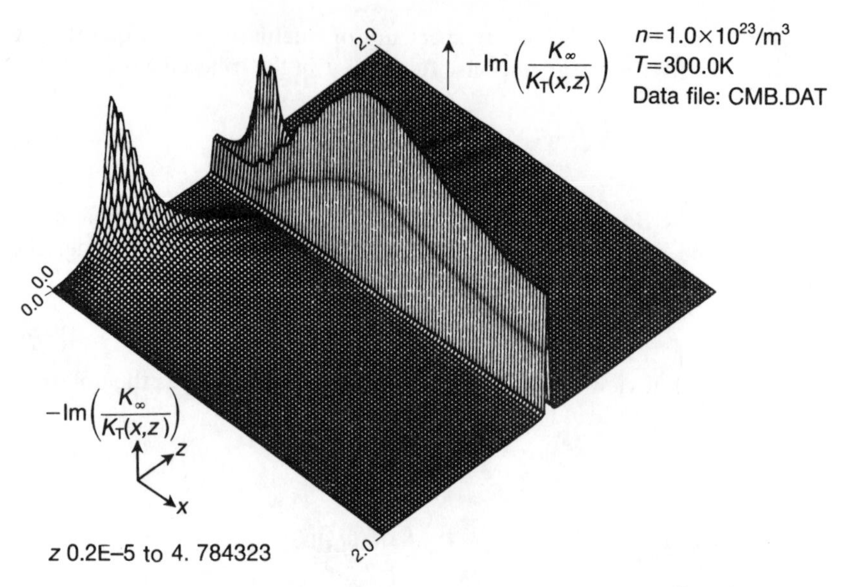

FIG. 9.7. Coupled-mode strengths in n-GaAs at 300 K for $n = 10^{17}\,\mathrm{cm}^{-3}$.*

Integration over ω gives the usual result. (Note that $n(-\omega)+1 = -n(\omega)$.)

Modes can now be defined in general by the maxima of $\mathrm{Im}(-1/\epsilon(q,\omega))$ with

$$\epsilon(q,\omega) = \epsilon_L + \epsilon_e - \epsilon_0 \tag{9.59}$$

where ϵ_L is given by eqn (9.56) and ϵ_e is given by eqn (9.44) (or eqn (9.48)). Examples of mode patterns are shown in Figs. 9.7 and 9.8.

The results support the general conclusions of our simple model. At small q the two modes are well-defined with maxima moving up in frequency with increasing q, as expected from the dispersion introduced by the plasmons. For the case of $n = 10^{18}\,\mathrm{cm}^{-3}$ the ω_- mode is rather weak and the ω_+ mode is very strong, just as our plot of R in Fig. 9.3 predicted. On entering the damping zone both modes rapidly lose amplitude and there is the growth of a mode at the longitudinal phonon frequency. This phonon-like mode is anti-screened over much of the interaction range at the lower electron density, but screened heavily over much of the range at the higher electron density. For $q > 2q_0$ ($\hbar^2 q_0^2/2m^+ = \hbar\omega_L$) the mode is essentially unscreened.

* The author is indebted to F. A. Riddoch for the data in Figs. 9.7 and 9.8.

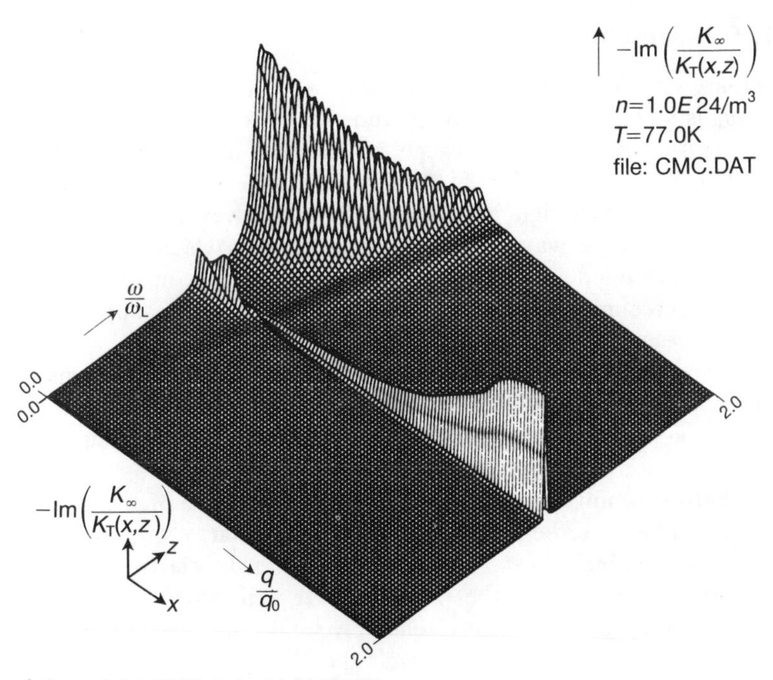

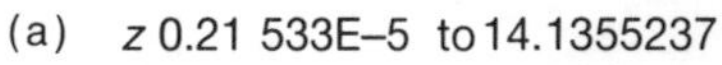

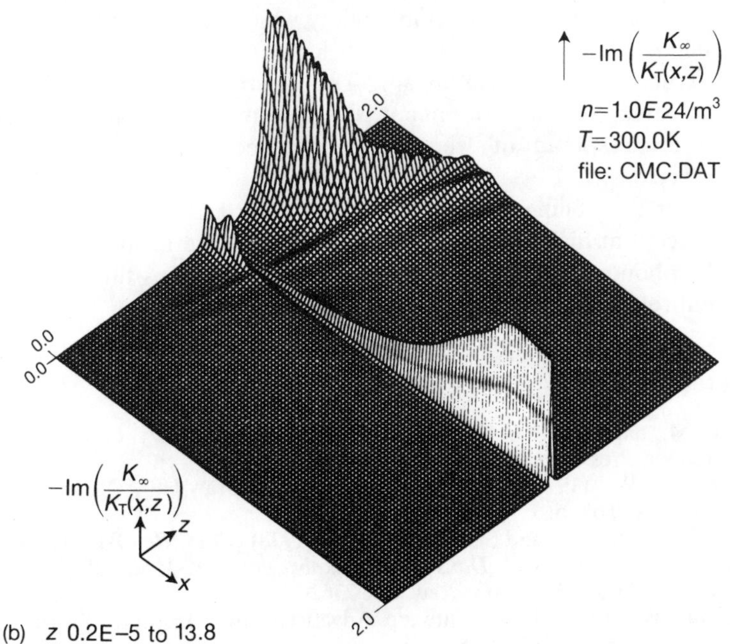

FIG. 9.8. Coupled mode strengths in n-GaAs for $n = 10^{18}\,\mathrm{cm}^{-3}$. (a) $T = 77\,\mathrm{K}$, (b) $T = 300\,\mathrm{K}$.

9.7. Screening regimes

Complex screening effects are important when the carrier density is such that ω_P is within an order of magnitude on either side of ω_L. In the case of n-GaAs this defines a range of electron densities roughly from 5×10^{16} cm^{-3} to 5×10^{18} cm^{-3}. Below this range screening is negligible; above it, it is well-nigh total for moderate wavevectors. (It is always negligible for large wavevectors.) The regime in which $\omega_P \approx \omega_L$ can be roughly subdivided into two—a lower density range in which $\omega_P < \omega_L$ with anti-screening as its characteristic feature, and an upper region in which $\omega_P > \omega_L$ and screening becomes dominant.

We can therefore summarize dynamic screening effects in terms of four regimes, as follows.

1. Weak-screening regime ($\omega_P \ll \omega_L$). Screening and coupled-mode effects are negligible. What screening there is, is anti-screening.
2. Anti-screening regime ($\omega_P \lesssim \omega_L$). The response of the electrons is too slow to screen at wavevectors such that $q < q_0$ and, instead, anti-screening occurs. The interaction with the electrons is therefore enhanced in exactly the range of wavevectors where it is strongest. Coupled-mode effects are relatively weak, but become important as ω_P approaches ω_L. As a result the 'phonon' energy becomes less well-defined as the density increases. In particular, the ω_- mode introduces a lower 'phonon' energy which may become important in determining energy relaxation at low temperatures.
3. Screening regime ($\omega_P \gtrsim \omega_L$). The electrons now move faster and screening becomes dominant. Coupled-mode effects are strong with a large spread of frequencies involved in the interaction with electrons.
4. Strong-screening regime ($\omega_P \gg \omega_L$). The LO frequency disappears (except at high q) and is replaced by ω_T. Interaction with electrons by phonon-like modes vanishes. Such interactions that do occur are entirely plasmon-like.

References

Born, M. and Huang, K. (1956). *The dynamical theory of crystal lattices.* Clarendon Press, Oxford.

Harrison, W. (1970). *Solid state theory.* McGraw-Hill, New York, and Kogakusha, Tokyo.

Kim, M. E., Das, A. and Senturia, S. D. (1978). *Phys. Rev* **B18,** 6890.

Lindhard, J. (1954). *Kgl. Danske Videnskab. Mat. Fys. Medd.* **28,** 8.

Mermin, N. D. (1970). *Phys. Rev.* **B1,** 2362.

Richter, W. (1982). Proc. Antwerp Advanced Study Institute, Priorij Corsendonk, Belgium, 26 July–5 August.

10. Phonon processes

10.1. Introduction

INTEREST in semiconductor physics generally focuses on properties consequent on the quantum transitions made by electrons and holes, but many of these transitions involve the emission of phonons, and therefore it is pertinent to ask what happens to these phonons. In Chapter 3 all the rates were calculated assuming that the phonon population was effectively described by the thermodynamic equilibrium distribution, but this implicitly assumes that any disturbances created by carrier emission processes decay instantaneously, which, of course, is unrealistic. Where electron and hole populations are dense, the emission of phonons can be rapid enough to create serious departures from equilibrium in those modes which interact most strongly, and when this occurs we often speak of 'hot phonons', metaphorically relating the increase in phonon numbers to an elevated temperature. The concept of a phonon temperature is sometimes valuable, even if it has to be limited to modes lying in a small section of the Brillouin zone, but 'temperature' implies randomization, and to justify its use it is necessary to know how rapidly phonon–phonon interactions occur. The presence of hot phonons will always slow down the rate of energy relaxation of carriers through the process of re-absorption, and so it is vital to know how long emitted phonons live. An emitted phonon in a normal process stores not only energy but also momentum. It is important to know how fast this momentum is relaxed since this determines thermal conductivity, but it is also important for electrical conductivity, for if momentum relaxation of the phonon is slow then re-absorption restores the momentum initially lost by the carriers (provided the same type of carrier is involved) and, in that case, hot phonons do not affect the drift of the carriers. It is clear, therefore, that even if our main interest is directed towards electronic or optoelectronic properties, we cannot avoid studying the processes that determine phonon lifetime and scattering rates.

Except at temperatures below roughly 40 K the most important electron–phonon interaction in semiconductors is that involving optical phonons. Compared with the literature on acoustic phonons, that on optical phonons is minuscule, and as a consequence the approach offered here has had to be rather unsophisticated, and for reasons of brevity the same approach has been adopted for the treatment of processes involving acoustic phonons. More rigorous and comprehensive accounts of processes involving acoustic phonons can be found in the books mentioned in the references at the end of this chapter. The aim here is to present the basic elements of phonon processes and so we adopt some simplifications such as assuming that elastic isotropy prevails and that all interactions are a consequence of a frequency shift induced by anharmonicity.

The Hamiltonian for the lattice vibrations is taken, following Klemens (1966), to be, in lowest order,

$$H = \sum_{\mathbf{r},i,j} M_i^{1/2}\omega_i M_j^{1/2}\omega_j \mathbf{u}_i \cdot \mathbf{u}_j. \tag{10.1}$$

Here $\mathbf{r}$ is the position coordinate of the unit cell, $M_{i,j}$ is the mass of the oscillator (total mass in the primitive unit cell for acoustic modes, reduced mass for optical modes), $\omega_{i,j}$ is the angular frequency, and $\mathbf{u}_{i,j}$ is the displacement (of the unit cell for acoustic modes, relative for optical modes). Expanding the spatial dependence of the displacement in a Fourier series we obtain

$$\mathbf{u} = \frac{1}{\sqrt{N}} Q_{\mathbf{q}}\mathbf{e}_{\mathbf{q}} \exp(\mathrm{i}\mathbf{q} \cdot \mathbf{r}) \tag{10.2}$$

where N is the number of unit cells, $\mathbf{q}$ is the wavevector and $\mathbf{e}_{\mathbf{q}}$ is the unit polarization vector. The convenient notation of second quantization can be exploited by the usual coordinate transformation

$$Q_q = \left(\frac{\hbar}{2M\omega}\right)^{1/2} (a_{\mathbf{q}}^+ + a_{-\mathbf{q}}) \tag{10.3}$$

where $a_{\mathbf{q}}$, $a_{\mathbf{q}}^+$ are the annihilation and creation operators which operate on the eigenstate ϕ consisting of a product of simple harmonic oscillator functions:

$$\phi = \pi_{n,\mathbf{q}} \frac{1}{\sqrt{n!}} (a_{\mathbf{q}}^+)^n \phi_0 \tag{10.4}$$

where ϕ_0 is the vacuum state. The displacement of eqn (10.2) becomes the operator

$$\hat{\mathbf{u}} = \sum_{\mathbf{q}} \left(\frac{\hbar}{2NM\omega}\right)^{1/2} \{\mathbf{e}_{\mathbf{q}}^* a_{\mathbf{q}}^+ \exp(\mathrm{i}\mathbf{q} \cdot \mathbf{r}) + \mathbf{e}_{\mathbf{q}} a_{-\mathbf{q}} \exp(-\mathrm{i}\mathbf{q} \cdot \mathbf{r})\}. \tag{10.5}$$

We assume most interactions are small perturbations which change frequencies by an amount $\delta\omega \ll \omega$, so that

$$H = H_0 + H_{\mathrm{i}} \tag{10.6}$$

$$H_0 = \sum_{\mathbf{q}i} \hbar\omega_{\mathbf{q}i}(a_{\mathbf{q}i}^+ a_{-\mathbf{q}i} + \tfrac{1}{2}) \tag{10.7}$$

$$H_{\mathrm{i}} = \sum_{\mathbf{r},i,j} M_i^{1/2}\omega_i M_j \omega_j \left(\frac{\delta\omega_i}{\omega_i} + \frac{\delta\omega_j}{\omega_j}\right) \mathbf{u}_i \cdot \mathbf{u}_j. \tag{10.8}$$

Rates are then given by

$$W = \frac{2\pi}{\hbar} \int |\langle \mathrm{f}| H_{\mathrm{i}} |\mathrm{i}\rangle \delta(E_{\phi} - E_{\mathrm{i}})\, \mathrm{d}N_{\mathrm{f}} \tag{10.9}$$

where f, i denote final and initial states and N_{f} is the number of final states. The only other type of interaction we will deal with is a Fröhlich interaction of longitudinally polarized optical modes with charged impurities which alters the energy electrostatically via the scalar potential of the mode.

The usual shorthand notation for vibrational modes will be adopted, namely LO, TO for longitudinally and transversely polarized optical modes and LA, TA for the equivalent acoustic modes.

10.2. Three-phonon processes

10.2.1. Coupling constants

Phonons interact via the lattice anharmonicity. The simplest approach is to assume that a fractional change of frequency is proportional to strain, in the case of acoustic modes, and to displacement, in the case of optical modes. This scheme is thus directly analogous to the deformation potentials introduced to describe the electron–phonon interaction. For an acoustic mode,

$$\left(\frac{\delta\omega}{\omega}\right)_{\mathrm{ac}} = \sum_{r,s} \gamma_{rs} S_{rs} \tag{10.10}$$

where S_{rs} are components of the strain tensor in reduced notation and γ_{rs} are the corresponding coupling constants. For optical modes,

$$\left(\frac{\delta\omega}{\omega}\right)_{\mathrm{op}} = \mathbf{\Gamma} \cdot \mathbf{u} \tag{10.11}$$

where Γ is the anharmonic coupling vector. As in the case of deformation potentials the γ_{rs} and Γ are hard to obtain from *ab initio* calculations. A mode-independent parameter, γ, the Grüneisen constant, can be obtained from the thermal expansion of the crystal and from frequency shifts; it has a magnitude typically between 1 and 2. A simplification of eqn (10.10) is therefore, for a travelling wave,

$$\left(\frac{\delta\omega}{\omega}\right)_{\mathrm{ac}} = \mathrm{i}\gamma q u. \tag{10.12}$$

In this approximation the wave produces the same anharmonic effect whatever its direction of polarization. For a travelling wave $S \sim qu$ and so we might expect $\Gamma \sim q\gamma$ with $q \sim \pi/a_0$, where a_0 is the dimension of the primitive unit cell. Thus with $\gamma = 1$ and $a_0 = 3$ Å we expect $\Gamma \sim 10^8\ \mathrm{cm}^{-1}$. An equivalent simplification for optical modes is to take the magnitude of Γ to be independent of direction, that is

$$\left(\frac{\delta\omega}{\omega}\right)_{\mathrm{op}} = \Gamma u. \tag{10.13}$$

The optical coupling constant, Γ, incorporates all modifications to the anharmonic Hamiltonian associated with the presence of an optical mode, and includes the effect due to the difference of force constants of the two atoms in the unit cell (which was manifested as a reduction factor, introduced by Klemens (1966)). In practice, we regard γ and Γ as phenomenological parameters to be determined by experiment.

10.2.2. Selection rules for acoustic phonons

When the fractional frequency change of eqn (10.12) is substituted in the interaction Hamiltonian, eqn (10.8), the triple product of displacements allows three-phonon events to occur. Thus

$$\begin{aligned} H_{\mathrm{i}} \equiv H_{3k} &= 2\mathrm{i}M\gamma \sum_{\mathbf{r}i,j} \omega_i\omega_j q_k u_k \mathbf{u}_i \cdot \mathbf{u}_j \\ &= \frac{\mathrm{i}\gamma\hbar^{3/2} q_k}{2^{1/2}N^{3/2}M^{1/2}\omega_k^{1/2}} \sum_{\mathbf{r}i,j} (\omega_i\omega_j)^{1/2} \\ &\quad \times \{a^+_{\mathbf{q}k}\exp(\mathrm{i}\mathbf{q}_k \cdot \mathbf{r}) + a_{-\mathbf{q}k}\exp(-\mathrm{i}\mathbf{q}k \cdot \mathbf{r})\} \\ &\quad \times \{\mathbf{e}^*_{\mathbf{q}i}a^+_{\mathbf{q}i}\exp(\mathrm{i}\mathbf{q}_{\mathrm{i}} \cdot \mathbf{r}) + \mathbf{e}_{\mathbf{q}i}a_{-\mathbf{q}i}\exp(-\mathrm{i}\mathbf{q}_i \cdot \mathbf{r})\} \\ &\quad \times \{\mathbf{e}^*_{\mathbf{q}j}a^+_{\mathbf{q}j}\exp(\mathrm{i}\mathbf{q}_i \cdot \mathbf{r}) + \mathbf{e}_{\mathbf{q}j}a_{-\mathbf{q}j}\exp(-\mathrm{i}\mathbf{q}j \cdot \mathbf{r})\}. \end{aligned} \tag{10.14}$$

Note that our simplifying assumption that the anharmonic effect is independent of the polarization means that we distinguish between the mode inducing the transition and the two modes which are thereby affected. Products consisting of all creation or all annihilation operators describe processes which violate energy conservation and so can be discarded. The rest describe processes of two types. In type 1 a phonon annihilates and creates two other phonons and in type 2 a phonon interacts with another with the result that both are annihilated and one phonon is created (Fig. 10.1). The phonon may or may not be the mode inducing the anharmonic frequency shift—the promoting mode, in brief.

It is sometimes helpful to make a distinction between those transverse modes with polarization vectors perpendicular to the plane containing the three wavevectors involved, the so-called s-modes, and those whose polarization lies in the scattering plane, the so-called p-modes (Fig. 10.2). The polarization vectors of the longitudinal modes always lie in the scattering plane. The following are the possible three-phonon processes

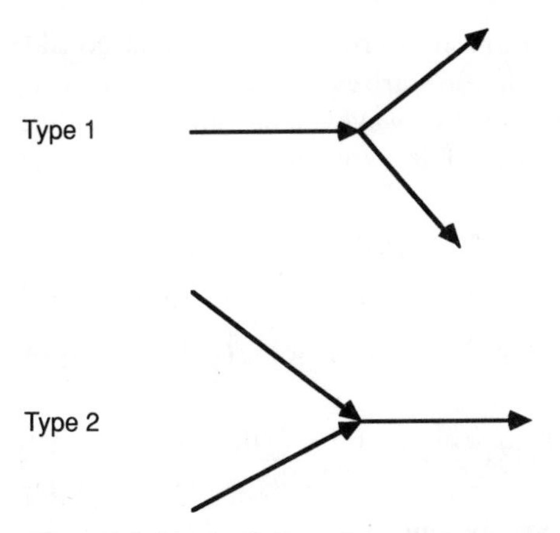

FIG. 10.1. Types of three-phonon processes.

of type 1 or type 2 involving only s and p TA modes and LA modes:

$$\begin{array}{llll} T_{s,p} = T_{s,p} + T_{s,p} & (8) & L = L + L & (1) \\ T_{s,p} = T_{s,p} + L & (4) & L = L + T_{s,p} & (2) \\ T_{s,p} = L + L & (2) & L = T_{s,p} + T_{s,p} & (4). \end{array} \tag{10.15}$$

The number of processes in each case is given in brackets, 21 in all. This number is drastically reduced when the necessity of conserving energy

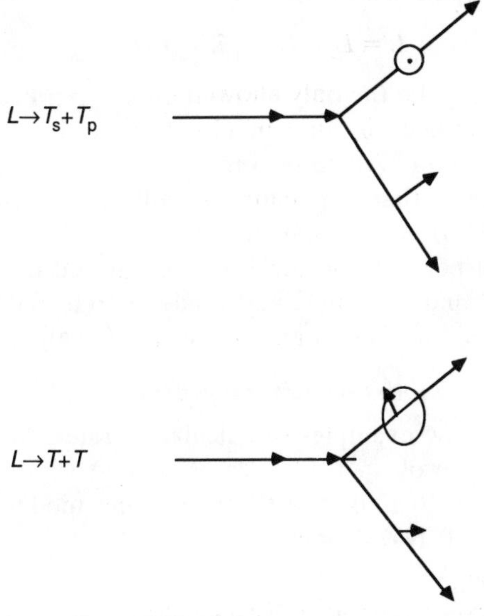

FIG. 10.2. Polarizations of TA modes.

and crystal momentum in normal processes is considered. (Note that conservation of momentum arises in the usual way from the sum over the lattice sites.) Labelling the wavevectors and frequencies with numerical subscripts 1, 2, 3 from left to right in one of the above equations we obtain the equations:

$$\omega_1 = \omega_2 + \omega_3 \tag{10.16a}$$

$$q_1^2 = q_2^2 + q_3^2 + 2q_2q_3 \cos\theta_{23} \tag{10.16b}$$

where θ_{23} is the angle between $\mathbf{q}_2$ and $\mathbf{q}_3$. In the absence of dispersion, $\omega = vq$, and hence

$$\cos\theta_{23} = \frac{2(v_2v_3/v_1^2)q_2q_3 - q_2^2\{1-(v_2/v_1)^2\} - q_3^2\{1-(v_3/v_1)^2\}}{2q_2q_3}. \tag{10.17}$$

If all three modes belong to the same branch, $v_1 = v_2 = v_3$, and $\cos\theta_{23} = 1$. This means that the process is allowed, but only if the modes are collinear; but this condition ensures that the normal rate is infinitesimal. Thus the processes $T_s = T_s + T_s$, $T_p = T_p + T_p$, $L = L + L$ are ruled out. In the absence of elastic anisotropy, so that the velocities of s and p polarization are equal, all processes involving transverse modes of orthogonal polarization (e.g. $T_s = T_s + T_p$) are also ruled out. In this way we can get rid of 9 of the 21 interactions depicted in eqn (10.15). A second culling can be made by noting that if v_1 is smaller than either v_2 or v_3 then $\cos\theta > 1$, which is impossible. This rules out all processes with $T_{s,p}$ on the left, and so the only allowed processes with finite rates are (with arbitrary TA polarizations):

$$L = L + T \qquad L = T + T. \tag{10.18}$$

These turn out also to be the only allowed umklapp processes. To see this it is sufficient to replace q_1 in eqn (10.16b) by $|\mathbf{q}_1 - \mathbf{g}|$, where $\mathbf{g}$ is a reciprocal lattice vector. All three vectors now lie in the first zone and provided it is assumed that dispersion can still be neglected the argument leading to eqn (10.18) proceeds as before.

When neither dispersion nor anisotropy is ignored the situation is very much complicated and the interested reader is referred to the books of Ziman (1960), Reissland (1973) and Srivastava (1990).

10.2.3. Rates for LA modes via normal processes

Let us now give a few examples of calculating rates. We begin with the three-phonon lifetime of an LA mode associated with type 1 processes, and suppose first of all that it is the promoting mode. The interaction Hamiltonian (eqn (10.14)) is now

$$H_i = \frac{i\gamma\hbar^{3/2}q_1}{2^{1/2}N^{1/2}M^{1/2}\omega_1^{1/2}}(\omega_2\omega_3)^{1/2}(\mathbf{e}_{q_2}^* \cdot \mathbf{e}_{q_3}^*)\delta_{\mathbf{q}_1-\mathbf{q}_2-\mathbf{q}_3,\mathbf{g}}a_{-\mathbf{q}_1}a_{\mathbf{q}_2}^+a_{\mathbf{q}_3}^+. \tag{10.19}$$

With $\omega = vq$, and noting that the annihilation operator introduces a factor $n^{1/2}$ and the creation operator a factor $(n+1)^{1/2}$, we obtain for the absorption rate,

$$W_a = \frac{\pi\gamma^2\hbar\omega_1}{NMv_L^2}\int |\mathbf{e}_2^* \cdot \mathbf{e}_3^*|^2\, \omega_2\omega_3 n_1(n_2+1)(n_3+1)\delta(\hbar\omega_1 - \hbar\omega_2 - \hbar\omega_2)\, \mathrm{d}N \tag{10.20}$$

and it is understood that the wavevectors are governed by a Kronecker delta. There will also be a reverse rate, W_e, identical to the above except for the phonon factor, which will be $(n_1+1)n_2n_3$. The net absorption rate will involve

$$n_1(n_2+1)(n_3+1) - (n_1+1)n_2n_3 = n_1(n_2+n_3+1) - n_2n_3. \tag{10.21}$$

We are usually interested in the rate at which a departure from thermodynamic equilibrium relaxes, and so we assume n_2 and n_3 are the equilibrium values, which we denote $\bar{n}_2$ and $\bar{n}_3$. But at equilibrium the net rate must be zero and thus

$$\bar{n}_1(\bar{n}_2+\bar{n}_3+1) - \bar{n}_2\bar{n}_3 = 0. \tag{10.22}$$

It follows that the relaxation rate which determines the lifetime of a disturbance involves the following phonon factor:

$$n_1(\bar{n}_2+\bar{n}_3+1) - \bar{n}_2\bar{n}_3 - \{\bar{n}_1(\bar{n}_2+\bar{n}_3+1) - \bar{n}_2\bar{n}_3\} = (n_1-\bar{n}_1)(\bar{n}_2+\bar{n}_3+1). \tag{10.23}$$

The reciprocal lifetime is thus

$$\frac{1}{\tau_L} = \frac{\pi\gamma^2\hbar^2\omega_1}{NMv_L^2}\int |\mathbf{e}_2^* \cdot \mathbf{e}_3^*|^2\, \omega_2\omega_3(\bar{n}_2+\bar{n}_3+1)\delta(\hbar\omega_1 - \hbar\omega_2 - \hbar\omega_3)\, \mathrm{d}N. \tag{10.24}$$

When equipartition prevails, $\bar{n}_2 \approx k_B T/\hbar\omega_2$, $\bar{n}_3 \approx k_B T/\hbar\omega_3$ and $\bar{n}_2 + \bar{n}_3 \gg 1$. Putting in these approximations and exploiting eqn (10.16a) leads to

$$\frac{1}{\tau_L} = \frac{\pi\gamma^2\hbar\omega_1^2 k_B T}{NMv_L^2}\int |\mathbf{e}_2^* \cdot \mathbf{e}_3^*|^2\, \delta(\hbar\omega_1 - \hbar\omega_2 - \hbar\omega_3)\, \mathrm{d}N. \tag{10.25}$$

The delta function can be transformed to eliminate, say, ω_2 and the integration taken over mode 3. Thus, for normal processes,

$$v_L q_1 = v_2 q_2 + v_3 q_3 \tag{10.26a}$$

$$q_2^2 = q_1^2 + q_3^2 - 2q_1q_3\cos\theta_{13}. \tag{10.26b}$$

This approach is convenient for $L = T + T$, when $v_2 = v_3 = v_T$, for then

$$q_3 = \frac{\{(v_L/v_T)^2 - 1\}}{2\{(v_L/v_T) - \cos\theta_{13}\}} q_1. \tag{10.27}$$

Putting $dN = (V/8\pi^3)q_3^2\, dq_3\, d(-\cos\theta_{13})\, d\phi$, performing the integrations, and replacing $|\mathbf{e}_2^* \cdot \mathbf{e}_3^*|^2$ by unity for s-polarized modes gives

$$\frac{1}{\tau_L} = \frac{\gamma^2 \omega_1^4 k_B T}{8\pi\rho v_L^2 v_T^3}\{1 - (v_T/v_L)^2\}. \tag{10.28}$$

When p-polarized TA modes are involved the polarization factor produces a more complicated angular dependence. This is also true for the transition $L = L + T_p$, and for all cases in which the LA mode is the passive rather than the promoting mode. Summing all the rates for type 1 normal processes leads to the form

$$\frac{1}{\tau_{L1}} = \frac{\gamma^2 \omega_1^4 k_B T}{8\pi\rho v_L^2 v_T^3} F_{L1}(v_T/v_L) \tag{10.29}$$

where $F(v_T/v_L)$ is a portmanteau function of v_T/v_L whose magnitude is of order unity and which conceals the results of the integrations of the polarization factors over the spherical angles, for arbitrary polarizations. (Note that, typically, $v_T/v_L \approx 0.5$.)

Type 2 processes involve the simultaneous absorption of two modes and the creation of a third. Instead of eqn (10.21) the net absorption rate for mode 1 is therefore

$$n_1 n_2(n_3 + 1) - (n_1 + 1)(n_2 + 1)n_3 \tag{10.30}$$

or with n_2 and n_3 taking thermodynamic equilibrium values, we end up with the factor

$$(n_1 - \bar{n}_1)(\bar{n}_2 - \bar{n}_3). \tag{10.31}$$

The calculation for $L = L + T$ proceeds as before and the result can be added to eqn (10.29) to give a rate of the same form. Thus the total rate for normal processes can be expressed as

$$\frac{1}{\tau_L} = \frac{\gamma^2 \omega_1^4 k_B T}{8\pi\rho v_L^2 v_T^3} F_L(v_T/v_L) \tag{10.32}$$

where $F_L(v_T/v_L)$ includes the results for both types of normal process.

LA phonons emitted by non-degenerate electrons in a semiconductor have wavevectors around $2 \times 10^6\ \text{cm}^{-1}$ corresponding to angular frequencies around $10^{12}\ \text{s}^{-1}$. The lifetime of these phonons with $\rho \sim 5\ \text{g cm}^{-3}$, $v_L = 5 \times 10^5\ \text{cm s}^{-1}$, $v_T = 2{\cdot}5 \times 10^5\ \text{cm s}^{-1}$ and $\gamma = 2$, according to eqn (10.32), is of order 3 μs, at room temperature, which is very long and reflects the weakness of coupling of comparatively low-frequency LA modes to the thermal bath of other phonons via normal processes. Umklapp processes, however, prove to be stronger, provided the temperature is not too low, as we will see.

At low temperatures when equipartition fails, type 2 processes cannot proceed and only type 1 survives. In this situation $\tau_L^{-1} \propto \omega^5$.

It is worth noting that our assumption of a directionally independent anharmonic effect merely simplifies the angular dependence from one involving all three polarization vectors to one involving $|\mathbf{e}_2 \cdot \mathbf{e}_3|^2$ but does not affect the result other than by a numerical factor of order unity. In view of uncertainties attending the assumption of a mode-independent Grüneisen constant our approach seems reasonably sensible.

10.2.4. Rates for TA modes via normal processes

The selection rules mean that TA modes are absorbed only via type 2 processes. In the interaction $T + T = L$ the frequencies of the modes are of comparable magnitude and the calculation of the rates proceeds along familiar lines, with the result which can be expressed as

$$\frac{1}{\tau_T} = \frac{2\gamma^2 \omega_1^4 k_B T}{\pi \rho v_L^2 v_T^3} F_T(v_T/v_L) \tag{10.33}$$

which is a larger rate than for LA modes, but still rather small for low-frequency modes.

A different result is obtained for the process $T + L = L$ because in this case both of the LA modes can have much higher frequencies than the frequency of the TA mode, and there is consequently a much higher density of states available for the scattering process. With the TA mode as the promoting mode we can write the rate as

$$\frac{1}{\tau_T} = \frac{\pi \gamma^2 \hbar^2 \omega_1}{N M v_T^2} \int |\mathbf{e}_2 \cdot \mathbf{e}_3^*|^2 \, \omega_2 \omega_3 (\bar{n}_2 - \bar{n}_3) \delta(\hbar\omega_1 + \hbar\omega_2 - \hbar\omega_3) \, \mathrm{d}N \tag{10.34}$$

and

$$q_3^2 = q_1^2 + q_2^2 + 2q_1 q_2 \cos\theta_{12}. \tag{10.35}$$

Restricting our attention to relatively small TA frequencies we can regard q_1 as small compared with either q_3 or q_2 and approximate eqn (10.35) as follows:

$$q_3 \approx q_2\left(1 + \frac{q_1}{q_2}\cos\theta_{12}\right) \tag{10.36}$$

and also $|\mathbf{e}_2 \cdot \mathbf{e}_3^*|^2 \approx 1$. Substituting eqn (10.36) into the delta function and integrating over the azimuthal angle and over $\cos\theta_{12}$ reduces eqn (10.34) to

$$\frac{1}{\tau_{\mathrm{T}}}=\frac{\gamma^2\hbar}{4\pi\rho v_{\mathrm{T}}v_{\mathrm{L}}}\int \omega_2^2(\bar{n}_2-\bar{n}_3)k_3^2\,\mathrm{d}k_3. \tag{10.37}$$

Because both LA modes often can be much higher in frequency than the typical TA mode involved in electron–phonon interactions we can make the approximation

$$\bar{n}_2-\bar{n}_3\approx\omega_1\frac{\mathrm{d}\bar{n}_3}{\mathrm{d}\omega_3} \tag{10.38}$$

and use the integration

$$\int_0^\infty \frac{x^4\mathrm{e}^x}{(\mathrm{e}^x-1)^2}\,\mathrm{d}x=\Gamma(5)\zeta(4) \tag{10.39}$$

where $\Gamma(z)$ is the gamma function and $\zeta(z)$ is Riemann's zeta function ($\zeta(4)=1.0823$). The rate is then

$$\frac{1}{\tau_{\mathrm{T}}}=\frac{6\gamma^2\hbar\omega_1(k_{\mathrm{B}}T)^4\zeta(4)}{\pi\rho v_{\mathrm{T}}v_{\mathrm{L}}^4\hbar^4} \tag{10.40}$$

which is appreciable—at room temperature, $\tau_{\mathrm{T}}\approx 10\,\mathrm{ps}$. This curious property, that only TA phonons can relax rapidly to 'thermal' phonons, was noted first by Landau and Rumer (1937), and the interaction described above is usually referred to as the Landau–Rumer process.

When one of the LA modes is promoting, the TA mode is p-polarized and this introduces some angular dependence. Approximating the polarization factor in each case by its spherical average leads to

$$\frac{1}{\tau_{\mathrm{T}}}=\frac{6\gamma^2\hbar\omega_1(k_{\mathrm{B}}T)^4\zeta(4)}{\pi\rho v_{\mathrm{T}}v_{\mathrm{L}}^4\hbar^4}\left(1+\frac{2}{3}\frac{v_{\mathrm{T}}}{v_{\mathrm{L}}}\right). \tag{10.41}$$

At room temperature, τ_{T} for a mode with wavevector $2\times 10^6\,\mathrm{cm}^{-1}$ is about 10 ps. This is to be compared with 3 μs for an equivalent LA mode!

At low temperatures the approximation leading to the integration in eqn (10.39) fails. In fact $\bar{n}_2$ and $\bar{n}_3$ become negligible and the type 2 process cannot proceed and $\tau_{\mathrm{T}}^{-1}\rightarrow 0$.

10.2.5. Rates for umklapp processes

In a type 1 process a phonon converts into two other phonons, and all the wavevectors must lie in the first Brillouin zone. Clearly, for modest wavevector this can never be an umklapp process. However, in a type 2 process phonons of large wavevector can be created and umklapp processes become possible (Fig. 10.3).

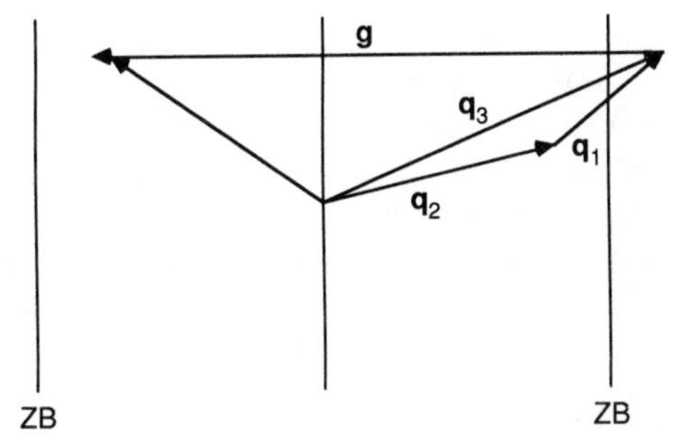

FIG. 10.3. Umklapp process.

Adding a reciprocal lattice vector **g** to the wavevector of the created mode in a type 2 process (which is always an LA mode) and ignoring dispersion allows us to write the conservation equations as follows:

$$v_1 q_1 + v_2 q_2 = v_{\mathrm{L}} \, |\mathbf{q}_3 + \mathbf{g}| \tag{10.42a}$$

$$|\mathbf{q}_3 + \mathbf{g}|^2 = q_1^2 + q_2^2 + 2q_1 q_2 \cos\theta_{12}. \tag{10.42b}$$

Restricting interest once more to the lifetime of comparatively low-frequency modes we can assume that maximum rates occur when $q_2 \approx |\mathbf{q}_3 + \mathbf{g}| \gg q_1$, i.e. when there is a large density of final states. Hence

$$|\mathbf{q}_3 + \mathbf{g}| \approx q_2\left(1 + \frac{q_1}{q_2}\cos\theta_{12}\right) \tag{10.43}$$

and therefore, approximately, $q_2 \approx |\mathbf{q}_3 + \mathbf{g}| \approx g/2$. The rate is thus

$$\frac{1}{\tau} = \frac{\gamma^2 \hbar \omega_1}{4\pi\rho v_{\mathrm{p}}^2 v_2} \int |\mathbf{e}_i \mathbf{e}_j^*|^2 \, \omega_2 \omega_3 (\bar{n}_2 - \bar{n}_3) \, \mathrm{d}(-\cos\theta_{12}) q_2^2 \tag{10.44}$$

where the subscript p denotes the promoting mode and the polarization vector subscripts denote the passive modes. Because ω_1 is small we can use

$$\bar{n}_2 - \bar{n}_3 = -\omega_1 \frac{\mathrm{d}\bar{n}_2}{\mathrm{d}\omega_2} = \frac{\hbar\omega_1}{k_{\mathrm{B}}T} \frac{\exp(\hbar\omega_2/k_{\mathrm{B}}T)}{\{\exp(\hbar\omega_2/k_{\mathrm{B}}T) - 1\}^2} \tag{10.45}$$

giving

$$\frac{1}{\tau} = \frac{\gamma^2 \hbar^2 \omega_1^2}{4\pi\rho v_{\mathrm{p}}^2 k_{\mathrm{B}} T} \left(v_{\mathrm{L}} \left(\frac{g}{2}\right)^4 \frac{\exp(\hbar v_{\mathrm{T}} g/2k_{\mathrm{B}}T)}{\{\exp(\hbar v_{\mathrm{T}} g/2k_{\mathrm{B}}T) - 1\}^2} \right) \int |\mathbf{e}_i \cdot \mathbf{e}_j^*|^2 \, \mathrm{d}(\cos\theta_{12}) \tag{10.46}$$

and finally

$$\frac{1}{\tau_{\mathrm{L}}} = \frac{\gamma^2 \hbar^2 \omega_1^2}{4\pi\rho v_{\mathrm{T}}^2 k_{\mathrm{B}} T} \left(v_{\mathrm{L}} \left(\frac{g}{2}\right)^4 \frac{\exp(\hbar v_{\mathrm{T}} g/2k_{\mathrm{B}}T)}{\{\exp(\hbar v_{\mathrm{T}} g/2k_{\mathrm{B}}T) - 1\}^2} \right) F_{\mathrm{v}}(v_{\mathrm{T}}/v_{\mathrm{L}}). \tag{10.47}$$

This rate, at high temperature, is bigger than that for LA normal processes (eqn (10.32)) by a factor $(v_{\mathrm{L}}g/\omega_1)^2(v_{\mathrm{L}}/v_{\mathrm{T}})$, and so umklapp processes allow LA modes to thermalize more rapidly. Rates for TA modes can be calculated in a similar way, and are of comparable magnitude, but because normal processes already give appreciable rates, umklapp rates are not as important as for LA modes. For the latter, with $\hbar v_{\mathrm{T}}g/2k_{\mathrm{B}}T=1$ and $\omega_1=10^{12}\,\mathrm{s}^{-1}$, T_{L} is about 200 ps. Obviously umklapp processes cannot proceed effectively at low temperatures.

10.2.6. Higher-order processes

The general features of third-order processes stem from the proportionality of $|H_3|^2$ to ω^3 and that of the density of states to ω^2, thus giving a basic rate proportional to ω^5. Some of the frequencies are replaced by $k_{\mathrm{B}}T/\hbar$, say n of them, and so the three-phonon rate is of the general form

$$\frac{1}{\tau}\sim\omega^{5-n}(k_{\mathrm{B}}T/\hbar)^n. \tag{10.48}$$

Fourth-order processes have $|H_4|^2$ proportional to ω^4, and the phase-space integrals add ω^5 (volume plus surface). Thus, in this case,

$$\frac{1}{\tau}\sim\omega^{9-n}(k_{\mathrm{B}}T/\hbar)^n. \tag{10.49}$$

Such processes actually contribute to the LA rate as significantly as third-order normal processes, but in most cases they can be ignored, and we will not consider them further.

10.2.7. Lifetime of optical phonons

Long-wavelength phonons decay by converting into two acoustic modes† (Fig. 10.4). This is always a type 1 process. With the optical phonon (LO or TO) as the promoting mode, the Hamiltonian is

$$H_{3,0}=2\Gamma\sum_{\mathbf{r},i,j}M\omega_i\omega_j\mathbf{u}_i\mathbf{u}_ju_0 \tag{10.50}$$

where M is the mass of the unit cell. The reciprocal lifetime is therefore

$$\frac{1}{\tau_0}=\frac{2\pi\Gamma^2\hbar^2}{NM\omega_0}\int|\mathbf{e}_2\cdot\mathbf{e}_3|^2\omega_2\omega_3(\bar{n}_2+\bar{n}+1)\delta(\hbar\omega_0-\hbar\omega_2-\hbar\omega_3)\,\mathrm{d}N. \tag{10.51}$$

The factor of 2 appears as a consequence of the reduced mass of the optical mode. Conservation of crystal momentum entails that

$$q_0^2=q_2^2+q_3^2+2q_2q_3\cos\theta_{23} \tag{10.52}$$

† LO modes can also decay into TO+LA or TO+TA. In GaN, for example, the process LO=LA+LA is impossible, because energy cannot be conserved.

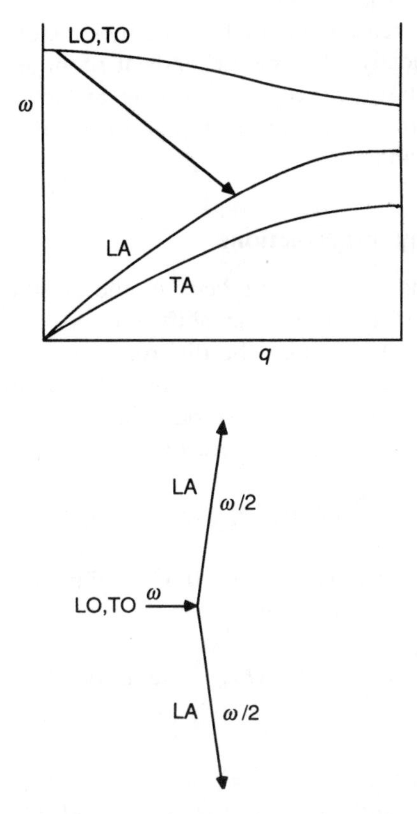

FIG. 10.4. Conversion of an optical phonon into two LA modes.

and since q_0 is small we can take $q_2 \approx q_3$, which also implies that $\omega_2 \approx \omega_3 \approx \omega_0/2$. When dispersion of the acoustic branches is taken into account it is usually the case that energy conservation can be satisfied only by LA modes. Since $\cos\theta_{23} \approx -1$ we can take $|\mathbf{e}_2 \cdot \mathbf{e}_3|^2 = 1$. The integrations are straightforward and we obtain, taking the optical oscillator mass to be $M/2$,

$$\frac{1}{\tau_0} = \frac{\Gamma^2 \hbar \omega_0^3 \{2n(\omega_0/2) + 1\}}{32\pi\rho v_L^3}. \tag{10.53}$$

The equation is the expression derived by Klemens (1966) if Γ is put equal to $(\frac{4}{3})^{1/2}\gamma\omega_0/v_L$.

When one of the LA modes is promoting, the interaction is practically ruled out for LO modes because the polarization factor is negligible, but it survives for TO modes. For this case we can replace the polarization factor by its spherical average and exchange Γ^2 for $\gamma^2\omega_0^2/4v_L^2$. Thus, for TO modes, eqn (10.53) holds but with Γ^2 replaced by $\Gamma^2 + \gamma^2\omega_0^2/12v_L^2$, amounting to an increase over the LO rate by a factor of roughly $(1+\frac{1}{12})$. The lifetime is typically a few picoseconds at room temperature.

Type 2, quasi-elastic processes involving an acoustic mode are also possible provided the dependence of ω_0 on q is taken into account, but the rates are very small.

10.3. Scattering by imperfections

Deviations from the perfect periodicity of the crystal lattice can scatter phonons by virtue of the frequency shift induced by differences in mass and force constants. Very often the difference in force constants can be neglected. Basically the interaction is a two-phonon process in which an incoming phonon is absorbed and an outgoing one is created (Fig. 10.5). From eqns (10.5) and (10.8) the perturbation is

$$H_2 = \frac{1}{N}\sum_{\bar{r},i,j} \hbar\delta\omega \mathbf{e}_i \cdot \mathbf{e}_j^* \exp\{i(\mathbf{q}_j - \mathbf{q}_i)\cdot \mathbf{r}\}a_i a_j^+ \tag{10.54}$$

in which we take $\delta\omega = 0$ except at $\mathbf{r} = 0$. Crystal momentum in this case is not conserved. The net scattering rate is

$$\frac{1}{\tau} = \frac{\hbar V(\delta\omega)^2}{4\pi^2 N^2}\int |\mathbf{e}_i \cdot \mathbf{e}_j^*|^2\, \delta(\hbar\omega_i - \hbar\omega_j) q_j^2\, \mathrm{d}q_j\, \mathrm{d}(-\cos\theta_{ij})\, \mathrm{d}\phi. \tag{10.55}$$

For LA and LO modes $|\mathbf{e}_i \cdot \mathbf{e}_j^*|^2 = \cos^2\theta_{ij}$, for s-polarized TA and TO modes $|\mathbf{e}_i \cdot \mathbf{e}_j^*|^2 = 1$, and for p-polarized TA and TO modes $|\mathbf{e}_i \cdot \mathbf{e}_j^*|^2 = \sin^2\theta_{ij}$. Introduction of the group velocity, $v_{gj} = \mathrm{d}\omega_j/\mathrm{d}q_j$ allows us to recast eqn (10.55) as follows:

$$\frac{1}{\tau} = \frac{V(\delta\omega)^2 q^2}{2\pi N^2 v_g}\int |\mathbf{e}_i \cdot \mathbf{e}_j|^2\, \mathrm{d}(-\cos\theta_{ij}) \tag{10.56}$$

where V is the volume of the cavity.

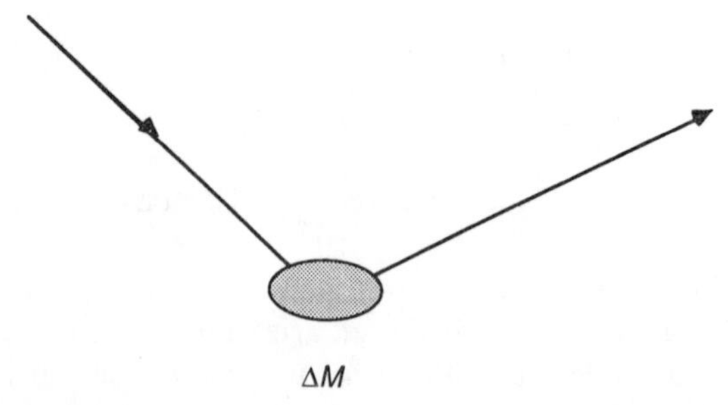

FIG. 10.5. Elastic scattering from an imperfection with mass difference ΔM.

Let us suppose that scattering sites are scattered randomly so that in fact we have an alloy of the chemical form A_xB_{1-x}. The mean square perturbation on each site is therefore

$$\langle \delta\omega \rangle^2 = (\Delta\omega_{AB})^2 x(1-x) \tag{10.57}$$

and so, taking $\frac{2}{3}$ for the integral, and with V_0 equal to the volume of a unit cell,

$$\frac{1}{\tau} = \frac{\Delta\omega_{AB}^2 q^2 V_0 x(1-x)}{3\pi v_g}. \tag{10.58}$$

This equation describes alloy scattering as well as dilute impurity scattering. Thus, $\Delta\omega_{AB}$ is the difference of frequencies between the two pure materials in an alloy, and it is related to a small mass difference δM in the case of impurity or isotope scattering according to $(\Delta\omega_{AB})^2 = \omega^2(\delta M/2M)^2$.

A momentum-relaxation rate can be defined in the usual way for elastic collisions by weighting the integral in eqn (10.56) by the factor $(1-\cos\theta_{ij})$, but because the scattering is random the factor averages to unity. There is no distinction between scattering and momentum-relaxation rates.

In the case of acoustic modes scattering off isolated imperfections whose density is N_I, eqn (10.58) is

$$\frac{1}{\tau} = \frac{(\delta M/2M)^2 V_0^2 \omega^4 N_I}{3\pi v_{L,T}^3} \tag{10.59}$$

which has the characteristic frequency dependence of Rayleigh scattering. For optical modes the group velocity is not, as in the case of dispersionless acoustic modes, equal to the phase velocity. Taking a quadratic dependence of frequency on wavevector near the zone centre, namely

$$\omega_0^2 = \omega_0^2 - v^2 q^2 \tag{10.60}$$

where v is of order of the acoustic velocity, we get

$$v_g = v^2 q/\omega_0 \tag{10.61}$$

and hence

$$\frac{1}{\tau} = \frac{(\delta M/2M)^2 V_0^2 \omega_0^3 q N_I}{3\pi v^2}. \tag{10.62}$$

This rate is much bigger than that for an acoustic mode of similar wavevector, but even so it is small compared with the reciprocal lifetime of an optical phonon for impurity concentrations of order $10^{18}\,\mathrm{cm}^{-3}$ and less. Substantial rates ($\gtrsim 10^{11}\,\mathrm{s}^{-1}$) generally require the large concentrations found in alloys or associated with isotopes, but large rates can also occur in the presence of impurity clusters. Clustering, in fact, can increase the rate by a factor equal to the number of atoms in the cluster.

Apart from mass and force-constant differences introduced by impurities one might think that there may be effects associated with the strain caused by the disparity in radii. A static displacement field surrounding the impurity is of the form (Love 1927, Landau and Lifshitz 1986)

$$u(r) = Ar + \frac{B}{r^2} \tag{10.63}$$

provided the effect of the impurity is like applying hydrostatic pressure to a spherical surface around the impurity. This field has the property $\operatorname{div} \mathbf{u} = 3A$, i.e. its divergence is a constant everywhere, which means the volume change is constant everywhere. Consequently this can have no effect on travelling elastic waves.

10.4. Scattering by charged impurities

An interaction quite different from the anharmonic interaction can occur in polar materials when the imperfection carries a charge and the phonon is an LO phonon (Fig. 10.6). Scattering occurs in a way analogous to the electron–LO phonon case, i.e. via the Fröhlich interaction (Ridley and Gupta 1991). The charge on the unit cell containing the impurity is the sum of the basic impurity charge, Ze, and the surface charge associated

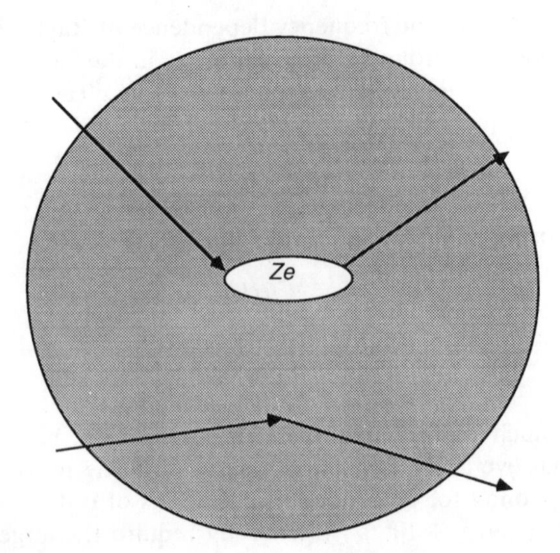

FIG. 10.6. Fröhlich scattering from charged impurity and from surrounding polarization.

with the surrounding polarization, that is,

$$e_{\mathrm{I}} = Ze + e_{\mathrm{s}}. \tag{10.64}$$

Taking a spherical cavity of radius r_0 we can write the surface charge as $e_{\mathrm{s}} = 4\pi r_0^2 \sigma_{\mathrm{s}}$ where σ_{s} is the surface density of charge. Standard theory equates this to $-P$, where P is the normal component of the polarization. Now $\mathbf{P}$ is related to the electric displacement $\mathbf{D}$ in the usual way, i.e. $\mathbf{D} = \epsilon_0 \mathbf{E} + \mathbf{P}$, and $D = Ze/4\pi r^2$, $E = Ze/4\pi\epsilon_{\mathrm{s}} r^2$, where ϵ_{s} is the static permittivity. Thus $P = (Ze/4\pi r^2)(1 - \epsilon_0/\epsilon_{\mathrm{s}})$ and hence $e_{\mathrm{s}} = -Ze(1 - \epsilon_0/\epsilon_{\mathrm{s}})$, and it follows that

$$e_{\mathrm{I}} = Ze(\epsilon_0/\epsilon_{\mathrm{s}}). \tag{10.65}$$

Since $\nabla \cdot \mathbf{P} = 0$, there is no bulk charge in the surrounding medium and so the interaction energy is simply $e_{\mathrm{I}}\phi$, where ϕ is the scalar potential associated with the LO mode, namely,

$$\phi = \frac{\mathrm{i}e^* u}{V_0 \epsilon_0 q N^{1/2}} \tag{10.66}$$

where e^* is the Callen effective charge (see Section 3.5).

The rate, to first order, is zero, because we assume that no energy can be transferred to the impurity. An elastic scattering process is allowed only in second order, thus

$$\frac{1}{\pi} = \frac{2\pi}{\hbar} \int \sum_n \left| \frac{\langle f|\, e_{\mathrm{I}}\phi \,|n\rangle \langle n|\, e_{\mathrm{I}}\phi \,|\mathrm{i}\rangle}{E_{\mathrm{i}} - E_n} \right|^2 \delta(E_{\mathrm{f}} - E_{\mathrm{I}})\, \mathrm{d}N_{\mathrm{f}}. \tag{10.67}$$

The initial state contains our LO mode of wavevector $\mathbf{q}$, the intermediate state corresponds to the phonon being absorbed, and the final state contains an emitted LO phonon of wavevector $\mathbf{q}'$. For a density N_{I} of charged impurities the scattering rate is

$$\frac{1}{\tau} = \frac{(Ze)^4 \omega N_{\mathrm{I}}}{4\pi(\epsilon_{\mathrm{s}}/\epsilon_0)^4 \hbar^2 v^2 q^3} \left(\frac{1}{\epsilon_\infty} - \frac{1}{\epsilon_{\mathrm{s}}} \right)^2. \tag{10.68}$$

As usual for polar interactions, the rate diverges at $q \to 0$. This divergence disappears in the presence of screening electrons. For GaAs with $Z = 1$ and $q = 2 \times 10^6\,\mathrm{cm}^{-1}$ (the typical wavevector involved in electron–LO interactions) the rate is about $10^{-7} N_{\mathrm{I}}\,\mathrm{s}^{-1}$, N_{I} in cm^{-3}. Compared with the electron-scattering rate this is small, but compared with the LO phonon lifetime it is appreciable when $N_{\mathrm{I}} \gtrsim 10^{18}\,\mathrm{cm}^{-3}$. However, the rate increases rapidly as q diminishes.

A similar interaction will be present for piezoelectric modes, but the effect is likely to be much weaker.

Associated with the polarization in a displacement field surrounding the impurity. The total polarization has electronic and ionic components, the electronic part being $(Ze/4\pi r^2)(1-\epsilon_0/\epsilon_\infty)$. Thus the ionic component is $(Ze/4\pi r^2)(\epsilon_0/\epsilon_\infty - \epsilon_0/\epsilon_s)$, where ϵ_∞, ϵ_s are the high-frequency and static permittivities, and since this is related to the relative ionic displacement via $e^*u(\mathbf{r})/V_0$ (see Section 3.5), we have a displacement field given by

$$u(\mathbf{r}) = (ZeV_0/4\pi r^2 e^*)(\epsilon_0/\epsilon_\infty - \epsilon_0/\epsilon_s). \tag{10.69}$$

This field promotes an anharmonic interaction via a radially varying frequency shift $\delta\omega/\omega = \Gamma u(\mathbf{r})$ which gives rise to first-order elastic scattering with the Hamiltonian given by eqn (10.54). Carrying out the sum over the lattice sites reduces the Hamiltonian to

$$H_2 = \frac{\pi Ze\Gamma\hbar\omega_0}{2e^*N\,|\mathbf{q}_j - \mathbf{q}_i|}(\epsilon_0/\epsilon_\infty - \epsilon_0/\epsilon_s)\cos\theta\, a_i a_j^+ \tag{10.70}$$

where $\mathbf{q}_i$, $\mathbf{q}_j$ are the incident and scattered wavevectors and θ is the scattering angle. Attempting to compute the rate in the usual way we run into a familiar problem—associated with Coulombic fields—that is, the rate is infinite because of the preponderance of small-angle scattering. This problem was encountered in the context of charged-impurity scattering of electrons, and the discussion of Section 4.2 is immediately relevant. Fortunately, the situation here is not as severe because the rate involves $|\mathbf{q}_j - \mathbf{q}_i|^{-2}$ rather than $|\mathbf{q}_j - \mathbf{q}_i|^{-4}$. The divergence can be removed by weighting the rate by the factor $(1-\cos\theta)$, since $|q_j - q_i|^2 = 2q^2(1-\cos\theta)$, and so using eqn (3.150) to substitute for e^*, and assuming a quadratic form for the dispersion (eqn (10.60), we obtain a momentum relaxation rate given by

$$\frac{1}{\tau_m} = \frac{\pi Z^2 e^2 \Gamma^2 \omega N_I}{24\epsilon_0 \rho v^2 q}(\epsilon_0/\epsilon_\infty - \epsilon_0/\epsilon_s). \tag{10.71}$$

For GaAs with $Z=1$ and $q = 2\times 10^6\ \mathrm{cm}^{-1}$, $\tau_m^{-1} = 5{\cdot}7\times 10^{-9} N_I\ \mathrm{s}^{-1}$, N_I in cm^{-3}, which is an order of magnitude less than the direct Fröhlich rate. This rate will be sensitive to screening by free carriers and by statistical screening (Section 4.2.4). The anharmonic interaction also allows TO modes to be scattered by charged impurities.

10.5. Scattering by electrons

The interaction of phonons with electrons and holes has been described in Chapter 3, with specific rates derived from the point of view of the electron. To obtain the equivalent rates for a phonon it is only necessary to integrate over the electron population, including spin degeneracy, instead of over phonon states, taking into account energy and momentum conservation. Here we will limit ourselves to quoting the result for the unscreened Fröhlich interaction (Section 3.5), which is

$$\frac{\mathrm{d}n(q)}{\mathrm{d}t}=\frac{1}{2\tau_0}\left(\frac{\hbar\omega}{E_q^3}\right)^{1/2}$$

$$\times\left(\int_{E_1}^{\infty} f(E)\{1-f(E-\hbar\omega)\}\,\mathrm{d}E-n(q)\int_{E_1-\hbar\omega}^{E_1} f(E)\{1-f(E+\hbar\omega)\}\delta E\right) \tag{10.72}$$

where τ_0 is the characteristic time-constant for LO-phonon emission, given by

$$\frac{1}{\tau_0}=\frac{e^2\omega}{2\pi\hbar}\left(\frac{m^*}{2\hbar\omega}\right)^{1/2}\left(\frac{1}{\epsilon_\infty}-\frac{1}{\epsilon_\mathrm{s}}\right) \tag{10.73a}$$

and

$$E_q=\frac{\hbar^2q^2}{2m^*} \qquad E_1=\frac{(\hbar\omega+E_q)^2}{4E_q} \tag{10.73b}$$

and $f(E)$ is the distribution function for electrons. When the latter is a Maxwellian characterized by an electron temperature T_e we obtain

$$\frac{\mathrm{d}n(q)}{\mathrm{d}t}=\frac{1}{2\tau_0}\left(\frac{\hbar\omega}{E_q^3}\right)^{1/2}\frac{nk_\mathrm{B}T_\mathrm{e}}{N_\mathrm{c}}\exp(-E_1/k_\mathrm{B}T_\mathrm{e})$$

$$\times[1+n(q)\{1-\exp(\hbar\omega/k_\mathrm{B}T_\mathrm{e}\}] \tag{10.74}$$

where n is the carrier density and N_c is the effective density of states. At low temperatures ($k_\mathrm{B}T_\mathrm{e}\ll\hbar\omega$) this rate maximizes when $E_\mathrm{q}=\hbar\omega$. Note that when the phonon occupation number, $n(q)$, is given by the

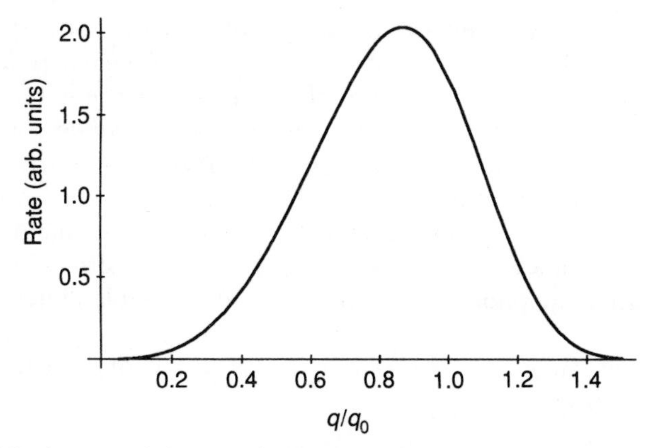

FIG. 10.7. LO phonon emission rate (arbitrary units) from a non-degenerate electron gas as a function of wavevector.

Bose–Einstein expression for a temperature T_e, the rate vanishes. Equation (10.72) is used in the description of hot-phonon effects in a polar semiconductor. As in the case of the Fröhlich interaction with charged impurities (eqn (10.68)) the basic rate is inversely proportional to the cube of the wavevector. Figure 10.7 shows the dependence on wavevector of the rate in eqn (10.74).

10.6. Other scattering mechanisms

In doped semiconductors phonons can interact with electrons bound in impurity levels in three ways. The phonon can be scattered elastically in a second-order process analogous to that discussed in Section 10.4, say via the deformation potential. Inelastic scattering may occur accompanied by a transition of an electron from one state to another—another second-order process. The phonon can interact with a second phonon as well as the electron in the impurity. The interaction can be particularly strong in the case of magnetic impurities such as Fe^{2+} which have a spread of levels allowing resonant phonon scattering to occur. The same is true of molecular impurities which can have vibrational levels which can interact resonantly with microwave phonons. One-phonon resonance rates are of the form

$$\frac{1}{\tau} \propto \frac{\Gamma}{(\omega - \omega_0)^2 + \Gamma^2} \tag{10.75}$$

where Γ is the level width and $\hbar\omega_0$ is the energy separation of the levels.

In the purest materials at low temperatures scattering is eventually determined by boundary scattering (Fig. 10.8). The efficacy of boundary scattering depends upon the particular properties of the surfaces—essentially how rough they are, a measure of which is not easy to obtain and in any case it will depend upon wavelength. Specular reflections do not count. However, boundary conditions, such as the requirement that a free surface is stress free, entail the phenomenon of mode conversion in which, say, an LA mode incident on a perfect surface triggers a TA mode, as well as reflecting non-specularly. This combination of surface roughness and mode conversion makes the detailed treatment of boundary scattering somewhat involved. The problem is often bypassed by simply equating the phonon mean free path to a suitable dimension of the specimen.

Dislocations induce elastic strain fields each of which have the approximate form

$$S(r) = \frac{b}{2\pi r}\begin{Bmatrix} \sin\theta \\ \cos\theta \end{Bmatrix} \tag{10.76}$$

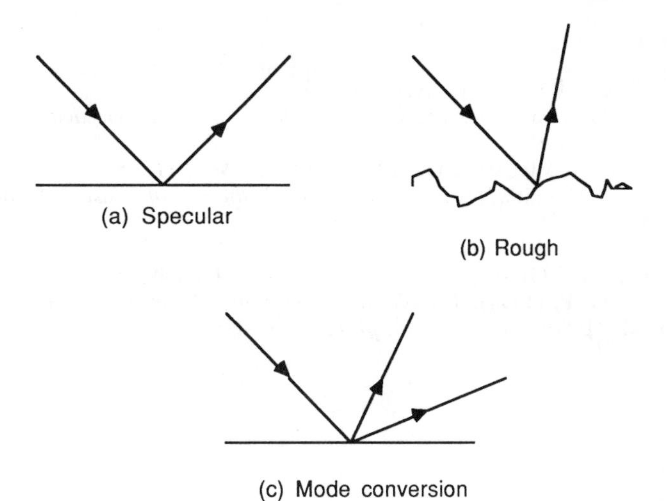

FIG. 10.8. Boundary scattering.

where b is the Burgers vector, and induce frequency shifts according to $\delta\omega/\omega = \gamma S(r)$. Computation of the first-order scattering rates for acoustic modes runs into the same difficulty as we have in Section 10.4 in that the rate diverges as $|\mathbf{q}_j - \mathbf{q}_i|^2$, which forces us to look at the momentum-relaxation rate. This is of the form

$$\frac{1}{\tau_\mathrm{m}} = Cb^2\omega N_\mathrm{d} \tag{10.77}$$

where N_d is the areal density of dislocations and C is a numerical constant. Note that this simple formula arises only by virtue (if it is a virtue) of neglecting the differences between edge and screw dislocations, between dilatation and shear strains, and between LA and TA phonons. It also neglects the fact that at the core of the dislocation the atomic displacements are so large that they fall outside of elasticity theory. Core effects might be expected to add a rate of the form (Ziman 1960).

$$\frac{1}{\tau_\mathrm{m}} = C'va(qa)^3 N_\mathrm{d} \tag{10.78}$$

where a is the core radius.

Stacking faults and grain boundaries also scatter phonons but are not likely to be present in experiments on semiconductor physics, given the remarkably high quality of crystals which are typically thought worth studying.

References

Klemens, P. G. (1966). *Phys. Rev.* **148,** 845.
Landau, L. D. and Lifshitz, E. M. (1986). *Theory of elasticity* (3rd edn), Pergamon, Oxford.
Landau, L. D. and Rumer, G. (1937). *Phys. Z. Sov.* **11,** 18.
Love, A. E. H. (1927). *The mathematical theory of elasticity* (4th edn), Cambridge.
Reissland, J. A. (1973). *The physics of phonons,* Wiley, London.
Ridley, B. K. and Gupta, R. (1991). *Phys. Rev.* **43B,** 4935.
Srivastava, G. P. (1990). *The physics of phonons,* Adam Hilger, Bristol.
Ziman, J. M. (1960). *Electrons and phonons,* Oxford.

11. Quantum transport

11.1. The density matrix

In practice, by far the majority of cases encountered are describable in terms of so-called 'pure' quantum states. Hitherto, in our description of scattering processes we have assumed that the electron begins in a well-defined initial state and ends up in an equally well-defined final state, and that the scattering process conserves energy and crystal momentum and occurs at a rate given by Fermi's Golden Rule. The transition rates involved Planck's constant, which made them quantum-theoretic, but otherwise the processes were basically classical, involving well-defined dynamic states. Such states can only be approximations to true quantum states, and we need to understand what their validity is. Moreover, for the most part we have focused on individual events and ignored the behaviour of populations. But we seldom observe individual events. We measure, for example, the electrical resistance of a piece of semiconductor that contains anything between, typically, 10^{10} and 10^{20} electrons or holes per cubic centimetre. These are huge numbers, implying huge numbers of dynamical states among which the carriers are distributed. As we can never know precisely the dynamic state of each particle, we need to resort to the concepts of statistical physics in order to describe the observable properties of the system. It is therefore necessary to incorporate both quantum and statistical concepts into our description of semiconductor physics. This can be achieved via the concept of the density matrix.

A comprehensive treatment involving many-body effects is beyond the scope of this book. Here we limit our account to the description of systems that can be reasonably well described in terms of single-particle states. A convenient shorthand for these states is the bra and ket notation, so that the Schrödinger equation can be written

$$\begin{aligned} i\hbar|\dot{\alpha}\rangle &= H|\alpha\rangle, \\ -i\hbar\langle\dot{\alpha}| &= \langle\alpha|H. \end{aligned} \tag{11.1}$$

The state $|\alpha\rangle$ can be taken to be a member of a normalized, orthonormal set that may describe, for example, the electronic eigenstates of a simple two-level system or, at the other extreme, the Bloch states of a conduction band.

We can introduce a statistical element by relating the probability of occupation, f_α, of the state $|\alpha\rangle$ to p_α, where

$$p_\alpha = \langle\alpha|f_\alpha|\alpha\rangle. \tag{11.2}$$

Obviously, in this case p and f are identical: p is the expectation value of the probability of occupation. This is fine for pure states, but it is a formalism that cannot deal with mixed states; i.e. when the electron in some non-classical sense occupies two states at once. In such a case we have to generalize the concept of occupation probability as the expectation value of some probability operator:

$$p_{\alpha'\alpha} = \langle \alpha' | \rho_{\alpha\alpha'} | \alpha \rangle. \tag{11.3}$$

This probability operator is what we call the single-particle density matrix element, and we define it as follows:

$$\rho_{\alpha\alpha'} = |\alpha\rangle f_{\alpha'\alpha} \langle \alpha'|. \tag{11.4}$$

The diagonal terms in this matrix give the occupation probabilities, but what do the non-diagonal terms mean? These are non-classical quantities whose nearest classical referents are the dipole moments associated with electrons oscillating between two states. They can be expected to assume importance at very short time intervals during which states couple coherently, and before that coherence is destroyed by perturbations within the system or by interactions with the rest of the universe.

The Schrödinger equation defines the time dependence:

$$i\hbar\dot{\rho}_{\alpha\alpha'} = H\rho_{\alpha\alpha'} - \rho_{\alpha\alpha'}H \tag{11.5}$$

with, in general, $H = H_0 + H_1$, where H_0 is a time-independent Hamiltonian and H_1 is a perturbation. Note that in the absence of any perturbation the states can be depicted by $|k\rangle$ and

$$\begin{aligned} i\hbar\dot{\rho}_{kk'} &= H_0|k\rangle f_{k'k}\langle k'| - |k\rangle f_{k'k}\langle k'|H_0 \\ &= E_k|k\rangle f_{k'k}\langle k'| - |k\rangle f_{k'k}\langle k'|E_{k'}, \\ i\hbar\langle k'|\dot{\rho}_{kk'}|k\rangle &= E_k f_{k'k}\delta_{k'k} - f_{k'k}E_{k'}\delta_{k'k} = 0 \end{aligned} \tag{11.6}$$

where E_k and $E_{k'}$ are eigenvalues of the unperturbed system. Furthermore,

$$\langle k'|\rho_{kk'}|k\rangle = \langle k'|k\rangle f_{k'k}\langle k'|k\rangle = f_{k'k}\delta_{k'k} \tag{11.7}$$

i.e. the density matrix reduces to its diagonal terms. At this point, a connection can be made with the electron density:

$$n = \frac{1}{\Omega}\sum_k f_k = \frac{1}{\Omega}\sum_k \langle k'|\rho_{kk'}|k\rangle \tag{11.8}$$

where Ω is the cavity volume.

Now consider the case when there exists a perturbation H_1 such that $H_1 \ll H_0$. We expand the perturbed states in terms of the unperturbed states (labelled r):

$$|\alpha\rangle = \sum_r |r\rangle\langle\alpha|r\rangle \tag{11.9}$$

and put

$$\rho_{\alpha\alpha'} = \rho_{0\alpha\alpha'} + \rho_{1\alpha\alpha'}, \qquad \rho_{1\alpha\alpha'} \ll \rho_{0\alpha\alpha'}. \tag{11.10}$$

It is straightforward to show, and the reader may care to verify, that the equation of motion for expectation values becomes

$$\begin{aligned} i\hbar\langle k'|\dot{\rho}_{1kk'}|k\rangle &= (f_k - f_{k'})\langle k'|H_1|k\rangle + (E_{k'} - E_k)\langle k'|\rho_{1kk'}|k\rangle \\ &\quad + \sum_r [\langle k'|H_1|r\rangle\langle r|\rho_{1kr}|k\rangle - \langle k'|\rho_{1kr}|r\rangle\langle r|H_1|k\rangle]. \end{aligned} \tag{11.11}$$

Note that the sum consists of second-order terms. These are important to retain, because for $k' = k$ the first-order terms vanish. Equation (11.11) can therefore be decomposed into two equations, one for the off-diagonal elements and one for the diagonal elements:

$$\begin{aligned} i\hbar\langle k'|\dot{\rho}_{1kk'}|k\rangle &\approx (f_k - f_{k'})\langle k'|H_1|k\rangle + (E_{k'} - E_k)\langle k'|\rho_{1kk'}|k\rangle, \quad k' \neq k \\ i\hbar\langle k|\dot{\rho}_{1kk}|k\rangle &= \sum_r [\langle k|H_1|r\rangle\langle r|\rho_{1kr}|k\rangle - \langle k|\rho_{1kr}|r\rangle\langle r|H_1|k\rangle]. \end{aligned} \tag{11.12}$$

The first of these equations describes the rate of change of the polarization of the electron population—a first-order effect—while the second describes the rate of change of occupancy—a second-order effect. The first response to a perturbation is therefore a mixing of states, which manifests itself in a polarization of the electron gas. Transitions from one state to another, the sort of scattering that we have been mostly concerned with, is a much weaker affair.

11.2. Screening

We gave an outline of the dielectric response of an electron gas in Section 9.5. Here we use the single-particle density-matrix approach to derive the Lindhard function.

Consider a model perturbation of the form

$$\begin{aligned} H_1 &= 2V_0 e^{\delta t}\cos(\omega t + \phi) \\ &= e^{\delta t}(Ve^{i\omega t} + V^* e^{-i\omega t}). \end{aligned} \tag{11.13}$$

The polarization or, in other words, the dielectric response, of the system is obtained from the first of eqn (11.12) and is given by

$$\langle k'|\rho_{1kk'}|k\rangle = (f_{k'} - f_k)e^{\delta t}\left[\frac{\langle k'|V|k\rangle e^{i\omega t}}{\hbar\omega_{k'k} + \hbar\omega - i\hbar\delta} + \frac{\langle k'|V^*|k\rangle e^{-i\omega t}}{\hbar\omega_{k'k} - \hbar\omega - i\hbar\delta}\right] \tag{11.14}$$

where $\hbar\omega_{k'k} = E_{k'} - E_k$. We can relate this to a perturbation of the electron density as follows:

$$n_1 = \frac{1}{\Omega}\sum_{k'k}\langle k'|\rho_{1kk'}|k\rangle. \tag{11.15}$$

If we imagine that there exists a fixed background positive charge that keeps the system neutral in the absence of the perturbation, this disturbance in the electron density gives rise to a space charge density and hence to a screening potential energy V_s related through Poisson's equation:

$$\nabla^2 V_s = -\frac{e^2 n_1}{\varepsilon} \tag{11.16}$$

where ε is the permittivity of the lattice. If V_0 is the unscreened perturbation, we have

$$V = V_0 + V_s \tag{11.17}$$

where V_s is linearly dependent on V.

We now specialize to the important case when the unperturbed states are the Bloch functions of the conduction band of a semiconductor. Making a Fourier expansion of the perturbation leads to a connection of states through the conservation of crystal momentum:

$$\begin{aligned} V &= \sum_{\mathbf{q}} V(\mathbf{q})e^{i\mathbf{q}\cdot\mathbf{r}}, \\ \langle k'|V|k\rangle &= V_{\mathbf{k}+\mathbf{q},\mathbf{k}}(\mathbf{q}). \end{aligned} \tag{11.18}$$

Also

$$V_s(\mathbf{q}) = \frac{e^2 n_1(\mathbf{q})}{\varepsilon q^2} \tag{11.19}$$

and hence, at $t=0$,

$$\begin{aligned} V_{\mathbf{k+q},\mathbf{k}}(\mathbf{q}) &= \frac{V_{0,\mathbf{k+q},\mathbf{k}}(\mathbf{q})}{\varepsilon(\omega,\mathbf{q})/\varepsilon}, \\ \frac{\varepsilon(\omega,\mathbf{q})}{\varepsilon} &= 1 - \frac{e^2}{\varepsilon q^2 \Omega} \sum_{\mathbf{k}} \frac{f_{\mathbf{k+q}} - f_{\mathbf{k}}}{E_{\mathbf{k+q}} - E_{\mathbf{k}} - \hbar\omega - \mathrm{i}\hbar\delta} \end{aligned} \tag{11.20}$$

which is the Lindhard expression used in eqn (9.43).

This can be evaluated with $\delta \to 0$ using

$$\frac{1}{z-\delta} \xrightarrow{\delta \to 0} P\left(\frac{1}{z}\right) + \mathrm{i}\pi\delta(z) \tag{11.21}$$

where P stands for the principal part, and the dielectric function has a real and an imaginary component:

$$\begin{aligned} \varepsilon(\omega,\mathbf{q}) &= \varepsilon_1(\omega,\mathbf{q}) + \mathrm{i}\varepsilon_2(\omega,\mathbf{q}), \\ \frac{\varepsilon_1(\omega,\mathbf{q})}{\varepsilon} &= 1 - \frac{e^2}{\varepsilon q^2 \Omega} \sum_{\mathbf{k}} \frac{f_{\mathbf{k+q}} - f_{\mathbf{k}}}{E_{\mathbf{k+q}} - E_{\mathbf{k}} - \hbar\omega}, \\ \frac{\varepsilon_2(\omega,\mathbf{q})}{\varepsilon} &= -\frac{\pi e^2}{\varepsilon q^2 \Omega} \sum_{\mathbf{k}} (f_{\mathbf{k+q}} - f_{\mathbf{k}})\delta(E_{\mathbf{k+q}} - E_{\mathbf{k}} - \hbar\omega). \end{aligned} \tag{11.22}$$

In general, the temperature dependence of these components precludes expressing them in simple analytical form; however, an informative analytical expression for the real part can be obtained for the case of a degenerate gas at $T=0\,\mathrm{K}$:

$$\begin{aligned} \frac{\varepsilon_1(\omega,q)}{\varepsilon} = 1 + \frac{e^2 N(E_F)}{2\varepsilon q^2}\Bigg\{1 - \frac{1}{4\eta^3}\Bigg[\{\gamma^2 - (1-2\gamma)\eta^2 + \eta^4\}\ln\left|\frac{\eta+\eta^2+\gamma}{\eta-\eta^2-\gamma}\right| \\ + \{\gamma^2 - (1+2\gamma)\eta^2 + \eta^4\}\ln\left|\frac{\eta+\eta^2-\gamma}{\eta-\eta^2+\gamma}\right|\Bigg]\Bigg\} \end{aligned} \tag{11.23}$$

where $N(E_F)$ is the density of states at the Fermi level, $\eta = q/2k_F$, k_F is the Fermi wavevector, and $\gamma = \hbar\omega/4E_F$. For slowly varying or static potentials $\gamma \to 0$ and

$$\frac{\varepsilon_1(0,q)}{\varepsilon} = 1 + \frac{e^2 N(E_F)}{2\varepsilon q^2}\left[1 + \left(\frac{k_F}{q} - \frac{q}{4k_F}\right)\ln\left|\frac{1+(q/2k_F)}{1-(q/2k_F)}\right|\right]. \tag{11.24}$$

This has the long-wavelength limit:

$$\frac{\varepsilon_1(0,q)}{\varepsilon} = 1 + \frac{q_0^2}{q^2}, \qquad q_0^2 = \frac{e^2 N(E_F)}{\varepsilon}. \tag{11.25}$$

where q_0 is the reciprocal of the Thomas–Fermi screening length. On the other hand, for ω finite, the long-wavelength result is (after carefully expanding the logarithms)

$$\frac{\varepsilon(\omega,0)}{\varepsilon} = 1 - \frac{\omega_p^2}{\omega^2}, \qquad \omega_p^2 = \frac{e^2 n}{\varepsilon m^*} \tag{11.26}$$

from which it may be seen that the dielectric function vanishes when $\omega = \omega_p$, the plasma frequency.

For a non-degenerate electron gas at finite temperatures, the calculation of the dielectric response function is much more difficult. The distribution function in eqn (11.20) can be taken to be the usual Maxwell–Boltzmann form and the dielectric function for electrons in a spherical, parabolic band can be expressed in the following terms:

$$\frac{\varepsilon(\omega,q)}{\varepsilon} = 1 - \frac{e^2 m^*}{4\pi^2 \varepsilon \hbar q^3 \lambda}\pi^{1/2}[Z(A_+) - Z(A_-)] \tag{11.27}$$

where $\lambda = (\hbar^2/2m^* k_B T)$ and $Z(w)$ is the plasma dispersion function:

$$Z(w) = \frac{1}{\sqrt{\pi}}\int_{-\infty}^{\infty} \frac{e^{-t^2}}{t-w}\,dt. \tag{11.28}$$

The complex arguments of Z are given by

$$A_\pm = \lambda^{1/2}\left(\frac{m^*\omega}{\hbar q} \pm \frac{q}{2} + i\delta\right). \tag{11.29}$$

The properties of $Z(w)$ have been explored in depth (Fried and Conte 1961). In numerical work $Z(w)$ is usually taken to be adequately represented by a so-called Padé approximant, which consists of a ratio of two polynomials:

$$Z(w) \approx \frac{p_0 + p_1 w}{1 + q_1 w + q_2 w^2}. \tag{11.30}$$

The trick is to choose the coefficients p_0, p_1, q_1, and q_2 to match the properties of the exact function as closely as possible. One such choice which has been shown to give good results (Lowe and Barker 1985) is

$$Z(w) = \frac{\mathrm{i}\sqrt{\pi} + (\pi - 2)w}{1 - \mathrm{i}\sqrt{\pi}w - (\pi - 2)w^2}. \tag{11.31}$$

Screening is basically a coherent process in which no energy is exchanged between the perturbing field and the system. However, the appearance of an imaginary component in the dielectric response alerts us to the inevitability of incoherent processes entering. The strength of these processes is usually fairly weak, so screening is not materially affected. Before looking at the rate at which incoherent processes occur it is useful to remain in a state of quantum-coherent bliss in order to briefly study the important case of the two-level system.

11.3. The two-level system

Consider two levels, labelled a and b, with equal probabilities of occupancy in the unperturbed state. (This will ensure that the system remains coherently connected to the perturbation.) Equation (11.12) becomes

$$\begin{aligned} \mathrm{i}\hbar\langle b|\dot{\rho}_{1ab}|a\rangle &= \hbar\omega_{ba}\langle b|\dot{\rho}_{1ab}|a\rangle \\ \mathrm{i}\hbar\langle a|\dot{\rho}_{1aa}|a\rangle &= \langle a|H_1|b\rangle\langle b|\rho_{1ab}|a\rangle - \langle a|\rho_{1ba}|b\rangle\langle b|H_1|a\rangle \end{aligned} \tag{11.32}$$

with a pair of similar equations with a and b interchanged. Converting the density matrix elements back into probability amplitudes c_a and c_b, i.e. $\langle b|\rho_{1ab}|a\rangle = c_a c_b^*$, and taking the perturbation to be of the form

$$H_1 = V_{\mathrm{e}}^{\mathrm{i}\omega t} + V^* \mathrm{e}^{-\mathrm{i}\omega t} \tag{11.33}$$

we obtain

$$\begin{aligned} \langle b|\rho_{ab}|a\rangle &= c_a c_b^* \mathrm{e}^{-\mathrm{i}\omega_{ba} t} \\ \langle b|\rho_{aa}|a\rangle &= -\frac{1}{\hbar}\left\{\frac{V_{ba}c_a c_b^*(\mathrm{e}^{\mathrm{i}(\omega-\omega_{ba})t} - 1) + V_{ab}^* c_b c_a^*(\mathrm{e}^{-\mathrm{i}(\omega-\omega_{ba})t} - 1)}{\omega - \omega_{ba}}\right. \\ &\qquad \left. - \frac{V_{ba}c_b c_a^*(\mathrm{e}^{\mathrm{i}(\omega+\omega_{ba})t} - 1) + V_{ab}^* c_a c_b^*(\mathrm{e}^{-\mathrm{i}(\omega+\omega_{ba})t} - 1)}{\omega + \omega_{ba}}\right\}. \end{aligned} \tag{11.34}$$

We now suppose that a is the state with the lower energy, so that ω_{ba} is positive. At and near the resonant condition $\omega = \omega_{ba}$ we need retain only the first term in the bracket, a choice known as the rotating-wave approximation, the name deriving from the study of magnetic resonance in which one component of the perturbation stays in phase with the Larmor precession. Thus

$$\begin{aligned} i\hbar(\dot{c}_a c_a^* + c_a \dot{c}_a^*) &= V_{ba} c_a c_b^* e^{i(\omega-\omega_{ba})t} - V_{ab}^* c_b c_a^* e^{-i(\omega-\omega_{ba})t} \\ &= V_{ba} c_a c_b^* - V_{ab}^* c_b c_a^*, \qquad \omega = \omega_{ba}. \end{aligned} \tag{11.35}$$

From this and the equivalent equation for level b we can deduce that

$$\begin{aligned} i\hbar\dot{c}_a &= -V_{ab}^* c_b \\ i\hbar\dot{c}_b &= -V_{ab} c_a \end{aligned} \tag{11.36}$$

with equivalent expressions for the complex conjugates. The equation of motion for the occupation probability is therefore

$$\begin{aligned} \ddot{c}_a &= -\frac{|V_{ab}|^2}{\hbar^2} c_a = -\frac{1}{4}\omega_R^2 c_a \\ \omega_R^2 &= 4\frac{|V_{ab}|^2}{\hbar^2} = \frac{|H_{1ab}|^2}{\hbar^2} \end{aligned} \tag{11.37}$$

where ω_R is the Rabi frequency, after Rabi (1937). What this means is that the occupation of the levels oscillates at a frequency determined by the strength of the perturbation. Thus, if $c_a = 1$ at $t = 0$,

$$\begin{aligned} c_a &= \cos(\tfrac{1}{2}\omega_R t), \qquad c_b = i\sin(\tfrac{1}{2}\omega_R t) \\ |c_a|^2 &= \tfrac{1}{2}(1 + \cos\omega_R t), \qquad |c_b|^2 = \tfrac{1}{2}(1 - \cos\omega_R t). \end{aligned} \tag{11.38}$$

A resonant perturbative pulse that lasts for a time t such that $\omega_R t = \pi/2$—called a $\pi/2$-pulse—produces a mixed state; a π-pulse flips the occupancy.

Two-level systems are of interest to the fields of quantum information, quantum cryptography, quantum teleportation, and quantum computing, where they are known as qubits. Manipulation is effected by using $\pi/2$- and π-pulses, but such manipulation is severely restricted to times less than the decoherence time. Decoherence occurs via the interaction with the system's surroundings; for example, via collisions with photons, phonons' or other electrons. As a result, information about phase is irretrievably lost.

11.4. Fermi's Golden Rule

To quantify the rate of decoherence we return to eqn (11.12). Substituting from eqn (11.14) into the second equation of eqn (11.12), and dispensing with second harmonic components that give negligible amplitude on integration, we obtain

$$i\hbar\langle k|\dot{\rho}_{1kk}|k\rangle = \sum_{\mathbf{k}'}(f_{\mathbf{k}'}-f_{\mathbf{k}})|V_{\mathbf{k}'\mathbf{k}}|^2\left\{\frac{1}{\hbar\omega_{\mathbf{k}'\mathbf{k}}+\hbar\omega-i\hbar\delta}+\frac{1}{\hbar\omega_{\mathbf{k}'\mathbf{k}}-\hbar\omega-i\hbar\delta} - \frac{1}{\hbar\omega_{\mathbf{k}'\mathbf{k}}-\hbar\omega+i\hbar\delta}-\frac{1}{\hbar\omega_{\mathbf{k}'\mathbf{k}}+\hbar\omega+i\hbar\delta}\right\}e^{\delta t}. \tag{11.39}$$

In the limit $\delta\to 0$, we can use eqn (11.21) to obtain

$$\langle k|\dot{\rho}_{1kk}|k\rangle = \frac{2\pi}{\hbar}\sum_{\mathbf{k}'}(f_{\mathbf{k}'}-f_{\mathbf{k}})|V_{\mathbf{k}'\mathbf{k}}|^2\{\delta(\hbar\omega_{\mathbf{k}'\mathbf{k}}-\hbar\omega)+\delta(\hbar\omega_{\mathbf{k}'\mathbf{k}}+\hbar\omega)\} \tag{11.40}$$

which is Fermi's Golden Rule (Fermi 1950). It quantifies the rate at which the occupation of the state k changes through transitions involving all other states in the system for which energy is conserved.

The symbol δ has been used as a mathematical convenience. The underlying assumption has been that the perturbation switches on at $t=0$ and (with $\delta\to 0$) stays on for ever. But in reality, other collisions occurs so any one interaction, if it is to have individuality, can have only a limited duration. A more physically meaningful time dependence would be to replace δ with $-\Gamma$, so that the interaction switches on at $t=0$ and decays exponentially with a time constant equal to Γ^{-1}. But this means that we cannot use eqn (11.21). Instead, we must replace the delta function by a Lorentzian:

$$\delta(\hbar\omega_{\mathbf{k}'\mathbf{k}}\pm\hbar\omega)\to\frac{1}{\pi}\frac{\hbar\Gamma}{(\hbar\omega_{\mathbf{k}'\mathbf{k}}\pm\hbar\omega)^2+\hbar^2\Gamma^2}. \tag{11.41}$$

Energy conservation becomes somewhat fuzzy.

It is clear that Fermi's Golden Rule will work only if Γ is very small. If Γ^{-1} is interpreted as being of order of the time between collisions, t_c, the condition is that it must be much longer than the duration, t_d, of the collision. A collision duration time can be defined as the time required for the effect of the energy uncertainty on the squared matrix element, Φ $(=|\mathbf{V}_{\mathbf{k}'\mathbf{k}}|^2)$, to be small. This puts the focus on the energy dependence of Φ. Thus, if Φ_0 is the squared matrix element for the case of perfect energy conservation, the value of Φ when there is an energy uncertainty of ΔE is

$$\Phi = \Phi_0 + \frac{d\Phi}{dE}\Delta E = \Phi_0\left(1 + \frac{1}{\Phi_0}\frac{d\Phi}{dE}\Delta E\right). \tag{11.42}$$

The expansion assumes that the variation with energy is not large. The uncertainty in the energy will not matter provided that the second term in the brackets is much smaller than unity. The Uncertainty Principle can be invoked to estimate the uncertainty in the energy within the average time between collisions, i.e. $\Delta E \approx \hbar / t_c$. The Fermi Golden Rule will be valid provided that

$$t_c \gg t_d, \qquad t_d = \hbar \frac{d \ln \Phi}{dE} \tag{11.43}$$

where we have identified an expression for the collision duration time. This the expression quoted in eqn (3.2). It turns out that the condition in eqn (11.43) is usually well-satisfied.

11.5. Wannier–Stark states

One of the most important interactions experienced by electrons in a semiconductor is with a uniform electric field associated with either an applied voltage or a long electromagnetic wave. In order to describe the quantum theory of this interaction, we must choose an appropriate gauge to describe the modification to the electron Hamiltonian brought about by the electric field. There are two obvious choices, either (1) to modify the kinetics via a vector potential $\mathbf{A}$, with ϕ, the scalar potential, equal to zero and $\nabla \cdot \mathbf{A} = 0$, or (2) to modify the potential energy via a scalar potential with $\mathbf{A} = 0$. We have already described the latter approach in Section 2.2, regarding the scalar potential as a perturbation. We will first adopt the kinetic approach, especially as this will apply to all field strengths.

The electric field is described in terms of a time-dependent vector potential:

$$\mathbf{F} = -\frac{d\mathbf{A}}{dt} \tag{11.44}$$

and the Schrödinger equation is

$$\left\{\frac{1}{2m}(\mathbf{p} - e\mathbf{A}(t))^2 + V(\mathbf{r})\right\}\Psi_n(\mathbf{r}, t) = E_n \Psi_n(\mathbf{r}, t). \tag{11.45}$$

Here m is the free-electron mass, $V(\mathbf{r})$ is the periodic potential of the lattice and n is a band index. This equation actually has an exact solution:

$$\begin{gathered}\Psi_n(\mathbf{r},t) = \Phi_n(\mathbf{r})e^{i(e/\hbar)\mathbf{A}(t)\cdot\mathbf{r}} \\ \left\{-\frac{\hbar^2}{2m}\nabla^2 + V(\mathbf{r})\right\}\Phi_n(\mathbf{r}) = E_n\Phi_n(\mathbf{r}).\end{gathered} \tag{11.46}$$

This shows that the field does not affect the band structure, but merely adds a phase to the Bloch function. The Bloch function for, say, the conduction band, has its usual periodic form:

$$\Phi_{\mathbf{k}} = u_{\mathbf{k}}(\mathbf{r})e^{i\mathbf{k}\cdot\mathbf{r}}. \tag{11.47}$$

In this formulation $\mathbf{k}$ is a good quantum number, independent of the electric field, but it is now associated with the total momentum $\mathbf{p}$. However, the velocity of the electron is given by

$$\mathbf{v} = \nabla_{\mathbf{p}}H = \{\mathbf{p} - e\mathbf{A}(t)\}/m. \tag{11.48}$$

An electron remaining in the state $\mathbf{p}_0$ will appear to move as if it had made the transition from a field-free state $\mathbf{k}_0$ to a field-free state $\mathbf{k}$ given by

$$\begin{gathered}\mathbf{k} = \mathbf{k}_0 - e\mathbf{A}(t)/\hbar \\ \therefore \dot{\mathbf{k}} = e\mathbf{F}/\hbar.\end{gathered} \tag{11.49}$$

This is the acceleration theorem, and it is clearly true for all field strengths. It predicts that unfettered motion of the electron in a conduction band will result in Bloch oscillations as described in Section 2.3 provided that the tunnelling probability to other bands is reasonably small. This in turn implies that the electron becomes localized about a particular unit cell in a Wannier–Stark state whose wavefunction will be a combination of time-dependent, free-field, Bloch states:

$$\Psi(\mathbf{r},t) = \sum_{n\mathbf{k}_0} c_{n\mathbf{k}_0}\Phi_{n\mathbf{k}(t)}(\mathbf{r})\exp\left\{-(i/\hbar)\int_{t_0}^{t} E_{n\mathbf{k}(t')}dt'\right\} \tag{11.50}$$

where $\mathbf{k}_0 = \mathbf{k}(t_0)$. For fields low enough for Zener tunnelling to other bands to be negligible, a Wannier–Stark (W–S) wavefunction for the Bloch oscillation state in a given band can be obtained from eqn (11.50) for an eigenvalue $E_{\mathbf{k}_\perp n}$ by distinguishing motion parallel and perpendicular to the field:

$$\mathbf{k} = (\mathbf{k}_\perp, k_\parallel), \qquad E_{\mathbf{k}} = E_{\mathbf{k}_\perp} + E_{k_\parallel(t)} \tag{11.51}$$

converting time dependence to $\mathbf{k}_{\|}$ dependence and solving for the $c_{\mathbf{k}}$:

$$\psi_{\mathbf{k}_\perp n}(\mathbf{r}) = \frac{a}{2\pi} \int_{-\pi/a}^{+\pi/a} \Phi_{\mathbf{k}_\perp, k_\|}(\mathbf{r}) \exp\left\{ \frac{-\mathrm{i}}{eF} \int_0^{k_\|} (E_{\mathbf{k}_\perp k'_\|} - E_{\mathbf{k}_\perp n}) \mathrm{d}k'_\| \right\} \mathrm{d}k_\| \qquad (11.52)$$

where n now labels the W–S state in the ladder of states each separated from its neighbour by the energy eFa, where a is the lattice constant:

$$E_{\mathbf{k}_\perp n} = E_{\mathbf{k}_\perp 0} + neFa. \qquad (11.53)$$

To the best of the author's knowledge, Wannier–Stark states have never been observed in bulk material. There are perfectly good reasons why this is the case, not least is the stringent condition on the scattering time τ:

$$\omega_{\mathrm{B}}\tau \geqslant 1, \qquad \omega_{\mathrm{B}} = eFa/\hbar. \qquad (11.54)$$

where ω_{B} is the Bloch angular frequency (see Section 2.3). The necessity for the field to be very high means that electrical breakdown is not avoidable in most semiconductors. The increasing purity of wide-bandgap materials such as AlN, GaN, ZnO, ZnS, and so on may allow W–S states to be observed. At the time of writing they have been observed only in the miniband of a semiconductor superlattice. A good account of this and related topics can be found in the book edited by Schöll (1998).

11.6. The intracollisional field effect

As discussed in Section 11.4, the duration of a collision is usually short compared with the period between collisions. Nevertheless, we have just seen that in the presence of an electric field the field-free Bloch functions become time-dependent, so that during a collision the initial and final states are not constant. The influence of this on the scattering rate is known as the intracollisional field effect.

There are two significant changes induced by the presence of a field. One is a change of energy of the electron during the collision which, for example, may enable an electron to emit an optical phonon or transfer to an upper valley even though its initial energy was too low. The second is that, inevitably, energy conservation becomes fuzzy. As a result, the energy-conserving delta function in the expression for the transition rate must be replaced by

$$\begin{aligned} \delta(\hbar\omega_{\mathbf{k}'\mathbf{k}} \pm \hbar\omega) &\to \frac{1}{\pi} \frac{\hbar\Gamma}{(E_{\mathbf{k}'_\perp, k'_\| + eFt_{\mathrm{d}}/\hbar} - E_{\mathbf{k}_\perp, k_\| + eFt_{\mathrm{d}}/\hbar} \pm \hbar\omega)^2 + \hbar^2\Gamma^2} \\ \hbar\Gamma &= eF(v' - v)_\| t_{\mathrm{d}} \\ \therefore \Gamma &= eF(k'_\| - k_\|)t_{\mathrm{d}}. \end{aligned} \qquad (11.55)$$

Alternatively, one may use Wannier–Stark wavefunctions instead of time-dependent Bloch functions, in which case the intracollisional field effect would be automatically accounted for.

If it is supposed that $t_d \approx 10^{-15}$ s it would require fields of at least 1 MV/cm for there to be significant effects.

11.7. The semi-classical approximation

The physics of phenomena with characteristic time scales much longer than the decoherence times of the underlying quantum phenomena can be very satisfactorily dealt with using the concept of long-lived pure states. Mixed states can be ignored and the Fermi Golden Rule, accomodating level broadening if necessary, can be used to describe interactions in which the collision duration time is much shorter than the average time between collisions. The common feature that is retained (common at any rate in our random-phase, one-particle theory, described in the previous sections) is the statistical nature of the description involving the Markovian approximation that collisions are uncorrelated. It is also assumed that only binary collisions occur.

One advantage that these simplifications afford is that spatially non-uniform systems can readily be treated, provided that the distribution function does not vary too rapidly over an electron wavelength.

References

FERMI, E. (1950). *Nuclear physics*, p. 142. University of Chicago Press, Chicago.

FRIED, B. D. and CONTE, S. D. (1961). *The plasma dispersion function*, Academic Press, London.

LOWE, D. and BARKER, J. (1985). *J. Phys. C: Solid State Phys.* **18,** 2507.

RABI, L. (1937). *Phys. Rev.* **51,** 652.

SCHÖLL, E. (1998). *Theory of transport properties of semiconductor nanostructures*, Chapman and Hall, London.

12. Semi-classical transport

12.1. The Boltzmann equation

As we have seen in Chapter 11, a comprehensive account of the response of electrons in semiconductors to external fields has to take into account their quantum nature. This is particularly necessary in order to describe phenomena at ultrashort length and time scales and high field strengths, when the electron cannot be regarded as being in an eigenstate of the unperturbed crystal, and when the effects of the uncertainty principle blur momentum and energy conservation. There is, however, a vast range of conditions over which the behaviour of electrons can be treated semi-classically, quantum effects entering only through band structure and through scattering, and this is what we will assume in what follows. This means that we can take the electron to be in one of the Bloch states of the conduction band with a well-defined energy, E, and a well-defined wavevector, $\mathbf{k}$, as we have done throughout the book, and we can exploit all the scattering rates discussed in previous chapters.

What we have to do that is different is to take account of the fact that in all experiments involving bulk material, typically, a large number of electrons is involved and only average quantities are ever measured. Statistics therefore enters. Electrons move between states in response to applied fields and scattering mechanisms, and it is necessary to have a book-keeping operation to follow the net occupancy of states. The Boltzmann equation provides just such a book-keeping operation. If $f(\mathbf{k})$ is the probability that the state is occupied, then its rate of change with time is determined by a conventional continuity equation consisting of a volume rate and a divergence of a probability current:

$$\frac{\partial f(\mathbf{k})}{\partial t} = \left(\frac{\partial f(\mathbf{k})}{\partial t}\right)_{\text{vol}} - \nabla \cdot (\mathbf{v}(\mathbf{k}) f(\mathbf{k})) \tag{12.1}$$

where $\mathbf{v}(\mathbf{k})$ is the group velocity. The volume rate is the sum of individual rates associated with applied fields, scattering, generation, and recombination, viz.:

$$\left(\frac{\partial f(\mathbf{k})}{\partial t}\right)_{\text{vol}} = \left(\frac{\partial f(\mathbf{k})}{\partial t}\right)_{\text{fields}} + \left(\frac{\partial f(\mathbf{k})}{\partial t}\right)_{\text{scat}} + \left(\frac{\partial f(\mathbf{k})}{\partial t}\right)_{\text{gen}} + \left(\frac{\partial f(\mathbf{k})}{\partial t}\right)_{\text{recomb}}. \tag{12.2}$$

The field term can be related to a probability current flow in $\mathbf{k}$-space:

$$\left(\frac{\partial f(\mathbf{k})}{\partial t}\right)_{\text{fields}} = -\nabla_{\mathbf{k}} \cdot \mathbf{j}_{\mathbf{k}} = -\nabla_{\mathbf{k}} \cdot \left(\frac{d\mathbf{k}}{dt} f(\mathbf{k})\right) = -\frac{d\mathbf{k}}{dt} \cdot \nabla_{\mathbf{k}} f(\mathbf{k}). \tag{12.3}$$

The rate of change of wavevector is just proportional to the force, $\mathcal{F}$:

$$\frac{d\mathbf{k}}{dt} = \frac{\mathcal{F}}{\hbar} \tag{12.4}$$

where the force in the presence of an electric field, **F**, and a magnetic field, **B** is

$$\mathcal{F} = \bar{e}(\mathbf{F} + \mathbf{v} \times \mathbf{B}). \tag{12.5}$$

In addition to the accelerating effect of the applied fields, there is also a polarizing effect that mixes states with the same **k** in different bands and induces inter-band transitions, such as occur in the Zener Effect. We will assume that the field strength is too small for tunnelling between bands to be important. Once more, we define $\bar{e}$ to carry the sign of the charge.

The volume scattering rate is given by eqn (8.1), which for convenience we reproduce here:

$$\begin{aligned}\left(\frac{\partial f(\mathbf{k})}{\partial t}\right)_{\text{scat}} = &\int [W(\mathbf{k}', \mathbf{k}) f(\mathbf{k}')\{1 - f(\mathbf{k})\} - W(\mathbf{k}, \mathbf{k}') f(\mathbf{k})\{1 - f(\mathbf{k}')\}] \\ &\times \delta(E_{\mathbf{k}'} - E_{\mathbf{k}} - \hbar\omega_q)\, d\mathbf{k}' \\ &+ \int [W(\mathbf{k}'', \mathbf{k}) f(\mathbf{k}'')\{1 - f(\mathbf{k})\} - W(\mathbf{k}, \mathbf{k}'') f(\mathbf{k})\{1 - f(\mathbf{k}'')\}] \\ &\times \delta(E_{\mathbf{k}''} - E_{\mathbf{k}} + \hbar\omega_q)\, d\mathbf{k}''.\end{aligned} \tag{12.6}$$

The integrals are over the final electron states, with conservation of crystal momentum having been taken into account. This rate describes the situation when a quantum of energy $\hbar\omega_{\mathbf{q}}$ is absorbed or emitted during the scattering event. For elastic processes such as charged-impurity scattering, alloy scattering, and so on, the energy of the quantum can be put to zero and the second integral eliminated. Equation (12.6) is also the form for radiative generation and recombination processes. It does not, however, describe carrier–carrier scattering, Auger, or impact ionization rates. The net rate for carrier–carrier scattering will be described later.

Here, we will focus on charge transport in the presence of an applied electric field, with the aim of describing some of the non-linear effects that make semiconductors so fascinating. (Transport in a magnetic field was discussed in Chapter 7.)

In the absence of non-uniformity in the carrier distribution, the steady-state Boltzmann equation is

$$\frac{\bar{e}\mathbf{F}}{\hbar} \cdot \nabla_{\mathbf{k}} f(\mathbf{k}) = \left(\frac{\partial f(\mathbf{k})}{\partial t}\right)_{\text{scat}}. \tag{12.7}$$

The presence of the field will induce both symmetric and antisymmetric disturbances to the distribution function. In order to model this analytically, we first expand the distribution function in Legendre polynomials:

$$f(\mathbf{k}) = \sum_{j=0}^{\infty} f_j(E)P_j(x) \tag{12.8}$$

where $P_j(x)$ is the Legendre polynomial of order j and $x = \cos\theta$, where θ is the angle between $\mathbf{k}$ and $\mathbf{F}$. The components of $\nabla_{\mathbf{k}} f(\mathbf{k})$ are

$$\nabla_{\mathbf{k}} f(\mathbf{k}) = \left(\frac{\partial f(\mathbf{k})}{\partial k}, \frac{1}{k}\frac{\partial f(\mathbf{k})}{\partial \theta}\right) = \left(\hbar v(E)\frac{\partial f(\mathbf{k})}{\partial E}, \frac{1}{k}\frac{\partial f(\mathbf{k})}{\partial \theta}\right). \tag{12.9}$$

Thus

$$\frac{\bar{e}\mathbf{F}}{\hbar} \cdot \nabla_{\mathbf{k}} f(\mathbf{k}) = \bar{e}Fv(E)x\sum_j P_j(x)\frac{\partial f_j(E)}{\partial E} + \frac{eF}{m^* v(E)}(1-x^2)\sum_j f_j(E)\frac{\mathrm{d}P_j(x)}{\mathrm{d}x} \tag{12.10}$$

where $v(E)$ is the magnitude of the velocity. Using the relations

$$\begin{aligned} (1-x^2)\frac{\mathrm{d}P_j(x)}{\mathrm{d}x} &= jP_{j-1}(x) - jxP_j(x) \\ xP_j(x) &= \frac{j+1}{2j+1}P_{j+1}(x) + \frac{j}{2j+1}P_{j-1}(x) \end{aligned} \tag{12.11}$$

we obtain

$$\begin{aligned} \frac{\bar{e}\mathbf{F}}{\hbar} \cdot \nabla_{\mathbf{k}} f(\mathbf{k}) = \bar{e}F\sum_j \Bigg[&\left(\frac{j+1}{2j+1}\frac{f_j(E)}{m^* v(E)} + \frac{v(E)}{2j+1}\frac{\partial f_j(E)}{\partial E}\right) jP_{j-1}(x) \\ &+ \left(v(E)\frac{\partial f_j(E)}{\partial E} - \frac{j}{m^* v(E)}f_j(E)\right)\frac{j+1}{2j+1}P_{j+1}(x)\Bigg]. \end{aligned} \tag{12.12}$$

As long as the field is not too strong, and as long as the scattering rate is reasonably high, we can expect the antisymmetric and non-spherical parts of the distribution function to be relatively small in amplitude. Thus, retaining only the zero-order symmetrical component of the distribution function, $f_0(E)$, and the first-order antisymmetrical component, $f_1(E)$, we obtain

$$\begin{aligned} \frac{\bar{e}\mathbf{F}}{\hbar} \cdot \nabla_{\mathbf{k}} f(\mathbf{k}) = \bar{e}F\Bigg[\frac{1}{3}\bigg(&\frac{2f_1(E)}{m^* v(E)}(P_0(x) - P_2(x)) + v(E)\frac{\partial f_1(E)}{\partial E}(P_0(x) \\ &+ 2P_2(x))\bigg) + v(E)\frac{\partial f_0(E)}{\partial E}P_1(x)\Bigg]. \end{aligned} \tag{12.13}$$

The second-order Legendre function, $P_2(x)$, can be averaged over all directions to convert it into a spherically symmetrical form.

The scattering rate can likewise be expanded in Legendre polynomials, and only terms in $f_0(E)$ and $f_1(E)$ retained. Then, equating symmetric and antisymmetric coefficients, we obtain

$$\frac{\bar{e}F}{3}\left[\frac{2f_1(E)}{m^*v(E)}(1-\langle P_2(x)\rangle)+v(E)\frac{\partial f_1(E)}{\partial E}(1+2\langle P_2(x)\rangle)\right]$$
$$=\left[\int[W(\mathbf{k}',\mathbf{k})f_0(E')\{1-f_0(E)\}-W(\mathbf{k},\mathbf{k}')f_0(E)\{1-f_0(E')\}]\right.$$
$$\times\,\delta(E'-E-\hbar\omega_q)\,\mathrm{d}\mathbf{k}'+\int[W(\mathbf{k}'',\mathbf{k})f_0(E'')\{1-f_0(E)\}$$
$$\left.-\,W(\mathbf{k},\mathbf{k}'')f_0(E)\{1-f_0(E'')\}]\delta(E''-E+\hbar\omega_q)\,\mathrm{d}\mathbf{k}''\right] \quad (12.14)$$

$$\bar{e}Fv(E)\frac{\partial f_0(E)}{\partial E}$$
$$=\left[\int\begin{bmatrix}W(\mathbf{k}',\mathbf{k})[f_1(E')\frac{x'}{x}\{1-f_0(E)\}-f_1(E)f_0(E')]\\-W(\mathbf{k},\mathbf{k}')[f_1(E)\{1-f_0(E')\}-f_1(E')\frac{x'}{x}f_0(E)]\end{bmatrix}\delta(E'-E-\hbar\omega_q)\,\mathrm{d}\mathbf{k}'\right.$$
$$+\int\begin{bmatrix}W(\mathbf{k}'',\mathbf{k})[f_1(E'')\frac{x''}{x}\{1-f_0(E)\}-f_1(E)f_0(E'')]\\-W(\mathbf{k},\mathbf{k}'')[f_1(E)\{1-f_0(E'')\}-f_1(E'')\frac{x''}{x}f_0(E)]\end{bmatrix}$$
$$\left.\times\,\delta(E''-E+\hbar\omega_q)\,\mathrm{d}\mathbf{k}''\right]. \quad (12.15)$$

The products of the antisymmetric components have been assumed to be negligibly small. Here $x'=\cos\theta'$ where θ' is the angle between $\mathbf{k}'$ and $\mathbf{F}$, and similarly x''. Now, in general,

$$\cos\theta'=\cos\alpha'\cos\theta+\sin\alpha'\sin\theta\sin\phi \quad (12.16)$$

where ϕ is the azimuthal angle which is to be integrated over, and α' is the angle between $\mathbf{k}'$ and $\mathbf{k}$. Nothing in the integrand depends upon ϕ except for the second term in eqn (12.16). Integration over ϕ between the limits 0 and 2π gives zero; thus, effectively, $x'/x=\cos\alpha'$ and $x''/x=\cos\alpha''$. Both energy and momentum are conserved. Therefore, for a parabolic band,

$$\text{(a) } E'-E-\hbar\omega_q=\frac{\hbar^2k'^2}{2m^*}-\frac{\hbar^2k^2}{2m^*}-\hbar\omega_q=\frac{\hbar^2q^2}{2m^*}+\frac{\hbar^2kq}{m^*}\cos\theta_q-\hbar\omega_q=0$$
$$\text{(b) } E''-E+\hbar\omega_q=\frac{\hbar^2k''^2}{2m^*}-\frac{\hbar^2k^2}{2m^*}+\hbar\omega_q=\frac{\hbar^2q^2}{2m^*}-\frac{\hbar^2kq}{m^*}\cos\theta_q+\hbar\omega_q=0. \quad (12.17)$$

In eqn (12.17a) $q_1 \leq q \leq q_2$, and in eqn (12.17b) $q_3 \leq q \leq q_4$, where

$$\begin{aligned} q_1 &= k[(1+\{\hbar\omega_q/E\})^{1/2} - 1], \qquad q_3 = k[1-(1-\{\hbar\omega_q/E\})^{1/2}] \\ q_2 &= k[(1+\{\hbar\omega_q/E\})^{1/2} + 1], \qquad q_4 = k[1+(1-\{\hbar\omega_q/E\})^{1/2}] \end{aligned} \tag{12.18}$$

and it is understood that the integral involving q_3 and q_4 vanish if $E < \hbar\omega_q$. These relations are familiar from our discussion in Chapter 3. Note that if $\hbar\omega$ is a function of q, as it is for acoustic modes, the limits on the phonon wavevector will be modified (unless $\hbar\omega \ll E$, in which case the collisions are quasi-elastic). The cosine factors are

$$\begin{aligned} \frac{x'}{x} &= \cos\alpha' = \frac{1-(q^2/2k^2)+(\hbar\omega_q/2E)}{(1+\{\hbar\omega_q/E\})^{1/2}} \\ \frac{x''}{x} &= \cos\alpha'' = \frac{1-(q^2/2k^2)-(\hbar\omega_q/2E)}{(1-\{\hbar\omega_q/E\})^{1/2}}. \end{aligned} \tag{12.19}$$

Finally, in the 3-D spherical symmetry assumed, one can take $\langle P_2(x)\rangle = 0$. (Keeping this average explicit in the equation is useful for making extensions to low dimensionality (Ridley 1997).) In general, the scattering coefficients, $W(\mathbf{k}',\mathbf{k})$, have a dependence on crystallographic direction—an example being the piezoelectric interaction. When this is the case, it is convenient to take a spherical average so that the integration over the azimuthal angle can be carried out straightforwardly. It is usually convenient to convert the integration over final electron states to an integration over phonon states, exploiting the one-to-one correspondence between $\mathbf{k}'$ and $\mathbf{q}$ via the conservation of crystal momentum, and to carry out the integration over $\cos\theta_q$ with the help of the energy-conserving delta function. The result is as follows:

$$\begin{aligned} &\text{(a)}\ \frac{\bar{e}F}{3}\left[\frac{2f_1(E)}{m^* v(E)} + v(E)\frac{\partial f_1(E)}{\partial E}\right] = \hat{I}_0 \\ &\text{(b)}\ \bar{e}Fv(E)\frac{\partial f_0(E)}{\partial E} = \hat{I}_1 \end{aligned} \tag{12.20}$$

where the collision integrals are given by

$$\begin{aligned} \hat{I}_0 &= \frac{2\pi m^*}{\hbar^2 k}\int_{q_i}^{q_{i+1}} [W(\mathbf{k}',\mathbf{k})f_0(E')\{1-f_0(E)\} - W(\mathbf{k},\mathbf{k}')f_0(E)\{1-f_0(E')\}]q\,dq \\ \hat{I}_1 &= \frac{2\pi m^*}{\hbar^2 k}\int_{q_i}^{q_{i+1}} [W(\mathbf{k}',\mathbf{k})\{f_1(E')\{1-f_0(E)\}\cos\alpha' - f_1(E)f_0(E')\} \\ &\qquad - W(\mathbf{k},\mathbf{k}')\{f_1(E)\{1-f_0(E')\} - f_1(E')f_0(E)\cos\alpha'\,]q\,dq \end{aligned} \tag{12.21}$$

and $i = 1$ or 3 (eqn 12.18), with $\cos\alpha'$ chosen appropriately (eqn 12.19) and $E' = E \pm \hbar\omega_q$.

The solutions now depend on the particular scattering mechanisms and on the strength of the field.

12.2. Weak electric fields

In the absence of a field and with thermodynamic equilibrium prevailing, it is clear that

$$\begin{aligned}&(a)\ f_1(E) = 0\\ &(b)\ W(\mathbf{k}',\mathbf{k})f_0(E')\{1 - f_0(E)\} - W(\mathbf{k},\mathbf{k}')f_0(E)\{1 - f_0(E')\} = 0\end{aligned} \tag{12.22}$$

for all scattering mechanisms, with $f_0(E)$ given by the Fermi–Dirac function. When there is a weak electric field, $f_1(E)$ is non-zero but it is of the same order as the field and so the left-hand side of eqn (12.20a) is of second order and can be neglected. This means that the spherical component of the distribution function is unaffected by the field and eqn (12.22b) continues to hold good. Exploiting this in eqn (12.20b), we obtain

$$\hat{I}_1 = \frac{2\pi m^*}{\hbar^2 k}\int_{q_i}^{q_{i+1}} W(\mathbf{k}',\mathbf{k})f_0(E')\{1 - f_0(E)\} \times \left(\frac{f_1(E')\cos\alpha'}{f_0(E')\{1 - f_0(E')\}} - \frac{f_1(E)}{f_0(E)\{1 - f_0(E)\}}\right) q\,\mathrm{d}q. \tag{12.23}$$

In the case of elastic collisions, $E' = E$ and $q_i = 0$, $q_{i+1} = 2k$ (from eqn (12.18), and we obtain

$$\hat{I}_1 = \frac{2\pi m^*}{\hbar^2 k} f_1(E)\int_0^{2k} W(\mathbf{k}',\mathbf{k})\{\cos\alpha' - 1\}q\,\mathrm{d}q \tag{12.24}$$

leading to the definition of a momentum relaxation time in the familiar way (see Chapter 4). Table 4.1 lists the momentum-relaxation times for impurity scattering. A momentum relaxation time can also be defined for acoustic-phonon scattering in the quasi-elastic approximation ($\hbar\omega_q \ll E$) (see Chapter 3). If the momentum relaxation time is τ_m then the solution of the Boltzmann equation can be expressed as follows:

$$f_1(E) = -\bar{e}v(E)\tau_\mathrm{m}(E)F\frac{\partial f_0(E)}{\partial E}. \tag{12.25}$$

When the collisions are inelastic, a relaxation time cannot always be defined. It is nevertheless useful to define an effective relaxation time, $\tau^*(E)$, using the form of eqn (12.25), which just means replacing τ_m by τ^*. Noting that

$$\frac{\partial f_0(E)}{\partial E} = -\frac{f_0(E)\{1 - f_0(E)\}}{k_B T} \tag{12.26}$$

the collision integral is then

$$\hat{I}_1 = \frac{2\pi \bar{e} F}{k_B T \hbar k} \int_{q_i}^{q_{i+1}} W(\mathbf{k}', \mathbf{k}) f_0(E')\{1 - f_0(E)\}[\tau^*(E')k' \cos\alpha' - \tau^*(E)k] q \, dq. \tag{12.27}$$

The effective time constant for each energy must be obtained from eqn (12.20b):

$$E + \pi \int_{q_i}^{q_{i+1}} W(\mathbf{k}', \mathbf{k}) \frac{f_0(E')}{f_0(E)} [\tau^*(E')k' \cos\alpha' - \tau^*(E)k] q \, dq = 0. \tag{12.28}$$

The important case here is that for optical-phonon scattering, in which a q-independent frequency can be assumed. Equation (12.28) is awkward in that the effective time constant at energy E is related to the time constants at energies $E \pm \hbar\omega$, except for an energy less than the phonon energy, in which case the only connection is with the state at $E + \hbar\omega$. Nevertheless, an exact solution is possible for non-polar modes in the general case and for polar modes provided that we know $\tau^*(E)$ at energies such that $E \gg \hbar\omega$; that is, where a relaxation time can be defined. Such a relaxation time is then straightforward to obtain for non-polar and for polar optical mode scattering (Table 12.1). Working backwards down the ladder of states (Fig. 12.1) eventually allows us to establish the effective time constant in the intervals of E where E is less than or of the order $\hbar\omega$ with arbitrary accuracy. This method, originally devized by Delves (1959), is superior to variational methods of solution, in that it is exact and illustrative of the quantum nature of the scattering process (Fig. 12.2). A good account can be found in Fletcher and Butcher (1972).

TABLE 12.1
High-energy relaxation times for optical-phonon scattering

Non-polar	$\dfrac{4\pi\rho\omega\hbar^3}{D_0^2(2m^*)^{3/2}\{2n(\omega)+1\}E^{1/2}}$
Polar	$\dfrac{2E^{1/2}}{W_0(\hbar\omega)^{1/2}\{2n(\omega)+1\}}$

D_0 = optical deformation constant, ρ = density, $n(\omega)$ = phonon factor, m^* = effective mass, $W_0 = (e^2/4\pi\hbar\varepsilon_0)(2m^*\omega/\hbar)^{1/2}(1/\kappa_\infty - 1/\kappa_S)$, ε_0 = permittivity of the vacuum, κ = dielectric constant.

One of the striking features of optical-phonon scattering is the sudden decrease of the effective time constant at the first threshold of emission. At temperatures low enough for the population of states above this threshold to be small, this rapid decrease allows us to define an approximate relaxation

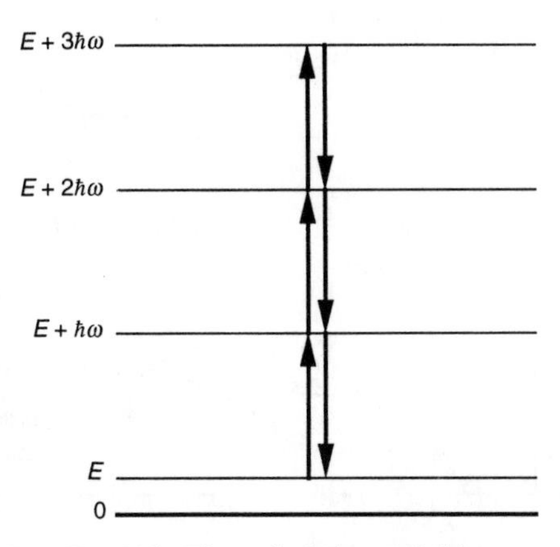

FIG. 12.1. The optical-phonon ladder.

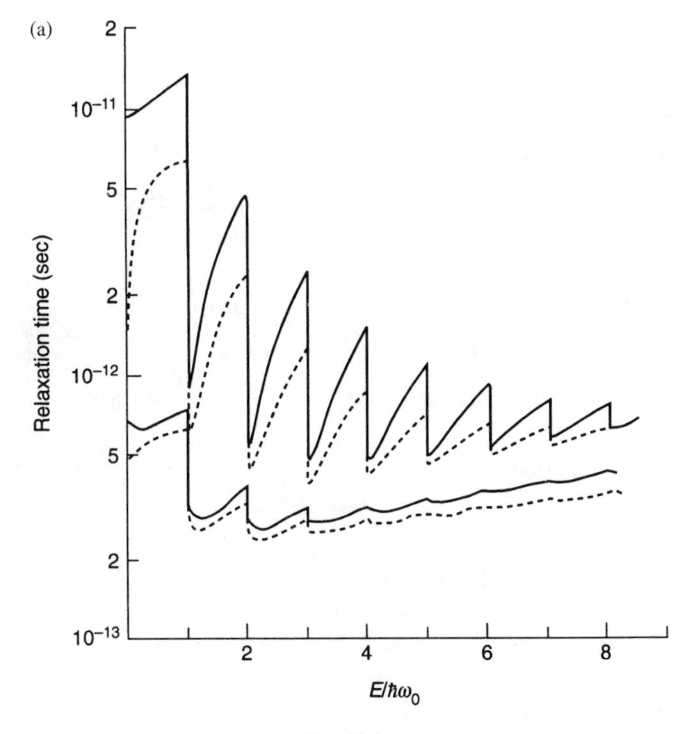

FIG. 12.2a.

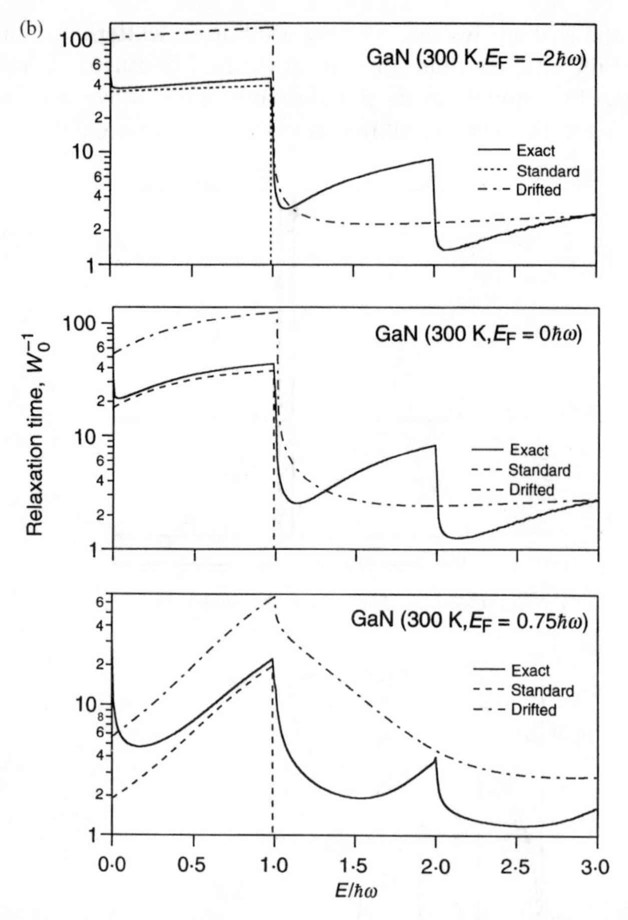

FIG. 12.2. Exact solutions of the Boltzmann equation. (a) Effective relaxation time in GaAs: continuous lines, polar mechanisms; upper curves at 100 K, lower at 300 K (after Fletcher and Butcher 1972). (b) Effective relaxation time in GaN at 300 K for three electron densities as measured by the Fermi level relative to the conduction bandedge. The exact solution is compared with the standard and drifted models $W_0 = 1.2 \times 10^{14}\mathrm{s}^{-1}$ (after Ridley 1998*a*).

time as that associated solely with the absorption process. Basically, this assumes that $\tau^*(\hbar\omega) \approx 0$. Thus we can obtain

$$\frac{1}{\tau_m} \approx \begin{cases} W_0\left(\dfrac{\hbar\omega}{E}\right)^{1/2} n(\omega)\sinh\left(\dfrac{E}{\hbar\omega}\right)^{1/2} & \text{polar} \\ \dfrac{\pi D_0^2}{\rho\omega} n(\omega) N(E+\hbar\omega), & \text{non-polar} \end{cases} \tag{12.29}$$

where $N(E)$ is the density of states and the various quantities are defined in Table 12.1.

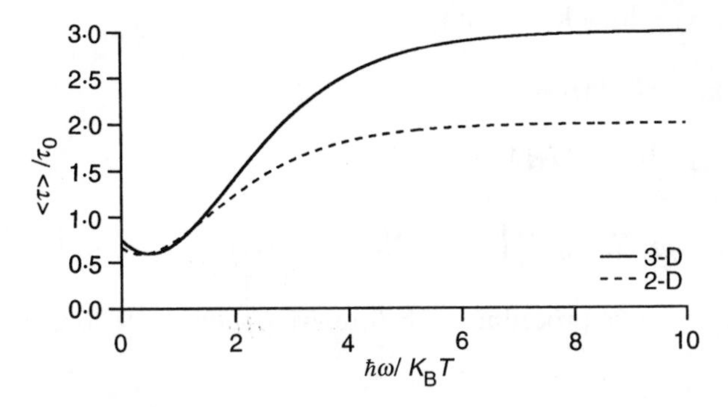

FIG. 12.3. Ratio of averaged momentum relaxation times for the drifted and standard models in 3-D and in 2-D (drifted time divided by standard time). (After Ridley 1998*b*).

The momentum relaxation rate for polar optical-phonon scattering just quoted can be contrasted with that derived by Callen (1949) (eqn 3.156). The latter turns out to be valid in the limit of rapid electron–electron scattering when a drifted Fermi–Dirac distribution function is established. This has the effect of transferring momentum back to the low-energy states by the emission process, thereby reducing the overall momentum relaxation rate. A comparison of the two rates is shown in Fig. 12.3. For $\hbar\omega \approx k_B T$, as is the case for GaAs at room temperature, the predictions of the two models coincide, but when $\hbar\omega \gg k_B T$, as is the case for GaN at room temperature, the 'drifted' model predicts a significantly longer momentum relaxation time.

12.3. Electron–electron scattering

It is appropriate at this point to consider the role of electron–electron scattering. By itself it cannot, of course, relax momentum or energy gained from the field, since interelectronic collisions conserve both quantities, but it can redistribute them over the electron states. Thus, both the symmetric and antisymmetric components of the distribution function can be affected.

The kinetics of carrier–carrier scattering were described in Sections 4.6 and 4.7. The net scattering rate associated with the state with wavevector $\mathbf{k}_1$ and energy E_1 is

$$\left(\frac{\mathrm{d}f(\mathbf{k}_1)}{\mathrm{d}t}\right)_{\mathrm{ee}} = \iiint [W(\mathbf{k}_1', \mathbf{k}_2', \mathbf{k}_1, \mathbf{k}_2) f(\mathbf{k}_1') f(\mathbf{k}_2')\{1 - f(\mathbf{k})_1)\}\{1 - f(\mathbf{k}_2)\}$$
$$- W(\mathbf{k}_1, \mathbf{k}_2, \mathbf{k}_1', \mathbf{k}_2') f(\mathbf{k}_1) f(\mathbf{k}_2)\{1 - f(\mathbf{k}_1')\}$$
$$\times \{1 - f(\mathbf{k}_2')\}]\, \mathrm{d}\mathbf{k}_2\, \mathrm{d}\mathbf{k}_1'\, \mathrm{d}\mathbf{k}_2' \tag{12.30}$$

where $\mathbf{k}_1' + \mathbf{k}_2' - \mathbf{k}_1 - \mathbf{k}_2 = 0$, and

$$W(\mathbf{k}_1\mathbf{k}_2\mathbf{k}_1'\mathbf{k}_2') = \frac{2\pi}{\hbar}|M|^2\delta(E_1' + E_2' - E_1 - E_2)$$
$$|M^2| = M_{12}^2 + M_{21}^2 - M_{12}M_{21}$$
$$M_{12}^2 = \left(\frac{e^2}{4\Omega\varepsilon K_{12}^2\sin^2(\theta/2)}\right)^2, \qquad M_{21}^2 = \left(\frac{e^2}{4\Omega\varepsilon K_{12}^2\cos^2(\theta/2)}\right)^2. \tag{12.31}$$

Following the same procedure as before, we obtain the following scattering integrals:

$$\hat{I}_0 = \iiint [W(\mathbf{k}_1'\mathbf{k}_2'\mathbf{k}_1\mathbf{k}_2)f_0(E_1')f_0(E_2')\{1 - f_0(E_1)\}\{1 - f_0(E_2)\} \\ - W(\mathbf{k}_1\mathbf{k}_2\mathbf{k}_1'\mathbf{k}_2')f_0(E_1)f_0(E_2)\{1 - f_0(E_1')\}\{1 - f_0(E_2')\}]\,\mathrm{d}\mathbf{k}_2\,\mathrm{d}\mathbf{k}_1'\mathrm{d}\mathbf{k}_2' \tag{12.32}$$

$$\hat{I}_1 = \frac{\bar{e}F\hbar}{m^*k_\mathrm{B}T}\iiint V(\mathbf{k}_1'\mathbf{k}_2'\mathbf{k}_1\mathbf{k}_2)\{\tau^*(E_1')\mathbf{k}_1' + \tau^*(E_2')\mathbf{k}_2' \\ - \tau^*(E_1)\mathbf{k}_1 - \tau^*(E_2)\mathbf{k}_2\}\,\mathrm{d}\mathbf{k}_2\,\mathrm{d}\mathbf{k}_1'\,\mathrm{d}\mathbf{k}_2' \tag{12.33}$$

where

$$V(\mathbf{k}_1'\mathbf{k}_2'\mathbf{k}_1\mathbf{k}_2) = V(\mathbf{k}_1\mathbf{k}_2\mathbf{k}_1'\mathbf{k}_2') \\ = W(\mathbf{k}_1'\mathbf{k}_2'\mathbf{k}_1\mathbf{k}_2)f_0(E_1')f_0(E_2')\{1 - f_0(E_1)\}\{1 - f_0(E_2)\}. \tag{12.34}$$

For brevity, we have quoted the antisymmetric scattering integral only for the case when the symmetric integral is zero, which allows us to simplify the expression considerably. This corresponds to the situation when the symmetric component of the distribution is of the form of a Fermi–Dirac function, which occurs either in the weak-field case or when electron–electron scattering is strong enough to randomize the energy and to define an electron temperature. These, in fact, are just the cases when it is interesting to look at the antisymmetric contribution of electron–electron scattering.

The permittivity that enters these equations is the appropriate dielectric function for the collision. This is the sum of a lattice term and a screening term (Chapter 9):

$$\varepsilon(q,\omega) = \varepsilon_\mathrm{L}(\omega) + \varepsilon_\mathrm{e}(q,\omega)$$
$$\varepsilon_\mathrm{L}(\omega) = \varepsilon_\infty\frac{\omega_\mathrm{LO}^2 - \omega^2}{\omega_\mathrm{TO}^2 - \omega^2}$$
$$\varepsilon_\mathrm{e}(q,\omega) = \frac{e^2}{q^2\Omega}\sum_\mathbf{k} f(\mathbf{k})\left[\frac{1}{E_{\mathbf{k}-\mathbf{q}} - E_\mathbf{k} + \hbar\omega + i\hbar\alpha} + \frac{1}{E_{\mathbf{k}+\mathbf{q}} - E_\mathbf{k} - \hbar\omega - i\hbar\alpha}\right]. \tag{12.35}$$

The frequency is $\mathbf{q}\cdot\mathbf{v}_{\mathrm{cm}}$, where $\mathbf{q} = \mathbf{K}'_{12} - \mathbf{K}_{12}$ and $\mathbf{v}_{\mathrm{cm}}$ is the velocity of the centre of mass. An assumption often made is that since the coulombic interaction favours small exchanges of momentum a static-screening approximation can be made, and so

$$\varepsilon(q,\omega) \approx \varepsilon_{\mathrm{s}}(1 + q_0^2/q^2) \tag{12.36}$$

where q_0 is the reciprocal screening length. A restriction of the collisions to purely binary encounters should also be made (Section 4.2.4). (For an example of the application of statistical screening to electron–electron scattering see Ridley (1998a).)

Conservation of crystal momentum converts the triple integral of eqn (12.33) to a double integral over $\mathbf{k}_2$ and either $\mathbf{k}'_1$ or $\mathbf{k}'_2$, but little further progress can be made without resorting to numerical techniques. Nevertheless, some general observations can be made regarding the effect of electron–electron collisions on momentum relaxation:

1. The occupation factors favour collisions in which all of the electrons are within $k_{\mathrm{B}}T$ of the band-edge in the non-degenerate case, or within $k_{\mathrm{B}}T$ of the Fermi level in the degenerate case.

2. Small momentum changes are favoured.

3. Electron–electron scattering acts to redistribute momentum among the energy states. Unless the momentum relaxation in the absence of electron–electron collisions is heavily energy-dependent, the effect of redistributing momentum will be small. Charged-impurity scattering in a non-degenerate population at low temperatures does produce a momentum relaxation time that is markedly energy-dependent and the corresponding mobility is found to be affected significantly. Other elastic processes are only weakly affected.

4. In optical-phonon scattering, the variation with energy of momentum relaxation is large only at the emission threshold. It follows that any significant effect can come from those less frequent electron–electron collisions involving at least one electron with energy above $\hbar\omega$. An estimate of the collision rate for these processes can be made which gives, ignoring screening

$$W_{\mathrm{ee}} = (2\pi n)^{1/3} v(\hbar\omega) f_0(\hbar\omega) \tag{12.37}$$

where n is the electron density and $v(\hbar\omega)$ is the group velocity of electrons at the phonon energy.

5. In a fully drifted distribution the effective relaxation times at each energy are equal to one another, so the net electron–electron scattering rate vanishes. This will occur when momentum redistribution dominates everything, which will be a rare occurrence.

6. On the other hand, the rate of energy redistribution at high electron concentrations can be an important factor, since the rate of energy relaxation is often much smaller than the rate for momentum relaxation. In this case the spherical part of the distribution is Fermi–Dirac with an electron temperature usually larger than the lattice temperature. If this can be assumed it makes the quantitative description of transport at high fields much simpler.

Carrier–carrier scattering also includes electron–hole scattering, but this is a very different affair, since this *always* contributes to the energy and momentum relaxation of the particular population, electron, or hole. Electron–hole scattering is treated in Section 4.6.

12.4. Hot electrons

With increasing strength of electric field the electrons become hot in the sense that their mean energy rises above that at thermodynamic equilibrium. Of course, that happens at all field strengths, but it becomes experimentally noticable by measurements of mobility, and by more spectacular manifestations such as negative differential resistance (NDR) effects and breakdown.

Theoretical work on hot electrons dates back 1930s. The prime motive was to understand electrical breakdown in insulators—so-called dielectric breakdown—and that is still a problem of some topical interest for large-gap semiconductors such as ZnS and GaN. Why the study of hot electrons should begin with the study of conduction processes in insulators is not hard to explain. The cause of breakdown was seen to have its origin in either a thermal runaway through excessive Joule heating or in purely electrical effects. The most straightforward way of creating hot electrons is to apply a strong electric field, but if that is done to a metal—or even a semiconductor—it simply melts, Joule heating being very powerful. Having very few electrons, an insulator does not get hot, although the few electrons might get very hot indeed. The study of the purely electrical effects in steady, high electric fields can be done only in insulators, and then only up to the breakdown field (10^6–10^7 V/cm).

An early theory of breakdown was that of Zener (1934), who described it in terms of quantum-mechanical tunnelling between valence and conduction bands, but the breakdown fields predicted proved to be much larger than any observed in the alkali halides. Other mechanisms proposed drew on the familiar field of breakdown in gases and emphasized impact ionization and subsequent avalanching.

The principal energy-relaxing mechanism for electrons in a solid is the emission of optical phonons, and the analysis by Fröhlich (1937) of the collision rate via the polar interaction stands today. Not quite as equally valid today is the idea of von Hippel (1937) that breakdown is associated with electrons gaining more energy from the field than they can dissipate via inelastic collisions, until they initiate an avalanche. This concept of runaway stems from the character of the polar interaction, whose strength weakens with increasing electron energy. Above a critical field no energy balance is possible with this particular scattering mechanism, so an energy runaway occurs and this was thought to be the cause of breakdown. Now we know that there are non-polar interactions that stop this runaway, so the mechanism of breakdown is not so simple, but the idea is interesting in providing the first example of a possible instability associated with hot electrons.

As in the case of a good deal of solid-state physics, a tremendous boost came with the advent of the transistor around 1947. If it were true that the study of hot electrons in metals was difficult, this was not the case for semiconductors, provided that fast electrical pulses were used. In those days, microsecond pulses were fast and were sufficient. In non-degenerate semiconductors, as in insulators, the thermal energy of the electron is ($\frac{3}{2}k_{\mathrm{B}}T$, a matter of 25 meV at room temperature, and far less than the Fermi energy of a typical metal. Thus, even though semiconductors had higher electron concentrations than insulators, it was just as easy to engineer changes in the average energy as it was in insulators, and moreover, the higher electron concentration facilitated the observation of hot-electron effects. In fact, semiconductors were the ideal materials, and as the control of impurities improved, quantitative studies began to emerge.

A significant difference between the study in insulators and that in semiconductors was that the paradigm material for transistors was germanium, which is, unlike most insulators that were studied, non-polar. The Fröhlich interaction did not apply, but, rather, the deformation-potential interaction. Consideration of non-polar scattering by acoustic phonons led to the quantitative prediction of a deviation from Ohm's Law by Shockley (1951) and, ultimately, a saturation of drift velocity at around 10^7 cm/s, which was determined primarily by the interaction with optical modes. These effects were observed in germanium, first by Ryder (1953) and in more detail by Gunn (1956). Shibuya (1955) predicted that the spheroidal valley structure of the group IV semiconductors would lead to anisotropic conduction at high fields as electrons in some valleys became hotter than those in others, and this was observed by Sasaki, Shibuya, and Mizuguchi (1958). The same effect was also responsible for the crystals becoming birefringent. These developments were reviewed by Gunn (1957), Reik (1962), Schmidt-Tiedermann (1962), Paige (1964) and Conwell (1967).

In order to account theoretically for hot-electron effects it is necessary to tackle the problem of the electron distribution function, which deviates from the thermodynamic-equilibrium Maxwell–Boltzmann or Fermi–Dirac forms in the presence of a high field. This can sometimes be done analytically by solving the Boltzmann equation. Solutions were obtained when non-polar phonon scattering is dominant: for acoustic phonons, Davidov (1937); and, for optical phonons, Reik and Risken (1961). Assuming electron–electron scattering to be dominant, Fröhlich and Paranjape (1956) used a drifted Maxwellian characterized by an electron temperature. In many cases, however, an electron temperature cannot be defined and only the average energy is meaningful. Nevertheless, where an electron temperature can be defined its value can be determined by a number of techniques, such as, for example, measuring the spectral dependence of photoluminescence or at low temperatures, relating the lattice-temperature dependence to the field dependence of Shubnikov–de Haas oscillations. The topic has been reviewed by Bauer (1974). The more general approach to obtaining the distribution function using the numerical techniques of the Monte Carlo simulation following Kurosawa (1965) have been increasingly used by Fawcett, Boardman, and Swain (1970), Rees (1972) and, more recently, Jacoboni and Reggiani (1983).

The increase in energy of an electron in the presence of an electric field is heavily influenced by the scattering mechanisms and, since these are dependent upon electron energy, the dynamic situation is fundamentally non-linear and possibly unstable.

That this is so can be most simply illustrated by the basic dynamical equations that describe the momentum and energy conservation of a representative electron:

$$\begin{aligned} \frac{\mathrm{d}(m^* v)}{\mathrm{d}t} &= eF - \frac{m^* v}{\tau_\mathrm{m}} \\ \frac{\mathrm{d}E}{\mathrm{d}t} &= eFv - \frac{E - E_0}{\tau_\mathrm{E}} \end{aligned} \tag{12.38}$$

where m^* is the effective mass, v is the velocity, F is the electric field, E is the energy, E_0 is the energy at zero field, τ_m is the momentum relaxation time, and τ_E is the energy relaxation time, and we overlook the problem that in some cases relaxation times cannot strictly be defined (most notoriously in the important case of polar-optical-mode scattering). These equations are directly coupled through the velocity and indirectly coupled through the energy dependences of m^* (non-parabolicity) and the time constants. In the steady state,

$$\begin{aligned} v &= \frac{e\tau_\mathrm{m}}{m^*} F = \mu F \\ E &= E_0 + \frac{(eF)^2 \tau_\mathrm{m} \tau_\mathrm{E}}{m^*} \end{aligned} \tag{12.39}$$

where μ is the mobility, which is field-dependent through the energy dependences of the momentum relaxation time and the effective mass. An estimate of the threshold field for hot-electron effects to appear is therefore

$$F_{\text{th}} = \sqrt{\frac{E_0}{e\mu(E_0)\tau_{\text{E}}(E_0)}}. \tag{12.40}$$

Typical values are $\mu = 1\ \text{m}^2/\text{Vs}$, $\tau_{\text{E}} = 1$ ps and $E_0 = 25$ meV, giving $F_{\text{th}} \approx 1\ \text{kV/cm}$.

The field dependences of energy and drift velocity can most simply be illustrated by assuming that the conduction band remains parabolic and that the energy dependence of the time constants can be represented by

$$\tau_{\text{m}} = AE^p, \qquad \tau_{\text{E}} = BE^q. \tag{12.41}$$

We then obtain

$$E = \left(\frac{(eF)^2 AB}{m^*}\right)^{1/(1-p-q)}$$

$$v = \frac{eFA}{m^*}\left(\frac{(eF)^2 AB}{m^*}\right)^{p/(1-p-q)}. \tag{12.42}$$

There are two things to note here. One is that the field dependences are crucially determined by the energy dependences of the scattering time constants. The other is that no steady-state solution is possible unless

$$1 - p - q > 0. \tag{12.43}$$

For the polar interaction with phonons at high enough energies for the momentum and energy time constants to be well defined, both p and q are positive quantities and this condition is violated; consequently, there is no steady state. However, as the polar interaction weakens towards high energies non-polar interactions, always present, take over if they have not been dominant before and, as for these p is negative and $q = 1 + p$ for optical modes and $q = p$ for equipartitioned acoustic modes, the stability condition can be met.

The non-polar interaction with phonons, particularly with optical and short-wavelength acoustic phonons, is therefore of great importance for determining the behaviour of very hot electrons. Usually, the energy dependence of the momentum relaxation time is that of the density-of-states function of the electronic band structure. For a simple parabolic band in 3-D this means that $p = -1/2$ and $q = 1/2$. Equation (12.42) then predicts that the energy increases with the square of the electric field and the drift velocity saturates.

The time dependence of hot-electron effects can best be appreciated by imagining an electron being accelerated from zero velocity in an electric field. Figure 12.4 gives a schematic picture of the various zones of behaviour delineated by imaginary vertical lines for the non-polar interaction.

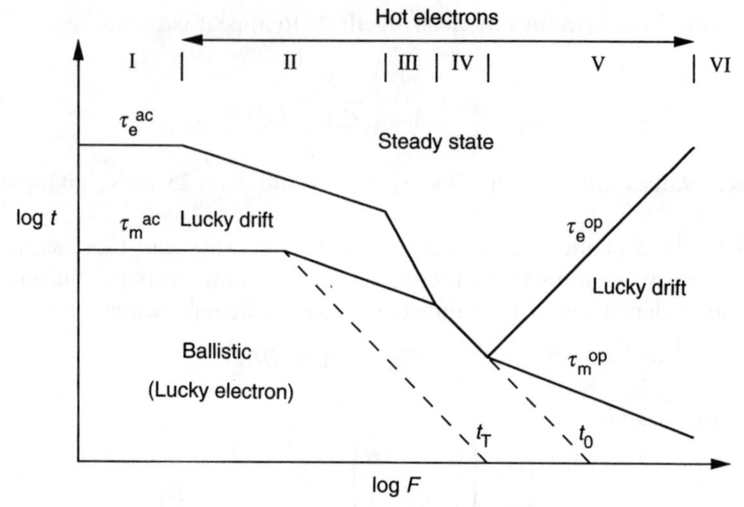

FIG. 12.4. A schematic 'phase' diagram for hot electrons: non-polar phonon scattering time versus electric field. I, Ohmic; II, warm electron; III, hot electron; IV, streaming; V, saturation drift; VI, impact ionization. $t_T = m^* v_T/eF$, $t_0 = m^* v_0/eF$; v_T is the average thermal velocity and v_0 is the group velocity of the electron when its energy equals the optical phonon energy. (After Ridley 1986.)

12.5. Hot electron distribution functions

The quantitative problem is to establish the distribution function as a function of applied electric field for arbitrary field strengths by solving the Boltzmann equation. It is usually reasonable to assume that the distribution of hot electrons is non-degenerate, so that classical statistics can be used. For the present we will assume that electron–electron scattering is too weak for a Fermi–Dirac distribution to be formed. The problem, in practice, is usually solved numerically using Monte Carlo techniques, but there are a few cases when analytical solutions can be found. As these are of considerable importance for understanding the consequences of the complex dynamics of energy and momentum input and energy and momentum relaxation, they are worth describing here, even though they generally involve the unrealistic assumption that only one scattering mechanism is operative. And, of course, if only one scattering mechanism is operative it has to be associated with phonons, since phonon mechanisms provide the necessary passage of energy to, ultimately, the thermal bath which determines the lattice temperature. In this case, the collision integrals take the form

$$\hat{I}_0 = \frac{2\pi m^*}{\hbar^2 k}\int_{q_1}^{q_2} W[\{n(\omega)+1\}f_0(E+\hbar\omega)\{1-f_0(E)\} - n(\omega)f_0(E)\{1-f_0(E+\hbar\omega)\}]q\,\mathrm{d}q$$

$$+\frac{2\pi m^*}{\hbar^2 k}\int_{q_3}^{q_4} W[n(\omega)f_0(E-\hbar\omega)\{1-f_0(E)\} - \{n(\omega)+1\}f_0(E)\{1-f_0(E-\hbar\omega)\}]q\,\mathrm{d}q$$

$$\hat{I}_1 = \frac{2\pi m^*}{\hbar^2 k}\int_{q_1}^{q_2} W[\{n(\omega)+1\}\{f_1(E+\hbar\omega)\{1-f_0(E)\}\cos\alpha' - f_1(E)f_0(E+\hbar\omega)\}$$

$$- n(\omega)\{f_1(E)\{1-f_0(E+\hbar\omega\} - f_1(E+\hbar\omega)f_0(E)\cos\alpha']q\,\mathrm{d}q$$

$$+\frac{2\pi m^*}{\hbar^2 k}\int_{q_3}^{q_4} W[n(\omega)\{f_1(E-\hbar\omega)\{1-f_0(E)\}\cos\alpha'' - f_1(E)f_0(E-\hbar\omega)\}$$

$$- \{n(\omega)+1\}\{f_1(E)\{1-f_0(E-\hbar\omega)\} - f_1(E-\hbar\omega)f_0(E)\cos\alpha'']q\,\mathrm{d}q \qquad (12.44)$$

where, for unscreened interactions,

$$\begin{aligned} W = W_\mathrm{a} &= \frac{\Xi^2 q^2}{8\pi^2\rho\omega} && \text{non-polar acoustic} \\ = W_\mathrm{pz} &= \frac{e^2 K_\mathrm{av}^2 c_\mathrm{av}}{8\pi^2\varepsilon_\mathrm{s}\rho\omega} && \text{piezoelectric} \\ = W_\mathrm{o} &= \frac{D_0^2}{8\pi^2\rho\omega} && \text{non-polar optical} \\ = W_\mathrm{po} &= \frac{e^2\omega}{8\pi^2 q^2}\left(\frac{1}{\varepsilon_\infty} - \frac{1}{\varepsilon_\mathrm{s}}\right) && \text{polar optical} \end{aligned} \qquad (12.45)$$

Except at very low temperatures in non-degenerate material, the acousti-phonon scattering can be taken to be quasi-elastic in the sense that $\hbar\omega \ll E$ for most collision processes. The validity of this assumption improves as the electrons get hot, and for extremely hot electrons it can even apply to the case of optical-phonon scattering. In the case when we can adopt the quasi-elastic approximation it is possible to expand the distribution function in a Taylor series and retain only the leading terms, i.e.

$$\begin{aligned} f_0(E \pm \hbar\omega) &\approx f_0(E) \pm \hbar\omega\frac{\mathrm{d}f_0(E)}{\mathrm{d}E} + \frac{1}{2}(\hbar\omega)^2\frac{\mathrm{d}^2 f_0(E)}{\mathrm{d}E^2} \\ f_1(E \pm \hbar\omega) &\approx f_1(E) \pm \hbar\omega\frac{\mathrm{d}f_1(E)}{\mathrm{d}E} \end{aligned} \qquad (12.46)$$

After some manipulation, we obtain the quasi-elastic collision integrals:

$$\hat{I}_0 = W\frac{4\pi m^* \omega k}{\hbar E}\left[f_0(E)\{1 - f(E)\} + n(\omega)\hbar\omega\frac{\mathrm{d}f_0(E)}{\mathrm{d}E}\right]_{q=2k}$$

$$+ \int_0^{2k} W\frac{2\pi m^* \omega}{\hbar k}\left[\frac{\mathrm{d}f_0(E)}{\mathrm{d}E}\{1 - 2f_0(E)\} + \frac{\hbar\omega}{2}\{2n(\omega) + 1\}\frac{\mathrm{d}^2 f_0(E)}{\mathrm{d}E^2}\right] q\,\mathrm{d}q$$

$$\hat{I}_1 = -\int_0^{2k} W\frac{\pi m^* q^3}{\hbar^2 k^3} f_1(E)\{2n(\omega) + 1\}\,\mathrm{d}q. \tag{12.47}$$

We have already seen from our simple model that a steady state at high fields is possible only for non-polar scattering (eqn 12.43). Thus we need only investigate solutions of the Boltzmann equation for non-polar interactions.

12.5.1. *Scattering by non-polar acoustic phonons*

Because acoustic-phonon scattering is the dominant energy-relaxing process at low temperatures, we will not assume that the distribution is non-degenerate. We will, however, assume that the phonon occupation obeys classical equipartition, i.e. $n(\omega) = k_B T/\hbar\omega \gg 1$, and that the long-wavelength dispersion relation $\omega = v_s q$ applies. Putting $W = W_a$, we obtain:

$$\hat{I}_0 = \frac{\Xi^2 (2m^*)^{5/2} E^{1/2}}{2\pi\rho\hbar^4}\left[\begin{array}{l} 2f_0(E)\{1 - f_0(E)\} + [E\{1 - 2f_0(E)\} + 2k_B T]\dfrac{\mathrm{d}f_0(E)}{\mathrm{d}E} \\ + E k_B T\dfrac{\mathrm{d}^2 f_0(E)}{\mathrm{d}E^2} \end{array}\right]$$

$$\hat{I}_1 = -\frac{\Xi^2 (2m^*)^{3/2} k_B T E^{1/2}}{2\pi\rho v_s^2 \hbar^4} f_1(E). \tag{12.48}$$

Equating these to the corresponding field terms in eqn (12.20), we obtain the following equation for the distribution function:

$$\frac{\mathrm{d}^2 f_0(x)}{\mathrm{d}x^2}(x + s) + \frac{\mathrm{d}f_0(x)}{\mathrm{d}x}\left[x\{1 - 2f_0(x)\} + 2 + \frac{x}{s}\right] + 2f_0(x)\{1 - f_0(x)\} = 0$$

$$x = \frac{E}{k_B T}, \qquad s = \frac{(eF\lambda)^2}{6m^* v_s^2 k_B T}, \qquad \lambda = \frac{\pi\rho v_s^2 \hbar^4}{\Xi^2 m^* k_B T}. \tag{12.49}$$

Here, λ is the energy-independent mean-free path. Liboff and Schenter (1986) (see also Schenter and Liboff 1987) have shown that the solution is (Fig. 12.5)

$$f_0(x) = \frac{1}{1 + e^{-\mu/k_B T}[s/(x+s)]^s e^x}. \tag{12.50}$$

For zero field ($s=0$), the distribution function becomes the Fermi–Dirac expression, as it should. For the case of non-degeneracy ($\mu < 0$),

$$f_0(x) = A(x+s)^s e^{-x}, \qquad A = e^{\mu/k_B T}/s^s. \tag{12.51}$$

This is the form first given by Davidov (1937). A is essentially a field-dependent normalizing constant maintaining a constant electron density. At high fields, the distribution becomes Gaussian:

$$f_0(E) = e^{\mu/k_B T} e^{-x^2/2s} \tag{12.52}$$

a distribution known as the Druyvestyn distribution (Druyvestyn 1930).

These results were obtained with the assumption of equipartition, but this can fail at low temperatures and high fields when there are few phonons of high enough energy to participate in scattering processes with high-energy electrons. As we noted in Section 3.3.2, the scattering rate is modified when $n(\omega) \ll 1$, with the momentum relaxation time being given by eqn (3.109). When this is the case, the equation for the spherical component becomes

$$\begin{aligned} &\frac{d^2 f_0(E)}{dE^2} + \frac{1}{2}\left(\frac{1}{E} + 5\beta E^{3/2}\right)\frac{df_0(E)}{dE} + 5\beta E^{1/2} f_0(E) = 0 \\ &\beta = \frac{8(2m^* v_s^2)^{1/2}}{(k_B T)^3 s}, \qquad s \gg \frac{m^* v_s^2}{k_B T} x^2. \end{aligned} \tag{12.53}$$

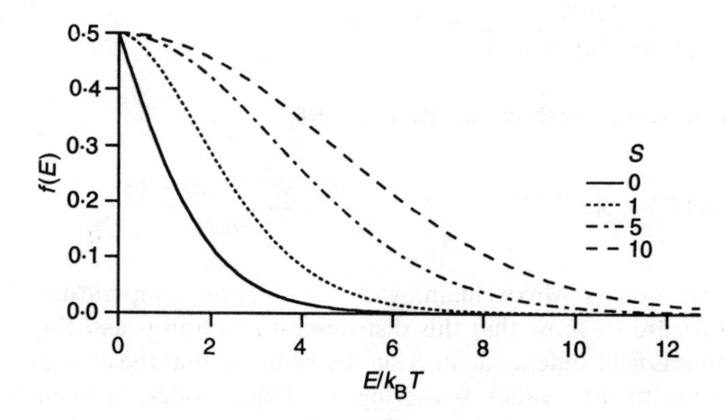

FIG. 12.5. Schecter–Liboff distribution function (see text).

the solution of which (Stratton 1957) is

$$f_0(E) = Ae^{-\beta E^{5/2}}. \tag{12.54}$$

Note that β is independent of lattice temperature, since s is inversely proportional to T^3.

12.5.2. Scattering by non-polar optical modes

An analytical solution can be found only for the high-energy regime where the quasi-elastic approximation can be made and $n(\omega)$ is no longer very much greater than unity. The collision integrals then become

$$\hat{I}_0 = \frac{D_0^2(2m^*)^{3/2}}{4\pi\rho\hbar^2 E^{1/2}}\left[\begin{array}{l} f_0(E)\{1-f_0(E)\} + E\{1-2f_0(E)\}\dfrac{df_0(E)}{dE} \\ +\frac{1}{2}E\hbar\omega\{2n(\omega)+1\}\dfrac{d^2f_0(E)}{dE^2} \end{array}\right]$$

$$\hat{I}_1 = \frac{-D_0^2(2m^*)^{3/2}\{2n(\omega)+1\}E^{1/2}}{4\pi\rho\hbar^3\omega}f_1(E) \tag{12.55}$$

At high energies it is a good approximation in most situations to assume a non-degenerate distribution, i.e. $f_0(E) \ll 1$. For high fields the equation for the spherical component is

$$\frac{d^2f_0(x)}{dx^2}x\alpha s + \frac{df_0(x)}{dx}(x+\alpha s) + f_0(x) = 0$$

$$x = \frac{E}{k_B T}, \qquad \alpha = \frac{\hbar\omega\{2n(\omega)+1\}}{2k_B T}, \qquad s = \frac{2(eF\lambda)^2}{3(\hbar\omega)^2} \gg 1,$$

$$\lambda = \frac{2\pi\rho\omega\hbar^3}{D_0^2 m^{*2}\{2n(\omega)+1\}}. \tag{12.56}$$

This has the solution (Reik and Risken 1961)

$$f_0(E) = Ae^{-E/k_B T_e}, \qquad k_B T_e = \frac{(eF\lambda)^2\{2n(\omega)+1\}}{3\hbar\omega}. \tag{12.57}$$

The distribution is Maxwelliain, with an electron temperature T_e. It is straightforward to show that this distribution function is associated with a current that is field-independent. It should be noted that the above treatment is applicable to inter-valley scattering involving modes, both optical and acoustic, in the region of flat dispersion with wavevectors near the zone edge.

12.5.3. The drifted Maxwellian

The analytical solutions of the Boltzmann equation just described were obtained with the assumption that the conduction band was spherical and parabolic. An extension to the case of spheroidal valleys is straightforward using the transformations described in Chapter 3, but any non-parabolic energy dependence invalidates the solutions. A different analytical approach to the hot-electron problem is to bypass the Boltzmann equation and assume a form for the distribution function which satisfies momentum and energy-balance equations. This can then be used even in the case of non-parabolic bands to describe hot-electron transport. The simplest choice is the drifted Maxwellian:

$$\begin{aligned} f(\mathbf{k}, E) &= A\mathrm{e}^{-E(\mathbf{k}-\mathbf{k}_D)/k_{\mathrm{B}}T_{\mathrm{e}}} \\ f(\mathbf{k}, E) &\approx f_0(E) - \mathbf{k}_D \cdot \nabla_{\mathbf{k}} f_0(E) \end{aligned} \tag{12.58}$$

where $\mathbf{k}_D$ is the shift in wavevector, common to all electron states, due to the field, and T_{e} is the electron temperature. This is physically possible only if electron–electron collisions are extremely frequent, so that drift momentum is randomized. Extra energy picked up from the field will also be randomized and will lead to the Maxwell–Boltzmann form. Since the rate of energy relaxation in phonon collisions is smaller than the rate of momentum relaxation, it is easier for electron–electron scattering to randomize energy than it is to randomize momentum. Thus, it is often possible to justify the use of an electron temperature at electron concentrations typically of $10^{17}\,\mathrm{cm}^{-3}$ and above. Justifying a common wavevector shift is less easy, but because a sum over all states is always involved, choosing a common shift, and regarding it as a kind of pre-emptive averaging, might be expected to produce little error. At high electron concentrations the drifted Maxwellian distribution is therefore a reasonable choice, even though complete momentum randomization is not to be expected.

The quantities $\mathbf{k}_D$ and T_{e} are derived from equations describing the momentum and energy balance:

$$\begin{aligned} \left\langle \frac{\mathrm{d}(\hbar\mathbf{k})}{\mathrm{d}t} \right\rangle &= \bar{e}\,\mathbf{F} \\ \left\langle \frac{\mathrm{d}E}{\mathrm{d}t} \right\rangle &= \bar{e}\langle \mathbf{v}_D \cdot \mathbf{F} \rangle \end{aligned} \tag{12.59}$$

where the angular brackets denote averages over the distribution function of the loss rates associated with the scattering mechanism. General expressions for these averages have been obtained for the case of spheroidal, non-parabolic bands, with scattering dominated by non-polar acoustic phonons obeying equipartition or by polar optical phonons (Harris and Ridley 1973). The non-parabolicity is taken to be of the form

$$\frac{\hbar^2 k_l^2}{2m_l^*} + \frac{\hbar^2 k_t^2}{2m_t^*} = E\left(1 + \frac{E}{E_g^*}\right) \tag{12.60}$$

where E_g^* is an effective bandgap, and the subscripts l and t denote longitudinal and transverse.

For acoustic-phonon scattering,

$$\begin{aligned}\left\langle \frac{d(\hbar k_\alpha)}{dt} \right\rangle &= \frac{3\Xi_\alpha^2 m_d^* k_B T (m_d^* k_B T_e)^{1/2} \hbar k_{D\alpha}}{2^{1/2} \pi \hbar^4 c_l} \frac{({}^0M_1^2)}{({}^0M_0^{3/2})} \\ \left\langle \frac{dE}{dt} \right\rangle &= \frac{3\Xi_0^2 m_t^* (2m_d^* k_B T_e)^{3/2} (1 - T/T_e)}{\pi \hbar^4 \rho} \frac{({}^0M_3^1) + 2\beta({}^0M_1^2)}{({}^0M_0^{3/2})}\end{aligned} \tag{12.61}$$

where Ξ_α, Ξ_0 and m_d^* are defined in Section 3.3.3, $\beta = k_B T_e / E_g^*$, and the M factors are the Maxwellian limits of the non-parabolic integrals (Kolodziecjak and Zukotynski 1964):

$${}^kM_n^m = \int_0^\infty e^{-x} x^k (x + \beta x^2)^m (1 + 2\beta x)^n \, dx \tag{12.62}$$

where $x = E/k_B T_e$.

For polar optical modes,

$$\begin{aligned}\left\langle \frac{d(\hbar k_\alpha)}{dt} \right\rangle &= \tfrac{1}{2} W_0 \left(\frac{\hbar\omega}{k_B T_e}\right)^{1/2} \frac{n(\omega) Y_\alpha \hbar k_{D\alpha}}{({}^0M_0^{3/2})} \{J_1 + J_2 e^{\lambda_0 - \lambda}\} \\ \left\langle \frac{dE}{dt} \right\rangle &= \tfrac{3}{2} W_0 \left(\frac{\hbar\omega}{k_B T_e}\right)^{1/2} \frac{n(\omega) Y_0 \hbar\omega}{({}^0M_0^{3/2})} I\{e^{\lambda_0 - \lambda} - 1\}\end{aligned} \tag{12.63}$$

where

$$W_0 = \frac{e^2}{4\pi\hbar} \left(\frac{2m_d^*\omega}{\hbar}\right) \left(\frac{1}{\varepsilon_\infty} - \frac{1}{\varepsilon_s}\right), \qquad \lambda_0 = \frac{\hbar\omega}{k_B T}, \qquad \lambda = \frac{\hbar\omega}{k_B T_e} \tag{12.64}$$

and the anisotropy factors are

$$\begin{aligned}Y_l &= 3(t+1)(t^{1/2} - \tan^{-1} t^{1/2}) t^{-3/2} \\ Y_t &= (3/2)\{(t+1)\tan^{-1} t^{1/2} - t^{1/2}\} t^{-3/2} \\ Y_0 &= t^{-1/2}(t+1)^{1/3} \tan^{-1} t^{1/2} \\ t &= \frac{m_l^*}{m_t^*} - 1.\end{aligned} \tag{12.65}$$

The other parameters are the integrals

$$I = \int_0^\infty e^{-x}(1 + 2\beta x)\{1 + 2\beta(x + \lambda)\} \sinh^{-1} \chi^{1/2}\, dx$$

$$J_1 = \int_0^\infty e^{-x}\{1 + 2\beta(x + \lambda)\}[\psi - \lambda\{1 + \beta(2x + \lambda)\} \sinh^{-1} \chi^{1/2}]\, dx$$

$$J_2 = \int_0^\infty e^{-x}(1 + 2\beta x)[\psi + \lambda\{1 + \beta(2x + \lambda)\} \sinh^{-1} \chi^{1/2}]\, dx \tag{12.66}$$

where

$$\chi = \frac{x + \beta x^2}{\lambda\{1 + \beta(2x + \lambda)\}}, \qquad \psi = (x + \beta x^2)^{1/2}\{x + \lambda + \beta(x + \lambda)^2\}^{1/2}. \tag{12.67}$$

The mobility can be expressed in terms of $\mathbf{k}_D$ via the current density:

$$j_\alpha = \bar{e}\int v_\alpha f(\mathbf{k}, E)\mathrm{d}\mathbf{k} = \bar{e}\mu_\alpha F_\alpha \int f(\mathbf{k}, E)\, \mathrm{d}\mathbf{k}$$

$$\therefore \mu_\alpha = \frac{\hbar k_{D\alpha}({}^0M_{-1}^{3/2})}{m_\alpha^* F_\alpha({}^0M_0^{3/2})}. \tag{12.68}$$

From eqns (12.59), (12.61), and (12.63), $\mathbf{k}_D$ is proportional to the field, and so the field can be eliminated from eqn (12.68): we obtain the mobility as a function of electron temperature. Putting $v_{D\alpha} = \mu_\alpha F_\alpha$ in eqn (12.59), we can obtain from the energy rate equations the relation between field and temperature. The variation of mobility and field with temperature can be deduced for either scattering mechanism or a mixture of both provided that the rates can be regarded as additive. Figure 12.6 shows the results for a few cases in which the ratio of optical to acoustic scattering rates and the non-parabolicity is varied. The runaway for purely polar mode scattering is exhibited, and it is also shown that a small degree of non-parabolicity removes this feature entirely, but it also produces a negative differential resistance (NDR) above a certain threshold field. This feature (NDR) remains when there is an equal amount of acoustic phonon scattering, even when the band is parabolic.

In a real material such as GaAs the role of the L and X valleys in the conduction band becomes important. Electron transfer to one or more of these valleys reduces the average mobility and produces a strong NDR (Ridley and Watkins 1961, Hilsum 1962), which is the basis of the Gunn effect (Gunn 1963). In general, NDR at high fields can be caused by a combination of mixed scattering, non-parabolicity, and inter-valley transfer.

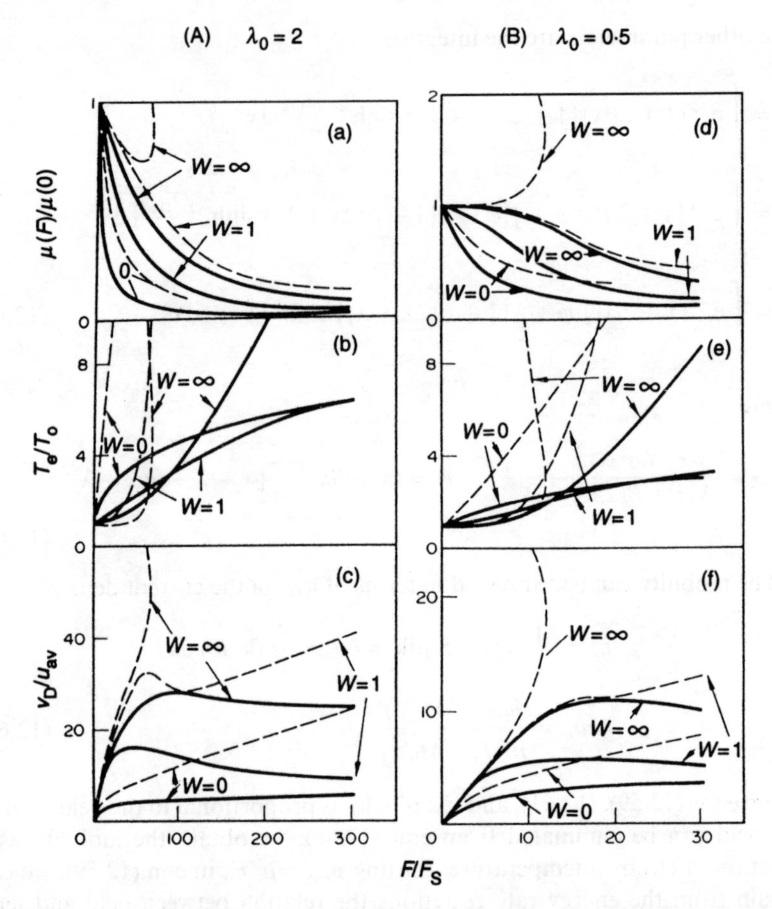

FIG. 12.6. Variation of mobility, electron temperature and drift velocity with field for different ratios of acoustic to optical phonon mobilities (W) assuming a drifted Maxwellian. $\lambda = \hbar\omega/k_BT$. Continuous lines, non-parabolic model; broken lines, parabolic model. $\mu(0)$ is the mobility at zero field; F_s is the field at which the ohmic drift velocity equals the average velocity of sound.

In the case of spherical, parabolic bands the anisotropy factors are all unity and the integrals in eqns (12.62) and (12.66) are standard:

$$
\begin{gathered}
{}^0M_0^0 = {}^0M_0^1 = 1, \qquad {}^0M_0^2 = 2, \qquad {}^0M_0^{3/2} = 3\sqrt{\pi}/4 \\
\int_0^\infty \mathrm{e}^{-x} \sinh^{-1}\left(\frac{x}{\lambda}\right)^{1/2} \mathrm{d}x = \tfrac{1}{2}\mathrm{e}^{\lambda/2} K_0(\lambda/2) \\
\int_0^\infty \mathrm{e}^{-x} x^{1/2}(x+\lambda)^{1/2} \mathrm{d}x = \tfrac{\lambda}{2}\mathrm{e}^{\lambda/2} K_1(\lambda/2)
\end{gathered}
\tag{12.69}
$$

where $K_n(x)$ is the modified Bessel function. For acoustic-phonon scattering,

$$\left\langle \frac{\mathrm{d}(\hbar k_\alpha)}{\mathrm{d}t} \right\rangle = \frac{8\Xi_\mathrm{d}^2 m^* k_\mathrm{B} T (m^* k_\mathrm{B} T_\mathrm{e})^{1/2} \hbar k_{D\alpha}}{2^{1/2}\pi^{3/2}\hbar^4 c_1}$$
$$\left\langle \frac{\mathrm{d}E}{\mathrm{d}t} \right\rangle = \frac{4\Xi_\mathrm{d}^2 m^* (2m^* k_\mathrm{B} T_\mathrm{e})^{3/2}(1 - T/T_\mathrm{e})}{\pi^{3/2}\hbar^4 \rho} \tag{12.70}$$

For polar optical-phonon scattering,

$$\left\langle \frac{\mathrm{d}(\hbar k_\alpha)}{\mathrm{d}t} \right\rangle = W_0 \left(\frac{\hbar\omega}{k_\mathrm{B} T_\mathrm{e}}\right)^{3/2} \frac{n(\omega)\hbar k_{D\alpha}}{3\pi^{1/2}} \begin{bmatrix} K_1(\lambda/2) - K_0(\lambda/2) \\ + \{K_1(\lambda/2) + K_0(\lambda/2)\}\mathrm{e}^{\lambda_0 - \lambda} \end{bmatrix} \mathrm{e}^{\lambda/2}$$
$$\left\langle \frac{\mathrm{d}E}{\mathrm{d}t} \right\rangle = W_0 \left(\frac{\hbar\omega}{k_\mathrm{B} T_\mathrm{e}}\right)^{1/2} \frac{n(\omega)\hbar\omega}{\pi^{1/2}} \{\mathrm{e}^{\lambda_0 - \lambda} - 1\}\mathrm{e}^{\lambda/2} K_0(\lambda/2). \tag{12.71}$$

Even after the adoption of a simple form for the distribution function and a simple form for the band structure, we are still faced with finding the mobility as a function of field from the transcendental equations that describe momentum and energy balance, which requires numerical work. In view of this and the unsatisfactory feature of estimating the form of the distribution function, it has become common practice to deal with the problem numerically right from the start using Monte Carlo techniques. This has the advantage that the best model band structure can be used and it can be applied to inhomogeneous semiconductors with all scattering mechanisms included. It has the disadvantage that understanding how the result comes about is more difficult than in a quasi-analytical approach. The Monte Carlo method usually produces a particular result for a particular system more accurately than an analytical method. The analytical method, although less accurate for a given system, produces a result that encompasses an infinite number of situations, and it therefore provides a general insight into the problem. The two methods are complementary to one another and, ideally, should be used side by side.

The Monte Carlo method was first used by Kurosawa (1965). There are several general accounts of the method (see Price 1979, Jacoboni and Reggiani 1983, Binder 1984).

References

Bauer, G. (1974). *Solid State Phys.* **74,** 1.
Binder, K. (1984). *Applications of the Monte Carlo method*, Springer, Berlin.
Callen, H. (1949). *Phys. Rev.* **76,** 1394.
Conwell, E. M. (1967). *High field transport in semiconductors*, Academic Press, New York.
Davidov, B. (1937). *Physik Z. Sov.* **12,** 269.

DELVES, R. T. (1959). *Proc. Phys. Soc.* **73,** 572.
DRUYVESTYN, M. J. (1930). *Physica* **10,** 61.
FAWCETT, W., BOARDMAN, A. D., and SWAIN, S. (1976). *J. Phys. Chem. Solids* **31,** 1963.
FLETCHER, K. and BUTCHER, P. N. (1972). *J. Phys. C: Solid State Phys.* **5,** 212.
FRÖHLICH, H. (1937). *Proc. R. Soc.* **A160,** 230.
—— and PARANJAPE, B. V. (1956). *Proc. Phys. Soc.* **B69,** 21.
GUNN, J. B. (1956). *J. Electron.* **2,** 87.
—— (1957). In *Progress in semiconductors, vol. 2* (ed. A. F. Gibson), p. 213. John Wiley, New York.
—— (1963). *Solid State Commun.* **1,** 88.
HARRIS, J. J. and RIDLEY, B. K. (1973). *J. Phys. Chem. Solids* **34,** 197.
HILSUM, C. (1962). *Proc. IRE,* **50,** 185.
JACOBONI, C. and REGGIANI, L. (1983). *Rev. Mod. Phys.* **55,** 645.
KOLODZIECJAK, J. and ZUKOTYNSKI, S. (1964). *Phys. Status Solidi* **5,** 145.
KUROSAWA, T. (1965). *J. Phys. Soc. Japan* **20,** 937.
LIBOFF, R. L. and SCHENTER, G. L. (1986). *Phys. Rev.* **34,** 7063.
PAIGE, E. G. S. (1964). *Progress in Semiconductors, vol. 2* (ed. A. F. Gibson), p. 1. John Wiley, New York.
PRICE, P. J. (1979). *Semiconductors and semimetals vol. 14*, p. 249. Academic Press, New York.
REES, H. D. (1972). *J. Phys. C: Solid State Phys.* **30,** 64.
REIK, H. G. (1962). *Festkorperprobleme* **1,** 89.
—— and RISKEN, H. (1961). *Phys. Rev.* **124,** 777; (1962). ibid., **126,** 1737.
RIDLEY, B. K. and WATKINS, T. B. (1961). *Proc. Phys. Soc.* **78,** 293.
—— (1998*a*). *J. Phys. Condens. Matter* **10,** 6717.
—— (1998*b*). *Semicond. Sci. Technol.* **13,** 480.
—— (1997). *Electrons and phonons in semiconductor multilayers*, p. 294. Cambridge University Press, Cambridge.
—— (1986). *Sci. Progr.* **70,** 425.
RYDER, E. J. (1953). *Phys. Rev.* **90,** 766.
SASAKI, W., SHIBUYA, M., and MIZUGUCHI, K. (1958). *J. Phys. Soc. Japan* **13,** 456.
SCHENTER, G. K. and LIBOFF, R. L. (1987). *J. Appl. Phys.* **62,** 177.
SCHMIDT-TIEDERMANN, K. J. (1962). *Festkorperprobleme* **1,** 122.
SHIBUYA, M. (1955) *Phys. Rev.* **99,** 1189.
SHOCKLEY, W. (1951). *Bell Systems Tech. J.* **30,** 990.
STRATTON, R. (1957). *Proc. R. Soc.* **A242,** 355.
VON HIPPEL, A. (1937). *J. Appl. Phys.* **8,** 815.
ZENER, C. (1934). *Proc. R. Soc.* **145,** 523.

13. Space-charge waves

13.1. Phenomenological equations

THE time constants that describe energy and momentum relaxation are typically picoseconds or less. At times longer than this, one can often assume that average energies and drift velocities have become established and measurable quantities, such as the mobility and the diffusion constant, have become well-defined. At even longer times the generally slower processes of trapping and recombination become characterized by time constants typically longer than a nanosecond. These distinctions based on characteristic times can frequently be made, but there are situations where this would be invalid; for example, in the case of an injection laser, where thermalization and stimulated emission times may be comparable. Such cases apart, it is possible to describe the various phenomena of charge transport in terms of average, rather than microscopical, quantities, relating the two through equations such as eqn (12.68), with electric and magnetic fields described by Maxwell's equations.

Defining statistical averages and moments from the Boltzmann equation leads to the so-called hydrodynamic model of transport, which is very often used in device simulation. In many practical cases the speed of momentum and energy relaxation is relatively so fast that simple phenomenological quantities such as the mobility and the diffusion constant can be used. This leads to the simplest transport model—the drift-diffusion model (Fig. 13.1). In what follows, we look at a selection of special transport phenomena amenable to a drift-diffusion model description, and limit our discussion to the case of non-degenerate distributions of electrons in a conduction band and holes in a valence band of a piezoelectric semiconductor. Our selection has taken into account that stable transport has been adequately treated by the drift-diffusion model in a number of standard texts; we therefore focus on phenomena that involve some kind of instability, since they are more exciting. (Certainly, the experimental investigation of transport instabilities can be regarded quite often as a complicated way of destroying specimens!)

We begin with Maxwell's equations, which take the form

$$\begin{aligned}\nabla \times \mathbf{H} &= \mathbf{j} + \frac{\partial \mathbf{D}}{\partial t}\\ \nabla \times \mathbf{E} &= -\frac{\partial \mathbf{B}}{\partial t}\\ \nabla . \mathbf{B} &= 0\\ \nabla . \mathbf{D} &= \rho.\end{aligned} \tag{13.1}$$

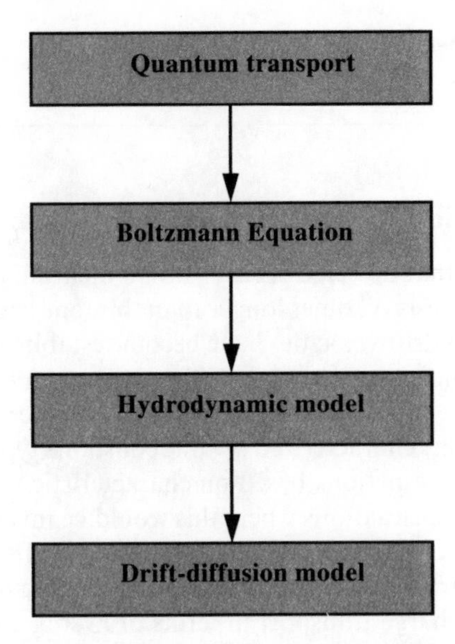

FIG. 13.1. A hierarchy of theoretical models of transport phenomena.

We limit our discussion to cases in which there is an absence of temperature gradients and flow of energy and focus on charge and particle transport. Since we will not need to refer specifically to energy in what follows, we use the usual convention of using **E** to denote the electric field. The electric displacement is related to the field and to the local elastic strain via the permittivity and piezoelectric tensors, and the field is related to the elastic strain and the stress tensor via the piezoelectric and the elastic-constant tensors:

$$D_i = \sum_j \varepsilon_{ij} E_j + \sum_{kl} e_{ikl} S_{kl}$$
$$T_{ij} = \sum_{kl} c_{ijkl} S_{kl} - \sum_k e_{ijk} E_k. \tag{13.2}$$

The current density, with $\mathbf{B} = 0$ for simplicity, is composed of drift and diffusion components:

$$j_i = e \sum_j \left((n\mu_{nij} + p\mu_{pij}) E_j + D_{nij} \frac{\partial n}{\partial x_j} - D_{pij} \frac{\partial p}{\partial x_j} \right) \tag{13.3}$$

where e is the magnitude of the electron charge, n and p are the electron and hole concentrations, $\mu_{n,pij}$ is the mobility tensor, and $D_{n,pij}$ is the diffusion-constant tensor. (Note that the presence of the subscript n or p will distinguish the diffusion constant from the electric displacement.) The space-charge density is composed of deviations from equilibrium concentrations of electrons and holes in bands and traps:

$$\rho = e\left[p - p_0 - (n - n_0) - \sum_t (n_t - n_{t0})\right] \tag{13.4}$$

where n_t is the electron concentration in the trap, and the subscript zero denotes equilibrium values. The particle continuity equation for electrons is

$$\begin{aligned} \frac{\partial n}{\partial t} &= \left(\frac{\partial n}{\partial t}\right)_{\text{cap/gen}} - \nabla \cdot \mathbf{j}_n \\ \mathbf{j}_n &= -n\hat{\mu}_n \mathbf{E} - \hat{D}_n \nabla n \end{aligned} \tag{13.5}$$

where the superscript denotes a tensor quantity. A similar equation can be written down for holes, with appropriate sign changes. The capture/generation rate for electrons has four components:

$$\left(\frac{\partial n}{\partial t}\right)_{\text{cap/gen}} = \left(\frac{\partial n}{\partial t}\right)_{\text{Auger}} + \left(\frac{\partial n}{\partial t}\right)_{\text{rad}} + \left(\frac{\partial n}{\partial t}\right)_{\text{exc}} + \left(\frac{\partial n}{\partial t}\right)_{\text{trap}}. \tag{13.6}$$

Auger processes are carrier–carrier interactions that include recombination, trapping, and impact ionization; radiative processes involve the emission and absorption of a photon; excitonic processes are the formation and disruption of excitons; and trapping is the capture into and generation out of one or more localized states. All of these processes are comprehensibly described in the book by Landsberg (1991). Here we will limit our attention to the trapping of electrons at a single deep level. (For an account of trapping at a spread of trapping levels, see Rose (1951) and Ridley and Leach (1977).) The trapping rate is

$$\begin{aligned} \left(\frac{\partial n}{\partial t}\right)_{\text{trap}} &= -c_n p_t n + e_n N_c n_t \\ n_t + p_t &= N_t \end{aligned} \tag{13.7}$$

where c_n is the volume trapping rate and e_n is the volume emission rate; N_c is the effective density of states in the conduction band (non-degeneracy assumed), and N_t is the density of traps.

We now apply these equations to some cases in which the transport is anomalous, but interesting.

13.2. Space-charge and acoustoelectric waves

The first and fourth of Maxwell's equations can be combined to give the equation for current continuity in the form

$$\frac{\partial \rho}{\partial t} = -\nabla \cdot \mathbf{j} \tag{13.8}$$

which, for one type of carrier in the case that spatial variations are limited to one dimension, becomes

$$\frac{\partial \rho}{\partial t} = -\bar{e}v\frac{\partial n}{\partial x} - \bar{e}n\frac{\partial v}{\partial x} + \bar{e}\frac{\partial D_\mathrm{n}}{\partial x}\frac{\partial n}{\partial x} + \bar{e}D_\mathrm{n}\frac{\partial^2 n}{\partial x^2} \tag{13.9}$$

where v is the drift velocity and, to avoid possible confusion arising out of the convention for sign of current and field, we use the symbol for electron charge which implies that it carries the sign of the charge. Effectively, we regard the electrons in the manipulation as positively charged and leave the insertion of the sign to the end of the calculation. Within the time hierarchy assumed here, v is an instantaneous function of the field; thus

$$\frac{\partial v}{\partial x} = \frac{\mathrm{d}v}{\mathrm{d}E}\frac{\partial E}{\partial x} = \mu_\mathrm{d}\frac{\partial E}{\partial x} \tag{13.10}$$

where μ_d is the differential mobility at the field E. If the deviations of electron density, and therefore space charge, are small, μ_d can be taken to be a constant and we can therefore ignore the spatial variation of D_n, since it is related to the mobility by the same scattering mechanisms at the same average energy. The field is related to the electric displacement, and hence to the space-charge density via eqn (13.2) but, in piezoelectric materials, the relationship involves the strain and hence the stress. The latter can be eliminated by using the equation of motion of the lattice (see eqn 3.196). For variations in one direction only, the equation of motion reduces to

$$\begin{aligned}\frac{\partial T}{\partial x} &= \rho_\mathrm{m}\frac{\partial^2 u}{\partial t^2}\\ \frac{\partial^2 T}{\partial x^2} &= \rho_\mathrm{m}\frac{\partial^2 S}{\partial x^2}\end{aligned} \tag{13.11}$$

where ρ_m is the mass density, and T and S are the components that vary along that direction. From eqn (13.2), we obtain

$$\begin{aligned}\frac{\partial^2 S}{\partial t^2} &= v_\mathrm{s}^2\frac{\partial^2 S}{\partial x^2} - v_\mathrm{s}^2\frac{\varepsilon}{e_\mathrm{p}}K^2\frac{\partial^2 E}{\partial x^2}\\ v_s^2 &= \frac{c}{\rho_\mathrm{m}}, \qquad K^2 = \frac{e_\mathrm{p}^2}{\varepsilon c}\end{aligned} \tag{13.12}$$

where v_s is the velocity of sound, K^2 is the dimensionless electromechanical coupling coefficient, and c, e_p, and ε are the appropriate components of the elastic-constant, piezoelectric, and permittivity tensors. For a travelling plane wave of the form $\exp\{i(kx - \omega t)\}$ the relation between S and E is

$$e_p S = -\frac{\varepsilon v_s^2 K^2 k^2}{\omega^2 - v_s^2 k^2} E \tag{13.13}$$

and this allows us to define an effective permittivity as follows:

$$\begin{aligned} D &= \varepsilon^* E \\ \varepsilon^* &= \varepsilon \frac{\omega^2 - v_p^2 k^2}{\omega^2 - v_s^2 k^2}, \qquad v_p^2 = v_s^2(1 + K^2). \end{aligned} \tag{13.14}$$

For semiconductors that crystallize in the sphalerite structure, the electromechanical coupling coefficient is rather small (see Tables 3.4 and 3.5), so unless the velocity of the wave is extremely close to the velocity of sound, the effective permittivity can be replaced by its usual value. Piezoelectric effects are significantly stronger in wurtzite structures, so a larger range of wave velocities is affected.

We obtain the general relation between field and space charge:

$$\frac{\partial E}{\partial x} = \frac{\rho}{\varepsilon^*}. \tag{13.15}$$

Equation (13.9) becomes

$$\begin{aligned} \frac{\partial \rho}{\partial t} &= -\frac{\bar{e} n \mu_d}{\varepsilon^*} \rho - v \frac{\partial \rho}{\partial x} + D_n \frac{\partial^2 \rho}{\partial x^2} \\ \rho &= \bar{e}(n - n_0). \end{aligned} \tag{13.16}$$

This is a non-linear equation through the first term on the right-hand side. For small deviations from space-charge neutrality, n can be replaced by n_0, and the linearized equation can be solved in terms of travelling space-charge waves:

$$\begin{aligned} \rho &= \rho_0 \exp\{i(kx - \omega t)\} \\ -i\omega &= -\omega_c^* - ivk - D_n k^2 \\ \omega_c^* &= \frac{\bar{e} n_0 \mu_d}{\varepsilon^*} = \omega_c \frac{\varepsilon}{\varepsilon^*}, \qquad \omega_c = \frac{\bar{e} n_0 \mu_d}{\varepsilon} \end{aligned} \tag{13.17}$$

where ω_c is the differential conductivity frequency. We can cast the solution in terms of temporally attenuating travelling waves, by letting $\omega \to \omega - i/\tau$, or in terms of spatially attenuating waves, by letting $k \to k + i\alpha$, and equating real and imaginary quantities. The frequency dependence of the effective permittivity generates three distinct solutions: a simple space-charge wave; and two acoustoelectric waves, one travelling forwards and the other backwards.

To be specific, we will give the solutions in terms of spatially attenuating waves. The space-charge wave is obtained when the condition $v \neq v_s$ is met:

$$\rho = \rho_0 e^{-\alpha x} e^{i(kx - \omega t)}$$
$$\omega = vk$$
$$\alpha = \frac{\omega_c + D_n k^2}{v}, \qquad \alpha \ll k. \tag{13.18}$$

In the Ohmic regime, any space-charge fluctuation dissipates at a rate determined by the conductivity and by diffusion. In the hot-electron regime, the differential conductivity can be negative, i.e. $\omega_c < 0$. In this case, the waves grow when $\omega_c + D_n k^2 < 0$ and the system is electrically unstable.

When $v \approx v_s$ one obtains the acoustoelectric solutions:

$$\omega^2 = v_s^2\left(1 - \frac{2\alpha(1 + k^2L^2)}{\Gamma k^2 L^2}\right)k^2, \qquad \alpha \ll k$$
$$2\alpha = -\frac{K^2 \Gamma k^2 L^2}{(1 + k^2L^2)^2 + \Gamma^2 k^2 L^4} \tag{13.19}$$
$$L^2 = \frac{D_n}{\omega_c}, \qquad \Gamma = \frac{(v - v_s)}{D_n}$$

Here, L is the Debye screening length. The waves attenuate when they travel against the drift, i.e. $v_s < 0$. They grow when $v > v_s$ (Fig. 13.2). (Actually, the condition for amplifying acoustic waves is more stringent than this, since the non-electronic losses described in Chapter 10 have to be overcome.) The acoustoelectric effect was first described by Hutson and White (1962), White (1962), and Gurevitch (1962). The gain is a maximum for $k^2L^2 = 1$ and $\Gamma L = 2$, in which case $2\alpha L = K^2/4$.

In the presence of a negative differential resistance (NDR), L^2 is negative and the conditions for acoustic amplification are reversed. Since NDR depends on hot-electron non-linearities which appear at drift velocities well above the velocity of sound, the effects are generally expected to be small. However, it is not the drift velocity of the carriers that matters but, rather, the drift velocity of the space charge. In the presence of trapping, these two velocities are not the same; in fact, the drift of trapped space charge can be much slower. There exists, therefore, the possibility of NDR when the drift velocity of space charge is in the vicinity of the velocity of sound.

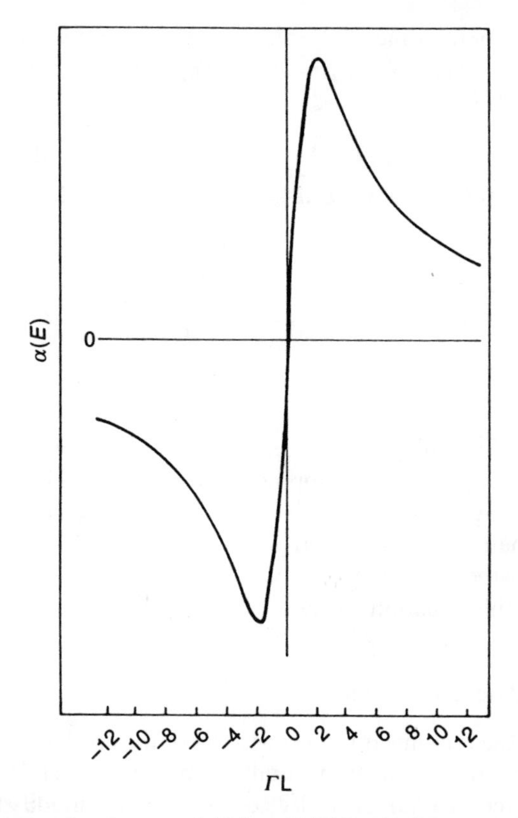

FIG. 13.2. Acoustoelectric gain versus field (eqn 13.19) for the frequency of maximum gain.

The trapping rate is given by eqn (13.7) and the space-charge density by eqn (13.4). The total rate of change of free electron density is therefore

$$\begin{aligned}\frac{\partial n}{\partial t} &= -c_{\mathrm{n}}(N_{\mathrm{t}} - n_{\mathrm{t}})n + e_{\mathrm{n}}N_{\mathrm{c}}n_{\mathrm{t}} + \frac{\partial N}{\partial t} \\ &\approx -\frac{n}{\tau} + \frac{N}{\tau_{\mathrm{g}}} + \frac{\partial N}{\partial t}\end{aligned} \tag{13.20}$$

where N is the total electron density free and trapped. Equation (13.7) is recovered if the total density remains fixed, but this is not the case when there is space-charge build-up. Without trapping the rate of change of free electron density would equal that of the total. Introduction of the frequency of the wave leads to the following relation between free and total densities:

$$\begin{aligned}n &= \frac{f - \mathrm{i}\omega\tau}{1 - \mathrm{i}\omega\tau}N \\ f &= \frac{\tau}{\tau_{\mathrm{g}}}\end{aligned} \tag{13.21}$$

Here, τ is the trapping time constant, τ_g is the generation time constant and f is the trapping factor. With $\rho = \bar{e}(N - N_0)$, the attenuation coefficient is (Uchida *et al.* 1964)

$$2\alpha = K^2 k \frac{[kL^2(\Gamma_0 fb - \Gamma_s) - a(k^2L^2fb + k\Gamma_s L^2 a)]}{[1 + k^2L^2fb + k\Gamma_s L^2 a]^2 + k^2L^4[\Gamma_0 fb - \Gamma_s + (a/kL^2)]} \tag{13.22}$$

where

$$a = \frac{(1-f)\omega\tau}{f + \omega^2\tau^2}, \qquad b = \frac{f^2 + \omega^2\tau^2}{f(f + \omega^2\tau^2)}$$
$$\Gamma_0 = \frac{v}{D_n}, \qquad \Gamma_s = \frac{v_s}{D_n} \tag{13.23}$$

In the absence of trapping ($\tau \to \infty$), $a = 0$, and $b = 1/f$, and we return to eqn (13.19). In the case of fast trapping $\tau \to 0$, $a = 0$, and $b = 1$, we obtain eqn (13.19) once again, but this time the effective drift velocity is fv, which can be much less than typical hot-electron drift velocities. Thus, it is possible for NDR and acoustoelectric effects to be coincident (Ridley 1974). As far as the author knows, this situation has never been investigated.

13.3. Parametric processes

In eqn (13.16), the non-linear term consisting of the product of carrier density and space-charge deviations from equilibrium was ignored. When there is an instability, this term cannot be neglected. Its effect is to modify the wave being amplified by engendering a second-harmonic component and producing a frequency shift. This self-interaction is not as important as the principal effect of the non-linearity, which is to couple waves to one another via parametric processes. A parametric process involves the transfer of energy from one wave, termed the pump, to two other waves, known (unpoetically) as bucket modes, which are phase matched to the pump. The pump is typically one of the waves whose growth is favoured by the amplification process, and the bucket modes are waves that may not be amplified very much, or even at all, in the linear regime.

If ρ_1 is the amplitude of the space-charge of the pump and ρ_2 and ρ_3 are the corresponding amplitudes of the bucket modes, the set of equations describing growth, neglecting diffusion for simplicity, is as follows:

$$\frac{\partial \rho_1}{\partial t} = \alpha_1\rho_1 - \beta_1\rho_2\rho_3$$
$$\frac{\partial \rho_2}{\partial t} = \alpha_2\rho_2 + \beta_2\rho_3^*\rho_1$$
$$\frac{\partial \rho_3}{\partial t} = \alpha_3\rho_3 + \beta_3\rho_1\rho_2^* \tag{13.24}$$

plus corresponding equations for the complex conjugates. The waves obey the phase-matching conditions:

$$\begin{aligned} k_1 &= k_2 + k_3 \\ \omega_1 &= \omega_2 + \omega_3. \end{aligned} \tag{13.25}$$

In the case of the acoustoelectric effect, the growth coefficients (in time) in the weak activity regime (Γ small) are

$$\begin{aligned} \alpha_i &= \frac{K^2}{2}\frac{v_s k_i^2 L^2 \Gamma}{(1+k_i^2L^2)^2} - \alpha_{Li} \\ \beta_i &= \frac{K^2}{2}\frac{v_s k_i^2 L^2}{(1+k_i^2L^2)^2}\frac{ek_jk_kL^2}{k_BT\varepsilon}. \end{aligned} \tag{13.26}$$

The linear growth coefficient differs from that in eqn (13.19) in describing the growth of amplitude, not intensity (which accounts for the factor of 2), growth in time rather than space (which accounts for the appearance of the velocity of sound, and it includes a lattice-loss term. Parametric acoustoelectric processes have been treated by Ganguli and Conwell (1969), Reik, Schirmer, and Hinkelmann (1969), and Ridley (1971). The expression for the parametric gain is taken from the last reference.

The process described above can be described as (p,bb) in obvious notation, and it is a downconversion process; that is, energy is transferred to lower frequencies. The rate of downconversion maximizes for the subharmonic, $\omega_2 = \omega_3 = \omega_1/2$. Another downconversion process is (p,pb). Upconversion processes are (pp,b) and (pb,b), the latter being weak.

We have described the interaction between three waves but, in reality, there are many waves, and a bucket mode will receive energy from a wide spectrum. The net result, as revealed experimentally by Brillouin scattering, is a huge shift in frequency downwards (Fig. 13.3).

13.4. Domains and filaments

Ultimately, the situation becomes determined by the inherent non-linear dynamics of the growth process, and new physical features emerge. In the case of NDR-generated growth, the emergent phenomenon is either a propagating high-field domain or a stationary high-current filament (Ridley 1963). In the case of the acoustoelectric effect, it is a domain of both high field and high acoustic flux that propagates at the speed of sound.

The high-field domain is a creature of so-called N-type NDR. The latter is characterized by the current falling beyond a certain field—to be distinguished from the S-type NDR, in which the field falls beyond a certain current

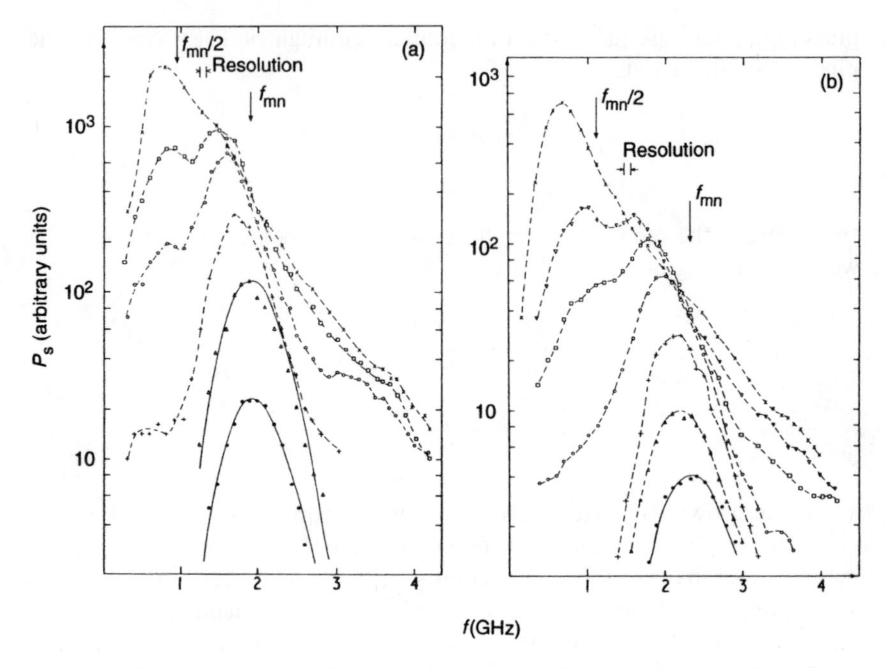

FIG. 13.3. The frequency distribution of acoustic flux in n-GaAs as a function of time for an applied field of 573 V/cm. (a) $\rho = 3.6\,\Omega\,$cm. Time in μs: •, 2.5; △, 2.8; +, 3.1; ○, 3.48; □, 3.7; ×, 4.2. (b) $\rho = 2.6\,\Omega\,$cm. Time in μs: •, 2.7; △, 2.9; +, 3.1; ○, 3.38; □, 3.47; ▽, 3.66; ×, 4.2. Full curves are linear theory curves. (Sussmann and Ridley, 1974)

(Fig. 13.4). The domain is a dipole layer propagating from cathode to anode; that is, in the direction of the electron drift, with a leading depletion layer and a trailing accumulation layer. Whereas accumulation is unlimited, depletion is limited by the electron density. In a completely depleted layer the space charge is constant and the field drops linearly, giving a triangular shape to the field profile (Fig. 13.5). When the domain reaches the anode it collapses: a new one is formed near the cathode and the process is accompanied by a current spike. Thus, current pulses are produced with a period equal to the transit time of the domain and these can be exploited as a source of microwave power. Domain formation requires that there be sufficient electron density in the sample to form a dipole layer. This imposes a criterion on the product of electron density and sample length, l: in GaAs, $nl > 10^{11}\ \mathrm{cm}^{-2}$. When this is not satisfied, a stationary domain forms at the anode. If situated in a resonant cavity, space-charge build-up can be controlled to inhibit domain formation. In this way, microwave frequencies up to 100 GHz can be generated, the limit being the rate at which electrons return from the upper valley to the central valley.

There are a number of mechanisms that produce a NDR, the main ones being electron transfer to an upper valley (Ridley and Watkins 1961*a*),

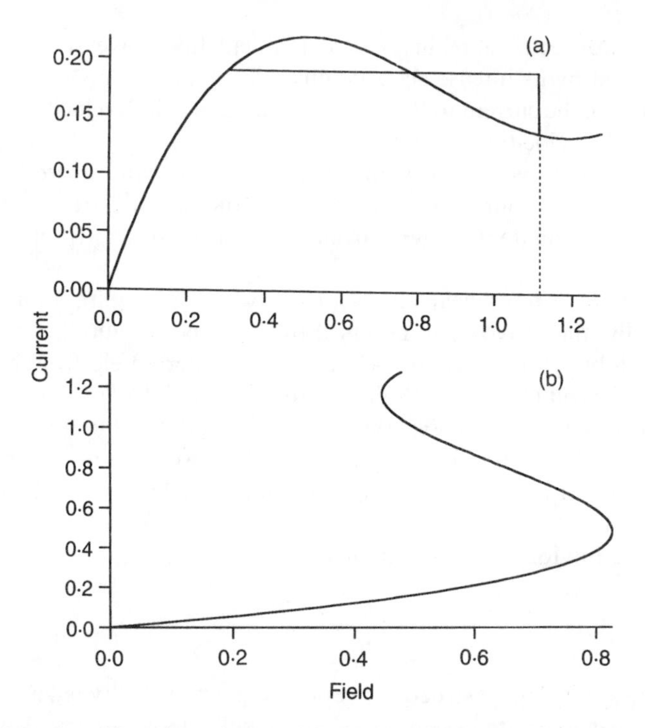

FIG. 13.4. Negative differential resistance (NDR). (a) Voltage-controlled NDR with equal-areas rule shown; (b) current-controlled NDR.

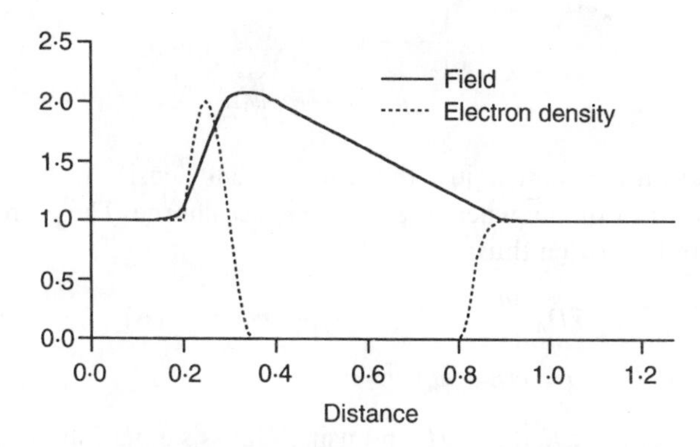

FIG. 13.5. The field and density profiles of a dipole domain propagating to the right.

scattering induced (Hilsum and Welborn 1966; see also Fig. 12.6), field-enhanced capture over an impurity barrier (Ridley and Watkins 1961*b*), impact ionization (Schöll 1981), and double injection (Stafeev 1959). For a survey of these mechanisms in 3-D and in 2-D, see Ridley (1993).

The acoustoelectric domain is also a high-field dipole layer, but its structure is determined by its intense acoustic flux, which acts simply as a high resistance, limiting the current to that corresponding to a drift velocity near to and just above the velocity of sound.

S-type NDR is associated with impact ionization and it produces more complicated behaviour involving both longitudinal and transverse charge instabilities. Ultimately, a high-current filament is formed from cathode to anode.

To treat these non-linear effects properly would require another book. Individually, they have been discussed by a number of authors. A far from complete bibliography might include: for transferred-electron NDR and domains, Carroll (1970) and Bulman, Hobson, and Taylor (1972); for filaments, Gaa, Kunz, and Schöll (1996) and Schöll (1998); and for the acoustoelectric effect, Meyer and Jörgensen (1970). Here, we focus our attention on a simple model of a propagating field domain (Fig. 13.3) associated with an N-type NDR.

The equation for current continuity is

$$\frac{\partial \rho}{\partial t} = -\frac{\partial j}{\partial x}. \tag{13.27}$$

In a stable space-charge structure propagating with velocity v_{D}, a conversion to a moving coordinate system, $x \to x - v_{\mathrm{D}}t$, will remove all time dependence. Thus,

$$\begin{gathered} \frac{\partial}{\partial t} \to -v_{\mathrm{D}}\frac{\mathrm{d}}{\mathrm{d}x} \\ \therefore\ v_{\mathrm{D}}\rho = j - \bar{e}n_0 v_0. \end{gathered} \tag{13.28}$$

The integration constant is just the drift-current density in the regions on either side of the domain where there is no space charge. The current-density equation can be written thus:

$$\begin{aligned} \bar{e}D_{\mathrm{n}}\frac{\mathrm{d}n}{\mathrm{d}x} &= \bar{e}n(v - v_{\mathrm{D}}) - \bar{e}n_0(v_0 - v_{\mathrm{D}}) \\ \rho &= \bar{e}(n - n_0). \end{aligned} \tag{13.29}$$

Replacing $\mathrm{d}/\mathrm{d}x$ by $(\mathrm{d}E/\mathrm{d}x)\mathrm{d}/\mathrm{d}E$ and using Gauss's equation, we obtain

$$\begin{aligned} (\bar{e}D_{\mathrm{n}}/\varepsilon^*)(n - n_0)\frac{\mathrm{d}n}{\mathrm{d}E} &= n(v - v_{\mathrm{D}}) - n_0(v_0 - v_{\mathrm{D}}) \\ \varepsilon^* &= \varepsilon\frac{v_{\mathrm{p}}^2 - v_{\mathrm{D}}^2}{v_{\mathrm{s}}^2 - v_{\mathrm{D}}^2}. \end{aligned} \tag{13.30}$$

Assuming that the diffusion coefficient is field-independent, the equation can be integrated to give

$$\frac{n}{n_0} - \ln\frac{n}{n_0} - 1 = \frac{\varepsilon^*}{\bar{e}D_n n_0}\int_{E_0}^{E}\left[(v - v_D) - \frac{n}{n_0}(v_0 - v_D)\right]\mathrm{d}E. \tag{13.31}$$

Now $n = n_0$ on either side of the domain where the field is E_0 and both sides of the equation vanish. But $n = n_0$ also in the interior of the domain when the field is at its maximum value E_D, and since the left-hand side again vanishes, so must the integral with the upper limit equal to E_D. But this integral can be performed over either the acumulation layer or over the depletion layer, and the only way it can be zero in both cases is if $v_D = v_0$. Thus the domain velocity is the same as the drift velocity of electrons outside the domain. But v_D must also be such that (Butcher 1965)

$$\int_{E_0}^{E_D}(v - v_D)\mathrm{d}E = 0. \tag{13.32}$$

In other words, the domain velocity is such that the area under the curve of drift velocity and field above v_D must be equal to that below it (Fig. 13.4). This, in turn, determines v_0.

The domain dynamics can be repeated in the presence of fast trapping. Similar results are obtained with n replaced by the total electron density, free and trapped, N, and all velocities reduced by the trapping factor τ/τ_g. Domains in the Gunn effect travel at around 10^7 cm/s, but the velocity of slow domains that have been observed can be as low as 10^{-4} cm/s. NDR domains (as distinct from acoustoelectric domains) that travel near the velocity of sound have not been observed, to the author's knowledge, but they are predicted to have interesting properties (Ridley 1974).

13.5. Recombination waves

So far, we have considered processes involving a single type of carrier; we now look at an instability involving the recombination of electrons and holes. We need all the equations from eqns (13.1)–(13.8) but we do not need to consider Auger, radiative, or excitonic processes. Writing these explicitly, we have

$$\begin{aligned}
\frac{\partial D}{\partial t} &= e(\mu_p p_0 + \mu_n n_0)E_0 - e\left[(\mu_p p + \mu_n n)E - D_p\frac{\partial p}{\partial x} + D_n\frac{\partial n}{\partial x}\right]\\
\frac{\partial D}{\partial x} &= e(\delta p - \delta N)\\
\frac{\partial p}{\partial t} &= -\delta p\omega_p + g_p(\delta n - \delta N) - \frac{\partial}{\partial x}\left(\mu_p pE - D_p\frac{\partial p}{\partial x}\right)\\
\frac{\partial n}{\partial t} &= -\delta n\omega_n + g_n\delta N - \frac{\partial}{\partial x}\left(-\mu_n nE - D_n\frac{\partial n}{\partial x}\right)
\end{aligned} \tag{13.33}$$

$\omega_{p,n}$ and $g_{p,n}$ are capture and generation rates averaged over the traps, and N is the total electron density, free and trapped. We also have $D = \varepsilon^* E$. Linearizing in the usual way, we obtain four equations for the four unknowns δE, δp, δn and δN, each considered to be small compared with equilibrium value, and to vary as $\exp\{i(kx - \omega t)\}$.

Solutions exist provided that a certain dispersion relationship is satisfied. Deriving this dispersion relationship in the general case involves a tedious amount of algebra. In the situation we are interested in, certain approximations are allowable. We assume that the conductivity frequencies associated with the electrons and holes are much larger than the frequencies associated with the single deep centre through which the recombination traffic proceeds. Additionally, we regard the centre to be deep enough for the generation rates $g_{p,n}$ to be negligible. These approximations are actually applicable to most practical situation, so there is not a great loss in generality.

Applying them allows us to express the dispersion relationship as follows:

$$\begin{aligned}&[\omega + i\{\omega_c + k^2(D_p + D_n)\} - k(v_p - v_n)]\\&\times [\omega^2 + i\omega\{\omega_p + \omega_n + ik(\mu E_0 - ikD_a)\}\\&- \omega_p\omega_n - ik\{(\mu\omega)_t E_0 - ik(D\omega)_t\}] = 0\end{aligned} \tag{13.34}$$

where $\omega_c = e(\mu_p p_0 + \mu_n n_0)/\varepsilon^*$ is the conductivity frequency. In the second bracket, the terms μ and D_a are the ambipolar mobility and ambipolar diffusion constant respectively:

$$\begin{aligned}\mu &= \frac{\mu_p\mu_n(n_0 - p_0)}{\mu_p p_0 + \mu_n n_0}\\ D_a &= \frac{D_p D_n(n_0 + p_0)}{\mu_p p_0 + \mu_n n_0}.\end{aligned} \tag{13.35}$$

The terms in the last internal bracket are

$$\begin{aligned}(\mu\omega)_t &= \frac{\mu_p\mu_n(n_0\omega_n - p_0\omega_p)}{\mu_p p_0 + \mu_n n_0}\\ (D\omega)_t &= \frac{D_p D_n(n_0\omega_n + p_0\omega_p)}{\mu_p p_0 + \mu_n n_0}.\end{aligned} \tag{13.36}$$

The first bracket of eqn (13.34) equated to zero describes an ambipolar space-charge wave:

$$\omega = -i\{\omega_c + k^2(D_p + D_n)\} + k(v_p - v_n). \tag{13.37}$$

Since the conductivity frequency is large (and here assumed to be positive), this solution describes heavily damped space-charge waves travelling with velocity $v_p - v_n$. The slower solutions described by the second bracket of eqn (13.34) therefore describe waves in which electrical neutrality pertains, thereby accounting for the appearance of the ambipolar transport coefficients.

The character of these slower solutions is easily established for the case $k=0$, viz. $\omega = -\mathrm{i}\omega_p$ *or* $\omega = -\mathrm{i}\omega_n$. Fluctuations simply decay at the trapping rate for the individual carrier. However, for $k \neq 0$ and for large enough fields so that drift dominates diffusion, at least over some range of k, a fluctuation may grow. For example, in *n*-type material transport is determined by the minority carrier, as eqn (13.35) shows. In such a case it is also likely that the capture of holes is much more rapid than for electrons (since the deep centres will tend to be well-occupied by electrons).

We can explore this case by assuming that

$$\begin{aligned}
&\mu_n n_0 \gg \mu_p p_0 \\
&\therefore \mu \to \mu_p, \qquad D \to D_p \\
&(\mu\omega)_t \to \mu_p(\omega_n - \omega_p p_0/n_0) \\
&(D\omega)_t \to D_p(\omega_n + \omega_p p_0/n_0).
\end{aligned} \tag{13.38}$$

Inserting these into the second bracket of eqn (13.34) and equating to zero gives two solutions, which are

$$\omega_+ = -\mathrm{i}\{\omega_p(1+a) + k^2 D_p\} + kv_p(1+b)$$

$$\omega_- = -\mathrm{i}(\omega_n - \omega_p a) - kv_p b$$

$$a = \frac{p_0}{n_0}\frac{k^2 v_p^2}{k^2 v_p^2 + (\omega_p - \omega_n + k^2 D_p)^2}\left\{1 - \frac{D_p(\omega_p - \omega_n + k^2 D_p)}{v_p^2}\right\} \tag{13.39}$$

$$b = \frac{p_0}{n_0}\frac{\omega_p(\omega_p - \omega_n + 2k^2 D_p)}{k^2 v_p^2 + (\omega_p - \omega_n + k^2 D_p)^2}$$

where v_p is the drift velocity of holes. These are the solutions obtained by Konstantinov and Perel (1965). The first describes a damped hole wave travelling in the direction of the hole drift. The second describes an electron wave travelling in the direction of the electron drift, which grows in time if

$$\omega_p a > \omega_n. \tag{13.40}$$

This condition implies that the hole concentration and the drift length should both be high enough such that

$$\frac{p_0\omega_p}{n_0\omega_n} > 1$$
$$v_p\omega_p^{-1} > \sqrt{D_p\omega_p^{-1}}. \tag{13.41}$$

If these conditions are satisfied, then there will exist a range of wavelengths that are unstable. Current oscillations in silicon that may be caused by a recombination instability have been observed, for example, by Holonyak and Bevacouva (1964). One of the problems in identifying this instability is that it has to be distinguished from the instability caused by field-enhanced capture. In principle, there is no difficulty, since the latter is a hot-electron effect, but in practice a distinction may not be easy.

Physically, what is happening is that the instability is triggered off by a local increase, say, in the electron density which reduces the field (since the electrons are the majority carriers). As a result, holes pile up on the anode side of the fluctuation and get trapped rapidly. The resultant space-charge attracts more electrons which do not get trapped quickly, and this results in an enlargement of the field trough and a movement of the fluctuation towards the anode, i.e. in the direction of electron drift. Ultimately, stability will be established when the hole traps are so full that the electron trapping rate increases. A fuller discussion can be found in Konstantinov, Perel, and Tsarenkov (1967) in which the possibility of fast recombination waves, as distinct from the slow recombination waves that we have been considering, is explored. These waves travel in the direction of the minority holes and satisfy the condition

$$p_0\omega_n > n_0\omega_p \tag{13.42}$$

which can be satisfied only if the trapping of holes is extremely slow.

References

Bulman, P. J., Hobson, G. S., and Taylor, B. C. (1972). *Transferred electron devices*, Academic Press, London.
Butcher, P. N. (1965). *Phys. Lett.* **19,** 546.
Carroll, J. A. (1970). *Hot electron microwave generators*, Arnold, London.
Gaa, M., Kunz, R. E., and Schöll, E. (1996). *Phys. Rev.* **53,** 15971.
Ganguli, A. K. and Conwell, E. M. (1969). *Phys. Lett.* **29A,** 221.
Gurevitch, V. L. (1962). *Sov. Phys.—Solid State* **4,** 668.
Hilsum, C. and Welborn, J. (1966). *J. Phys. Soc. Japan* **21,** 532.
Holonyak, N. and Bevacouva, S. F. (1964). *Appl. Phys. Lett.* **2,** 4.
Hutson, A. R. and White, D. L. (1962). *J. Appl. Phys.* **33,** 40.
Konstantinov, O. V. and Perel, V. I. (1965). *Sov. Phys.—Solid State* **6,** 2691.

——, and Tsarenkov, G. V. (1967). *Sov. Phys.—Solid State* **9,** 1381.
Landsberg, P. T. (1991). *Recombination in semiconductors*, Cambridge University Press, Cambridge.
Meyer, N. I. and Jörgensen, M. H. (1970). *Festförperprobleme* **10,** 21.
Reik, H. G., Schirmer, R., and Hinkelmann, H. (1969). *Solid State Commun.* **7,** 1309.
Ridley, B. K. (1963). *Proc. Phys. Soc.* **82,** 954.
—— (1971). *Phys. Status Solidi B* **48,** 149.
—— (1974). *J. Phys. D: Appl. Phys.* **7,** 1555.
—— (1993). In *Negative differential resistance and instabilities in 2D semiconductors* (ed. N. Balkan, B. K. Ridley, and A. J. Vickers), p. 1. NATO ASI Series **B307,** Plenum Press, London.
—— and Leach, M. F. (1977). *J. Phys. C: Solid State Phys.* **10,** 2425.
—— and Watkins, T. B. (1961*a*). *Proc. Phys. Soc.* **78,** 293.
—— (1961*b*). *Proc. Phys. Soc.* **78,** 710.
Rose, A. (1951). *RCA Rev.* **12,** 362.
Schöll, E. (1981). *J. Physique* **42,** C7.
—— (1998). *Theory of transport properties of semiconductor nanostructures*, Chapman and Hall, London.
Stafeev, V. I. (1959). *Sov. Phys.—Solid State* **1,** 763.
Sussmann, R. S. and Ridley, B. K. (1974). *J. Phys. C: Solid State Phys.* **7,** 3941.
Uchida, I., Ishiguro, T., Sasaki, Y. and Suzuki, T. (1964). *J. Phys. Soc. Japan* **19,** 674.
White, D. L. (1962). *J. appl. Phys.* **33,** 2547.

Author index

Subject index